Desert Meteorology

Aridity prevails over more than one third of the land area of Earth, and over a significant fraction of the oceans as well. Yet to date there has been no comprehensive reference volume or textbook dealing with the weather processes that define the character of desert areas.

Desert Meteorology fills this gap by treating all aspects of desert weather, such as large-scale and local-scale causes of aridity, precipitation characteristics in deserts, dust storms, floods, climate change in deserts, precipitation processes, desertification, the land-surface physics of deserts, numerical modeling of desert atmospheres, and the effect of desert weather on humans. A summary is provided of the climates and surface properties of the desert areas of the world. The book is written with the assumption that the reader has only a basic knowledge of meteorology, physics, and calculus, making it useful to those in a wide range of disciplines. It includes review questions and problems for the student.

This comprehensive volume will satisfy all who need to know more about the weather and climate of arid lands. It will appeal especially to advanced students and researchers in environmental science, meteorology, physical geography, hydrology, and engineering.

TOM WARNER was a Professor in the Department of Meteorology at the Pennsylvania State University before accepting his current joint appointment with the National Center for Atmospheric Research and the University of Colorado in Boulder. Professor Warner's career has involved teaching and research in mesoscale meteorological processes and in numerical weather prediction, and he has published on these subjects in numerous professional journals. His recent research and teaching have focussed on atmospheric processes and operational weather prediction in arid areas.

Desert Meteorology

THOMAS T. WARNER

CAMBRIDGE
UNIVERSITY PRESS

CAMBRIDGE UNIVERSITY PRESS
Cambridge, New York, Melbourne, Madrid, Cape Town, Singapore, São Paulo, Delhi

Cambridge University Press
The Edinburgh Building, Cambridge CB2 8RU, UK

Published in the United States of America by Cambridge University Press, New York

www.cambridge.org
Information on this title: www.cambridge.org/9780521817981

© Thomas T. Warner 2004

First published 2004
This digitally printed version 2008

A catalogue record for this publication is available from the British Library

ISBN 978-0-521-81798-1 hardback
ISBN 978-0-521-10048-9 paperback

To my mother and father,
Dorothy and Tom,
who never lost faith,
   and
to my wife Susan

"Water, thou hast no taste, no color, no odor; canst not be defined, art relished while ever mysterious. Not necessary to life, but rather life itself, thou fillest us with a gratification that exceeds the delight of the senses."

Antoine de Saint-Exupéry, *Wind, Sand and Stars* (1939)

Contents

Preface

This book is intended as a text and as a reference book for students from a range of disciplines, not just the atmospheric sciences. However, it is expected that students will have had at least an introductory, undergraduate, non-technical course in weather. The equations will have to be interpreted by an instructor for those without preparation in physical sciences and mathematics with calculus. And the short refresher tutorials on various topics at the beginning of some chapters will likely be skipped by those with more rigorous backgrounds in atmospheric sciences. At the end of each chapter are lists of suggested general references for further reading, questions for review, and problems and exercises. The references represent a spectrum of difficulty levels; some are qualitative whereas others may be quite technically oriented. Students without a technical background will need to choose from the qualitative references, of which there are many. The review questions are provided as study aids. The technical discipline of each student, which has served as the motivation for studying this subject, will ultimately determine what material is most germane. Thus, the student and the instructor should add their own study questions to those provided. The problems and exercises are sometimes sufficiently challenging that background reading in other texts in atmospheric sciences may be required. A number of the problems require some mathematics, and these are most appropriate for students with such a background. Hints to solutions of some of the problems can be found at the back of the book.

Metric units will be used throughout the book. Exceptions are limited to providing near-surface temperatures in degrees Fahrenheit as well as Celsius. Technical words will often be printed in **bold** the first time that they are used in the text in order to emphasize that they are important and that their meaning should be remembered. If the bold words are not defined in the text or in a footnote, a brief definition will be found in the glossary in Appendix A. Also provided are lists of symbols employed and their meaning, abbreviations, and conversion factors and physical constants (Appendixes B–D). Maps with national boundaries and country names are included in Appendix E to serve as a temporary substitute for a good world atlas, which should be available and consulted frequently. Most city names used in the text are located on the maps.

The fact that this is one of the only books exclusively devoted to desert meteorology has served as motivation for trying to make it useful to a broad audience, with both technical and non-technical backgrounds. It has also been the aim to provide as even a treatment as possible for all of the world's deserts, to make it uniformly useful to students and researchers worldwide. This aim has been occasionally compromised because many good studies of desert meteorology are not readily accessible to a Western author, especially when they are available in report form only. It is nevertheless hoped that subject-area omissions that have resulted from the unavailability of some literature are minor.

Throughout the book are inset boxes containing short informal presentations of special topics that generally relate in some way to the subjects treated in the text. The material is primarily of human interest, and is intended to provide the reader with light treatments of some subjects related to arid-land meteorology that do not fit within the more formal text presentations.

Even though it has been the primary aim to provide a complete technical treatment of the various aspects of desert meteorology, it is hoped that the reader also develops a subjective impression of the desert environment in terms of its spiritual effects on people, the importance of protecting it from further degradation, and the excitement associated with exploring and appreciating one of the few remaining frontiers on the planet.

To facilitate using this book as a text in desert meteorology, course materials are provided free of charge on the author's web site, which can be accessed through http://publishing.cambridge.org/resources/0521817986/. Included are Microsoft PowerPoint ® files that are used by the author in a desert meteorology course at the University of Colorado, Boulder.

Acknowledgements

Numerous libraries provided reference material and human resources, including those of the National Center for Atmospheric Research; The University of Arizona, Office of Arid Lands Studies; and the University of Colorado, Boulder. Data were prepared by Hilary Justh, Hsiao-Ming Hsu, and Seth Linden. Technical discussions and manuscript reviews were provided by Elford Astling, Barbara Brown, Toby Carlson, Fei Chen, Brant Foote, Margaret LeMone, Ron Smith, and David Yates. Technical assistance with the manuscript preparation was provided by Inger Gallo, Dara Houliston, Candace Larsen, Carol Makowski, and Carol Park. The graphical design of figures was by Justin Kitsutaka. Encouragement and valuable assistance in many forms were provided by Matt Lloyd, Editor, Jayne Aldhouse, Production Editor, and Lynn Davy, Subeditor, all of Cambridge University Press.

Introduction

I shall never be able to express clearly whence comes this pleasure men take from aridity, but always and everywhere I have seen men attach themselves more stubbornly to barren lands than to any other. Men will die for a calcined, leafless stony mountain. The nomads will defend to the death their great store of sand as if it were a treasure of gold dust. And we, my comrades and I, we too have loved the desert to the point of feeling that it was there we had lived the best years of our lives.

Antoine de Saint Exupéry, French aviator and writer
Wind, Sand and Stars (1939)

If one is inclined to wonder at first how so many dwellers came to be in the loneliest land that ever came out of God's hands, what they do there and why stay, one does not wonder so much after having lived there. None other than this long brown land lays such a hold on the affections . . . once inhabiting there you always mean to go away without quite realizing that you have not done it.

Mary Austin, American naturalist and writer
The Land of Little Rain (1903)

Deserts, in spite of the popular perception of their uniformity over vast distances, often contain within them a complex mosaic of **microclimates** and local weather. Great contrasts also exist in the **climate** and surface characteristics from one desert to the next. There are "cold" deserts and hot deserts, deserts with winter precipitation and deserts with summer precipitation, deserts with virtually no precipitation, perpetually foggy coastal deserts and continental deserts with near the maximum possible sunshine, barren deserts and heavily vegetated deserts, sand-dune deserts and deserts with rocky plains. The existence of such variety and complexity in deserts, and their ubiquity as well, are important messages of this book.

The expression "desert weather" has a peculiar ring to mid-latitude-centric meteorologists. It even strikes some as a contradiction in terms. This

1

preconception is belied by the facts. When it does rain in the desert it sometimes results in a violence that is rarely matched in more temperate places.
Dry riverbeds that have lain parched for decades pass a wall of water, after which
they may lie parched again for years. In some places, the day–night temperature change is immense, and larger than that associated with most mid-latitude
frontal passages. And, most who have experienced a desert sand storm or dust
storm would likely tell you that the discomfort level and fear are much greater
than, for example, in snow storms.

The existence and distribution of desert climates have profoundly shaped
historical patterns of human travel, settlement, economic development, and communication, where the deserts seem to have curiously served as both barriers
and attractions. Most of the world's early great civilizations developed at the
margins of deserts. Virtually all of the world's great contemporary religions were
born in desert regions: Judaism, Islam, Christianity, Hinduism, and Buddhism.
The Aborigines of arid Australia are thought to have the oldest continuously
maintained culture and the world's oldest language family. And the world's
longest continuously inhabited human settlement is in the desert (Jericho,
Jordan). Deserts are still significant as physical barriers, even in this age when
technology has allowed many of us to become insulated from our environment.
In some respects, the greater deserts of the world have been, and still are, more
important than mountain barriers or seas in terms of inhibiting surface transportation and commerce. In addition, with a few notable exceptions in high-
technology societies, recent human habitation has not encroached significantly
into arid, especially highly arid, areas. It is thus arguable that the deserts represent one of the last terrestrial frontiers in terms of human dominance over, and
exploitation of, the natural environment.

Most humankind understandably feel culturally, emotionally, and geographically distant from desert environments. However, the green conditions that we
take for granted may be ephemeral on time scales shorter than we imagine.
Prolonged and unpredictable droughts are natural and frequent occurrences in
some areas, and human degradations of the environment have made vast once-
green places become desert. Our perceptions can quickly change when dust
fills the air, and water sources that have been taken for granted for centuries
are no longer available. Fig. 1.1 illustrates that only about 3% of Earth's water
is fresh (non-saline), and much of this is not readily usable because it is in the
form of permanent ice and snow, and permafrost. Of the remaining water, only
1% is available at the surface in rivers, lakes and vegetation. The remainder
is groundwater, some of which is not renewed by rainfall; i.e., when it is depleted to the point that it is no longer economically recoverable for greening the

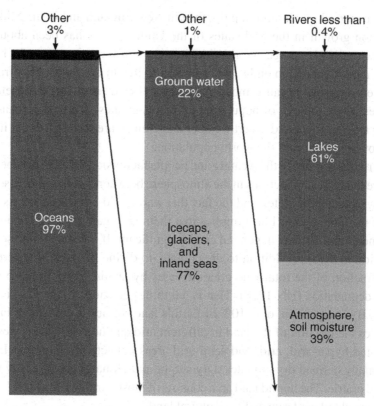

Fig. 1.1 Distribution of water on Earth, including fresh and salt water.

surface by irrigation, the resource is permanently lost. This very small slice of the hydrosphere that sustains our civilizations and keeps most of the land green is vulnerable to overuse and abuse, and to natural variability in climate, and it should not be taken for granted that the green parts of Earth will stay that way. The thinness of this veneer of water security is one motivation for studying the dynamics of desert formation and maintenance.

Desert weather and climate have historically been primarily of academic concern to all but the few local residents, even though some broader long-term interest has existed because of military activities, the exploitation of the desert's great natural resources, agricultural reclamation through irrigation, and the perception of some regional climatic trends toward desertification. However, more recently, population growth in arid areas has often surpassed that in more temperate zones, the attractions being unpolluted air, abundant sunshine, beautiful landscapes, and endless open space. In North America, for example, population growth in urban areas of the Sonoran Desert in the second half of the twentieth century occurred much more rapidly than in New England and in the Midwest. Population growth in the arid states of the United States has been about five times the national average. Worldwide, in excess of three-quarters of a billion people, more than one in eight, are estimated to live in dry lands. This fraction will likely increase because most inhabitants of arid lands are in developing countries where the rate of population growth is greatest. Thus, deserts are becoming less deserted, and desert weather and climate are having a direct impact on a growing fraction of the world population.

Deserts also influence the weather and people far outside their borders because there are dynamic connections in the atmosphere between weather processes that are geographically remote. And the fact that warm, arid climates comprise over 30% of the land area of Earth implies that their aggregate effect on non-desert areas and global climate in general can be significant. If the state of the soil and vegetation is used, in addition to the climate, to define the area of deserts, the desert fraction of the total land area increases by another 10% because of the human degradation (UN 1978). That is, all things considered, the total is close to 40%. By comparison, only 10% of Earth's land surface is cultivated. Fig. 1.2 illustrates the fraction of the land in different arid and non-arid categories. The designated hyper-arid, arid, and semi-arid areas collectively represent Earth's traditionally defined dry climates. Dry sub-humid climates are typically grassland and prairie. The humid land area consists of mostly tropical and mid-latitude forest, grassland, and rain-fed agricultural land.

This book is motivated by the fact that, in spite of the historical and present importance of the world deserts, there is no available in-depth review of their meteorology and climate. There are numerous good *general* references on the subjects of weather (Ahrens 2000), regional climate (Yoshino 1975), land-surface physics and micrometeorology (Geiger 1966; Oke 1987; Yoshino 1975), planetary boundary-layer physics (Stull 1988; Garratt 1992a), and mesoscale dynamics (Atkinson 1981), but none provides a complete treatment of such processes for desert environments. There are also excellent standard references on the physical and biological environments of deserts (McGinnies *et al.* 1968; Petrov 1976; Mares 1999), some of which review the meteorology, but the meteorological processes are not the primary focus. In addition, desert geomorphology, the desert hydrologic system, and the desert thermal energy budget are all related to desert microclimate and weather, but the literature on these subjects is somewhat specialized and is generally not commonly read by students of atmospheric sciences. Making the study of desert meteorology especially challenging is the fact that a general knowledge of traditional meteorology may not extrapolate well to arid environments because the physical conditions in the desert are

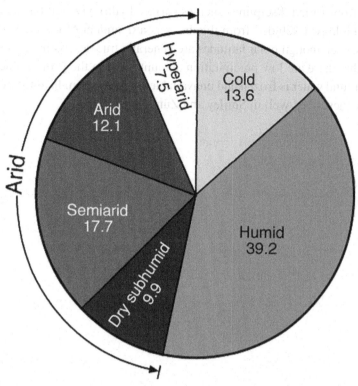

Fig. 1.2 Global land area by aridity zone (%). (Data from UNEP 1992.)

well outside our experience base: large horizontal temperature gradients, high temperatures, large diurnal oscillations, dry soils, low relative humidity, etc. This problem is compounded by the fact that there has not been a commensurate amount of experimental work for deserts, compared with forests and agricultural land (Jury *et al.* 1981), because of their perceived low economic significance and their geographic remoteness, and this has limited our understanding of the physical system.

As a step toward satisfying this need, this book describes the large-scale and mesoscale[1.1] meteorological processes and climates of the world's deserts. To properly address the near-surface atmospheric processes, the disciplines of micrometeorology and land-surface physics must be employed. These encompass the study of *surface* and *sub-surface* thermodynamic, hydrologic, and biospheric processes, as well as processes in the atmosphere very near the ground. Thus, our study of desert meteorology and climate will need to embrace a range

---

[1.1] Mesoscale, as the term is most often applied, simply refers to a broad range of horizontal space scales between the familiar "synoptic" scale, or weather-map scale, and the very small microscale of turbulence.

of topics from varied disciplines such as surface hydrology, soil physics, geology, and biology. Excluded from this book is a treatment of the polar regions as deserts, even though polar latitudes are generally true deserts from a climatic sense in that there are low precipitation amounts, and in the geomorphic sense that the ground water is frozen and unavailable for vegetation. In any case, polar deserts are described well in Smiley and Zumberge (1974).

2

The atmospheric dynamics of deserts

The Bedouin of the desert, born and grown up in it, had embraced with all his soul this nakedness too harsh for volunteers, for the reason, felt but inarticulate, that there he found himself indubitably free. He lost material ties, comforts, all superfluities and other complications to achieve a personal liberty which haunted starvation and death.

T. E. Lawrence, British writer and adventurer
Seven Pillars of Wisdom (1926)

And it is an almost terrifying magnificence . . . In a distance that is much clearer than usual earthly distances, mountain chains join and overlap. They are in regular arrangements that man has not interfered with since the creation of the world. And they have harsh brittle edges, never softened by the least vegetation. The closest row of mountains is a reddish brown; then, as they stand closer to the horizon, the mountains go through elegant violet, turning a deeper and deeper blue, until they are pure indigo in the farthest chain. And everything is empty, silent, and dead. Here you have the splendor of fixed perspectives, without the ephemeral attractions of forests, greeneries, and grasslands; it is also the splendor of almost eternal stuff, freed of life's instabilities. The geological splendor from before the Creation . . . .

Julien Viaud, French writer, soldier, painter, and acrobat
Le Désert (1895)

Because discussions in this chapter, and others in later chapters, will employ many concepts in basic atmospheric dynamics, the first section below will present some essential background review material. Then, the complex issue will be discussed of how we define "desert," and how we quantify degrees of aridity. There are definitions based on climatology, surface hydrology, plant communities, and soil types. Once the various definitions of aridity have been reviewed, the different dynamic causes of deserts will be described. These desert-forming and -maintaining processes will be seen to span scales from planetary to very local. There are also dynamic feedbacks between the atmosphere and surface that can contribute to the development and sustainment of deserts, and these will be discussed. Finally, the dynamics of desert heat lows will be addressed. Numerous other dynamic processes that occur in desert atmospheres are not related to maintaining the aridity, and these will be discussed in later chapters. The locations of deserts mentioned by name in this chapter can be found in Fig. 3.1.

Some basic concepts of atmospheric structure and dynamics

It was noted earlier that it is assumed that the reader already has a basic knowledge of atmospheric sciences. Nevertheless, some particular concepts that will be employed throughout this book will be summarized here for convenience. This will not be a comprehensive review, and non-technical texts on atmospheric sciences, such as Ahrens (2000), and technical ones on atmospheric dynamics, such as Holton (1992), should be used as additional resources as needed.

Scales of atmospheric motion

Atmospheric processes occur on a wide range of space and time scales. The horizontal space scale can be thought of as the typical horizontal dimension, or wavelength, of a phenomenon. On the large end of the scale are planetary waves, with only a few spanning the circumference of the planet. On the small end are turbulent eddies of air that may be only a few centimeters in size. Between these extremes, a variety of terms have been used to classify the different scales of motion. Below the largest planetary scale, there is what is labeled the **synoptic scale** by meteorologists. This is the typical weather-map scale that spans part or most of a continent. Smaller than the synoptic scale is the **mesoscale** (meso meaning middle), which is generally partitioned into three bands, mesoalpha, mesobeta and mesogamma, from large to small scales. Smaller than the mesogamma scale is the microscale, which includes small turbulent eddies. The following is a summary of the horizontal length scales associated with each term.

Planetary scale	> 5000 km
Synoptic scale	2000–5000 km
Mesoscale	2–2000 km
Microscale	< 2 km

Terminology used here will consistently adhere to the above definitions. In addition to these space scales, there are time scales associated with different atmospheric phenomena. Depending on usage, the term time scale can refer to the life time of an atmospheric feature, or the time period that it influences a particular location. For example, a thunderstorm complex may last most of a day as it

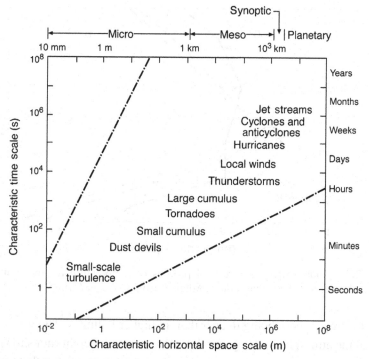

Fig. 2.1 Horizontal space and time scales of various atmospheric phenomena. (Adapted from Smagorinsky 1974.)

traverses several states, but specific locations may only be affected by the rainfall for a half hour. Fig. 2.1 depicts the approximate time and horizontal space scales for different atmospheric phenomena. Note that *vertical* length scales are often quite different from horizontal scales because of the relatively shallow depth of the atmosphere. Vertical scales and atmospheric structure will be discussed in the next section.

Vertical structure of the atmosphere

The atmosphere can be divided in the vertical into various layers according to a number of different criteria, such as the vertical temperature structure, turbulence characteristics, electrical properties, and chemical composition. The divisions based on temperature and turbulence will be important points of reference later, and are described here. Fig. 2.2 shows average vertical profiles of temperature in the winter season for different latitudes. The lower atmosphere where the temperature decreases with height is the troposphere, and above that, where the temperature increases with height, is the stratosphere. The point at which the temperature ceases to decrease rapidly with height is the tropopause, with

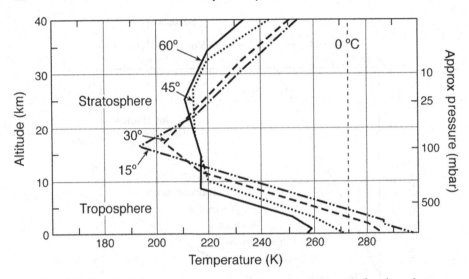

Fig. 2.2 Standard (average) vertical profiles of temperature for the winter atmosphere, for different latitudes. (Adapted from Cole *et al.* 1965.)

the tropical troposphere being deeper than in higher latitudes. The water-vapor content of the atmosphere also generally decreases with height (not shown), because its source is surface evaporation. It is important to remember that these are average conditions, and that there is much (1) day-to-day variability that reflects changing weather regimes at a particular location; (2) horizontal variability at a particular time, associated with the spatial distribution of weather regimes; and (3) diurnal variability near the ground in response to the regular surface heating and cooling cycle.

A departure from this average that is especially relevant to our study of deserts is related to the fact that a common condition over arid areas is one in which the tropospheric air is sinking, or subsiding. An important dynamic response to this **subsidence** is that the temperature no longer decreases as rapidly with height, and in extreme cases can even increase with height over shallow layers. An important consequence of this condition is that clouds and precipitation are suppressed. Conversely, in a layer of air that is rising, the temperature will decrease more rapidly with height.

At the lower boundary of the troposphere is the **turbulent** layer through which the influence of the surface is directly transmitted to the atmosphere above. Through this **boundary layer**, or **mixed layer**, turbulent eddies transport water vapor and heat upward from their source at the surface. Also, the frictional stress exerted by the surface on the atmospheric fluid is transmitted by the turbulence. There are two causes of turbulence, or sources of turbulent energy. One is the buoyancy that creates rising parcels of air, or convection, and the compensating subsidence, when the land surface is heated during the day. The other source is related to the rate of change of the horizontal wind speed with height, i.e. the vertical **shear** of the horizontal wind. When this shear is small, and there is no buoyancy, the flow is non-turbulent, or **laminar**. When the shear exceeds a threshold, the flow becomes turbulent, with the turbulent energy derived from the mean wind speed. The buoyancy-driven convective turbulence is dominant during the day, whereas the shear-driven turbulence is more common at night. Because the wind speed perpendicular to any surface, including the ground, must be zero, turbulence cannot exist at the surface and cannot transport heat or moisture. Thus, in a very shallow layer of a few molecules to a few millimeters above the surface, called the **laminar sublayer** (also, microlayer), transfers of heat, moisture and surface-frictional effects are through molecular processes. Thus, the laminar sublayer is the non-turbulent interface between the ground and the turbulent mixed layer, and the mixed layer is the turbulent interface between the laminar sublayer and the **free atmosphere** above the mixed layer. The lower 50–100 m of the mixed layer, where the turbulent transport of heat, moisture and momentum vary relatively little compared to the rest of the mixed layer above, is called the **surface layer**.

Fig. 2.3a is a schematic of the geometry of the daytime (convective) and nocturnal (stable) mixed-layer structure, and of the transitions between the two regimes. During the daylight hours, the mixed layer will increase in depth as the surface heating generates buoyancy-driven turbulence that erodes upward into the troposphere. The depth will typically reach about 1 km, but may span the entire about 10 km depth of the troposphere in strongly heated deserts. After sunset, the ground and the lower atmosphere cool, and the buoyant source of turbulent energy diminishes. The nocturnal, or stable, mixed layer derives its turbulent energy from the wind shear, with the depth of the layer being considerably less than in the daytime. In contrast to the daytime, at night the mixing can be intermittent. The shear will develop to a critical value; mixing will abruptly ensue and decrease the shear to a subcritical value, shutting off the mixing; the shear will then increase again; and so on. At the ground, this process is manifested as periods of calm that are occasionally interrupted when moderate or strong winds are briefly mixed downward from above. Above the stable, nocturnal boundary layer, in the **residual layer**, there exists residual turbulence from the daytime mixed layer, with the intensity decaying with time as a result of internal friction within the fluid.

Fig. 2.3b shows typical vertical daytime profiles of wind speed (u), **water-vapor density** (ρ_v), and **potential temperature** (θ). The potential temperature,

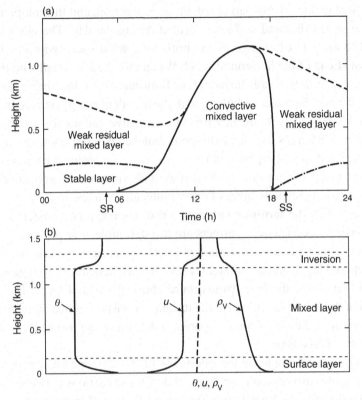

Fig. 2.3 Schematics of (a) the typical diurnal variation of the boundary-layer structure and (b) typical vertical profiles within the daytime boundary layer of potential temperature (θ), horizontal wind speed (u), and water-vapor density (ρ_v). The times of sunrise (SR) and sunset (SS) are shown in (a). The dashed line in (b) represents the wind speed that would exist without friction between the atmosphere and ground.

defined mathematically later in this chapter, is approximately uniform in the mixed layer. The temperature itself decreases at about 10K km^{-1} in this region. Within the surface layer, closer to the ground, the temperature decreases even more rapidly with height. At the top of the mixed layer is often a potential **temperature inversion**, within which potential temperature increases rapidly with height. The transition from uniform potential temperature to the inversion above is often used to define the depth of the mixed layer based on radiosonde soundings. Wind speed increases rapidly with height within the surface layer, from zero at the ground, and remains relatively uniform within the mixed layer. Throughout the troposphere above the mixed layer, the climatological north–south temperature contrast causes the wind speed to increase with height up to the tropopause. The dashed line represents the value that the wind speed would attain without the retarding effect of friction, that is transmitted through the

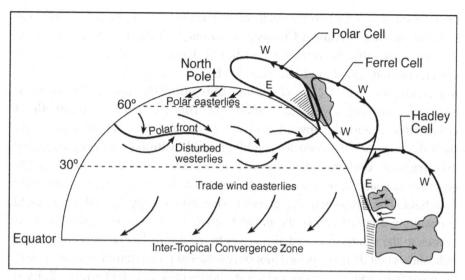

Fig. 2.4 Schematic of the different surface wind regimes, and the associated vertical circulations. The heavy straight line in the cross-section is the polar front, separating polar and tropical air masses, and the arrows in the cross-section indicate the direction of the north–south components of the air circulation. Shading shows general areas of cloud related to stable ascent along the polar front and convection in the tropics. Westerly and easterly winds are indicated with a W and E, respectively. (Adapted from Rossby 1941.)

boundary layer by turbulence. Above the mixed layer in the free atmosphere, where turbulence does not transmit the frictional stress of Earth's surface, the wind speed is greater. The water-vapor content, here defined in terms of the density of the water vapor, is fairly uniform within the mixed layer, but it does decrease somewhat with height because the source is at the surface.

Wind and pressure regimes of global weather

Earth's average weather can be described in terms of a simple conceptual model of the different weather regimes that prevail across the latitudes. There are numerous exceptions to this model that exist because of the profound and complex effects of the spatial distributions of the continents, mountains and ocean-surface temperatures. Even though these effects are what make atmospheric science so interesting and challenging, it is worth first understanding the average large-scale conditions so that the anomalies can be placed in context. Fig. 2.4 shows a schematic of the large-scale horizontal winds at the surface, and the vertical structure of the north–south circulations for the Northern Hemisphere. Between the Equator and about 30° latitude, the horizontal low-level flow is dominated by the very regular northeasterly **trade winds** to the north of the Equator and the southeasterly trades to the south of it. The area where the trades converge is called the **Inter-Tropical Convergence Zone** (ITCZ). The cross-section on the right shows that the vertical circulation is defined by the Hadley Cell, with upward motion, cloud, and rain near the Equator, and subsidence and cloud-free conditions near 30° latitude in the **subtropics**. The low-level trade-wind easterlies transition to westerlies with height. This is called a **thermally direct circulation** because the upward motion is co-located with the greatest heating (near the Equator). Between 30° and 60°, in **mid-latitudes**, eastward migrating **extratropical cyclones** (low-pressure anomalies) and anticyclones (high-pressure anomalies), with associated rotation in the winds, create a disturbed flow field. The extratropical cyclones derive their energy from the horizontal temperature contrast across the **polar front** (shown) that separates warm air masses to the south from cold air masses to the north. The cross-section to the right shows that there is upward motion, cloud and precipitation associated with the polar front and the cyclones along the northern side of the Ferrel Cell. Even though the wind direction is highly variable in this latitude belt, the average direction is westerly at low and, especially, high levels. In higher latitudes are polar easterlies at low levels, with a weak vertical circulation called the Polar Cell.

A complexity in the global circulation that is not reflected in the simple patterns of Fig. 2.4 is the existence of mean circulations along east–west-oriented vertical planes. Especially in equatorial latitudes, there are a number of circulation cells whose positions are related to longitudinal variations in sea-surface temperature, planetary **albedo**, and the atmospheric heat budget. At the equator, subsidence is found in the eastern Pacific Basin, the western Indian Ocean, and the eastern Atlantic Basin. Upward motion prevails over Indonesia, Africa, and South America. These vertical-motion patterns, referred to as the Walker Circulation (with numerous interpretations described in Hastenrath 1994), have an impact on the amount of rainfall and the aridity in near-equatorial climates.

The considerable degree of variability with longitude and season of each of the latitude regimes in Fig. 2.4 is shown in the January-average and July-average surface wind and sea-level pressure maps displayed in Fig. 2.5. First note the seasonal cycle in the north–south migration of the ITCZ, the trade winds, and the high-pressure belts of the subtropics. In general, the overall pattern shifts with the sun, being farther north in July and farther south in January. The seasonal ITCZ displacement is greater over land than over water because it is the land surface that experiences the greater seasonal change in temperature. Also, the subtropical high-pressure "belt" near 30° is really a series of highs, punctuated by semipermanent lows over land in the summer hemisphere, with the high-pressure centers being more intense in summer. A comparison of the Northern and Southern Hemisphere weather patterns in the figure clearly shows the importance

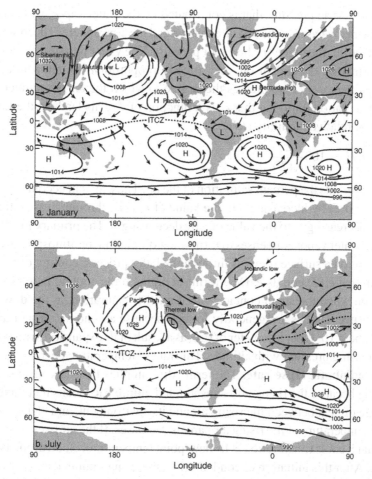

Fig. 2.5 Average January and July patterns of near-surface wind (arrows) and sea-level pressure (solid lines, millibars). The names of some of the different semi-permanent centers of high and low sea-level pressure are identified. (From Ahrens 2000.)

of the continents (with much greater area in the Northern Hemisphere) in breaking up the circulation. For example, compare the irregular Northern-Hemisphere pattern with the regular westerly flow in the mid-latitudes of the Southern Hemisphere. The transient low-pressure centers associated with storm systems tend to average out in the statistics.

Formation of fog, cloud, and precipitation

There are many measures of the water-vapor content of air, with the one most often employed here being the vapor pressure (e). This reflects the part of the total atmospheric pressure that is contributed by the water vapor, which is only one of the many gases in the mixture that we call the atmosphere. The more water vapor in the air, the higher is the vapor pressure. The vapor pressure at saturation (e_s) is primarily dependent on, and directly proportional to, temperature (T), so the quotient $e/e_s(T)$ expresses the degree to which the air is saturated, with the "(T)" notation implying that e_s is a function of T. This leads to the familiar concept of relative humidity (RH),

$$RH = 100\frac{e}{e_s(T)},\tag{2.1}$$

which represents the percent saturation of the air. Thus, air can become saturated ($RH = 100\%$) by increasing e to the value of e_s (say, through evaporation), or through decreasing e_s to the value of e by decreasing T. The primary mechanism by which water vapor condenses to form water droplets in the atmosphere (clouds and fog) or on a surface (dew) is through the latter process: cooling of the air to the point that $e_s(T)$ equals e. Dew forms at night on the cooling bare ground or vegetation when the air in contact with it reaches saturation, at the **dew-point temperature**. Clouds are generally formed because ascending, unsaturated air cools at about 10 °C per kilometer of rise, as a result of its expansion in response to the decreasing pressure with altitude. This rate of temperature decrease is called the **dry adiabatic lapse rate**. The term "dry" implies that no condensation is occurring because the rising air is unsaturated, **adiabatic** means that there is no exchange of heat between the rising air and its environment, and "lapse" means that the temperature of the rising air parcels decreases. If the air rises sufficiently far so that it cools to saturation at its dew-point temperature, cloud droplets begin to form. After this initiation of condensation, the rising saturated air cools at less than the dry adiabatic rate because the **latent heat of condensation**[2.1] is being added. The cooling is now at the **moist adiabatic lapse rate**, which varies from about 2 to 10 K km^{-1}, depending on temperature and pressure. The value is about 6 K km^{-1} at 800 mb and 0 °C. Once a cloud composed of liquid droplets and/or ice crystals is formed, processes within the cloud sometimes allow the growth or aggregation of the cloud water such that snow or raindrops are produced within the cloud. Especially in arid environments with dry air below the cloud base, raindrops often evaporate before reaching the ground. The visible rain or snow streaks that terminate below the cloud base are called virga.

Fog can develop through a number of processes. An **advection fog** forms when air moves over a cooler surface (land or water), and the air near the surface is cooled to its dew-point temperature. **Radiation fog** occurs at night when the

---

[2.1] The latent heat of condensation is released when water condenses from the vapor to the liquid phase, and is equal to 2.500×10^6 J kg^{-1} at 0 °C. The reverse process, evaporation, consumes energy at this rate.

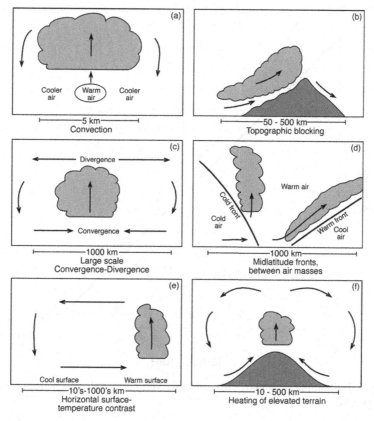

Fig. 2.6 Schematic of different basic mechanisms that can be responsible for the development of cloud and precipitation.

land cools by infrared emission to space, and the air near the ground is, in turn, cooled by the surface to its dew point. Lastly, a **mixing fog** can be produced when two unsaturated volumes of air, with different temperatures and vapor pressures, mix. If the resulting mixture is saturated, a fog forms. We will see later in this chapter, and in the next one, how advection fogs form over cold ocean currents and influence the climate of deserts that exist near the shore.

Air can be caused to rise, possibly leading to cloud formation, through the mechanisms depicted in the cross-sections of Fig. 2.6, with the ascent being categorized as either stable or unstable. Typical horizontal length scales for each process are also indicated. Unstable ascent, called **convection**, is illustrated in Fig. 2.6a, and results when a parcel of air is warmer, and therefore less dense, than its surrounding environment. This is the familiar concept of buoyancy, and can result in the production of *cumulus* clouds when the lower atmosphere is heated during the day and when the air is sufficiently moist (high e) so that

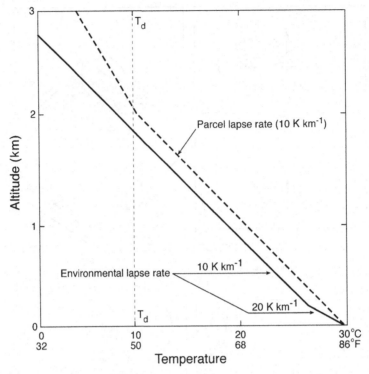

Fig. 2.7 Example of a buoyantly unstable vertical profile of environmental temperature.

condensation takes place before a rising parcel loses its buoyancy. When cloud is formed, the process is called moist convection. Fig. 2.7 shows how the existence of convection depends on the lapse rate of atmospheric temperature, as might be measured by a temperature sensor on an ascending radiosonde. This atmospheric, or environmental, lapse rate is distinct from the dry or moist adiabatic lapse rates that reflect the temperature change of a parcel of air that is rising (and expanding and cooling) through its environment. The example in the figure shows an absolutely unstable (superadiabatic) environmental lapse rate of 20 K km^{-1} (solid line) in the lowest 100–200 m, surmounted by an adiabatic lapse rate of 10 K km^{-1}. This would be typical of a situation with strong mid-day surface heating and a deep, turbulent boundary layer. If a parcel of the unsaturated air at the surface, with a temperature of 30 °C and a dew-point temperature of 10 °C, is displaced slightly upward, it will cool at 10 K km^{-1} (dashed line) and therefore become warmer and less dense than its environment. Because of the upward buoyant force, it will continue to rise. When its temperature decreases to 10 °C (the dew-point temperature), it is saturated, cumulus cloud will form, and it will cool less rapidly because of the release of the latent heat of condensation. The cloud will continue to grow upward until the rising air loses its buoyancy. In contrast to this unstable situation, consider one in which the surface temperature is the same, but the environmental lapse rate is less than the dry adiabatic value of 10 K km^{-1}. That is, the solid line would be to the right of the dashed line in the figure. Now, a parcel that is displaced upward will immediately become cooler and more dense than its environment, and the downward buoyant force will cause the parcel to return to its original position. This is a stable environmental lapse rate, and will not lead to the production of cloud through buoyancy. These examples are simplifications because real environmental lapse rates are generally more complex. Also, a parcel that is initially displaced upward may already be saturated, in which case the stability of the environmental lapse rate is determined by whether it is greater or less than the *moist* adiabatic lapse rate. Unstable lapse rates, or rapid decreases of environmental temperature with height, can be produced in a number of ways, but certainly a common one is through the heating of the lower atmosphere by the surface of Earth as it warms during the day. In contrast, the cooling of the lower atmosphere at night, as the ground cools, leads to stable lapse rates. This buoyancy-driven moist convection is the dominant source of rainfall in the tropics, and is responsible for rainfall that occasionally invades arid subtropical areas from lower equatorial latitudes.

Some of the other mechanisms shown in Fig. 2.6 for lifting air and forming clouds and precipitation represent stable ascent in that they do not rely on buoyant forces, and **stratus**, or layered, clouds are formed. However, in some cases the stable lifting of the layer destabilizes the lapse rates, and the resulting buoyant rise generates cumulus clouds. Fig. 2.6b illustrates topographic forcing of vertical motion. When air moving horizontally encounters mountain barriers, it can be potentially deflected over or around them. Air is more likely to rise over a barrier that represents a relatively continuous ridge that is perpendicular to the flow, and it is more likely to rise over than go around an isolated obstacle when the atmospheric lapse rate is not too stable or when the wind speed is high. If the air is sufficiently moist and the mountain sufficiently high to produce lifting of the air to the condensation level at the dew-point temperature, cloud will form on the upwind side, and dissipate on the downwind side where the air warms as it descends and compresses. This cloud and precipitation formation can occur over a wide range of scales, depending on the width of the mountain barrier. We will see that such topographic modulation of cloud and precipitation is important to desert dynamics in a couple of respects. On the one hand, precipitation over mountain barriers removes water vapor from the air, creating more-arid conditions downwind. Alternatively, mountainous areas within deserts can increase the precipitation locally and produce **altitude oases**.

Figure 2.6c shows the creation of upward motion and cloud as a result of upper-atmospheric horizontal divergence or low-level convergence of air on the synoptic, or weather-map, scale. Regardless of whether the forcing is from upper or lower levels, continuity of mass requires the development of the upward motion and the associated compensating subsidence that are depicted. The convergence or divergence results from upper- or lower-atmospheric disturbances.

Cold and warm fronts that represent boundaries between air masses also produce upward motion, and cloud and precipitation (Fig. 2.6d), as the warm, less dense, air rises over the colder, more dense, air. The clouds associated with cold fronts tend to be more convective and smaller in horizontal scale (cumulus clouds) than those caused by warm fronts (stratus clouds), but the combined scale of the cloud and precipitation pattern is similar to that of the synoptic-scale extratropical cyclones with which the fronts are associated. These cyclones are generally mid-latitude phenomena, and their intrusion into arid subtropical areas is responsible for some of the winter precipitation there.

Figure 2.6e illustrates thermally direct air circulations that result from horizontal contrasts in the heating of Earth's surface and lower atmosphere. For example, on small scales, horizontal pressure gradients[2.2] are caused by differences across coastlines in the diurnal heating and cooling of land and water. These diurnally reversing pressure gradients generate **sea-** or **lake-breeze circulations** that cause upward motion, and sometimes cloud and precipitation, over the warmer land during the daytime part of the cycle. The circulation reverses at night when the water surface is warmer than that of the land. On larger, continental scales, the greater summer warming of the land surface than the sea surface produces **monsoon** circulations directed at low levels from the cooler water to the warmer land. This produces warm-season upward motion, cloud and precipitation over land, with the circulation reversing in the cold season. Such summer monsoon rains can penetrate arid areas, and represent a large fraction of the total annual accumulation. Also, such thermally direct circulations will be shown later in this chapter to be intimately related to desert heat lows, or climatological surface low-pressure areas.

Lastly, another type of thermally direct circulation that can produce cloud and precipitation is caused by the daytime heating of elevated terrain (Fig. 2.6f). Relative to the surrounding atmosphere, the air over the high terrain is warmed, producing horizontal pressure gradients that generate the circulations. Upward motion and possible cloud and precipitation occur over the higher elevations.

[2.2] A gradient of a scalar property is the rate at which it changes value in the horizontal, in the direction of most rapid change.

Definition of desert

In spite of the frequency with which the term is applied, there is no universally accepted common or technical definition of "desert." It is perhaps one of the oldest written words (El-Baz 1983), having come to us from an Egyptian hieroglyph pronounced Tésert, and through the Latin words *desero* – to abandon, *desertum* – a waste place or wilderness, and *desertus* – abandoned, relinquished or forsaken. It is important to recognize that the various attempts described below to define deserts and quantify degrees of aridity are not of just academic interest. Rather, good objective measures are required to allow assessment of how people and climate may be changing our arid environments.

A characteristic of arid lands, and a common criterion for their designation as such, is that the surface water loss (the sum of the **potential evaporation** (PE) and the **transpiration**, called **evapotranspiration**) exceeds the surface water gain (precipitation). It is thus misleading to simply use annual precipitation amount as a metric of aridity because the water loss is just as important a component of the water budget. Thus, one intuitive definition of an arid area is that it is "the land of the empty bucket" (NSF 1977): if you place a bucket outside on the ground and it never fills up, you are in a desert. That is, evaporation exceeds precipitation. An illustration that both the supply and demand halves of the water balance are important is that a 40 cm annual precipitation supports only sparse vegetation in a hot climate, but permits the growth of a dense forest in a high-latitude cool climate having a much lower evaporation rate. The seasonal distribution of precipitation also has a bearing on the dryness of the climate. If most of the annual precipitation is concentrated in the warm summer months when the evaporation is greatest, the climate will be more arid than if the precipitation is distributed more evenly throughout the year. For example, in parts of western Australia the annual precipitation is only about 25 cm, but it falls reliably in the cooler season and supports good crops of wheat (Miller 1961). In contrast, Hillet Doleib, Sudan, receives over 75 cm of rain annually, but primarily during the summer, which has **daily-average temperatures**[2.3] of over 26 °C (80 °F). As a result, it appears just as arid as other places in North Africa that have one-tenth the precipitation. Note that cold-season precipitation is only beneficial in promoting vegetation if the temperature is sufficiently high.

Soil properties also influence the degree of aridity. Tundra, in high latitudes, which may have precipitation of only 25 cm, has a subsurface permafrost in summer and low evaporation, which cause the moisture to be retained near the surface, leading to a distinctly non-arid surface that is soggy or muddy when it

---

[2.3] Daily-average temperature is defined as the mean of the maximum and minimum temperature during the 24 h period from midnight to midnight local time.

is not frozen. Other natural subsurface barriers to rapid precipitation loss into the deep soil include rock and clay layers. A man-made subsurface barrier is the "plow-pan" of compressed soil produced by repeated plowing. *Surface* barriers to water penetration, however, can *contribute* to aridity. There are arid climates with more than 50 cm of annual rain that falls in intense thunderstorms on hard soils or rock, and the water runs off horizontally or is evaporated quickly by the sun and hot, dry air. At the other extreme, there are arid-land soils that are extremely porous, and this same annual precipitation drains through them so rapidly that it is virtually unavailable to vegetation.

There are still more factors that make annual precipitation a poor reflection of aridity, or whether we call a place desert. What of lushly vegetated areas with very low precipitation, where the moisture is supplied by very high water tables, or by abundant water from the floodplain of a perennial river such as the Nile. What of virtually rainless areas where vegetation is nourished by coastal fog? Are they deserts, nevertheless?

In spite of these complexities, an annual precipitation threshold is sometimes still used as a simple criterion for desert; a traditional geographer's definition is 25 cm, others use 5 cm. Many regard 50 cm as the limit of the desert in hot climates and 25 cm in cold climates (Miller 1961). Some define as "true deserts" only those in which there have been recorded 12 consecutive months without precipitation (McGinnies *et al.* 1968). Howe *et al.* (1968) employ for their criterion the number of precipitation days (> 2.5 mm) per month, irrespective of the total amount of precipitation. The terminology is further complicated because various authorities have used many descriptors such as desert, sub-desert, semi-desert, absolute desert, true desert, superarid, hyperarid, arid, semi-arid, extremely arid, demi-arid, hemi-arid, etc.

Miller (1961) has incorporated both annual precipitation and temperature, a surrogate for the evaporation rate, in a definition of desert. This criterion for a desert is that $R < T/5$, where R is annual precipitation in inches and T is **annual-average temperature**[2.4] in degrees Fahrenheit. That is, for higher annual mean temperatures, more precipitation is required to offset the greater associated evaporation rate. For an annual mean temperature of 22 °C (72 °F), the area is classified as desert if the mean annual precipitation is less than 36 cm (14.2 in). Fig. 2.8 shows how this formula was defined graphically, based on prevailing vegetation. The $R = T/5$ line in annual-temperature versus annual-precipitation space roughly separates desert and non-desert climates. There are numerous other algorithms for defining the limits of deserts in terms of mean

---

[2.4] Daily mean temperature is the average of the daily maximum and minimum temperatures. Annual mean temperature is the average of the daily means for the calendar year.

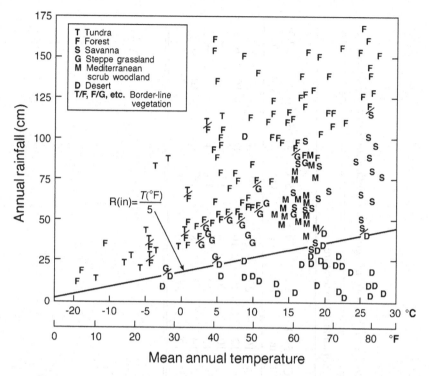

Fig. 2.8 Temperature and rainfall associated with various climate regions that are defined in terms of vegetation, showing the line separating desert and non-desert climates $(R(in) = T(°F)/5)$. (Adapted from Miller 1961.)

annual precipitation and temperature. Table 2.1 lists some of these algorithms, and shows that, for a given temperature, the precipitation thresholds that separate desert and non-desert sometimes differ by 50–100%. Unfortunately, other meteorological variables often have a much larger correlation with evaporation than does temperature, even though the latter is a convenient quantity to use because it is frequently measured. Thompson (1975) shows the correlation of evaporation rate with solar radiation intensity, relative humidity, air temperature and wind speed. For one set of measurements, the correlations were approximately 0.77, 0.72, 0.46, and 0.27, respectively, so the radiant energy input at the surface and the relative humidity near the surface (not the air temperature) had the greatest correlation with water loss.

A number of different quantitative approaches exist for defining degrees of aridity, rather than just the threshold between desert and non-desert. Dzerdzeevskii (1958) presents 19 climate indices that can be used for this purpose, many of which have served as the basis for constructing bioclimate maps. For example, in one aridity index, the ratio of precipitation (*P*, water

Table 2.1 *Combinations of mean annual temperature and rainfall used by various authors to define the limits of deserts*

Annual rainfall defining the desert limit according to	Mean annual temperature											
	25 °C cm	77 °F in	20 °C cm	68 °F in	15 °C cm	59 °F in	10 °C cm	50 °F in	5 °C cm	41 °F in	0 °C cm	32 °F in
Köppen (1)[a]	32	13	29	11	26	10	23	9	20	8	16	6
Köppen (2)	41	16	36	14	31	12	26	10	21	8	16	6
Thornthwaite	37	15	30	12	25	10	20	8	15	6	—	—
De Martonne (1)	50	20	40	16	30	12	20	8	10	4	—	—
De Martonne (2)	35	14	30	12	25	10	20	8	15	6	10	4
Miller	40	15	35	14	30	12	25	10	20	8	15	6

[a] See Miller (1961) for details of references.
Source: From Miller (1961).

supply) to **potential evapotranspiration** (PET), is used (UN 1977), where the following ratios delineate areas with different degrees of aridity: hyperarid for $P:PET < 3:100$; arid for $P:PET$ between $3:100$ and $20:100$; semi-arid for $P:PET$ between $20:100$ and $50:100$; and subhumid for $P:PET$ between $50:100$ and $75:100$. The PET is calculated by using observed solar radiation and atmospheric wind speed and humidity. The classification does not apply to cold deserts (see Chapter 3 for definition), such as tundra regions and the high deserts of Tibet. A number of other classification systems also employ ratios of P and PET in their indices of aridity. In contrast, the Budyko index, a "radiational index of dryness," is the depth of water that could be evaporated by the observed total annual net radiation (net radiant energy at the ground), divided by the depth of the observed annual precipitation (Budyko 1958, 1986; Henning and Flohn 1977; El-Baz 1983). Local atmospheric conditions, such as temperature, wind speed, and humidity, are not accounted for as they are in the $P:PET$ ratio. Fig. 2.9 displays a comparison of three aridity indices or classifications, using the Arabian Peninsula as an example. Shown are the Budyko index, the Meigs classification (1953, 1957), and the UN (1977) $P:PET$ ratio, each computed from inevitably different meteorological input. Even though the large-scale patterns are similar, there are numerous differences in the maps that result from the definitions used for the aridity categories and the details of the input data. For example, both the Meigs and UN climate zones show the eastern Sahara and the Empty (southeastern) Quarter of Saudi Arabia to be similarly extremely arid or hyperarid, but the Budyko ratio is four times larger for the Sahara. Also, only the Budyko ratio (or the data used to calculate it) accounts for the existence of the less arid coastal areas in southeastern Yemen and Oman, but it is also the only analysis that does not show the wetter conditions in the mountains along the northeast coasts of Oman and the United Arab Emirates. A shortcoming of many simple aridity indices is that they use annual-average data as input. Thus, they cannot account for situations where there is sufficient water supply relative to demand in certain months of the year to permit a distinctly non-arid vegetated condition. In contrast, the climate classification schemes of Köppen (1931), Thornthwaite (1948), Trewartha (1968), Meigs (1953), and others, account for seasonal effects.

As we have seen, most methods for defining the degree of aridity rely on an estimate of long-term water demand, which should include transpiration by the vegetation and evaporation. This is much more difficult to obtain than the other half of the balance equation, the precipitation input. There are various complications such as the fact that actual water demand rarely equals the more easily estimated potential demand, such as PET, or simply evaporation from a pan. Also ignored is the situation where storage of an excess over demand within

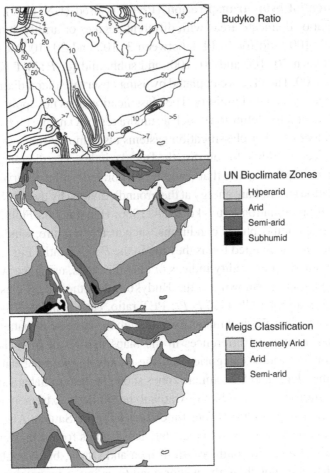

Fig. 2.9 Indices of aridity for the Arabian Peninsula based on the Budyko ratio (adapted from Henning and Flohn 1977), the ratio of precipitation (P) to potential evapotranspiration (PET) (adapted from UN 1977), and the Meigs classification (from Meigs 1953, 1957.)

the vegetation or the **substrate** may provide water for later periods of deficit. Such storage potential varies widely from place to place in arid environments. Because actual, or even potential, water demand are very difficult to measure or estimate through direct means, estimation techniques have proliferated, and are based on temperature, water-vapor deficit, wind speed, or net radiation. One early approach for estimating evaporation from temperature involves calculating regression equations based on selected measurements of both quantities. Thornthwaite's (1945) estimates using this approach are good for climate types that are similar to the areas for which the equations are defined, but for other areas the evaporation rate is poorly defined. Correction terms successfully reduce the error, but significant problems remain for some regions. Bailey (1958, 1981) improves upon Thornthwaite's approach by also inferring humidity from temperature. The aforementioned aridity estimates by the UN (1977), Meigs (1953) and Budyko-index approaches all employ different ways of obtaining water demand.

All of the above methods of defining deserts, and degrees of aridity, are climatologically oriented. Other approaches for defining aridity involve analysis of the hydrology, soils and vegetation cover. For example, de Martonne and Aufrère (1927) and de Martonne (1927) employ the fact that water drainage in deserts generally is internal. That is, the one-third of Earth's surface for which the rivers do not reach the sea corresponds closely with arid areas on maps that are based on climate and vegetation. Deserts are also defined solely in terms of the soil's ability to support vegetation. If it is too permeable to retain water for vegetation (coarse sand) or too impermeable to allow vegetation to root (rock), it is called an edaphic desert. Vegetation types have also been employed (Shantz 1956). These non-climatological approaches also have drawbacks that make them difficult to apply in a general way (Thames and Evans 1981). A typical classical qualitative definition of deserts is offered by Petrov (1976), and refers to the climate, vegetation and soils.

Deserts and semideserts are considered to be territories with an extremely arid climate (irregular precipitation, less than 200–250 mm annually, and excessive evaporation), sometimes with dried-up or intermittent streams, absence of regular surface runoff, with saline soils and crusts where sulfates and chlorides predominate. In these areas where organic life is restricted, the plant cover and the fauna are very meager, migration of saline solutions predominates over biogenic processes in the soil, and agriculture is possible only by means of irrigation.

Climatological causes of aridity

The simplest conceptual model of the large-scale, climatological distribution of deserts is based on the global-scale circulation of the atmosphere. The longitudinally averaged vertical cross-section of the pole to equator circulation shows, in each hemisphere, a broad region of subsidence between the convective upward motion of the near-equatorial ITCZ and the stable ascent associated with mid-latitude storms (Fig. 2.4). This band of rain-suppressing subsidence is centered on approximately 30° latitude in the subtropics, where its latitudinal width and position, and its strength, depend on longitude, season and hemisphere. The subsidence zone is reflected at the surface as a subtropical high-pressure belt

characterized by light winds and clear skies. This latitudinal band has sometimes been referred to as the horse latitudes, presumably because prolonged,
clear, calm conditions at sea could force sailing vessels to lighten their cargo and
lessen their water requirement by offloading horses. This large-scale conceptual
model is clearly only a zero-order approximation for explaining the distribution
of arid lands, given the multitude of exceptions to it that are required to explain
the details of the Meigs (1953) aridity distribution in Fig. 3.1. Rain-producing
mechanisms from equatorial latitudes or mid-latitudes seasonally penetrate into
the belt of subtropical highs, reducing the aridity. Conversely, local factors, such
as the **orography** and the distance from maritime water-vapor sources, act to
create regional desert climates outside of this subtropical belt.

The first subsection below will describe the global or planetary forcing mechanisms, and the remaining ones will discuss local or regional factors. These
desert-generating influences have often had their names adopted to describe
the types of deserts that they create. The deserts caused by large-scale subsidence are referred to as **subtropical deserts**, because that is the latitude belt
where the subsidence occurs. Similarly, there are **rain-shadow deserts** caused
by orographic effects; **cool coastal deserts** that are adjacent to cold, coastal ocean
currents; and **continental-interior deserts** that exist because of their distance
from maritime moisture sources.

Planetary-scale circulations

Aridity can be caused by a variety of mechanisms, but most are related to the
circulation of the atmosphere and its heat and moisture budgets over a wide
range of scales. On the largest scale, the global circulation of the atmosphere
suppresses precipitation between about 10° and 30°–40° latitude because of
the widespread and persistent subsidence caused by the downward branch of the
Hadley circulation and the associated subtropical belt of high pressure. Most of
the world's great deserts are found within this latitude belt in both hemispheres:
The Sahara, Arabian, Iranian, Thar, Kalahari, Australian, Monte, Patagonian, and
parts of the Sonoran and Chihuahuan Deserts. Refer to Fig. 2.4 for a schematic
of the vertical circulation of the Hadley Cell, and to Fig. 2.5 for the climatological distribution of the related high-pressure centers in the subtropics. The latter
figure shows that the belt of high pressure is more continuous in the Southern
Hemisphere, with less of the interruption by low-pressure centers seen in the
north. This difference is related to the effect of the mountains, and the solar heating and nocturnal cooling, of the greater land area in the Northern Hemisphere.
The longitudinal average of the sea-level pressure variation with latitude is shown
in Fig. 2.10. Average high pressure prevails in the subtropics of both hemispheres,

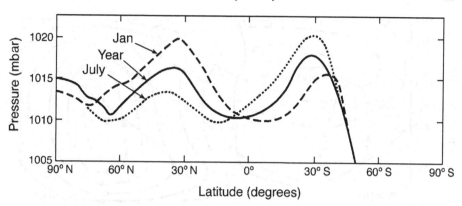

Fig. 2.10 Global average sea-level pressure (millibars) as a function of latitude in both hemispheres, for January, July and the annual mean. (Adapted from Godske *et al.* 1957.)

centered on about 30° latitude, in both the warm and cold seasons. The strength of the pressure maximum in the longitudinal average is greater during the cold season, even though Fig. 2.5 shows that the strengths of the individual subtropical maritime high-pressure centers are generally greater in the warm season. In the summer, the lower pressures over the continents more than compensate for the higher pressures of the maritime maxima. The climatological pressure minimum in mid-latitudes is related to the track of extratropical cyclones, and the minimum near the equator is related to the ITCZ and the associated convection. Parenthetically, the mid-latitude sea-level pressure gradient in the Southern Hemisphere is larger than that in the north, presumably because the fewer mountain ranges and land areas permit a less disturbed and stronger westerly flow. For graphical clarity, the low pressures in the high southern latitudes are not plotted.

The subsidence in the subtropical latitudes has multiple effects on climate: (1) it stabilizes the atmospheric lapse rate of temperature; (2) air with lower water-vapor content is transported downward from the upper troposphere; and (3) the relative humidity decreases because subsiding air warms. All of these effects suppress precipitation. Fig. 2.11 illustrates the area coverage and steadiness of this circulation. The abscissa is plotted as a sine-latitude to reflect the area coverage on the planet associated with each latitude band. This downwelling air that helps maintain many of the deserts of the world spans about one-third of the atmosphere. Examples at two times (night and day) of the vertical motion in the subsidence region are shown in Fig. 2.12, where the values are calculated kinematically from observed vertical profiles of the horizontal winds in the Rub Al Khali Desert in Saudi Arabia at about 20° latitude. The vertical motion, shown in pressure coordinates (Ω), is downward throughout the troposphere in

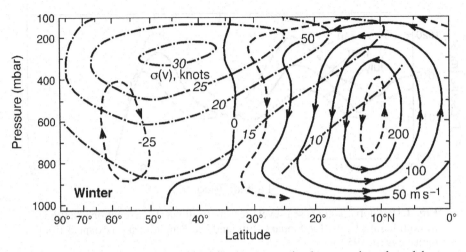

Fig. 2.11 Circulation along a latitudinal cross-section between the pole and the equator for winter. The solid lines indicate the streamlines associated with the Hadley circulation, which parallel the wind vectors in this cross-section plane. The streamlines are labeled in terms of mass transport in millions of short tons per second. The subsidence is shown between 10° and 35° latitude. The steadiness of this subtropical Hadley circulation, relative to the mid-latitude circulation, can be seen in terms of the dashed lines, which reflect the standard deviation (knots) of the wind component in this plane. The abscissa is plotted as a sine-latitude to reflect the area coverage on the planet associated with each latitude band. The downwelling air that helps maintain many of the deserts of the world spans nearly one third of the atmosphere. (From Palmen and Newton 1969.)

the nighttime (Fig. 2.12a), with horizontal divergence below about 600 mb and convergence above. The conditions during the day (Fig. 2.12b) are somewhat different in the low levels because of the effects of the heat low, the dynamics of which are discussed in the last section of this chapter.

The Hadley circulation should be viewed as only the zero-order cause of the subtropical aridity, with other superimposed large-scale processes enhancing or reducing the aridity in different longitudes. Exceptions to the conceptual model of the longitudinal-mean Hadley circulation are numerous. The Asian summer monsoon, which produces over 10 m of rain per year at some locations in India, prevails in the same subtropical latitudes as does the Sahara. Also, the longitudinal-mean subsidence at 25° N in the summer is relatively weak, but this is the driest period of the year at the longitudes of the northern Sahara. A number of mechanisms are responsible for this longitudinal irregularity. There are the seasonal monsoons, as well as the aforementioned Walker-type east–west circulations in equatorial latitudes. In addition, Rodwell and Hoskins (1996)

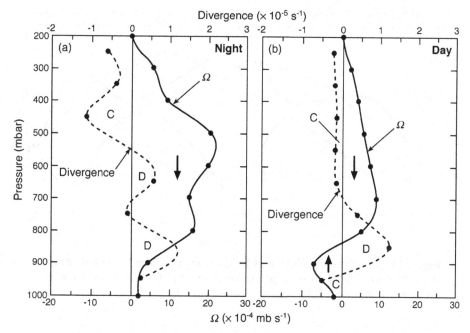

Fig. 2.12 July vertical profiles of vertical motion and horizontal divergence in the subtropics in the Rub Al Khali Desert in Saudi Arabia at about 20° latitude. The vertical motion (Ω) is shown in pressure coordinates (solid lines), and convergence and divergence are indicated as C and D, respectively, in terms of excursions in the dashed line. Vertical motion is depicted by the arrows. The right panel (b) is for daytime conditions and the left panel (a) is for nighttime. (Adapted from Blake *et al.* 1983.)

describe a mechanism by which the Asian summer monsoon may contribute to the summer aridity in the eastern Sahara Desert and in the Kyzylkum Desert of Asia. They speculate that other summer monsoon circulations worldwide may have similar remote effects of contributing to Mediterranean-type climates with dry summers.

As noted earlier, aridity should not be presumed simply because of the low precipitation associated with the subsidence. Fig. 2.13 shows the latitudinal distribution of both annual precipitation and evaporation. Even though these values should be viewed qualitatively because of uncertain conditions over the oceans, the plot shows an excess of evaporation over precipitation between about 10° and 40° latitude in both hemispheres. Again, the abscissa is plotted as a sine-latitude to depict the area involved.

Annual precipitation on the equatorial side of the belt of high pressure is generally dominated by warm-season convective processes, and passages of

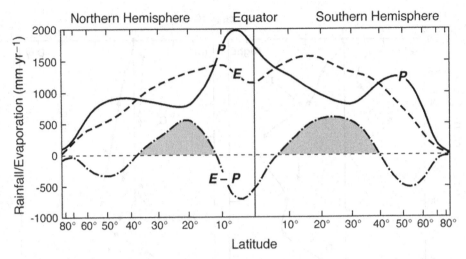

Fig. 2.13 Latitudinal distribution of annual precipitation and evaporation, and the difference. The *P* curve is precipitation and the *E* curve is evaporation. (Adapted from Sellers 1965.)

the ITCZ. On the poleward side, cold-season precipitation from extratropical cyclonic disturbances is responsible for much of the precipitation. As the high-pressure belt and the large-scale rain-suppressing subsidence is approached from the north or the south, the annual precipitation totals progressively decrease, sometimes to virtually zero. The greater the rain-suppressing strength of this immense area, the less the penetration by the occasional cold- or warm-season, precipitation-producing disturbances.

It is curious that the atmospheric heating, which is a first-order expression of these desert climates, can sometimes lead to influences that lessen the aridity. Heating that is localized, on a large or small scale, causes a redistribution of atmospheric mass between the column that is heated greatly (the arid land area in this case) and the surrounding atmosphere that is heated less (the less arid area). The consequence is a lowering of the surface pressure at the base of the atmosphere that is heated the most. The development of this **heat low**, especially in the summer, sometimes locally overcomes the high pressure associated with the global circulation. Fig. 3.12 shows this effect over the Sahara in North Africa. The anticyclone dominates in the winter, but lower pressure prevails in the summer. The low-level air inflow into this desert-scale low pressure can often augment the rain-bringing monsoon circulation associated with normal land–sea summer temperature contrasts. The result of this enhanced summer monsoon is deeper penetration of convective precipitation into the arid area. Thus, deserts in the summer subtropics can be areas of intense large-scale low pressure, rather

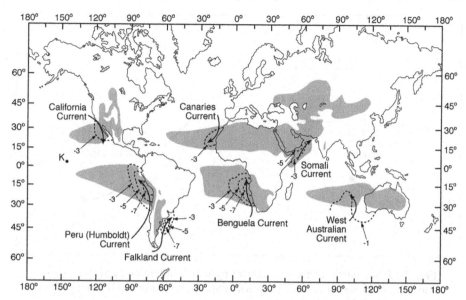

Fig. 2.14 The names and locations of cold, coastal ocean currents, and the approximate temperature anomaly (°C) of the surface water, relative to that of the water more distant from the coast. Also shown are oceanic extensions of the dry climates. K shows the location of Kiritimati Island (see p. 000). (Adapted from Logan 1968.)

than the converse. The dynamics of the heat low will be described in the last section of this chapter.

Our discussion of deserts has been, so far, limited to land areas. But, it is important to recognize that the large-scale subtropical deserts also cover vast areas of the oceans. Because of a lack of regular precipitation observations, the limits of such ocean deserts are not well documented, but an estimate is shown in Fig. 2.14. Their existence is solely related to the position of the subtropical subsidence and high-pressure belts. Fig. 15.10 illustrates the precipitation for Kiritimati Island (formerly Christmas Island) in the central Pacific Ocean. It is clearly arid, except during El Niño years when the circulation is disturbed.

Orographic effects

The effects of orography on aridity are paradoxical. On the one hand, large-scale elevated terrain features can wring water from the air passing over them (Fig. 2.6b), and the dried and latent-heat-warmed air may contribute to the existence of deserts on the downwind side. However, desert rain is often focussed on the cooler high altitudes of mountains that are within the deserts and on their fringes. This lessens the aridity, and the areas are called **altitude oases**. There is

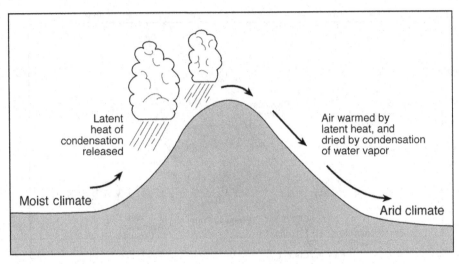

Latent
heat of
condensation
released

Air warmed by
latent heat, and
dried by condensation
of water vapor

Moist climate

Arid climate

Fig. 2.15 Schematic of the mechanism by which mountain barriers can produce rain shadows, or arid areas, downwind. Precipitation reaches the ground on the upwind side of the barrier.

no real difference in the physical processes in the two situations. However, the desert-causing mountains are often (but not always) larger in scale than the oasis-causing mountains. In both cases, the mountains force the uplift of air, causing more precipitation over the higher elevations and a possible deficit downwind. The difference is that, in otherwise-rainy areas, the rain deficit is viewed as an anomaly and the place is called arid. But, in an arid area the downwind deficit is little noticed, and the rainfall excess is the anomaly. The desert-causing influences of terrain features are treated here; Chapter 13 describes orographic causes of desert rain. Among the deserts of the world, the following are wholly or partly caused by different orographic effects: the Eurasian Deserts from Russia to Mongolia, the Monte Desert, the Patagonian Desert, all the deserts of North America, and the semi-arid Great Plains of North America. Thus, in North America, South America, and Asia, the largest arid regions are immediately downwind of the major mountain ranges. These arid areas tend to be in higher latitudes than the subtropical deserts.

A few causes of aridity are related to the geographic distribution of mountain ranges. First, mountains can block the flow of moist air from a source region such as the oceans, or cause winter storms to weaken. More commonly, some of the air flows over the mountains, but in the process moisture is lost though precipitation. The schematic in Fig. 2.15 illustrates this mechanism. As the mountain barrier forces the air upward, the air cools to its dew-point temperature and clouds form.

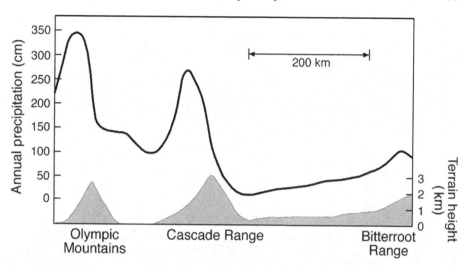

Fig. 2.16 Annual precipitation along an east–west line between the Washington State coastline and western Montana in North America, illustrating the influence of the orography on the distribution of wet and arid mesoclimates. (From Neuberger and Cahir 1969.)

This releases the latent heat of condensation into the air, and removes water vapor. If precipitation does not form within the cloud, or if the precipitation does not reach the ground, the liquid water will eventually evaporate into the airstream as the subsiding air warms on the downstream side of the mountain, replacing the water vapor lost and consuming the heat gained earlier in the condensation process. However, if the condensed water reaches the ground, the air-drying and -heating process cannot be reversed downwind. Thus, the air that subsides on the lee side has been dried and warmed, and within that air it is more difficult to produce precipitation by stable or buoyant processes. A factor that may lessen the efficiency of this process is that air heated on the upwind side may not very effectively run down the lee slope. If higher mid-level parcels of air descend instead, this may produce less of an effect on downwind rainfall because this air has not been warmed and dried to the same degree. On the other hand, mid-level air is drier because the moisture content decreases with height, so the effect on diminishing the lee side rainfall may be the same. This rain-shadow effect can modulate the precipitation climatology and the aridity on the scale of large mountain ranges, or on smaller scales. For example, Fig. 2.16 shows a plot of the annual precipitation along a line extending eastward from the Pacific coast of Washington State of the United States, crossing the Olympic Mountains on the Olympic Peninsula, the Cascade Mountains in western Washington State, and the Bitterroot Mountains in eastern Idaho and western Montana in North

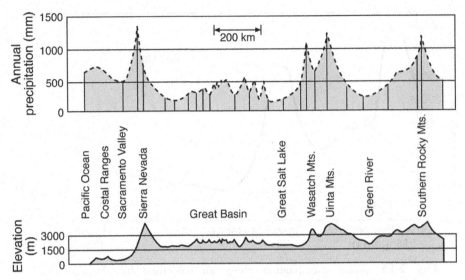

Fig. 2.17 Annual precipitation and terrain elevation along an east–west line from San Francisco on the Pacific Ocean, across the Great Basin Desert, to Denver to the east of the Rocky Mountains in North America. The driest areas are on the east, or downwind, side of the highest mountains. (Adapted from Bailey 1941.)

America. Each time the prevailing westerly-flowing air crosses a mountain range, it releases some of its Pacific-acquired moisture. The subsiding air on the lee side is warmer than it was upwind because of the latent heat released with the condensation, it is drier, and it is more stable. This leads to a rain shadow on the lee side that is characterized by lower precipitation than if the mountain had not been there to modify the atmosphere. Every time the flow goes over the next range of mountains, more moisture is squeezed out and the regional climate on the lee side becomes even drier. Additional examples of this rain-shadow effect are seen in Figs. 2.17 and 2.18. Fig. 2.17 is on a larger scale, extending from the Pacific Ocean to the east of the Rocky Mountains of North America. The correlation of the annual precipitation with the major mountains ranges is clear, with the more-arid areas prevailing on the leeward, or eastern, sides. This rain-shadow effect in the northwestern United States can also be seen in Figs. 3.1 and 3.29. Fig. 2.18 shows the rapid decrease in precipitation to the east of the Andes Mountains in the Monte Desert.

On much smaller scales, mountainous tropical islands in rainy climates can exhibit distinct rain shadows, with the lee side containing small local deserts. Examples include the island of Hispaniola, and islands in the Hawaiian and Lesser Antilles chains. The aridity on the southwest side of the island of Madagascar is also associated with a rain shadow. Fig. 2.19 shows that the

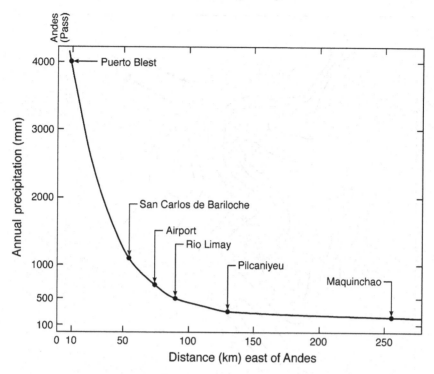

Fig. 2.18 Annual precipitation as a function of distance eastward into the Monte Desert from the peak of the Andes Mountains. (From Soriano 1983.)

upwind east side of the island of Hawaii, which faces the trade winds, has an annual rainfall in excess of 800 cm, whereas parts of the downwind coast and interior are arid. The exact locations of the arid microclimates are likely a function of the way in which the low-level trade winds are channeled by the local orography. The large-scale average annual rainfall, unaffected by the islands, is 65–75 cm. Irrigation channels have even been constructed from the wet side to the arid side of some island mountains. This orographic effect on the climate of these islands is so pronounced because they are in the trade wind regime that has a fairly steady direction. Similarly, mountainous locations within a steady seasonal monsoon flow are also prone to rain-shadow effects. In contrast, if the wind direction is highly variable throughout the year, there is going to be a less noticeable effect on the climate in the area surrounding the mountains. Small-scale rain-shadow effects can also be seen where hydrologically closed basins are encircled by high mountains. For example, in the Monte Desert such basins have a very arid nucleus that results from the rain shadow of the mountains to the west. In one case (Mares *et al.* 1985), rainfall varies from 344 mm to only 77 mm per year over a 50 km distance.

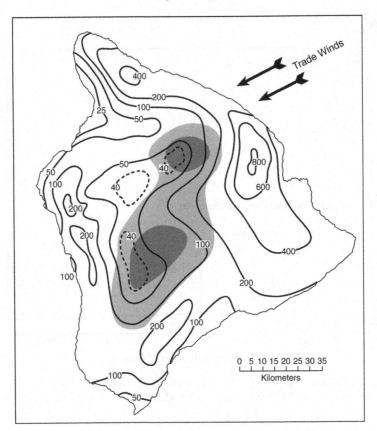

Fig. 2.19 Annual average rainfall for the island of Hawaii in the Pacific Ocean trade winds. The isohyets are plotted at an irregular interval, and labeled in centimeters. Terrain elevations in excess of 2 km are shaded, with elevations greater than 3 km in dark gray. The terrain begins sloping upward at the coastline. The ambient rainfall over the surrounding ocean is 60–75 cm. Arid areas of 25–50 cm per year are localized on the lee side of the island's mountains. (Adapted from Armstrong 1983.)

Another mechanism by which large mountain ranges may affect aridity can be understood in the context of the conservation of *potential vorticity*, Z_θ, where

$$Z_\theta = (\zeta_\theta + f)\frac{\partial \theta}{\partial p}. \tag{2.2}$$

The quantity $\zeta_\theta + f$ is the vertical component of the *absolute vorticity*, which is a measure of the rotation of the atmosphere at a point. This rotation is a sum of the rotation of the Earth–atmosphere planetary system (f, unrelated to air motion relative to the planet) and the rotation of the atmosphere relative to the surface of Earth (ζ_θ, the *relative vorticity*). The Coriolis parameter, f, represents the vertical component of Earth's vorticity, or rotation, and is defined as

$$f = 2\Omega \sin \phi, \tag{2.3}$$

where Ω is the rotational frequency of Earth ($2\pi/86\,400$ s^{-1}), and ϕ is the latitude. The potential temperature, θ, is defined as the temperature that a parcel of air would have if it were moved adiabatically to a pressure of 1000 mbar. A parcel of air for which the adiabatic assumption applies (no exchange of heat with the environment) conserves its potential temperature as it moves horizontally and vertically. The potential temperature of a parcel is defined by Poisson's equation, which is derived from the First Law of Thermodynamics, as

$$\theta = T \left(\frac{p}{1000} \right)^{R/c_p}, \tag{2.4}$$

with R the gas constant (287 J kg^{-1} K^{-1}), c_p the specific heat at constant pressure (1005 J kg^{-1} K^{-1}), p the pressure in millibars (mbar), and T the temperature in degrees Kelvin. The θ subscripts on Z and ζ in Eqn. 2.2 imply that horizontal derivatives are evaluated at constant θ.

The potential vorticity of a parcel of air is conserved in **inviscid** (no internal friction, or viscosity), adiabatic, **hydrostatic** flow. As a layer of westerly tropospheric air flows up the western slope of a north–south-oriented mountain range (Fig. 2.20), the air remains confined within the layer of depth $\Delta\theta$ because the potential temperatures of the parcels at the top and bottom of the layer do not change in adiabatic motion. Near the surface, the flow follows the topography, and it is assumed that there is a relatively undisturbed layer at some greater distance above the ground. Because $\Delta\theta$ does not change, but Δp becomes smaller in magnitude as the layer of air decreases in depth toward the top of the mountain, $\partial\theta/\partial p$ in Eqn. 2.2 becomes more negative. But, because Z_θ must be conserved, the absolute vorticity, $\zeta_\theta + f$, must decrease in magnitude. For f positive, ζ_θ must develop less positive or more negative values. If we assume that the horizontal shear does not change and affect the vorticity, the flow must thus acquire anticyclonic curvature (i.e. negative relative vorticity, ζ_θ) by turning equatorward. However, f decreases with southward displacement (Eqn. 2.3), reducing the magnitude required of the developing anticyclonic curvature. At the top of the mountain, $\partial\theta/\partial p$ has its maximum negative value, and therefore the absolute vorticity must be a minimum. If we assume only a small southward displacement, and therefore a small decrease in f, the anticyclonic curvature will be a maximum there. As Δp returns to its original values on the lee side, f is now smaller in magnitude and the flow will develop cyclonic curvature and turn toward the north. Associated with the anticyclonic curvature, a ridge of high pressure prevails over the mountains, and a low-pressure trough with cyclonic

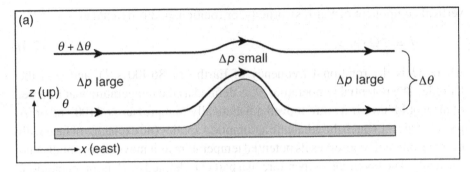

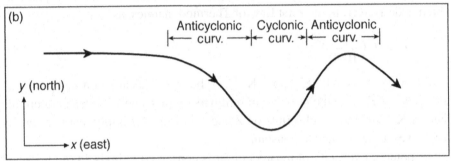

Fig. 2.20 Schematic showing how the conservation of potential vorticity causes the development of regions of cyclonic and anticyclonic flow over and downwind of a mountain barrier in a westerly low-level flow. (Adapted from Holton 1992.)

curvature exists downwind (Riehl *et al.* 1954). Even though this argument is a simplification of the actual dynamics (Smith 1979b), it may help explain how large mountain ranges such as the Rockies, the Alps, the Himalayas, and the Andes create very long stationary waves in the deep tropospheric westerlies (the waves are of smaller amplitude in the summer season). In North America and Asia, the ridges are centered over the Rocky Mountains and the Tibetan Plateau, and the troughs are located near the east coasts of the continents (van Loon *et al.* 1973). Fig. 2.21 illustrates the climatology of the wave in the 500 mbar heights and in the sea-level pressure, for the 40–50° N latitude belt. The wave pattern results from the thermal effects of the land masses, the thermal effects of the mountains (Fig. 2.6f), and the barrier effect of the mountains. The figure shows the wave to be much more prominent in the winter season. There are at least two dynamic effects associated with this mountain-caused wave pattern that can contribute to aridity. Where the flow is anticyclonic near the surface, there is low-level divergence and subsidence, which suppresses precipitation. Also, it can be shown (Holton 1992; Dutton 1976) that the region between the

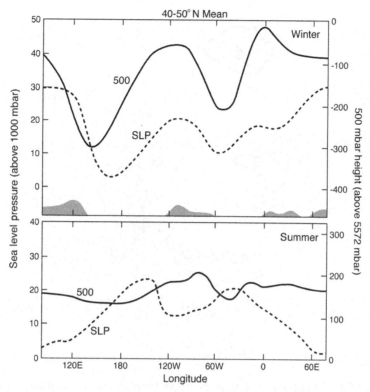

Fig. 2.21 Stationary atmospheric waves in the winter (upper) and summer (lower) in terms of the 40–50° N latitude, sea-level pressure and 500 mbar height variation as a function of longitude. (Adapted from Hoskins *et al.* 1989.)

upper-air ridge over the mountains and the trough to the east is conducive to the development of anticyclones (high-pressure centers), not cyclonic storms, and that this region is characterized by subsidence. Thus, precipitation related to extratropical cyclones is suppressed by the mountain-induced wave in this wide longitude belt. The dryness of the North American Plains, which continue to decrease in elevation for many hundreds of kilometers to the east of the Rocky Mountains, is possibly at least partly attributable to this effect. And the aridity of central Asia is relatable to the long wave generated by the Tibetan Plateau. Similar perturbation of the westerly large-scale flow by the Andes has been documented in southern South America (Boffi 1949), and may contribute to the dryness of the Patagonian and Monte Deserts.

This effect on aridity of the long waves forced by orography is documented by Manabe and Broccoli (1990) and Broccoli and Manabe (1992) in numerical global-model simulations of the atmosphere over a full annual cycle. In a

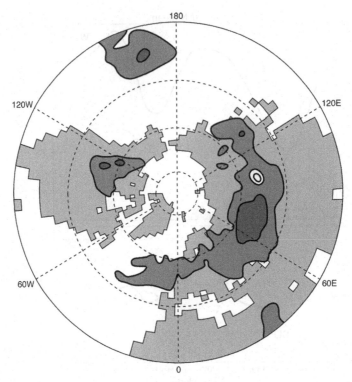

Fig. 2.22 The difference between Northern Hemisphere annual precipitation simulated by a global numerical model, with and without mountains. Light gray shows land areas, medium gray indicates the areas where the rainfall is less with the mountains by an amount greater than 1 mm d^{-1}, and darker gray delineates areas of mountain drying of greater than 2 mm d^{-1}. (Adapted from Broccoli and Manabe 1992.)

simulation with mountains, the cyclonic-storm activity was suppressed over the mid-latitude arid regions of Asia and North America. Without the mountains represented in the model, the mid-latitude storm activity and related precipitation was as great in these areas as in other longitudes. Fig. 2.22 shows the simulated effect of the mountains in decreasing the precipitation. Areas in which precipitation was enhanced by the orography are not shown, for graphical clarity. The model simulated a mountain drying effect in the southern North American Great Plains, but the amount is less than the large 1 mm d^{-1} plotting threshold. Farther north, to the east of the Canadian Rocky Mountains, the annual rainfall in the prairie was about a factor of three greater without the mountains. The seasonal effect of the mountains on the simulated rainfall for an area in central Asia is illustrated in Fig. 2.23. There is clearly a significant influence of the mountains in all seasons. It should be kept in mind that the model physics represents all of the possible effects of the mountains on precipitation: rain shadows, the effect

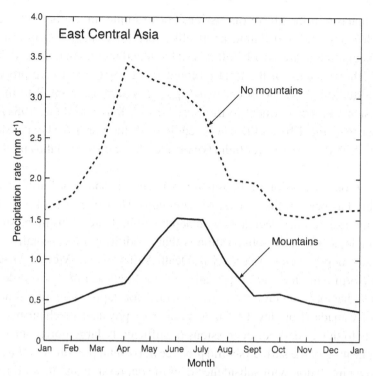

Fig. 2.23 Seasonal variation in model-simulated monthly precipitation (mm d^{-1}) in east-central Asia, with and without mountains. These are the same simulations that provided the data for Fig. 2.22. (Adapted from Broccoli and Manabe 1992.)

of the long waves on the downstream suppression of cyclones, and the blocking of flow from upstream moisture sources. Even though the model has coarse horizontal resolution and poorly resolved orography, the figures clearly show the strong aggregate effect of all orographic processes on the suppression of precipitation. The fact that Asia does not have a north–south-oriented, upwind, blocking ridge, as exists in North America, perhaps implies that the effect of the orography there is more through the development of long waves in the westerlies rather than through the rain-shadow effect.

A complicating issue is that the lee sides of mountain ranges are sometimes favored regions (along with coastlines) for the genesis of cyclonic storms in mid-latitudes. Horizontal temperature contrasts can be greater near the slopes, and such storms derive their energy from these gradients. Apparently the precipitation from such incipient storms is not sufficient to negate the otherwise aridifying effect of the mountains on the downwind climate.

It is important to understand that the dominant effect of large-scale mountains on precipitation in arid lands is not only the development of deficits, in spite of the emphasis in the above discussion. For example, monsoon circulations, on which many arid and humid areas alike rely for a large fraction of their annual precipitation, are greatly influenced by inland mountain ranges. In North America, the existence of the Rocky Mountains is important to the properties of the North American monsoon, which provides summer rainfall to the arid southwest part of the continent (Barlow *et al.* 1998; Tang and Reiter 1984). And the presence of the Tibetan Plateau is required for the northward expansion of the south-Asian monsoon over India (Hahn and Manabe 1975; Hahn and Shukla 1976).

Thus far, our discussion of the relation between the shape of Earth's surface and aridity has been in the context of mountains. However, it is observed that precipitation sometimes decreases with depth within depressions in the surface. For example, a qualitative observation is that conditions are especially arid at the greatest depths below sea level in Death Valley in the Mojave Desert in western North America and in the valley of the Dead Sea in the Middle East. This effect has been quantitatively documented for depressions in a semi-arid prairie of Canada (Longley 1975). A number of physical mechanisms could contribute to this effect. For depressions of sufficiently large horizontal extent, the solar heating of the surface should produce horizontal pressure contrasts, and a resulting circulation with subsidence over the depression and upward motion over the surrounding slopes and plains. This daytime subsidence should suppress precipitation over the lower elevations. Also, over depressions, precipitation must fall through a deeper column of atmosphere, and greater evaporation can occur. It is also arguable that the warming associated with the subsidence could result in higher daytime temperatures in the depressions.

Geographic remoteness from moisture sources

Most of the atmosphere's water vapor originates as evaporation from the oceans. Thus, the more remote a location from this source, the less likely that the water vapor will survive the long transit without being precipitated in a storm of some sort. For example, this is likely a contributing reason for the existence of the vast deserts of central Asia (see Fig. 3.1). In winter, the flow there is northerly, associated with the Siberian high-pressure system, so the air originates in cold high latitudes and its water-vapor content is low. In the summer, these same areas experience dry northwesterly winds that are associated with the Azores high and that have had a long trajectory across northern Asia. Only in the eastern Taklamakan Desert do the monthly precipitation totals betray the influence of the summer monsoon flow from the Pacific Ocean. Thus, remoteness from moisture

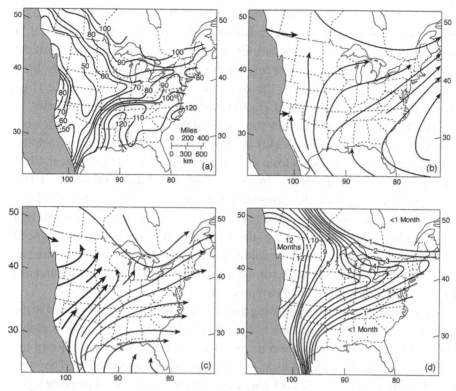

Fig. 2.24 (a) Percent of normal rainfall in North America to the east of the Rocky Mountains during drought years. (b) Streamlines of July-mean at the top of the mixed layer for non-drought years. The heavier streamlines show airflow that originates at the base of the mountains. (c) Same as (b), except for July 1934, a drought year. (d) Average number of months per year with a mean transport of air from the eastern base of the Rocky Mountains. (Adapted from Borchert 1950.)

sources must naturally be viewed in terms of the prevailing wind direction. In North America, the Rocky Mountain barrier in the west is sufficiently wide and high to insulate the central and eastern continent from most Pacific moisture, but there is considerable available moisture from airstreams with their origin over the warm Gulf of Mexico. Thus, variability in the source regions of air on time scales of weeks to years can cause the degree of aridity in a region to vary substantially. For example, the High Plains prairie to the east of the Rocky Mountains is well known for the frequency with which it experiences extended droughts. These droughts are correlated with periods in which the air over the Plains persistently originates only from the mountains and continental interior to the west. Fig. 2.24a shows the percent of the normal July precipitation observed during years in which a major drought has prevailed in the area. A tongue of precipitation deficit extends eastward across the central continent in the High Plains. Also shown are the airflow patterns, at a level near the top of the boundary layer, based on average July conditions (Fig. 2.24b) and for July 1934 (Fig. 2.24c), which was a major drought year. The low-level air over the High Plains during the drought year originated more from the continental interior, in contrast to a normal year in which the source of the air was the Gulf of Mexico. Another revealing map (Fig. 2.24d) shows the number of months in which the monthly mean flow originates from the base of the Rockies. The area near 40° N latitude with a dominant airflow from this direction of the continental interior corresponds closely with the eastward extension of the semi-arid High Plains prairie. Thus, even during non-drought years, the distribution of the normal High Plains aridity seems to be related to the degree to which the air has experienced the long traverse of the western continent. It is also likely that a rain-shadow effect may be contributing to the drought when the flow is from the mountains, but it is difficult to separate the two processes.

A revealing method for diagnosing the degree to which aridity is related to the lack of upwind water-vapor sources over the ocean is through the use of global models to trace water vapor from its origin as evaporation at the surface of the ocean or land to its removal from the atmosphere as precipitation. The most complete approach for such tracking of water vapor is described and employed by Bosilovich and Schubert (2002). This method was used for the creation of Fig. 2.25, which illustrates the percent of local precipitation that results from water vapor originating as evaporation at the land surface. With some exceptions, there is a good general correspondence between arid climates and areas for which a small percent of the annual precipitation originates as ocean water. The prevailing direction of the wind circulation, the blocking effect of mountains on the airstream, and the distance from oceans conspire to shield these areas from the most prolific sources of water vapor: the oceans.

Coastal effects

In spite of the fact that the oceans are the greatest source of atmospheric moisture, barren and dry environments bounded by the sea are not uncommon. Such areas extend for great distances along the west coasts of North and South America, Australia and Africa; along the north, west and east coasts of the Sahara; along almost the entire coast of the Arabian Peninsula; and along the southwest coasts of Asia. Many smaller coastal areas with arid climates exist, including numerous desert islands. Amiran and Wilson (1973), Meigs (1966) and Lydolph (1973)

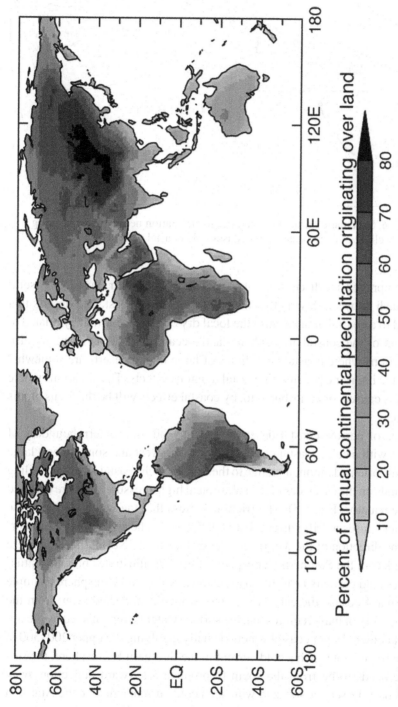

Fig. 2.25 The percent of local precipitation that results from water vapor originating as evaporation at the land surface, calculated using the approach described by Bosilovich and Schubert (2002). (Courtesy Michael Bosilovich, National Aeronautics and Space Administration, Goddard Space Right Center, Data Assimilation Office.)

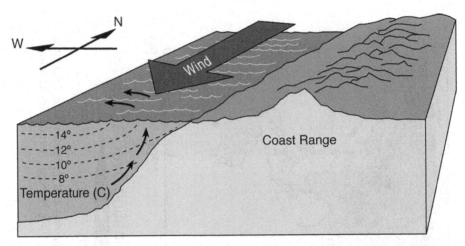

Fig. 2.26 Schematic of the wind and ocean circulation near a coast, showing how upwelling of deep ocean water causes cold coastal currents.

provide comprehensive discussions of coastal deserts and their causes. It is useful to distinguish between deserts that happen to include a coast, and deserts that are limited to a coastal area because the local dry climate is intimately related to some aspect of the coastal atmosphere–land–ocean interaction. There are also some situations where coastal boundaries of large inland deserts are somewhat drier than the interiors, because of coastal dynamic effects. The situations where the aridity is caused, or contributed to, by coastal effects will be the focus of this discussion.

One type of narrow coastal desert is typically on the western boundary of continents, with a cold, coastal ocean current to the west and sometimes a large north–south-oriented mountain range to the east. In such coastal areas, prevailing surface winds on the east sides of the subtropical high-pressure areas are oriented toward the equator (Fig. 2.5). The frictional stress that the winds impart to the ocean water causes a drift current that is deflected to the right (facing equatorward, in the direction of the wind) near the surface in the Northern Hemisphere and to the left in the Southern Hemisphere. Fig. 2.26 illustrates how this situation creates cold currents near the coast, using a Northern Hemisphere example of California. Because the current near the surface is deflected away from the coast, continuity of mass requires that the surface water be replenished by the upwelling of colder, deeper coastal waters (usually involving the upper 200–300 m of water). To generate such a cold current in the Southern Hemisphere, the surface wind needs to be from the south to produce a leftward deflection away from the coast. Deserts associated with such cold currents include the Atacama and Peruvian Deserts of South America, and the Namib Desert of southwestern

Africa. Other cold currents with nearby coastal deserts are depicted in Fig. 2.14, along with an indication of the typical temperature anomaly of the surface water. Also included are the coastal Sahara in northwest Africa, the coast of southwestern North America, the coasts of the Arabian Peninsula and the Horn of Africa, and the western coast of Australia. The coldness of the Falkland Current along the coast of southeastern South America is not so much associated with upwelling, but it is nevertheless cold because of its origin in the Weddell Sea in more polar latitudes. Icebergs are even transported equatorward of 40° latitude in this current, adjacent to the Patagonian Desert. Even though precipitation data are sparse in most of these areas, there is some evidence that the precipitation is a minimum at the coast, and increases to the seaward and landward. All these coastal deserts lie at similar latitudes, are commonly on the western boundaries of continents, are influenced by subtropical high-pressure cells, and (with the possible exception of Australia) have a nearby cold ocean current. Lydolph (1973) suggests that the Somali Desert of east Africa might be included in this type of coastal desert because of the cold water of the Somali Current near the coast. To the north, the southeast coast of Oman is also a classic coastal desert with dense fog, an annual variation of monthly average temperatures of as little as 5 °C, and a coastal current with water temperatures as low as 16 °C (Price *et al.* 1985).

Figure 2.27 depicts the latitudinal variation in annual precipitation for some of the above west-coastal deserts. The minima generally prevail between 15° and 30° latitude, in the vicinity of the strongest subsidence from the subtropical high-pressure belt (Fig. 2.11). However, other factors must cause the aridity maximum at the coast. For example, the stabilizing effect of the upwind cold ocean current on the temperature lapse rate in the lower troposphere would contribute to a coastal minimum in precipitation over land. That is, the stability would be greatest at the coastline because the eastward movement of the marine air layer over land that is warmer than the sea would cause the layer to become less stable and more supportive of moist convection. Fig. 2.28 shows an example of such high stability in the temperature profile for the coasts of Peru and Chile. The radiosonde temperature profile is a climatological average for coastal Lima, Peru, and Antofagasta, Chile, and reflects a temperature inversion between 500 m and 1500 m above the surface, at the top of the cool marine air layer. Daytime temperature measurements (open circles) along highways between coastal Mollendo, Peru, and higher-elevation inland locations also show the existence of a strong inversion. For the stable, marine air mass to move over the coastal area and influence the climate, it is necessary to suppose the dominance of onshore winds. This situation is, in fact, realized a large fraction of the time as a result of monsoon and sea breeze winds that are forced by the warmer continental

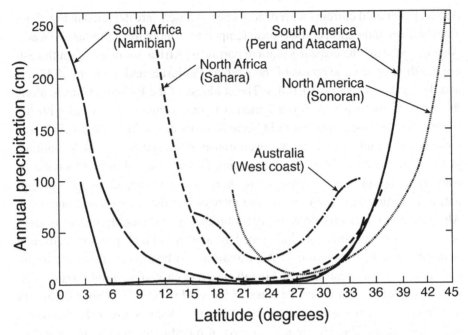

Fig. 2.27 Annual rainfall as a function of latitude (north and south) for five coastal deserts. (From Lydolph 1973.)

interior. These onshore circulations near the surface sufficiently deflect the planetary-scale flow so that the movement of the cool marine air is onshore. In the summer, this thermal onshore flow, and consequently the marine influence, is greatest. During the winter, higher pressures over land cause more frequent offshore winds.

There is some evidence that irregular coastline geometries that deflect the cold current away from the coast, and allow higher sea-surface temperatures to develop near the shore, can lead to higher local precipitation amounts and reduced aridity. For example, sharp increases in rainfall occur near coastal capes or protrusions in northern Peru and along the west coast of Africa.

Orographic barriers near the coastal zone, such as occur in North America and South America, may have multiple contributing effects on the aridity. First, the orography blocks the intrusion of most precipitation-producing weather systems from the east. Secondly, the orography represents an eastern terminus of the semi-permanent or eastward-migrating subtropical high-pressure systems. The effects of the blocking orography on anchoring the high pressure and anticyclonic wind field may cause enhanced subsidence to the west of the mountains over the coastal desert. Lastly, any upward motion over the heated, elevated terrain (Fig. 2.6f)

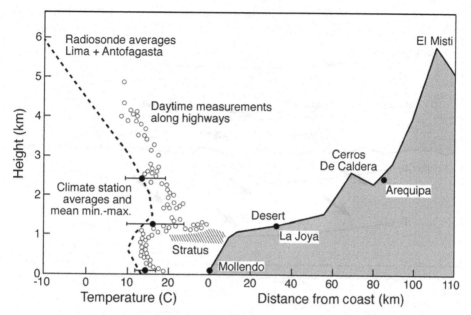

Fig. 2.28 The climatological average vertical temperature profile for coastal Lima, Peru, and Antofagasta, Chile (dashed line). The open circles are daytime temperature measurements along highways between coastal Mollendo, Peru, and higher-elevation inland locations. The horizontal bars show the averages and mean maximum and minimum temperatures for the three climate stations shown (Mollendo, La Joya, and Arequipa). (Adapted from Lettau 1978a.)

will have compensating subsidence on both sides, including the already arid coastal strip to the west.

Another mechanism by which coastlines may contribute to low precipitation is related to the fact that the frictional drag of Earth's surface on the atmosphere causes the near-surface winds to be deflected toward low pressure; for example, to the left of the **geostrophic wind**[2.5] in the Northern Hemisphere. The magnitude of this deflection is proportional to the magnitude of the frictional drag, which is greater over the land surface than over the smoother sea surface. Thus, if the large-scale wind is parallel to a coast in the Northern Hemisphere, with the land on the left looking downwind, the wind on the landward side of the coastline will be deflected more to the left than will the wind to seaward. The resulting low-level divergence will induce subsidence centered on the coastline, and this

[2.5] The geostrophic wind is a hypothetical wind that results from a balance of the pressure gradient and Coriolis forces. It is oriented parallel to the isobars, with low pressure to the left in the Northern Hemisphere, looking in the direction of the wind vector, and the wind magnitude is proportional to the pressure gradient.

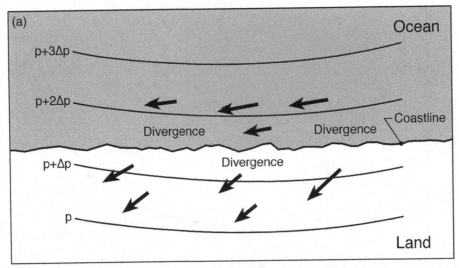

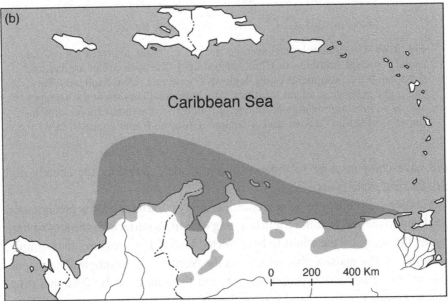

Fig. 2.29 (a) The development of low-level divergence when the large-scale, low-level wind is parallel to a coast (Northern Hemisphere example). The frictional stress of the land surface on the air is greater than that of the sea surface, creating a force balance such that the surface wind over land is deflected more across the isobars toward low pressure. This creates low-level divergence near the coast, and conservation of mass requires compensating subsidence. The Δp signifies a positive pressure increment. (b) Semi-arid area (dark shading) along the Caribbean coast of Venezuela and Colombia (latitude about 10–12° N), possibly caused by this mechanism (Lahey 1973). The trade winds are persistently easterly and parallel to the coast.

subsidence will inhibit precipitation. Fig. 2.29a shows this situation schematically. If this coast-parallel wind is a prevailing part of the local climatology, so will be the tendency for the subsidence and lower precipitation over the coastline. In the Southern Hemisphere, a similar effect will exist when the large-scale wind is parallel to the coast, with the land to the right looking downwind. Lahey (1973) ascribes the moderate aridity along the north coasts of Venezuela and Colombia, and in the southern Caribbean, to this effect (Fig 2.29b). Here, the prevailing easterly trade winds are roughly parallel to the coast.

Dynamic feedback mechanisms that may cause and sustain deserts

Surface–atmosphere feedback mechanisms are discussed in different contexts in this book, but here the focus will be on those dynamic processes that may develop, enhance, or sustain arid areas. A feedback process is one in which an initial perturbation to a system causes a system response such that the departure from equilibrium either increases or decreases. If the system causes an increased departure from the original state, it is a positive feedback. If the departure lessens, a negative feedback is at work. The coupled system that includes desert atmospheric, biospheric, and surface-hydrologic processes is so complex that there are myriad positive and negative feedbacks, many of which are too weak to be of much consequence most of the time.

Feedback processes that affect precipitation can act through the vertical motion or the atmospheric moisture content. That is, both upward motion and sufficient water vapor to permit saturation are required for the production of precipitation, and thus processes that affect either can change the degree of aridity. Both large-scale processes, such as the El Niño–Southern Oscillation (ENSO) and the North Atlantic Oscillation, and local processes, may be involved. In fact, it is reasonable that large-scale processes might initiate a precipitation anomaly, and local feedbacks could then enhance or mitigate the large-scale influence. For example, a vegetation reduction could result from large-scale causes of a precipitation deficit, and then a local vegetation–albedo feedback, or one of the others described below, might enhance the departure from the normal. Some possible feedback mechanisms are summarized below. Their importance, and even the sign of the feedback in some cases, is a matter of debate.

When discussing feedback mechanisms that might lead to a more desert-like regional or global climate, it is useful to keep in mind the distinction between **transitive** and **intransitive** climate systems (Lorenz 1968). A transitive climate system is one in which there is only one permitted set of long-term climate statistics: that is, given a particular set of external forcing parameters for the atmosphere, such as the orography, the solar input, Earth's rotation rate, etc., there is only one stable long-term climate. In contrast, an intransitive system has more than one possible stable climate, with the prevailing climate determined by the present state of the system. A special type of system is an **almost-intransitive** one, in which the climate remains within a regime for a finite time, with the system then migrating into another equally acceptable regime without any change in the external forcing. In other words, climate regimes have sufficient "inertia," in a dynamic sense, to be self-perpetuating for a period of time. It will be seen in the following discussion that it might be possible for positive atmosphere–surface feedbacks to lead to fairly stable regional climates that can endure for a significant period. For example, when a drought occurs in a region, feedbacks that perpetuate that condition could lead to a fairly stable (say, metastable) regional climate. This is consistent with the observed situation where distinct periods of prolonged regional drought can transition abruptly into periods with normal or abundant precipitation.

Vegetation – albedo/transpiration feedback

A vegetated desert surface is generally less reflective than one that is unvegetated. That is, the foliage has a lower albedo than does the bare desert surface. If a drought causes more of the ground surface to be bared, because vegetation either dies or becomes covered with dust or sand, the albedo will increase. The argument offered is that less solar energy will be absorbed by the ground, the lower atmosphere will be heated less, the atmospheric lapse rate of temperature will be more stable, and there will possibly be less convection to produce precipitation. Less precipitation will lead to further reduced vegetation, representing a positive vegetation–albedo feedback. However, surface temperatures are often observed to be higher than normal (not lower) during droughts, perhaps as a result of the fact that less solar heat is used for direct evaporation of soil water and transpiration by vegetation. Thus, other processes may be at work. In particular, the decreased transpiration that results from decreased vegetation during a drought represents another positive feedback. Less vegetation means less release of water vapor into the atmosphere by transpiration, which might lead to less precipitation in the same area and a further decrease in vegetation, etc. This feedback mechanism is analogous to the one discussed in Chapter 18 wherein deforestation may begin a feedback process that leads to desertification. Important to such a vegetation–transpiration feedback is the amount of local precipitation recycling that occurs (discussed later in this chapter). Discussion of these feedbacks, on both sides of the argument, can be found in Charney (1975), Charney *et al.* (1975, 1977), Xue and Shukla (1993), Nicholson (1989b),

Otterman (1974, 1989), Berkofsky (1976), Ellsaesser *et al.* (1976), Sud and Fennessey (1982), Jackson and Idso (1975), Ripley (1976a,b), Idso (1977), Idso and Deardorff (1978), Wendler and Eaton (1983), Laval and Picon (1986), Sud and Smith (1985), Zheng and Eltahir (1997), and Courel *et al.* (1984). It is worth noting that the sensitivity of natural vegetation productivity to rainfall amount is generally greater in the desert than in more humid climates. There are dozens of examples of this correlation, but Seely (1978) shows it for Namib desert grasses, and Tucker *et al.* (1991) illustrate it for the varied Sahel vegetation.

Dust–radiation feedback

If a drought causes desiccation of the soil, and the loss of vegetation that normally helps prevent erosion of even dry soil, dust can become elevated into the atmosphere. The radiative effects of desert dust result from the scattering and absorption of incoming solar radiation and from the absorption of outgoing terrestrial infrared radiation. If the net effect of the dust is to locally cool the atmosphere (which will enhance subsidence) or stabilize it to convection, this will inhibit precipitation and the feedback will be positive. If local atmospheric warming occurs (which will contribute to upward motion) or the atmosphere is destabilized, the feedback will be negative. Duce (1995) has a good list of references on this subject of radiative forcing by mineral dust. There is still considerable uncertainty about the local atmospheric effects of airborne mineral dust because it depends on many factors, such as the particle size distribution, the altitude of the dust, the particle composition, and the albedo of the underlying surface. Kellogg (1979) and Fouquart *et al.* (1987) suggest that these factors determine whether the net effect will be atmospheric cooling or warming, and what the vertical distribution will be of the temperature effect. Kellogg (1979) states that, if the aerosol is carried out over the ocean, the result will be cooling, whereas over land surfaces the local effect will be warming. Carlson and Benjamin (1980) have a similar conclusion: Saharan dust over the Atlantic appeared to cool the surface and warm the atmosphere, producing a more stable vertical profile of temperature. Bryson and Baerreis (1967) and Bryson *et al.* (1964) describe a study near the Thar Desert of India and Pakistan, for which they showed that the dust had the effect of increasing the mid-tropospheric diabatic cooling rate, which increased the subsidence rate by about 50%. The enhanced subsidence decreased the depth of the monsoon layer and reduced the monsoon rainfall's penetration into the desert. Thus, a positive feedback prevailed. Hansen *et al.* (1980) have additional discussion of the radiative effects of mineral dust. More will be found in Chapter 4 about the effect of mineral dust on the desert radiation budget.

Soil-moisture feedback

Because the soil is the repository for precipitation that does not run off into streams, and the soil moisture represents a significant source of the water vapor (through direct evaporation, and transpiration by vegetation) for production of clouds and precipitation, it would seem that a feedback might be possible. For example, if a precipitation deficit of sufficient duration occurs, the soil will dry, and the local source of water vapor to fuel the production of cloud by convective or non-convective disturbances will decrease. Thus, an initial perturbation toward less precipitation may be reinforced, and the anomaly may persist. Conversely, a moist period might be sustained because the precipitation will be recycled through soil moisture back to subsequent storms. The importance of this feedback is partially determined by the degree to which water is locally recycled; that is, how much of the precipitation in an area is associated with vapor from local evapotranspiration in contrast to vapor that has been imported from upwind land areas or oceans. Brubaker *et al.* (1993) compute the degree of local water recycling for areas having dimensions of about 1000–2000 km, and find that 10–30% of the precipitation is typically derived from local evapotranspiration, depending on the season and geographic area. The percent is much higher for some seasons and locations. Trenberth (1999) produced global maps of the recycling rate, with the values depending greatly on the size of the area considered and the location. In addition, Chapter 14 cites a number of studies that describe a positive correlation between irrigation in arid areas and rainfall, which is further evidence of local water recycling.

There is much evidence, especially from numerical modeling studies, that this feedback can be important. Walker and Rowntree (1977) show that, when the Sahara in West Africa was replaced by moist land in a model, precipitation related to easterly wave depressions persisted for weeks after, and maintained the moist soil. In contrast, when the experiment was repeated with the typically dry surface, there was much less rainfall and the aridity was maintained. Model simulations by Hong and Kalnay (2000) show that a spring drought over North America was prolonged throughout the summer because of the dry land surface. In an analysis of 65 years of data for North America, Namias (1991) shows that spring seasons that were warmer and drier than normal were likely to be followed by summers that were also hotter and drier than normal. Wetter than normal summers followed cool and wet springs. On the smaller scale of individual convective events, Taylor *et al.* (1997) analyze data obtained in southwest Niger during the Hydrological Atmospheric Pilot Experiment in the Sahel (HAPEX-Sahel), and describe how a soil–moisture feedback explains how an initial convective event influenced the development of further convection in subsequent large-scale disturbances.

Broccoli and Manabe (1992) used a model to study orographic effects on the maintenance of dry climates, and demonstrated that the soil–moisture feedback accounted for over 50% of the drying effect in the summer for Asia and North America. Other studies showing evidence of such positive feedback are those of Rowell and Blondin (1990), Xue and Shukla (1993), and Lare and Nicholson (1994) for the Sahel; Cunnington and Rowntree (1986) for the Sahara; and Oglesby and Erickson (1989) for North America. Importantly, it is evident from many of these studies that the relationship of soil moisture to precipitation is more than simply through its role as a moisture source. Rather, the dependence of the surface heat fluxes to the atmosphere on the amount of surface evaporation allows the soil moisture to significantly influence the pressure distribution and the wind field on a variety of scales.

Upwelling – coastal-desert feedback

We have seen that cold coastal waters may help maintain the aridity of nearby deserts. In these situations, if the surface of the arid land near the coast is warm, there is a considerable difference between the surface temperatures of the land and the water. We will see that this temperature contrast can help maintain the upwelling of cold water: a positive feedback that contributes to the aridity. Specifically, the lower surface pressure over land relative to that over the water, which results from the temperature contrast, will produce a near-surface geostrophic wind field that is oriented approximately parallel to the coast. In the Northern Hemisphere, the Coriolis force will cause the wind to be oriented with the warm land to the left, facing with the wind. In the Southern Hemisphere, the land will be to the right. This effect of the temperature contrast on the development or enhancement of the coast-parallel wind will enhance the upwelling of cold water, through the mechanism described earlier in this chapter, and this should contribute to the maintenance of the coastal desert and the temperature contrast. This is a positive feedback mechanism. Such coast-parallel, low-level jets forced by coastal temperature contrasts are ubiquitous, and have been documented along many coasts, especially those bounding cold currents and arid lands. For example, Lettau (1978b) and Enfield (1981) describe the equatorward-directed jet and the above feedback in the context of the Peruvian and Atacama Deserts, while Holt (1996), Doyle (1997), and Burk and Thompson (1996) document such jets along the California coast.

Vegetation–substrate feedback

Loss of vegetation leads to substrate changes that can further reduce vegetation amount, representing a positive feedback. When vegetation is lost, say through a drought, the lack of foliage for interception of rainfall means that the kinetic energy of raindrops falling on the bare ground will increase water erosion and wash away the upper soil or organic-matter layer. This eroded condition is less supportive of vegetation because nutrients are lost, the soil is compacted, and the soil moisture is lower because of the loss of organic matter. The lack of vegetation also means that the wind speed is greater near the surface, which results in a drier soil and greater wind erosion.

Dust–biogeochemical feedbacks

There are numerous ways in which mineral dust that is elevated and transported from desiccated desert surfaces can influence Earth's biological and chemical processes on the land and in the atmosphere and oceans. Some of these effects may involve positive or negative feedbacks. For example, all mineral dust has some iron content, and a significant amount of the airborne dust is deposited in the oceans. The growth of the phytoplankton biomass is stimulated by the iron content of the water, and the response of the system to increased dissolved iron would be a reduction of the carbon dioxide content of the atmosphere (Moore *et al.* 2001). This reduction in atmospheric carbon dioxide may decrease global warming effects. Depending on the response of regional rainfall patterns to reduced global warming, there may be an increase in the area and severity of some desert climates, and therefore in dust production: a positive feedback.

The dynamics of desert heat lows

Heat lows, also known as thermal lows or thermal troughs, are climatological features of many arid areas. The name refers to the fact that the surface pressure is low relative to that of the surrounding area by about 3–10 mbar or more. See Fig. 3.12 for an example of the heat low over the summer Sahara, and Fig. 2.5 for the positions of other large-scale heat lows. The feature results from the stronger sensible heating of the atmosphere over the desert, and thus is most distinct in the summer season and in lower latitudes where the solar input is greater. It is a ubiquitous property of desert environments, and has been thoroughly documented as follows.

- Northern and southwestern Africa (Ramage 1971; Pedgley 1972; Griffiths and Soliman 1972)
- West Pakistan and northern India (Ramage 1971; Chang 1972; Joshi and Desai 1985)
- The Qinghai–Xizang Plateau of China (Junning *et al.* 1986)
- Southwestern North America (Sellers and Hill 1974; Rowson and Colucci 1992; Douglas and Li 1996)

- The Arabian Peninsula (Ackerman and Cox 1982; Blake *et al.* 1983; Smith 1986a,b; Bitan and Sa'aroni 1992; Mohalfi *et al.* 1998)
- Spain (Uriarte 1980; Gaertner *et al.* 1993; Alonso *et al.* 1994; Portelo and Castro 1996)
- Northwestern and northeastern Australia (Moriarty 1955; Leslie 1980; Leighton and Deslandes 1991)

Because of the paucity of data in deserts, many studies of the heat-low phenomenon have relied on model simulations to produce surrogate data (see, for example, Rácz and Smith 1999). In general, the observed data and the model simulations are consistent, and collectively provide us with a good description of the processes. The general cause of the low-pressure anomaly is fairly simple. The greater sensible heating of the column of atmosphere over the desert surface, relative to that over the surroundings, causes a high-pressure anomaly on a horizontal surface near the top of the heated column. This forces the propagation of gravity waves that remove mass from the warm column, which creates the lower surface pressure. The associated vertical-motion field is one in which there is upward motion over the heated area, and surrounding subsiding motion. If the heating of the boundary layer over the desert did not vary diurnally, the winds would respond to the steady pressure gradient and the Coriolis force and assume a general cyclonic rotation around the surface low-pressure anomaly and an anticyclonic rotation around the upper high-pressure anomaly. But, because of the diurnal variation in the surface heating and cooling, the pressure anomaly and the related winds also vary diurnally. In particular, the surface pressure in the low is a minimum in the later afternoon and a maximum in the morning, with the amplitude of the diurnal variation being about 1–10 mbar. Daytime winds near the surface are generally directed inward, converging toward the developing low pressure. The Coriolis force will progressively impart greater cyclonic rotation (counter-clockwise in the Northern Hemisphere, and clockwise in the Southern Hemisphere) to the converging surface wind as the day progresses into evening, turning the wind more parallel to the edge of the desert. Conversely, the diverging winds near the top of the heated column will be turned anticyclonically. Upward motion in the center of the low is strongest in the late evening when the convergence of the near-surface horizontal winds is greatest. The near-surface cyclonic rotation is strongest late in the night, with some analysts referring to the core of highest winds as a low-level jet.

The vertical and horizontal motion patterns become more complex when the effects of the desert-scale heat low are superimposed on the planetary-scale subtropical subsidence of the north–south Hadley circulation that exists over some deserts, and east–west Walker-type circulations. Figure 2.12, shown earlier, of the day and night divergence and vertical motion measured in the subtropics in the Rub Al Khali Desert on the Arabian Peninsula, illustrates this point. At night, when the heat-low-related winds are mostly rotational and not convergent at low levels over the desert, the associated vertical motion is small and there is generally Hadley-cell-related lower tropospheric divergence, upper tropospheric convergence, and subsidence throughout the entire column. During the day, however, when the upward vertical motion and the low-level convergence from the heat low are well developed, the subtropical subsidence and divergence are reversed at low levels.

A further complication to the dynamics of the heat low is the effect of dust. Over the desert, the dust in the mixed layer is heated during the day through absorption of solar energy, and this heat is imparted to the atmosphere. Also, less heat thus reaches the surface, and there are effects of the absorption by the dust of terrestrial infrared radiation. These radiative effects are weaker in less-dusty areas outside of the desert region. A modeling study of the effect of dust on the heat low in the Rub Al Khali Desert shows that only through including the dust effects can a heat low of sufficient strength be simulated so that it corresponds to observations (Mohalfi *et al.* 1998). Bounoua and Krishnamurti (1991) also discuss the dynamic effects of dust on a five-day oscillation of the Saharan heat low.

Specific characteristics of the different heat lows over Earth's deserts will depend on many factors. These include the size and configuration of the desert, its latitude, the orography, and the large-scale flow.

Suggested general references for further reading

Ahrens, C. D., 2000: *Meteorology Today: An Introduction to Weather, Climate, and the Environment* – a general non-technical survey of atmospheric science for readers of all backgrounds.

Amiran, H. K., and A. W. Wilson (Eds.), 1973: *Coastal Deserts: Their Natural and Human Environments* – a selection of papers on coastal deserts and their causes.

Holton, J. R., 1992: *An Introduction to Dynamic Meteorology* – a general reference on atmospheric dynamics for readers with a physical-science background.

Lashof, D. A., *et al.*, 1997: *Terrestrial ecosystem feedbacks to global climate change* – a summary of some land–atmosphere feedbacks.

Meigs, P., 1966: *Geography of Coastal Deserts* – a summary of the climates and surface conditions of all the coastal deserts.

Otterman, J., and D. O'C. Starr, 1995: *Alternative regimes of surface and climate conditions in sandy arid regions: possible relevance to Mesopotamian drought 2200–1900 B.C.* – a discussion of arid-land surface–atmosphere feedbacks.

Sikka, D. R., 1997: *Desert climate and its dynamics* – a summary of desert climate dynamics.

Smith, R. B., 1979a: *The influence of mountains on the atmosphere* – a technically oriented review of all the effects of orography on the atmosphere.

Whiteman, C. D., 2000: *Mountain Meteorology: Fundamentals and Applications* – a non-technical review of the various effects of mountains on weather.

Questions for review

(1) Explain the difference between the Budyko index and the P : PET ratio for defining the degree of aridity.
(2) What is the mechanism by which fog forms over the sea near coastal deserts?
(3) What are the three mechanisms by which subsiding air creates an environment that is less conducive to precipitation formation?
(4) What meteorological factors control water loss at the surface?
(5) In order for a rain shadow to form, why must precipitation reach the ground when air rises over a mountain barrier?
(6) Review the possible causes that contribute to the existence of coastal deserts.
(7) By what mechanisms do mountain barriers affect downwind climate?
(8) Describe the diurnal cycle of the heat low.
(9) Discuss the relationship between the heat low and monsoon processes.
(10) By what different criteria (e.g. climate) can places be defined as desert?
(11) Summarize the different feedback mechanisms that can contribute to the development of deserts.
(12) What factors influence the relative humidity as an airstream flows over a mountain to create a rain shadow?
(13) To produce coastal upwelling in the Southern Hemisphere, what must the orientation of the near-surface wind be relative to the coast?

Problems and exercises

(1) Propose and discuss additional atmosphere–surface feedback mechanisms that might affect the stability of regional desert climates.
(2) Explain, dynamically, how desert heat lows form.
(3) You are tasked with defining whether a location is becoming more or less arid, and with quantifying any natural change. Develop and describe a useful procedure.
(4) Hypothesize some potential mechanisms by which soil moisture can influence large-scale circulations that affect precipitation, and whether a positive or negative soil–moisture feedback would result.
(5) Describe how a depression in the terrain elevation might generate horizontal pressure contrasts in the atmosphere above a heated surface, and cause subsidence over the depression that can suppress precipitation and produce a positive temperature anomaly.

3

The climates of the world deserts

It would be well-nigh impossible for one who has heeded the call of the vast Desert solitudes to pass back through the Gateway to the Desert [Tripoli] without a special tribute to the insidious charm of that great land of sand and silence which lies behind it. South, the interminable African main drifts on to the Sudan; west to east it sweeps the whole width of Africa. Even at the Red Sea it merely pauses for a moment at the brink, then dips beneath the limpid waters and continues across Arabia, Persia, and into northern India. . . .

It is little wonder that the ancients saw in the Sahara, dark dotted with oases, the graphic simile of the leopard's skin. The call of those limitless reaches is as subtle and insidious as must be the snow fields of the Arctic. Listening to it, one is beguiled onward against the gentle pressure of its capricious south-east breezes, under which date-palms nod their graceful crests over the murmoring oases . . . No sound but the soft scuff of our horses and the creaking saddle leathers breaks the stillness; no shadows except our own paint splashes of azure upon the orange sand. Again, white-walled, bastioned Tripoli lies many miles behind me on the edge of the coast like a great silver shell in a stretch of golden sand, and I feel that somehow I have again drawn back the veil of the ages.

Charles Wellington Furlong, American adventurer and writer
The Gateway to the Sahara (1914)

According to the well-known Köppen (1931) classification of climates, in which climate types are defined primarily in terms of differences in vegetation, the dry regions of the world are divided into two climate types: the arid and semi-arid areas. Collectively these types occupy 26% of the total land area of the world, more than any other major climatic type. The arid component represents 12% of the land area. However, the Meigs (1953, 1957) classification system is thought to be one of the best suited for the definition of arid lands. In terms of percent of the world's land surface, the following categories are defined: extremely arid, 4%; arid, 15%; semi-arid, 14.6%; with the total being 33.6%. The extremely arid and arid components represent $29\,000\,000$ km^2. Table 3.1 defines

Table 3.1 *Area (million km^2) by continent, of arid lands*

Continent	Extremely arid	Arid	Semi-arid	Extremely arid plus arid	Total area	Percent of arid land
Africa	4.56	7.30	6.08	11.86	17.94	36.7
Asia	1.05	7.91	7.52	8.96	16.48	31.7
Australia	—	3.86	2.52	3.86	6.38	10.8
North America	0.03	1.28	2.66	1.31	3.97	12.0
South America	0.17	1.22	1.63	1.39	3.01	8.8

Source: From Meigs (1957).

the area, by continent, represented by each of these categories. Fig. 3.1 shows the distribution of these three arid climate types. Note that the classification code includes information on the precipitation, its seasonal distribution, and the temperature, all factors that are related to the degree of aridity. The figure shows that the dry climates occur in regions of relatively continuous aridity that are made up of a number of individually named deserts. The North Africa–Eurasia area sweeps from the Atlantic coast in North Africa, through the Arabian Peninsula and into southern Asia. This arid region covers more land area than that of all of the other deserts of the world combined. In southern Africa, the narrow coastal Namib Desert exists on the west, with the Kalahari and Karoo Deserts adjacent to the east. In east central Africa is the Somali–Chalbi Desert, which might be viewed as part of the North Africa–Eurasia desert region. The South American dry areas include the Peruvian and Atacama Deserts on the west coast, and the Patagonian and Monte Deserts to the east of the Andes in the south. There are small semi-arid areas in eastern Brazil and northern coastal Venezuela and Colombia. The North American Desert area consists of four separate deserts: the Sonoran, Chihuahuan, Great Basin and Mojave. The Australian deserts, the Great Sandy, the Gibson, the Great Victoria, the Tanami, and the Simpson, which includes the Sturt Stony, cover two-thirds of the continent, a larger fraction of their continent than do any other deserts of the world.

In excess of 80% of Earth's total arid lands are found on three continents: Africa (37%), Asia (32%) and Australia (11%). Fifty-five of the sixty-six countries affected by aridity are found on these continents, as are all of the countries with 75–100% of their land area in either the arid or the semi-arid category. The above statistics about desert areas apply only to land. Fig. 2.14 shows that the oceans have roughly as large an area of dry climates as does the land.

It is clear from the maps in Fig. 3.1 of the worldwide distribution of aridity that there is much mesoscale structure to the pattern. In fact, except for the immense Sahara, most deserts show considerable irregularity in the boundaries between

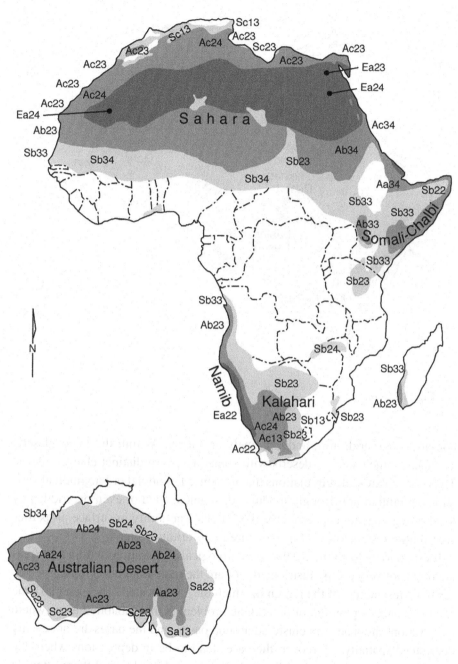

Fig. 3.1 Geographic distribution of extremely arid, arid, and semi-arid land. Note from the key that the classification system includes information about precipitation and temperature. (From Meigs 1953, 1957.)

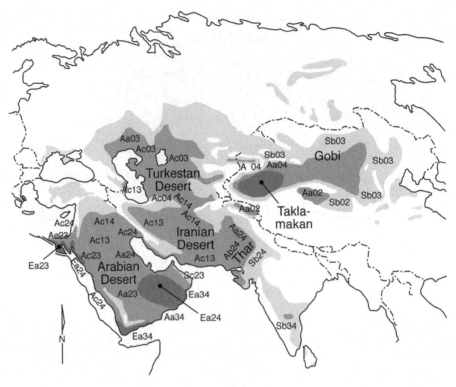

Fig. 3.1 (*cont.*)

non-arid, semi-arid, arid, and extremely arid areas. Within the major deserts are many named smaller deserts with sometimes very distinct characteristics. Thus, the global-scale circulations that influence latitudinal or longitudinal variations in rainfall are strongly modulated, supplemented or even overridden by local or regional atmospheric processes. Local variability in surface properties contributes to some of this fine structure in the aridity pattern.

It is important to recognize that meteorological observations within arid areas are often not very evenly distributed, nor are the individual observations necessarily representative of the region in which they are located. This can make the weather-analysis problem difficult. One problem is that inhabited areas where observations are made are clustered around oases, and the oases are not evenly distributed spatially. Moreover, the oases tend to be in depressions where the water is near the surface, and the meteorological data obtained there often do not reflect the broader-scale conditions outside of the depression.

The following geographical and meteorological descriptions of the major deserts serve as a brief summary only. The characteristics of the surface, including the orography, the substrates, and the vegetation, are briefly discussed

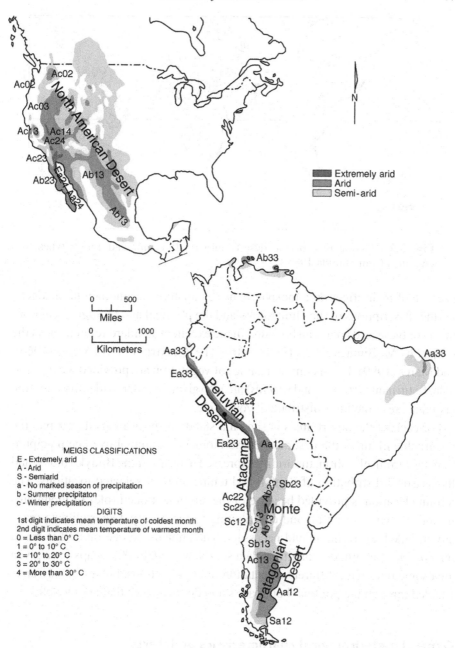

Fig. 3.1 *(cont.)*

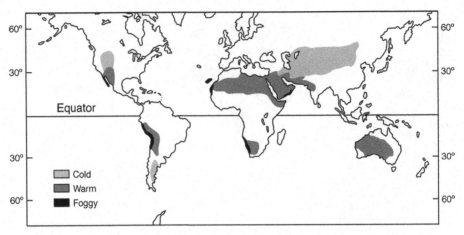

Fig. 3.2 Classification of the deserts into cold, warm, and foggy types. (Adapted from Shmida 1985.)

because of their effect on atmospheric heat, moisture and momentum budgets. Further descriptions of the climatology and the physical and biological environments of these arid areas can be found in the excellent standard references on the subject by McGinnies *et al.* (1968), Petrov (1976), Arritt (1993), Meigs (1966), and Mares (1999). Local common names of vegetation are provided when available. Latin names (*Genus* and *species*) are also given in order to identify common taxa that are sometimes distributed worldwide.

For each of the desert areas is shown a standard set of maps that depict the geographic limits of the deserts and arid biomes, the annual-average precipitation, the terrain elevation, the aridity expressed in terms of the Budyko index, and the seasonal distribution of rainfall and temperature at selected locations. The terrain elevation is provided because of the aforementioned roles of mountains in both increasing rainfall and contributing to the aridity. Other climate maps are provided, as available and necessary. The annual-average rainfall is based on a 0.5 degree latitude–longitude data set (New *et al.* 2000). Maps for a given area sometimes have different orientations and projections, but a comparison is enabled through the political boundaries and desert names that are plotted.

General meteorological characteristics of deserts

Chapter 2 describes the various dynamical processes that can lead to the low rainfall that is part of many definitions of desert. However, there has been little discussion as yet of the general characteristics of the sensible (sensed by humans) weather variables that would be experienced near the surface in deserts. This section will serve as a preparation and introduction for the following sections,

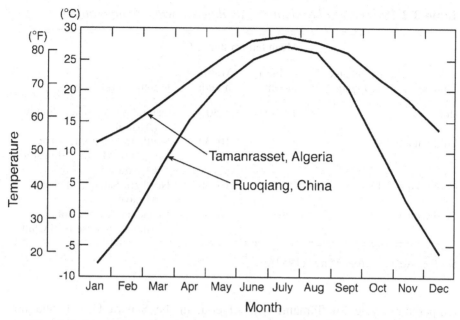

Fig. 3.3 Monthly mean temperatures for Ruoqiang, China (annual mean rainfall of 23 mm), in the eastern part of the Taklamakan Desert, and for Tamanrasset, Algeria (annual mean rainfall of 44 mm) in the Sahara Desert. The Taklamakan is a cold desert and the Sahara is a hot desert.

which document the specific meteorological properties of the different individual deserts of the world.

Deserts may be classified according to whether they have a distinct cold season. All have summers that are hot, so reference to them as "cold" or "warm" deserts refers only to the winter temperatures, or the annual-average temperature that reflects the coldness of the winters. Fig. 3.2 shows which of the world's deserts are generally defined as cold, warm, and foggy. Cold deserts are generally at higher latitudes, which is not surprising because the seasonal variation in solar-energy input increases with latitude and causes more definable seasons for non-deserts and deserts alike. Cold deserts include the central Asian deserts, the Great Basin Desert of North America, and the Patagonian Desert of South America. In spite of the differences in latitude, and therefore of solar energy input, of cold and hot deserts, the energy budgets are such that summer temperatures are not greatly different. For example, Luktchun, in the Taklamakan Desert of Asia at latitude 43° N, is as hot in the summer as most places in the Sahara, even though it is 20° latitude farther north. The Luktchun winter, however, has four months with monthly mean temperatures below freezing (Miller 1961). This distinction between cold and hot deserts is shown in Fig. 3.3 in terms of the annual

Table 3.2 *Percentage of arid lands with different winter temperatures*

Climate	Percentage of arid lands	Mean temperature (°C)		Example arid zones
		Coldest month	Warmest month	
Hot	43	10–30	>30	Central Sahara, Great Sandy (Australia)
Mild winter	18	10–20	10–30	Southern Sahara, Kalahari, Deserts of Mexico, Simpson (Australia)
Cool winter	15	0–10	10–30	Northern Sahara, Atacama, Mojave
Cold winter	24	<0	10–30	Canadian Prairie, Gobi, Turkestan, deserts of China

Source: Adapted from Meigs (1952).

temperature cycle for Tamanrasset, Algeria, in the Sahara Desert, and for Ruoqiang, China, in the Taklamakan Desert, with the latter location being 15°–20° higher in latitude. The highest monthly average summer temperatures are similar, but the January means are about 20 K (36 °F) different. Cold deserts exhibit some of the largest annual ranges in temperature on Earth. Table 3.2 further refines the classification of deserts in terms of their winter temperatures. About 60% of desert areas have hot or mild winters, and about 25% have cold winters.

The marine, or coastal, desert is a special case of the hot desert. The coastal parts of the Peruvian, Atacama, and Namibian Deserts, for example, have such high humidity and low diurnal and annual temperature ranges that their climates are essentially maritime, except for the lack of rainfall. The annual range of monthly average temperatures can be less than 5 K (9 °F). In terms of discomfort, however, the high relative humidity can more than compensate for the more-moderate temperatures. For example, Walvis Bay on the coast of the Namibian Desert has a January (summer) mean relative humidity of 85%, and a nearly as high July value of 77%. At Cape Juby, on the Atlantic coast of the Sahara, the mean relative humidity is 82% in January (winter) and 91% in July. With increasing distance from the coast, the deserts become more typical of the continental hot desert. In the Namibian Desert, Windhoek is 650 km inland of coastal Swakopmund at the same latitude, and the mean annual temperature is 4 K (7 °F) warmer in spite of the fact that it is 1500 m higher in elevation. See Box 3.1 for additional information on the climates of coastal deserts.

Box 3.1
Coastal climates of deserts

Sometimes deserts are caused by their proximity to ocean coastlines, and sometimes they just happen to abut coastlines. In either case, many aspects of the climate of the desert near the coast are strongly affected by the nearby body of water. The sea breeze, with a low-level landward wind during the day and seaward wind at night, is forced by the pressure gradient caused by the coastal temperature contrast. The temperature over the land near the coast, is moderated by the fact that the water temperature, and therefore the temperature of the atmosphere in contact with it, changes only very slowly. This means that daily and seasonal swings in temperature are smaller than would be experienced farther inland. The humidity will also be higher because of the nearby source of water vapor. If the temperature is also high, this will increase the discomfort level and decrease the body's ability to maintain a heat balance through evaporative cooling (Chapter 19). Higher relative humidity and the existence of sea-salt condensation nuclei also will cause the atmosphere to be hazier. And, if a cold ocean current exists near the coast, a landward wind will bring fog with it, greatly affecting many aspects of the microclimate in terms of water available for vegetation, the amount of sunshine, and the temperature.

In such coastal deserts, the annual rainfall statistic may misrepresent the amount of moisture available to flora and fauna. Here, fog may occur on over 200 days per year, and wet the ground with more moisture than the rainfall (see Chapter 13). Fig. 3.4 shows the diurnal variation of temperature and humidity for three locations near the coast in the Namibian Desert, and illustrates the effect of distance from the coast on how greatly fog impacts the local climate. On this typically foggy summer day, at Swakopmund the solar radiation was only briefly able to begin to dissipate the fog and allow the temperature to rise. At Rössing, the fog evaporated between morning and evening, allowing the temperature to rise significantly. Even farther inland at Lintvelt, the atmosphere remained unsaturated throughout the diurnal cycle. Plots are very similar for a foggy winter day (Walter 1986).

There are also distinctions among deserts, and among subareas within deserts, in terms of the seasonality of the rainfall. Fig. 3.5 illustrates, for arid areas, where summer and winter rain are dominant, and the transition areas where there is no seasonal dominance. An example of the seasonality is shown more quantitatively in Fig. 3.6 for Australia. Both figures show that latitude is the most strongly controlling factor, with rainfall concentrated in the summer months in more tropical latitudes and in the winter months as mid-latitudes are approached. Coastal, monsoonal, and orographic factors also play a role, however.

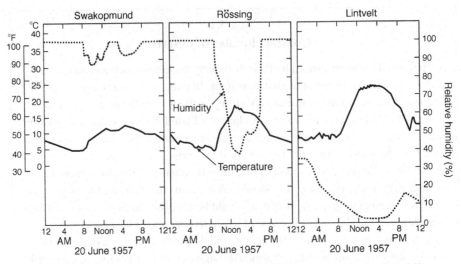

Fig. 3.4 Diurnal variation in temperature (solid line) and relative humidity (dashed line) on a typically foggy day near the coast in the Namib Desert. The Swakopmund observation is 5 km from the coast, Rössing is 30 km inland, and Lintvelt is yet farther inland on the escarpment that rises from the coastal plain. (Adapted from Walter 1986.)

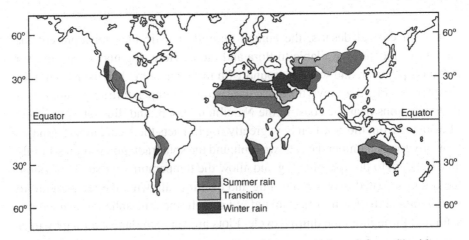

Fig. 3.5 Seasonality of precipitation in arid areas. (Adapted from Shmida 1985.)

The diurnal range of the air temperature near the ground in deserts can be larger than is typical in non-deserts, for reasons that will be described later in terms of the surface energy budget and the land-surface physics. The 2 m air temperature in summer may reach in excess of 50 °C (122 °F) during the afternoon, and approach freezing at night. A diurnal variation of 56 K (101 °F) has been observed in Tucson in the northern Sonoran Desert. Ground temperature

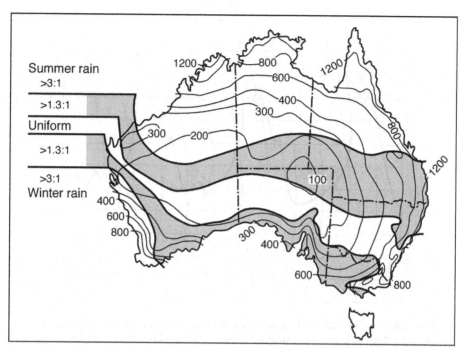

Fig. 3.6 Seasonality of precipitation in Australia. The lines indicate annual precipitation in millimeters, and are plotted at an irregular interval. (From AUSLIG 1992.)

diurnal amplitudes can exceed 80 K (144 °F). However, when the ground has physical properties that allow it to efficiently store daytime heat that becomes available in the night, the night temperatures of the ground or air may not drop greatly from the daytime maximum. Temperature minima of 38 °C (100 °F) are not uncommon.

The desert atmosphere is often subjectively considered to be "dry" in some sense, presumably because of the lack of rain, and it often is. But we need to be more specific than this. Because of the often near-zero rate of evaporation from the dry surface, and the small amount of water transpired by meager amounts of vegetation, there are definitely few local sources of water vapor in many inland desert climates. This can, indeed, lead to low water-vapor pressures near the surface. The combined effect of the low vapor pressure and the high mid-day saturation vapor pressure (resulting from high temperatures) causes the relative humidity to be low, sometimes as low as a few percent. But some deserts experience dew deposition at night when the temperature falls precipitously and the dew-point temperature is reached. Thus, the relative humidity can have large diurnal swings. Fig. 3.7 illustrates an example of the diurnal relative humidity variation in Khartoum, Sudan (southeastern Sahara), and its dependence on

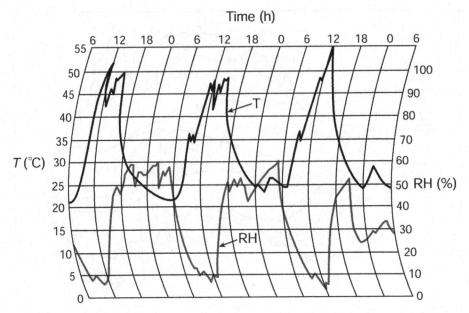

Fig. 3.7 Near-surface relative humidity and temperature for Khartoum, Sudan, during a three-day period. (Adapted from Cloudsley-Thompson 1977.)

temperature. Evenari (1985a) shows a more extreme example (see Fig. 12.9, this volume), where the diurnal variation of relative humidity is from 2% to 100% on two of three consecutive days in the Negev Desert. There can also be large annual variations in the vapor pressure and the relative humidity. In the warm season, when the land heats up and the heat lows develop, monsoon circulations can draw moist maritime air into arid areas.

The references above to the low water-vapor content of desert air need to be interpreted cautiously. Compared with locations with comparably high temperatures, such as the humid tropics, the desert atmosphere indeed does contain less water vapor. But because the water-vapor pressure at saturation is dependent on temperature, desert air at 35 °C (95 °F) and 15% relative humidity has *more* water vapor than does saturated air at 0 °C (32 °F) in a winter mid-latitude storm. This effect can be seen in the water-vapor climatology of many regions. For example, the dew-point temperature, which is an absolute measure of the amount of water vapor in the air, has a December average in the warm, northern Sonoran Desert that is the same as over the cool humid-climate Appalachian Mountains in eastern North America at the same latitude (NOAA 1983).

It is also risky to generalize about desert winds. Even though the high pressures that sometimes characterize subtropical deserts are typically associated with weak horizontal pressure gradients, and therefore weak large-scale winds, there are many weather disturbances that can intrude on the subtropics from tropical and mid-latitudes and cause high wind speeds. Also, the strong convective vertical overturning of air that occurs within deep desert boundary layers during the day can mix higher wind speeds from aloft down to the surface. The importance and prevalence of these higher-wind conditions are evidenced by the dust storms that represent an important part of the desert climate. Because of the positive correlation between wind speed and evaporation rate, high wind speeds can enhance the desiccating influence of the high temperature and low relative humidity. For example, Evenari (1985a), referring to a strong spring wind in Israel, states

After the first spring sharav has blown for a few hours, the annual vegetation looks as if boiling water has been poured over it.

The wind speed also affects the diurnal range of temperature, and it controls the exchange of heat with the soil. The wind direction near the desert's edge, where temperature contrasts can be large, can very strongly affect the air temperature through advection.

Even though advective fogs along coasts, and inland radiation fogs, can significantly reduce visibility, visibility impairment in deserts generally results from elevated dust. When the humidity is high and salt particles exist in the air near coastlines, the haze is especially heavy. Even when there is no significant mean wind, the daytime turbulent motions will elevate fine dust particles and reduce visibility to much less than a kilometer. This omnipresent haze of mineral dust is sometimes referred to as "desert fog." Even when convective turbulent mixing of the dust ceases to be forced by surface heating (after sunset), the fine particles can remain suspended through the night. The greatest reduction of visibility, sometimes to near zero, is naturally a result of dust storms that are generally caused by high winds forced by strong synoptic-scale pressure gradients or associated with cold-air outflows from convective storms.

General physiographic characteristics of deserts

The desert surface has been carved and shaped by millennia of high wind and flood water acting on the geological structures. It can be one of the most uninterrupted expanses of flat land on Earth. And it can be the most rugged of possible otherworldly landscapes that has been tortured by moving water cutting deeply through soil and rock. Significant fractions of some deserts are composed of

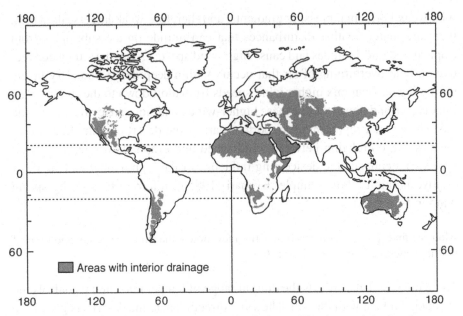

Fig. 3.8 Areas of Earth with interior water drainage. That is, precipitation does not reach the sea. (From de Martonne and Aufrère 1927.)

sand dunes, some of which are "mobile," while others are "stable" with the sand anchored by vegetation. Other surfaces can be immense areas of bare rock, or desert pavement composed of a matrix of pebbles that have been cemented together by chemical processes. Also, there are salt flats that can have a variety of properties, including moist surfaces from high water tables. Many of the above landforms are unique, being found only in arid regions. It was mentioned in Chapter 2 that much of the water drainage in desert areas is internal, in that water never reaches the sea because the drainage is to interior basins. Fig. 3.8 maps the areas of Earth with such internal drainage, and the correspondence with the arid areas is clear. In any case, the shape of the surface and its physical composition strongly affect the microclimate through roughness and surface water- and energy-budget effects.

Within a single named desert, a mosaic of different types of surfaces may be found. Fig. 3.9 illustrates this variability for the Australian deserts, which include: mountainous surfaces; clay plains; stony surfaces; sandy surfaces with dunes; and bare rock shields. There are many good references on the geomorphology of arid zones (see, for example, Thomas 1997a) and on desert soils (see, for example, Dunkerley and Brown 1997). Box 3.2 describes the effects of one type of physiographic feature, mountains, on the desert climate.

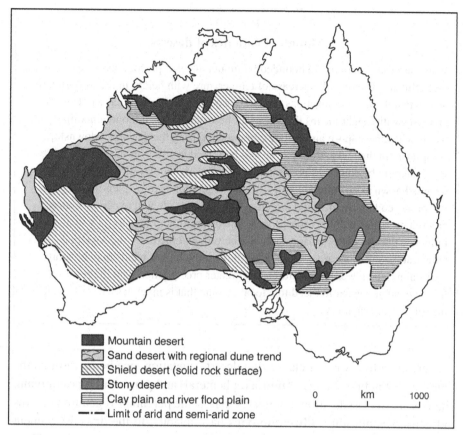

Fig. 3.9 The variability of the substrate properties in the deserts of Australia. (Adapted from Goudie and Wilkinson 1977.)

General vegetative characteristics of deserts

When vegetation exists in arid areas, it can be more diverse than in some temperate ecosystems. It primarily consists of lichens, shrubs, herbs, grasses, succulents, and small trees. Microphytic (small or microscopic) vegetation is especially important, with surfaces often being covered with mosses, liverworts, algae, lichens, fungi, and bacteria. In extremely dry environments that appear to be barren of vegetation, microphytes may be the only vegetation present over vast areas. And even then, they sometimes can only exist in special niches. For example, algae need moisture and light to photosynthesize, and one location where those conditions can co-exist in the desert is beneath translucent stones (e.g. calcite, selenite, or quartz) where any water is protected from evaporation and where sunlight can penetrate. Thus, a favorite habitat for desert algae is in these natural "greenhouses."

> ## Box 3.2
> ## Mountain climates of deserts
>
> Many deserts contain or are bounded by mountains that profoundly influence the local climate. The overall decrease of temperature with height in the troposphere means that the higher elevations tend to be cooler, and the fact that wind speed increases with height means that they will be windier. Local mountain-valley winds will superimpose an upslope component during the day and a downslope component in the night. Because air will be forced to rise over the barrier, there will generally be enhanced precipitation over the higher elevations. This leads to what is known as an "altitude oasis" where vegetation is more plentiful. If there is a preferred large-scale wind direction, the upwind side will experience more precipitation than will the other slopes. The orientation of the slopes will also influence the amount of solar radiation received, and therefore the substrate temperature and moisture content. Even though the higher winds may be more desiccating, the lower temperatures, the greater precipitation, and the resulting more abundant vegetation lead to a local climate that is much less desert-like than that of the surroundings.

There are a few ways of classifying desert vegetation, one being through the common distinction between **annual** (ephemeral) and **perennial** types of plant. There is also a classification that relates to how the perennial vegetation obtains and conserves moisture. **Phreatophytes** have adapted to the dry environment by developing long tap roots that can reach the water table, or close to it. They do not have special mechanisms for conserving water, with examples being mesquite (*Prosopis*), tamarisk (e.g. *Tamarix aphylla* and *T. gallica*) and the date palm (*Phoenix dactylifera*). In contrast, **xerophytes** have developed a variety of properties that promote conservation of water and reduction of the heat load: waxy or hairy leaf surfaces; leaves with high albedo; leaves that maintain their edge to the sun; small rather than large leaves; and leaves that roll up, wilt, drop from the plant, or close their stomata during the dry season or droughts. These processes reduce the rate of transpiration. Some types, such as the paloverde tree (*Cercidium*) have chlorophyll on the green woody trunks and branches so that photosynthesis can continue even after leaves have desiccated. Other plants, the succulents, sequester water in their leaves, roots and stems. These include the cacti and euphorbias. The foliage of some species (**nephelophytes**) can absorb dew and fog on the leaf surfaces, and store it (Monod 1973; Evenari 1985a). There is even disputed evidence that some species can transport water absorbed by the foliage into the soil through the roots, to be utilized later (Went 1975; Mooney *et al.* 1980). Annual plants, in contrast, have adapted to the aridity through rapid flowering and formation of seeds during the brief periods when rain and soil moisture suffice. The seeds of these **therophytes** (annuals) remain dormant in the soil during the hot, dry season. Because of the prevalence of alkaline soils in arid environments, some plant types have developed salt tolerance. The most well-adapted are called **halophytes**, of which the saltbush is an example. Plants that grow more typically in sandy soils are called **psammophytes**.

The terms steppe and savanna are often applied to semi-arid areas, and refer to the typical vegetation. Steppes are primarily grassland, without trees, and are generally in mid-latitudes. Savannas are also open grassland, but the grassland is punctuated by scattered shrubs and trees. Savanna is typically subtropical and represents a transition between the closed tropical forests of humid areas and the open grasslands of arid areas.

McGinnies *et al.* (1968), Evenari *et al.* (1985, 1986), West (1983a,b,c,d,e), Petrov (1976), and Mares (1999) have fairly complete lists of the myriad vegetation types that exist in the various deserts. In the following sections, the names of the more common grasses, trees, shrubs, and succulents are derived primarily from these sources. The reader should consult these same sources if more comprehensive listings of vegetation types are required.

African deserts

The continent of Africa is second only to Australia in terms of the fraction of its land surface that is arid (Fig. 3.10). The northern 40% of the continent is covered by the vast Sahara, straddling the Tropic of Cancer. Where the continent narrows in the south are the Kalahari, Namib, and Karoo Deserts, crossed by the Tropic of Capricorn. The eastern horn of Africa is largely covered by the Somali–Chalbi Desert. The island of Madagascar to the east of Africa also has a hot semi-arid area in its south. Collectively, these areas represent 35% of Earth's semi-arid or drier land, and 69% of the hyper-arid land (UNEP 1992). A review of the geomorphology of African deserts is found in Shaw (1997).

The Sahara Desert

The Sahara, the largest desert in the world, derives its name from the Arabic word *sahra*, meaning desert, empty area or wilderness. It includes areas that are among the hottest and driest on the planet. From the Atlas Mountains and the Mediterranean coast it extends southward over 1500 km, and spans the 5000 km between the Atlantic coast and the Red Sea (Figs. 3.1 and 3.10). The total area of extremely arid to semi-arid land represents over 40% of the continent of Africa, and is larger than the area of the continental United States. The

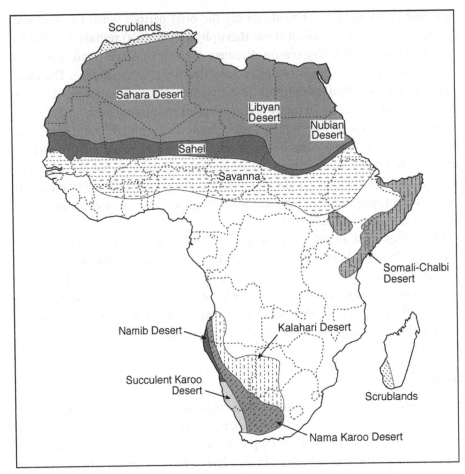

Fig. 3.10 The deserts of Africa. (Adapted from Mares 1999 and Meigs 1953, 1957.)

southern boundary of the Sahara is difficult to define. It is a region called the Sahel that experiences recurrent drought. Oases represent less than 2000 km². Other permanent surface water includes the longest river in the world, the Nile, which empties into the Mediterranean on the east edge, the Niger River along the southern edge, and a few rivers that drain out of the Atlas mountains in the northwest. All of these, and other, rivers originate outside of the Sahara, and are sometimes consumed by it. In some areas, especially in the north (e.g. Libya), there are large underground aquifers that are being depleted and in some instances becoming saline.

Although much of the Sahara is flat, there are some notable mountains (Fig. 3.11). The largest expanse is the Atlas Mountains near the Mediterranean

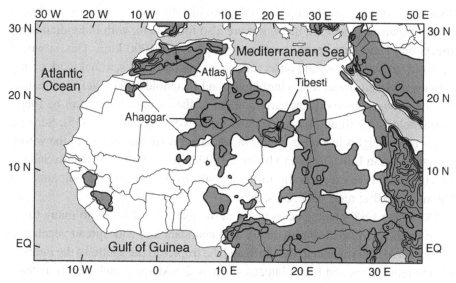

Fig. 3.11 Terrain elevation for the Sahara Desert region. The contour interval is 500 m, and heavy contours are used for 500 and 1000 m. Elevations greater than 500 m are shaded.

coasts of Algeria, Morocco and Tunisia, with higher elevations of 2000–3000 m. Over the rest of the Sahara, isolated small mountain ranges attain similar elevations. For example, in southern Algeria, the Ahagger Mountains reach elevations of over 2900 m. And to the east in northern Chad and southern Libya, the Tibesti Mountains reach 3450 m, the highest point in the Sahara. The intermountain plains have elevations between 500 m and 2000 m. There are also some significant closed depressions, such as the Qattara Depression in western Egypt (134 m below sea level). This is the Sahara's deepest salt basin. Some of the depressions are occupied by oases. Other relief is provided by the sand dunes, and by dry eroded watercourses called **wadis**. The wadis have their origins in the higher terrain, most having been formed during earlier **pluvial** (rainy) periods. They terminate at coastlines, or in intermountain depressions. The depths of young wadis that have not yet been filled by sand from wind erosion sometimes exceed 100 m (see cross-sections in Petrov 1976).

Over 80% of the Sahara is covered by rock, gravel or pebble, and sand surfaces, with the sand representing about 15–20% of the total. The smooth plains of dense, cemented fine gravel (called *regs* in the western Sahara and *serir* in the eastern desert) result from the winds carrying away the lighter material over the millennia. Similar processes have resulted in surfaces of bouldery terrain (called **hammadas,** or **hamadas**). The gravel plains stretch unbroken over vast distances. For example, a single gravel plain in Libya covers 880 000 km^2. The largest expanses of sand in the world are also found here, with individual **sand sheets** and **sand seas (ergs)** having areas of over 60 000 km^2. This is where Bagnold performed the field observations in support of his seminal work on the physics of desert dunes and windblown sand (Bagnold 1954). Very interesting first-hand observations of the environment of the eastern Sahara can be found in Bagnold's (1990) autobiography and other works (Bagnold 1935). Saharan dunes reach heights in excess of 300 m, and are among the highest in the world. One sand sea in Libya is said to be over 1000 m deep. For the northern Sahara, Le Houérou (1986) reports the following surface types: rocky mountains and hammadas, 10%; regs, 68%; ergs, 22%.

More than 1600 plant species occur in the interior Sahara, with many times more on the margins. The only areas with abundant and widespread vegetation border the Mediterranean to the north and the tropics to the south. In the interior, mountainous areas and high plateaus, such as the Ahaggar and Tibesti, receive sufficient rainfall to support open grassland. Apart from oases, wadi ecosystems are the only habitats where large bushes and trees grow. *Tamarix* and *Acacia* are widespread and can form dense communities in rocky beds of wadis. Over half of the sandy areas are without perennial vegetation. Dunes, although often completely devoid of vegetation, sometimes have a variety of different shrubs and grasses that are regionally common. Quézel (1965) notes that, in the one-quarter of the Sahara with rainfall of less than 20 mm, vegetation is practically absent. Where the precipitation is between 20 and 50 mm, the vegetation cover is still poor, but in wadis and depressions that receive runoff it is more plentiful. Where the rainfall is above 50 mm, the vegetation is sometimes rich, but extremely variable depending on the substrate. For additional information about North African vegetation, see Le Houérou (1986), Ayyad and Ghabbour (1986), and Monod (1986).

Figure 3.12 shows the major large-scale surface weather features over North Africa in both January and July. In January, the ITCZ is near the southernmost extreme of its seasonal migration, at 5° N. To its north over the Sahara are north-easterly trade winds, and the Sahara high, which is an extension of the Azores anticyclone, part of the global subtropical high-pressure belt. Near the north coast are westerlies that are to the south of low pressure over the Mediterranean. Traveling depressions cause these coastal winds to have a high variability. To the south of the eastern part of the ITCZ is the equatorial trough, from which, extending southward, is a line of convergence of Indian Ocean easterlies and South Atlantic Ocean westerlies. The only maritime airmass trajectories that enter the Sahara are from the northwest and from the Red Sea. Virtually all of the Sahara is influenced by large-scale subsidence from the Sahara high. By

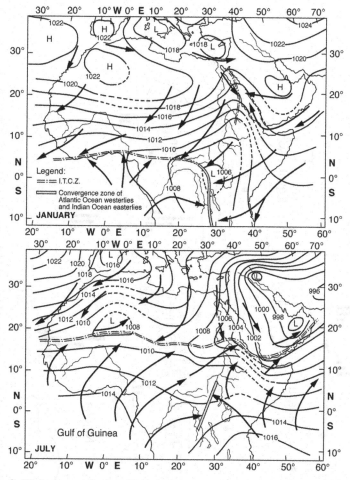

Fig. 3.12 Large-scale weather patterns over North Africa in January and July. The solid lines are sea-level pressure (millibars, plotted at a 2 mbar interval): arrows are streamlines of surface wind. See legend for other information. (From Griffiths and Soliman 1972.)

July, the ITCZ has moved northward to 15°–20° N, and the monsoon flow of the maritime Atlantic airmass to the south now penetrates well into the southern Sahara. The southwesterly monsoon winds can be viewed as "recurved" southeasterly trade winds from the south of the ITCZ, with the monsoon itself often being referred to as the Guinea monsoon because the winds are directed from the Gulf of Guinea. The ITCZ trough has replaced the Sahara high. To the north of the summer ITCZ, the trade winds still prevail and are known locally as the Etesian winds. The surface northerly to easterly trade winds in all seasons are also known as the harmattan. These surface winds of the large-scale trade circulation

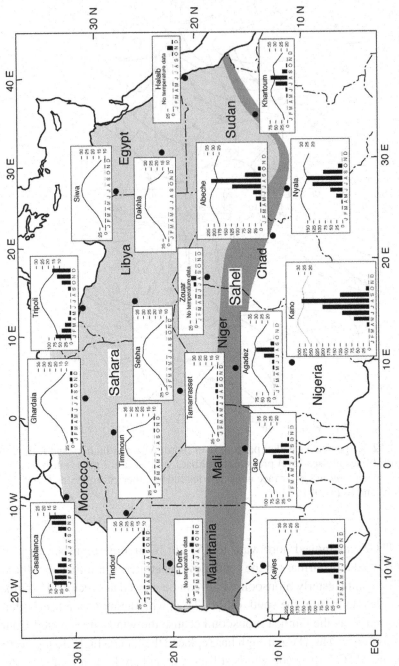

Fig. 3.13 Monthly distributions of rainfall (mm) and temperature (°C) for selected locations (dots) in and around the Sahara Desert. Light shading shows the approximate limits of the Sahara, and the dark shading is the Sahel.

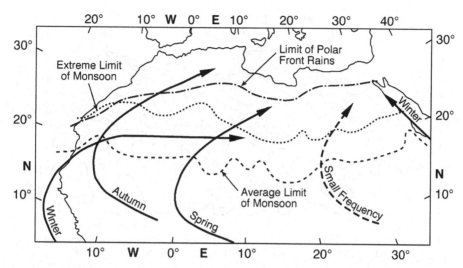

Fig. 3.14 The southern limit of the polar-front rains and the northern penetration of the monsoon in North Africa. The arrows show the seasonal changes in the paths of tropical weather disturbances. (From Griffiths and Soliman 1972.)

tend to become calmer in the evening as temperature inversions form near the surface and decouple the surface air from the winds above. They then commence again a couple of hours after sunrise as the inversion is eliminated by the surface heating. Because even a light wind will elevate fine mineral dust above the surface, the visibility is often extremely poor in the daytime boundary layer.

Over three-fourths of the Sahara has an annual-average rainfall of less than 10 cm and one-fourth has less than 2 cm, with Plate 1 illustrating the distribution. Mountains, near the desert margins and in the interior, cause local enhancements. The selected examples of the seasonal distributions of rainfall in Fig. 3.13 show three distinct rainfall climate regions.

- The northern Sahara, with a Mediterranean climate in which the greatest rainfall occurs in the winter season, associated with Atlantic cyclones, Mediterranean depressions, and the polar front that is displaced this far south. The summers are very dry.
- The central Sahara, with very scarce and irregular rainfall, and a mountain desert climate.
- The southern Sahara, with primarily summer rainfall that is related to the monsoon from the Gulf of Guinea and general, summer-season convection.

The interior highlands of the Sahara that trigger rain from large-scale disturbances allow us to determine how far the disturbances penetrate. For example, the Ahagger Mountains (Fig. 3.11) experience summer rain associated with the southwest monsoon as well as winter rains from westerly mid-latitude

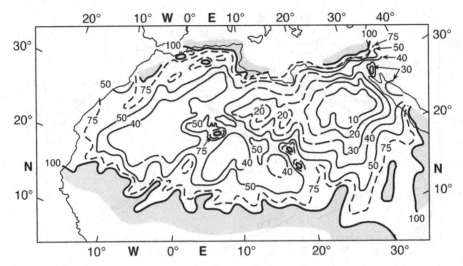

Fig. 3.15 Maximum observed 24 h rainfall (mm) in North Africa. The shading outlines the moist side of the 100 mm isohyet. (From Griffiths and Soliman 1972.)

disturbances, where the surrounding plains are virtually rainless in all seasons. Figure 3.14 further illustrates the dynamic causes of the geographic and seasonal divisions in the rainfall. Note the large area of the central Sahara that is to the south of the limit of polar-front rains and to the north of the normal limit of the monsoon. However, even in the most arid part of the Sahara, heavy rains do occasionally occur. Fig. 3.15 depicts the maximum recorded 24 h rainfall, and shows that a large fraction of the Sahara has experienced greater than 30 mm in a 24 h period. Heavy rains have caused extensive damage from rapid flooding of wadis. The small-scale maxima in the central Sahara are associated with the Ahaggar Mountains in southern Algeria and the Tibesti Mountains in northern Chad. Only the Wadi Halfa region in southern Egypt and northern Sudan fails to report any daily amounts in excess of 10 mm. Indeed, this area has never reported *annual* amounts in excess of 10 mm (Fig. 3.16, which also shows the Ahaggar and Tibesti Mountain effects). Where the desert reaches to the Mediterranean, the coast configuration also can greatly affect the rainfall climatology. For example, coastal Alexandria, Egypt has an annual average of 18 cm, while Port Said, also on the coast a little over 200 km to the east, averages only 8 cm.

Fig. 3.17 shows the Budyko index (defined in Chapter 2) for North Africa, indicating the expected substantial variability in the aridity. The maximum in the eastern Sahara is related to both the lack of rainfall as well as the large radiative input that results from the fact that this is one of the sunniest places on Earth

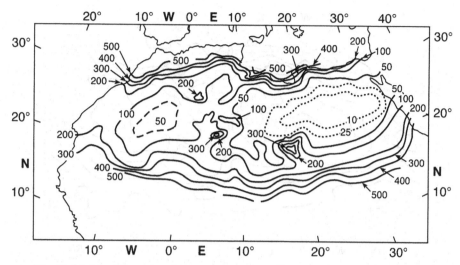

Fig. 3.16 Maximum observed annual rainfall (mm) in North Africa. (From Griffiths and Soliman 1972.)

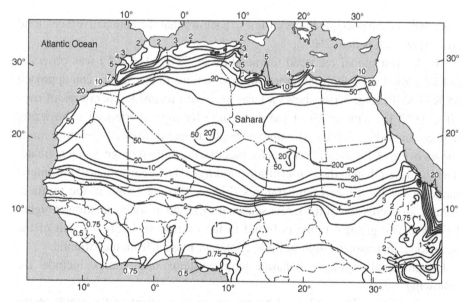

Fig. 3.17 For the Sahara Desert, the distribution of aridity based on the Budyko ratio. This is the depth of water that could be evaporated by the observed total annual net radiation, divided by the depth of the observed annual rainfall. (Adapted from Henning and Flohn 1977.)

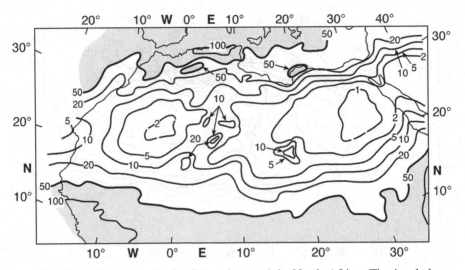

Fig. 3.18 Mean annual cloudiness (percent) in North Africa. The isopleth interval is irregular, and values greater than 50% are shaded. (From Griffiths and Soliman 1972.)

(over 90% of the possible hours). Fig. 3.18 shows that the annual average cloud cover there is less than 2%.

The highest global recorded temperature of 58 °C (136.4 °F) was observed in El Azizia, Libya. Diurnal temperature variations in the Sahara can approach 55 K (100 °F). Fig. 3.19a shows a large area with record temperatures of over 50 °C (122 °F), with much of the same area having experienced subfreezing temperatures (Fig. 3.19b). The map of mean annual diurnal temperature range shown in Fig. 3.19c clearly reflects the fact that the diurnal range is much greater over the desert than in the non-arid surroundings. The larger values of the annual range of the daily-mean temperatures (Fig. 3.19d) are found in the northern Sahara, an expression of the larger seasonal variation in solar radiation at higher latitudes. The greatest range is found in the northwest, likely a result of the occasional incursion of high-latitude cooler air masses from the Atlantic.

Geographically unique Saharan wind systems and names include the following.

• *harmattan* – This is forced by the pressure gradient to the south of the Sahara high in winter, and to some extent by the gradient to the southeast of the Azores high in summer (see Fig. 3.12). It is a dust-bearing northerly to easterly wind that originates in the desert and extends, at the surface, as far south as the ITCZ. The amount of dust is greater in the winter when the trajectory over the Sahara is longest. In the summer, the cooler, maritime, southerly monsoon flow

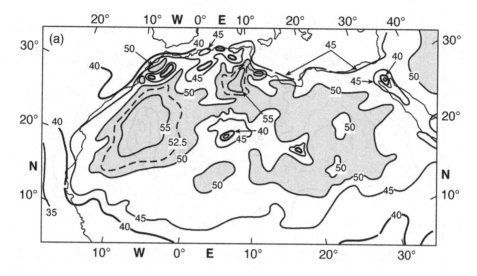

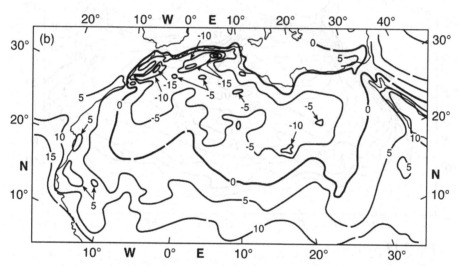

Fig. 3.19 The temperature (°C) climatology of North Africa (1926–1960); (a) Record maximum, with an isotherm interval of 5 °C, > 50 °C shaded; (b) record minimum, with an isotherm interval of 5 °C; (c) mean annual diurnal range with an isotherm interval of 5 °C and intermediate isotherms at 2.5 °C, > 20 °C shaded; and (d) annual range of daily mean, with an isotherm interval of 2 °C. (From Griffiths and Soliman 1972.)

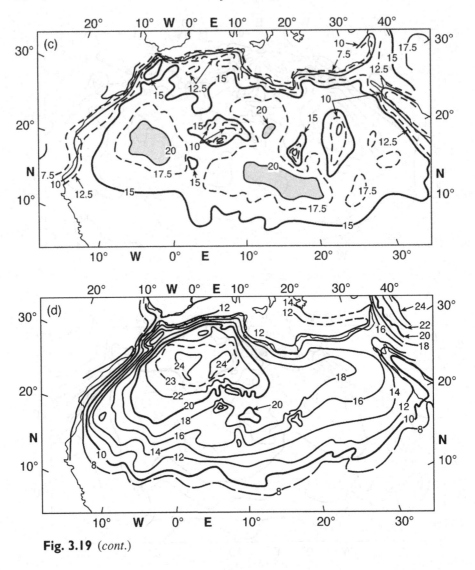

Fig. 3.19 (*cont.*)

undercuts the northerly harmattan, which continues to flow southward over the Gulf and Atlantic at higher elevations. The strength of the harmattan is controlled by synoptic-scale processes, which vary the pressure gradient, which is usually strongest from late December through early February. In some areas, the name is used generally to refer to the northeasterly trade winds.

• *sirocco* or *khamsin* (also chili, ghibli, leste, hamsin, sharav, samum, simoun, simoom (n), simm, zoboa) – This is a southerly or southeasterly wind along the Mediterranean coast of North Africa. It occurs primarily in the spring, and to a lesser degree in the fall, and consists of the continental tropical air that is in the warm sector to the south and east of eastward-moving depressions. During these seasons, the track of the depressions has moved south to the North African coast, or inland. A typical frequency is three to four depressions per month, representing a total of 20–60 days per year. Because the origins of this warm airmass are over the Sahara to the south, it is dusty and dry. Its temperatures may exceed 54 °C (130 °F) and the humidity may be below 10%. Even though the storms are smaller than the winter Mediterranean ones, the southerly winds can be strong and cause severe sandstorms, especially associated with cold-frontal passages. Sometimes the associated cyclones are given the names of these winds. In Italy, such storms produce strong southerly sirocco winds, rain laden with Saharan dust, high tides, and dirty snow in the Po Valley.

• *haboob* – This is an Arabic word meaning "to blow" or "strong wind," which is often used to refer to any high wind, or wind that raises dust or sand. Even though both synoptic-scale pressure gradients, and downdrafts and outflows associated with thunderstorms, can be responsible for the haboob, some African meteorologists limit the use of the term to the latter cause.

• *irifi* – In Western Sahara, this is a dreaded (Meigs 1966) easterly wind that arises suddenly and brings with it dust and extreme heat from the desert in the east. For example, in March 1941 it raised the temperature at Semara (177 km from the coast) from 18 °C (65 °F) at noon to 43 °C (109 °F) at 1600 LT. After 36 h of temperatures above 40 °C (103 °F), the northerly cool *alisio* wind returned the temperatures to the normal of near 16 °C (60 °F) (Meigs 1966; Diaz de Villegas *et al.* 1949).

The Kalahari, Namib, and Karoo Deserts

These deserts will be considered together here because they are of modest size and contiguous (Fig. 3.10). Their climates are distinct, however. The Namib Desert is a coastal desert that is washed by the cold Benguela Current of the Atlantic Ocean to the west, whereas the Kalahari is inland. The Karoo Desert's boundaries with the Kalahari and the Namib depend on whether the Karoo is defined by ecological characteristics, topography, or geology. On some maps it is not even shown as a distinct desert, with its area divided between the Namib and Kalahari. Sometimes the Karoo is shown as two deserts, the Nama Karoo and the Succulent Karoo (as in Fig. 3.10). Quite often the Namib Desert is defined as extending southward along the coast into South Africa, through the area defined in the figure as the Succulent Karoo.

The Kalahari and Nama Karoo Deserts exist largely in a sand-covered depression (the Kalahari Depression) in southern Africa's plateau, and are bordered by surrounding ridges and higher plateaus (Fig. 3.20). On the west they abut the

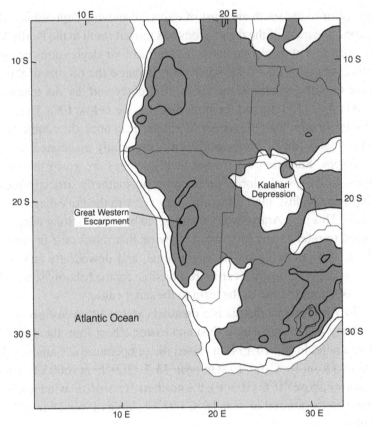

Fig. 3.20 Terrain elevation for the region of the Namib, Karoo, and Kalahari Deserts. The contour interval is 500 m, and heavy contours are used for 1000 and 1500 m. Elevations greater than 1000 m are shaded.

Namib Desert, and have an elevation of 900–1000 m. Most of the landscape is made up of gently undulating red sand dunes overlying deep layers of sand. The sands are interrupted by areas of exposed rock that are part of a vast sheet of sandstone that exists below. Many of the dunes form lines that are oriented parallel to the wind, but there are also barchan dunes in the southwest that achieve heights of 300 m. Sand-filled drainage channels are relics of former rivers, and collect water from streams that drain toward the center of the depression in the wet season, some into the great Okavango Swamps in northern Botswana. The water collects in basins with relatively impervious calcium-crusted surfaces, or pans, which serve to contain it for short periods. The pans of water are of a variety of sizes, with diameters reaching 5 km and depths of up to 10–15 m. Even though the pans number perhaps one thousand, there are currently only three permanent water holes in the Kalahari, and a few others that are perennial except during drought. The Kalahari Depression previously drained southward into the Orange River and to the sea, but during the past 1000 years that route has been blocked by sand. The fact that the sandy substrate allows rapid infiltration of rainwater contributes to the lack of water holes. Thus, the Kalahari and Namib are referred to as "thirstland."

In the local Nama language, Namib means "an area where there is nothing." The Namib Desert, when it is defined to include the Succulent Karoo, is a long and narrow coastal strip, more than 2000 km long and 50–150 km wide, between the rise to the east called the Great Western Escarpment and the Atlantic Ocean with the northward-flowing cold Benguela Current to the west. From Angola in the north, it extends southward through Namibia along the Skeleton Coast down to South Africa. In Namibia it merges eastward into the Nama Karoo and Kalahari Deserts that sit atop the escarpment, and in the south it merges with the less arid Nama Karoo. The extreme desert is limited to about half of this north–south distance, essentially the coast of Namibia. The northern part in Angola, also called "thirstland," does not possess a typical desert climate and vegetation. However, it does lack surface water because of deep layers of sub-surface sand.

Generally, riverbeds cut downslope from about 900 m at the Great Escarpment to the east, through the desert to the coastline. Only the Orange and Cunene Rivers in the south have perennial flow. To the north, the waters in the Omaruru and Swakop Rivers only mix with the sea after heavy rains have fallen on the interior plateau. The water of some rivers is caught by closed depressions containing saline evaporation ponds that are sometimes below sea level. The surface, northward from Walvis Bay in central Namibia, to Luanda, Angola, consists of bedrock and flat gravelly pavement, with scattered coastal dunes. The subsurface of the gravel pavement is often cemented by lime, limiting permeability to rain. South of Walvis Bay to Lüderitz, Namibia, are great sand dunes, 250 m high in places, and separated by valleys 1–2 km wide. From Lüderitz south to Saint Helena Bay, South Africa, is a rocky platform with some ridges, and sand and gravel.

Kalahari vegetation is rich for an arid area. It is generally savanna, with scattered grasses, shrubs, and trees that increase in density to the north. The grasses are dominated by *Aristida*, *Eragrostis*, and *Stipagrostis* and the trees by camelthorn (*Acacia erioloba*), grey camelthorn (*Acacia haematoxylon*) and shepherd's tree (*Boscia albitrunca*). The dunes have different species of grass growing on different parts, such as the slopes and the crests. The western coastal part of the Karoo, the succulent Karoo, has the highest plant species diversity of any desert in the world, with over 3500 species of succulent. The eastern Nama

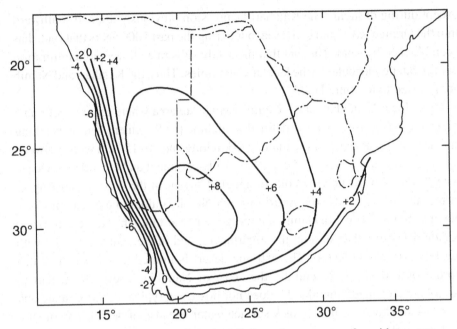

Fig. 3.21 The temperature anomaly (°C) from the area mean for mid-summer (January). The largest gradient is in the Namib Desert associated with the effects of the cold Benguela Current. The Namib in summer is about 12 K cooler than the Kalahari. (Adapted from Schulze 1972.)

Karoo has perennial grasses and dwarf shrubs, with extensive areas that have been overgrazed. The Namib has annual grasses on gravel plains where the annual rainfall is greater than 2 cm. Trees, such as *Acacia erioloba*, exist along drainage washes. The larger riverbeds are lined with large ana trees (*Faidherbia albida*) and tamarisk (*Tamarix usneoides*). The dunes in the sand seas contain nara plants (*Acanthosicyos horridus*) at the base of the dunes, perennial grasses (*Stipagrostic sabulicola*), and a few succulents (*Trianthema hereroensis*). The most well-known plant in the Namib is the *Welwitschia mirabilis*, a small evolutionary relic that can live in excess of 2000 years. For additional information about Namib Desert vegetation, see Walter (1986), and for Karoo and Kalahari Desert vegetation, refer to Werger (1986).

The Kalahari, Karoo, and Namib Deserts are relatively cool compared with other deserts at similar latitudes, such as in northern Africa and in Australia. The continent narrows here as it projects into the southern ocean, exposing the deserts to cool maritime air masses that are not greatly modified during their brief inland course. Also, the inland elevations are greater here. Nevertheless, there is considerable geographic variability in the mean temperature. Figure 3.21 shows

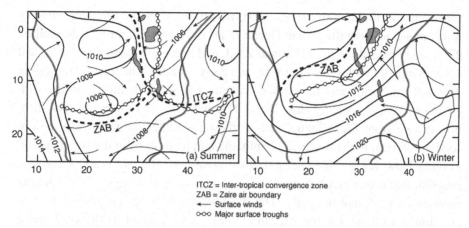

ITCZ = Inter-tropical convergence zone
ZAB = Zaïre air boundary
← Surface winds
ooo Major surface troughs

Fig. 3.22 Mean isobars (solid lines), streamlines of the surface winds (arrows), and positions of the Inter-Tropical Convergence Zone (ITCZ) and the Zaïre Air Boundary (ZAB) for summer and winter over southern Africa. The wide gray line outlines the continent. See the legend for other symbols. (Adapted from Tyson 1986.)

the temperature anomaly from the area mean for mid-summer (January), with the largest gradient being in the Namib Desert associated with the effects of the cold Benguela Current. The Namib in summer is about 12 K (22 °F) colder than the Kalahari, in spite of the higher elevation of the latter. This temperature gradient is similar to that found in the South American Atacama and Peruvian Deserts. In the winter, the temperature difference between the two areas is about 2 K (3.6 °F) (not shown). The Kalahari, being a continental rather than a maritime desert, experiences some of the highest and lowest monthly maximum and minimum mean temperatures in southern Africa. The diurnal amplitudes are also large.

Throughout the year, the subtropical highs dominate the surrounding maritime areas, off both east and west coasts, with the position of the ridge remaining near 30° S latitude (Fig. 2.5). In the summer, low pressure prevails over land, on average, whereas in the winter high pressure dominates. Schulze (1972) points out that the lower pressure in the summer contributes to an influx of moist tropical air, while in the winter the higher pressure inhibits the intrusion of maritime air. Fig. 3.22 shows the average sea-level pressure, winds, and ITCZ position for winter and summer for the latitudes of the northern Namib and Kalahari. Also shown is an extension of the ITCZ, the Zaïre Air Boundary (ZAB) (Tyson 1986). The convergence zones associated with the ITCZ and ZAB, and a strong low-pressure trough, prevail in these latitudes in summer.

The Kalahari annual rainfall ranges from less than 13 cm in the south and west to over 60 cm in the north and east, as shown in Plate 2. Most is convective and occurs in the summer (Fig. 3.23), associated with the monsoon flow of warm moist air from the Indian Ocean. The convective rainfall is enhanced by the convergence along the ITCZ and ZAB, and in the low-pressure trough, and occurs typically in the late afternoon and early evening. The meager rainfall in the southern Namib and the coastal Succulent Karoo falls mostly in the winter season (e.g. Alexander Bay, Fig. 3.23), and is related to the westerly flow of moisture from the Atlantic. This is the only winter-rainfall desert area in southern Africa. In the northern Namib, the rain tends to be in the summer (e.g. Lobito). However, for most coastal desert areas here, the rain is so meager and irregular that it is difficult to ascribe to any season. For example, Swakopmund receives only 15 mm annually. The dynamical causes of the Namib dryness are similar to those for the Atacama, with a cold current to the west and a subtropical latitude. The Namib, however, lacks the Atacama's large protective mountain range to the east. Because the latitudes of these deserts span the transition zone between tropical weather dominated by summer convective rainfall and mid-latitude weather with winter-cyclone precipitation, there is a tendency for summer rain to dominate to the north and winter rain to prevail in the south. Fig. 3.24 illustrates that the coastal Namib Desert is by far the most arid area in the region, in terms of the Budyko dryness index.

Figure 3.25 further illustrates the asymmetry in the rainfall and the mean wind field across the Namib and Kalahari. The easterly monsoon flow or trade winds prevail over the warm Mozambique current to the east of the continent. As the warm, moist flow encounters the Drakensberge Mountains (elevations to 3000 m high), much of the water vapor is precipitated. As the air flows westward over the Kalahari Depression, more water is precipitated through convective showers. This is reflected by the progressively decreasing average rainfall toward the west. The landward sea-breeze winds on the west coast do not produce precipitation as they encounter the Western Escarpment (elevations to 2000 m) because the air is cold and stable after traversing the cold Benguela Current. This cold and stable layer, with a temperature inversion 600 m deep, generally prevents the warm easterly flow from reaching close to the west coast. When the humid easterly trade or monsoon winds are sufficiently strong in the summer to penetrate to the west coast, a little convective rainfall can result there. This boundary between cool, humid air from the west, and warm, dry air from the east, shifts back and forth continuously, and results in the rapid changes in humidity and temperature than can occur in the Namib.

In common with other coastal deserts is the existence of fog, and its deposition on the surface, that provides some moisture to support vegetation. Swakopmund, at the coast, experiences 100–120 days per year with fog, with the frequency

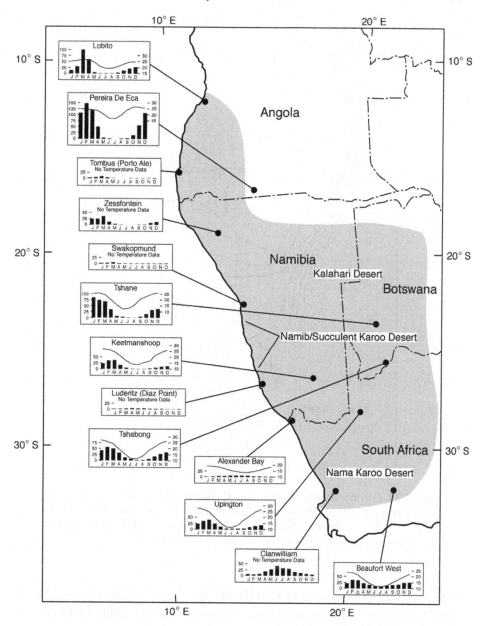

Fig. 3.23 Monthly distributions of rainfall (mm) and temperature (°C) for selected locations (dots) in and around the Namib, Karoo, and Kalahari Deserts. Shading indicates the approximate limits of these deserts.

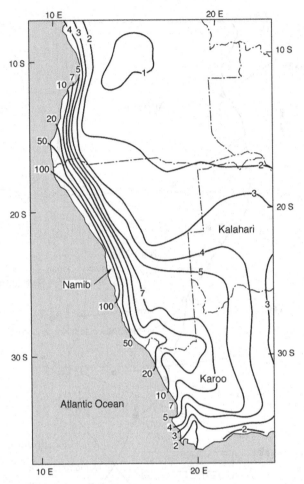

Fig. 3.24 For the Namib, Karoo, and Kalahari Deserts, the distribution of aridity based on the Budyko ratio. This is the depth of water that could be evaporated by the observed total annual net radiation, divided by the depth of the observed annual rainfall. (From Henning and Flohn 1977.)

decreasing to 5–10 days per year in the eastern Namib. Near the coast, Nagel (1962) estimates that the total fog water deposited is equivalent to 15 cm per year. Fig. 3.4 shows the effect of the fog, at and near the coast, on the diurnal temperature variation.

There are two maxima in mean annual sunshine hours in Africa. In addition to the large area over the eastern Sahara (Fig. 3.18), another covers the western Kalahari and the southern Namib, with its center in southern Namibia. Over this area, the mean annual sunshine hours are in excess of 3600, a value that is typical for a large part of the Sahara (Griffiths 1972a).

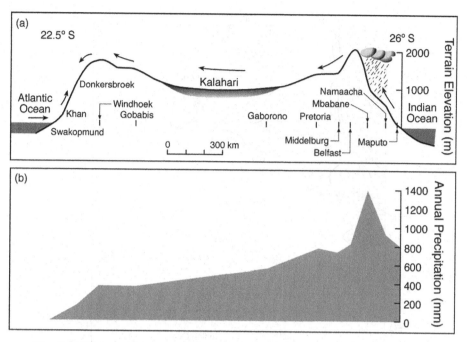

Fig. 3.25 Schematic cross-section through southern Africa along a transect from 22.5° S in the west to 26° S in the east. (a) Terrain elevation, the low-level wind field, and geographic place references. (b) The annual-average precipitation. (Adapted from Walter 1986.)

The winds over the Namib are often easterly in the winter, as cold airmasses and high pressure over the continent can produce a strong pressure gradient near the coast that forces downslope winds. The strong downslope winds, called *berg* winds, are warm and dry, and create dust storms that extend over the Atlantic. In the summer, west winds are more prevalent, as inland areas heat up.

Dust storms occur in significant number over these deserts, associated primarily with thunderstorms (the haboob of northeastern Africa). Great sandstorms have also occurred, especially along the coast associated with sustained large-scale winds. Schulze (1972) reports an event that closed the rail line between Swakopmund and Walvis Bay with 3–4 m drifts.

The Somali–Chalbi Desert

The southeastward extension of the aridity of the Sahara Desert is interrupted by the large area of orographic rainfall from the high Ethiopian Mountains (Plates 1 and 3). Without these mountains (Fig. 3.26), the aridity of North Africa and that of the Horn of Africa would be contiguous. The landscape includes stony plains,

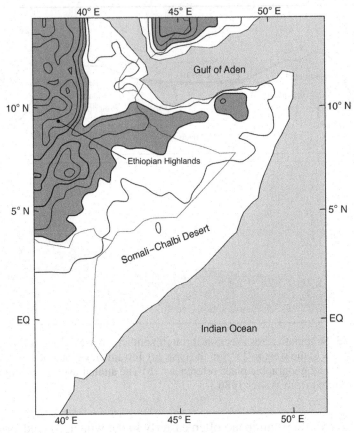

Fig. 3.26 Terrain elevation for the region of the Somali–Chalbi Desert. The contour interval is 500 m, and heavy contours are used for 1000 and 2000 m. Elevations greater than 1000 m are shaded.

with sand dunes generally limited to coastal areas. Inland are extensive areas with eolian and alluvial deposits. Vegetation is extremely limited where the annual rainfall is less than 10 cm. As the rainfall increases to 20 cm, grass, shrubs and tree savannas exist. Pichi-Sermolli (1955a,b) divides the vegetation in this area into eight categories: maritime vegetation (coastal areas), desert vegetation (the most arid part), subdesert shrub and grass, subdesert shrub with trees, subdesert shrub, subdesert bush and thicket, xerophytic open woodland, and vegetation where water is present (pools, wadis, etc., without running water). The desert-vegetation areas have scattered grass, shrubs, and stunted trees. *Acacia* trees are the most common, and some scattered palms occur (*Hyphaene* and *Phoenix*). In the more moist subdesert areas, phreatophytes such as *Acacia, Zizyphus, Ficus, Tamarix*, and species of palm occur.

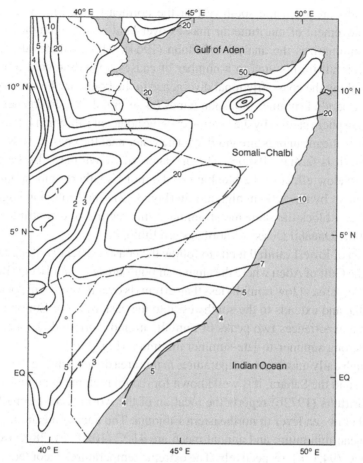

Fig. 3.27 For the Somali–Chalbi Desert, the distribution of aridity based on the Budyko ratio. This is the depth of water that could be evaporated by the observed total annual net radiation, divided by the depth of the observed annual rainfall. (From Henning and Flohn 1977.)

The entire Horn is semi-arid or arid, but the most arid parts are near the coasts of the Red Sea and the Gulf of Aden to the north, and the coast of the Indian Ocean to the east (Fig. 3.27). The Meigs (1953, 1957) aridity analysis in Fig. 3.1 also shows a narrow band of arid and extremely arid land along the coasts of the Red Sea and the Gulf of Aden, with a broader arid and semi-arid band extending southward along the Indian Ocean. Inland, an arid area exists around Lake Turkana in northern Kenya, and to the south below the Equator the semi-arid area extends through Kenya and into Tanzania.

The desert exists because of a combination of effects, and is unique as an equatorial desert on an east coast. The fact that the prevailing winds during most months are from the southwest or the northeast (see Fig. 3.12) means that the movement of maritime air masses over land is not common. Griffiths (1972b) summarizes the analysis of Flohn (1964), who attributes the lack of summer rainfall in this area to a number of causes, including cold upwelling (and fog) along the coast, frictional divergence in the boundary layer associated with coast-parallel summer-monsoon winds (as in Fig. 2.29), divergence in the lower troposphere caused by the heating over the adjacent Ethiopian Plateau, and variations in the mean pressure gradient. The water near the coast is 10 K (18 °F) colder than it is farther offshore. The dry lands of Kenya may also be related to a rain-shadow effect of the blocking of the moist southerly monsoonal flow in the summer by the 5200 m high terrain (Fig. 3.12). The Ethiopian Highlands may similarly block the same moist summer wind regime, and contribute to the aridity of the Danakil Desert of northeastern Ethiopia.

The area of lowest rainfall tends to follow the coast, especially along the Red Sea and the Gulf of Aden where the annual amounts are less than 10 cm (Plate 3). The broadest area of low rainfall (less than 20 cm) is over the end of the peninsula in Somalia, and extends to the southwest into the eastern tip of Ethiopia. Most of the area experiences two peaks of rainfall, around April and October, and a weak to strong summer-to-late-summer minimum (Fig. 3.28).

Although daily maximum temperatures in this area do not achieve the extreme values seen in the Sahara, it is well known for record high mean annual temperatures. Griffiths (1972b) reports the location of the world record to be Dallol, at 75 m below sea level in northeastern Ethiopia. The annual mean maximum, annual mean minimum, and annual mean are 41 °C (105.8 °F), 28 °C (82.5 °F) and 34.5 °C (94.1 °F), respectively. The extreme temperatures recorded here are 21 °C (69.8 °F) and 49 °C (120.2 °F). Griffiths (1972b) comments that Dallol does not have a high relative humidity, and he speculates that the nearby station of Massawa on the Eritrean Red Sea coast, with a mean annual temperature and relative humidity of 29.5 °C (85.1 °F) and 70%, respectively, is considerably more uncomfortable. Also, in the Strait of Bab el Mandab between Djibouti and Yemen is Perim Island (also Barim Island) with a mean June to September maximum of 37 °C (98.6 °F) and corresponding mean relative humidity of over 60%. Fig. 3.28 shows that most locations have very little seasonal variation in the temperature.

The winds are from the northeast in the cold winter-monsoon season, and become strong and southwesterly as the summer monsoon sets in (Fig. 3.12). In fact, during the summer monsoon these coastal surface southwesterlies represent one of the strongest and most-sustained wind systems on Earth (Ramage 1971). Gale-force winds (>14 m s^{-1}, 28 knots) are common in the summer in many locations. For example, Berbera in Somalia on the Gulf of Aden experiences over 50 days with gale-force surface winds from June through September (Griffiths

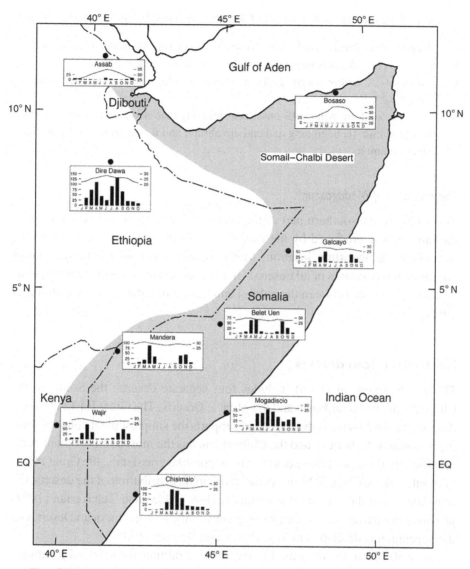

Fig. 3.28 Monthly distributions of rainfall (mm) and temperature (°C) for selected locations (dots) in and around the Somali–Chalbi Desert. Shading indicates the approximate boundary of the desert.

1972b). The July winds at Socotra Island to the east of Yemen exceed gale force over 28% of the time (Brody 1977). These winds are associated with the strong summer pressure gradient shown in Fig 3.12 over Somalia and northern Ethiopia, and produce severe dust storms along the northeast coast and the Gulf of Aden. The low-level jet over eastern, arid, equatorial Kenya (Findlater 1966, 1967, 1969) is related to these strong surface winds.

Some of the locally named winds of the area are listed below (Griffiths 1972b).

- *shamal* – A northerly wind in the winter that prevails for periods of 3–4 days.
- *kharif* – A hot, dry offshore wind with greatest intensity near dawn.
- *khamsin* – A northwesterly gusty wind that develops near noon and abates at sunset.
- *saba* – A westerly wind with onset near mid-day that is sometimes accompanied by a little rain. Its beginning and end are abrupt, and it is squally, with gale-force winds common.

The arid area of Madagascar

The aridity of the southern part of the island of Madagascar is likely related to the rain shadow produced by the island's terrain intercepting the trade winds. Because of its maritime location, Madagascar has one of the highest mean minimum temperatures of all deserts. Of all deserts, it also has the smallest temperature difference between the coldest and hottest months of the year (Evenari 1985a).

North American deserts

The North American Desert includes four separate deserts: the Sonoran, the Chihuahuan, the Great Basin and the Mojave Deserts. The Great Basin Desert is the coldest, the Mojave is the driest and supports the simplest plant communities, the Sonoran is the hottest, and the Chihuahuan has the most complex plant communities. To the east of these deserts are steppe-like grasslands, the Great Plains of North America. Fig. 3.29 shows the geographic distribution of the deserts, the grasslands, and the semi-arid scrublands. Hunt (1983) and Tchakerian (1997) provide a good overview of the physiography of the North American Desert, and the vegetation is described in MacMahon and Wagner (1985).

Fig. 3.30 illustrates the monthly partition of rainfall for selected locations, and the annual average rainfall is shown in Plate 4. The deserts generally prevail across the latitudes of the subtropical high-pressure belt, but extend northward as a result of rain shadows cast by high plateaus and mountains (Fig. 3.31). Four arid areas are a consequence of rain shadows: to the east of the Sierra Nevada and Cascade Ranges; to the east of the high plateaus of Utah on the Colorado Plateau; in the Wyoming Basin, to the east of the northern Rocky Mountains; and to the east of the Colorado Rocky Mountains on the Great Plains (see the orography in Fig. 3.31). The low-latitude areas are dominated most of the year by the subtropical belt of high pressure, with the North Pacific subtropical high to the west and the Bermuda high to the east. The Pacific high not only blocks

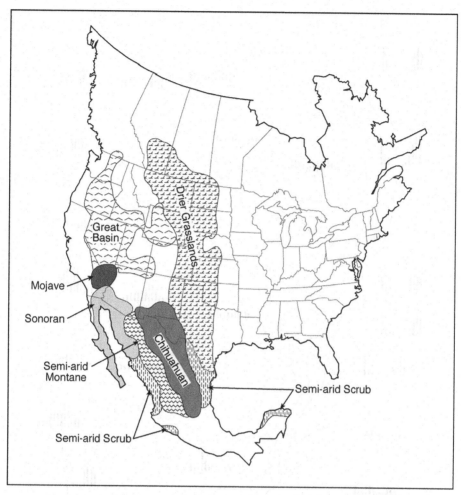

Fig. 3.29 The deserts and biomes of North America. (From Mares 1999.)

low-pressure systems from approaching the coast to its east, but the subsidence associated with it suppresses local moist convection. The high-pressure belt migrates with the season, with the Pacific high centered at about 38° N at the most northward point in its summer migration. In the winter, it is centered farther south off Baja California near 20°–25° N, and is considerably weaker (Fig. 2.5). Cyclone tracks flank the anticyclone on its poleward side.

Figure 3.32 shows the Budyko aridity index. By this measure, the most arid area is a narrow tongue that extends along the eastern side of the Sierra Nevada Range from the northern Sonoran Desert, through the Mojave Desert, and into the western Great Basin Desert, with the greatest aridity in the south. The northern Chihuahuan Desert is also quite arid.

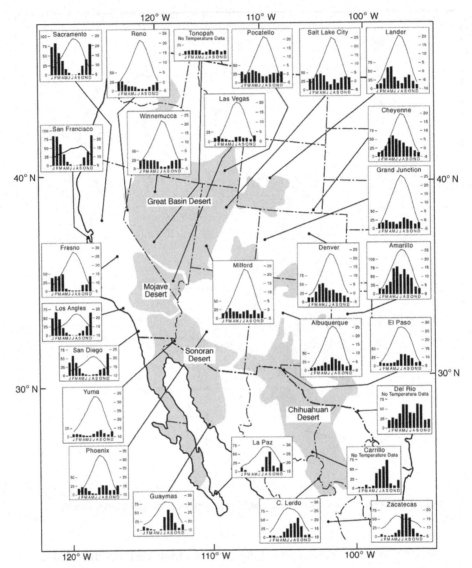

Fig. 3.30 The North American deserts (shaded), with monthly distributions of rainfall (mm) and temperature (°C) for selected locations.

The mountains near the Pacific coast – the Cascade and Sierra Nevada Ranges in the United States and the Sierra Madre Occidental Range in Mexico – have a few effects on the aridity. The Rocky Mountains to the east have similar effects on the aridity of the lands to their east.

- Winter storms entering the continent from the Pacific coast generally are weakened by the mountains as they move inland.

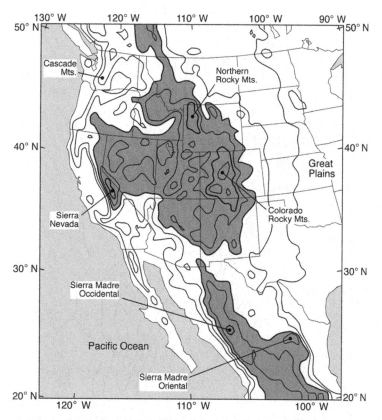

Fig. 3.31 Terrain elevation for the region of the North American deserts. The contour interval is 500 m; heavy contours are used for 1500 and 2500 m. Elevations greater than 1500 m are shaded.

- Penetration of moisture from the Pacific is physically blocked.
- The rain-shadow effect increases the stability of the downwind atmosphere and decreases the water vapor available for generation of precipitation. This effect allows the aridity to extend northward into Canada (Figs. 3.1 and 3.29).
- As noted earlier, the dynamic effect of the flow of air over a mountain range is the production of a ridge of high pressure. The anticyclonic curvature and northwesterly flow prevail for some distance downstream of the mountain ridge. This anticyclonic curvature in the upper-level flow can suppress storm development, and is at least partly responsible for the dry belt to the east of the Rocky Mountains (Borchert 1950).

The semi-arid California coast, with its Mediterranean climate, receives virtually all of its rainfall in the winter months because the North Pacific subtropical high-pressure system has retreated southward with the winter sun and no longer protects the coast from Pacific storms. For stations to the west of the orographic

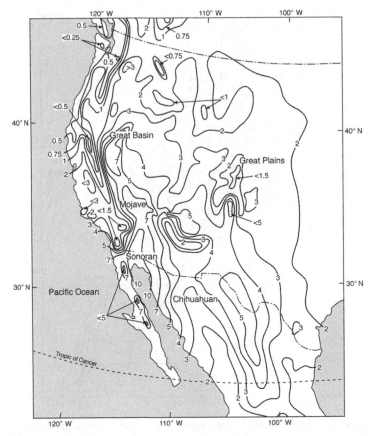

Fig. 3.32 For North America, the distribution of aridity based on the Budyko ratio. This is the depth of water that could be evaporated by the observed total annual net radiation, divided by the depth of the observed annual rainfall. (Adapted from Henning and Flohn 1977.)

influence of the Sierra Nevadas, the summer aridity is near absolute, and this area has virtually the only single-season rainfall climate in North America.

To the east of the coastal mountains in the southern deserts there tend to be two rainfall maxima per year (e.g. Phoenix in Fig. 3.30), one associated with weak winter storms that have traversed the mountains, and one that is a product of the North American summer monsoon (Douglas *et al.* 1993; Tang and Reiter 1984). The latter results from typical monsoon dynamics in which a warm-season, low-pressure center (the heat low) forms over strongly heated land, with a resulting low-level flow of moist air from the sea that supplies moisture for convection. In this case, the moist monsoon flow is associated with a southerly low-level jet from the Gulf of California and a southeasterly low-level jet from the Gulf of Mexico (Adams and Comrie 1997). This summer monsoon rainfall

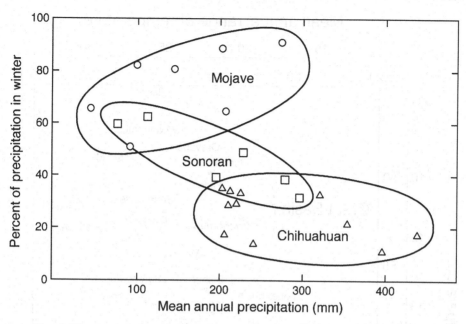

Fig. 3.33 Mean annual precipitation plotted against the percent of the precipitation occurring in the winter half of the year, for the hot deserts of North
America. (From MacMahon and Wagner 1985.)

affects the climates of the Sonoran Desert, the western Chihuahuan Desert, the
eastern Mojave Desert, and the western part of the semi-arid Great Plains to the
east of the Rocky Mountains. Barlow *et al.* (1998) show the July rainfall over
much of the Sonoran Desert to be over 4 mm d^{-1} greater than the June rainfall,
as a result of the onset of the monsoon, with the 1 mm d^{-1} change extending
as far north as 35° N. Note the large increase in rainfall from June to July at
Phoenix in Fig. 3.30. The dependability of the annual rainfall in these deserts
thus depends on the strength of the summer monsoon and the strength of the
winter Pacific storms after their penetration inland from the coast. A good summary of the mean annual precipitation and the seasonal distribution is shown in
Fig. 3.33.

Figure 3.34 illustrates the effects of latitude and other factors on the mean
annual temperature and the range of monthly mean temperatures for various locations in the northernmost three of the North American deserts: the Great Basin,
the Mojave, and the Sonoran Deserts. The Great Basin Desert has a mean annual
temperature that is about 5.6 K (10 °F) cooler than that of the Mojave and 11.1 K
(20 °F) cooler than that of the Sonoran. The Sonoran Desert is warmest because
its altitude is lower than that of the Mojave Desert, with its elevated intermountain basins, and that of the Chihuahuan Desert that is located on the Mexican

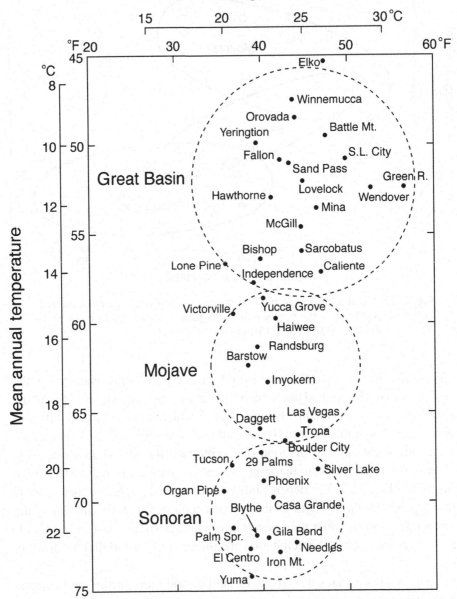

Fig. 3.34 Mean annual temperature range and mean annual temperature for various locations in the Great Basin, Mojave and Sonoran Deserts. (Adapted from Axelrod 1983.)

Plateau. The fact that the annual temperature range is generally greater in mid-latitudes than in the subtropics and lower latitudes accounts for the greater annual temperature range of Great Basin Desert locations. Fig. 3.30 also shows that the annual range in temperatures in these deserts is greatest at continental northern locations, with the range decreasing to the south and near the Pacific coast.

The Chihuahuan Desert

The Chihuahuan Desert extends from central Mexico into the southern United States, and in Mexico is flanked by the volcanic Sierra Madre Occidental to the west and the limestone Sierra Madre Oriental to the east. It increases in elevation from 600 m above sea level in the northern part to over 2000 m in the south. The desert surface generally is made up of mountain ranges, plains, basins, and **bajadas** (sloping aprons of rock waste on the flank of a mountain). Much of the surface is internally drained, with water flowing into closed basins called **bolsones**. Shallow saline lakes, or **playas**, near the centers of the bolsones are seasonally flooded. The primary external drainage is through the Rio Bravo (called the Rio Grande in the United States). Much of the desert overlies limestone that has formed from a sea that once prevailed there, and the soils are calcarious and often contain hardened layers of calcium carbonate (caliche). There are some gypsum soils, however, an example being the large area of gypsum sands in the far northern part of the desert in the Tularosa Basin in the United States. Other dune areas exist, an example being the enormous Samalayuca Dunes in the northern part of the State of Chihuahua in Mexico.

Vegetation is highly varied, and often abundant. Halophytic and gypsophilic (salt- and gypsum-tolerant, respectively) plants are common. The northern part of the desert was once a fairly lush grassland, with grass a meter high. However, overgrazing, primarily by cattle, has allowed the invasion by non-grass species, such as creosote bush, that are typical of more-arid parts of the desert. Shrubs include creosote bush (*Larrea tridentata*), lechuguilla (*Agave lechuguilla*), sotol (*Dasylirion wheeleri*), tarbush (*Flourensia cernua*), mesquite (*Prosopis juliflora*), mariola (*Parthenium incanum*), saltbushes (*Atriplex*), whitethorn acacia (*Acacia constricta*), allthorn (*Koeberlinia spinosa*), and soapweeds (*Yucca*). The creosote bush so dominates much of the landscape here (in addition to those of the Sonoran and Mojave Deserts to the west) that its common Mexican name is gobernador (governor). Some succulents are barrel cacti (*Ferocactus* and *Echinocactus*), prickly pears (*Opuntia*), chollas (*Opuntia*), and peyote (*Lophophora williamsii*). Rivers are bordered by riparian forests of willow (*Salix*), cottonwood (*Populus*) and sycamore. Grasses, which are much more common than in the adjacent Sonoran Desert, include grama grasses (*Bouteloua*), tobosa (*Hilaria mutica*), and fluffgrass (*Erioneuron*).

This is a subtropical desert, and is generally too far south to be greatly affected by the tracks of mid-latitude storms. Rainfall is primarily convective and occurs in the summer, with a significant impact from the moist southwesterly flow of the North American monsoon. Pacific storms provide a small amount of winter moisture (Fig. 3.30). Annual averages are typically 20–30 cm, but there is a great deal of interannual variability. The Sierra Madres, which bound the desert on the east and west, block the penetration of much of the potential moisture from the Pacific Ocean and the Gulf of Mexico. Summers are hot, but winter temperatures can be low, with subfreezing conditions prevailing sometimes for a few consecutive days.

The Great Basin Desert

The Great Basin Desert is about the same size as the Chihuahuan Desert, and like the Chihuahuan it is characterized by interior drainage basins. The surface is comprised of sandy or stony alluvial flats, salt- or clay-covered playas, lakes, **alluvial gravel fans**, lava flows, and mountains. The percent of the area that is covered by basin flats ranges between 10 and 40%, with the mountains and gravel fans representing most of the remainder. Small and large, shallow saline lakes are remnants of older ones that covered vast areas. There are more than 100 north–south-oriented mountain chains. The basins range in elevation from sea level to 1200–1500 m ASL, with many of the peaks exceeding 2700 m in elevation.

Vegetation consists mostly of shrubs and perennial grasses. The dominant shrubs are big sagebrush (*Artemisia tridentata*) and a saltbush, shadscale (*Atriplex confertifolia*), which leads to the fact that the Great Basin Desert is sometimes called the Sagebrush–Shadscale Desert. Other shrubs include blackbrush (*Coleogyne ramosissima*), hopsage (*Grayia spinosa*), greasewood (*Sarcobatus vermiculatus*), and rabbit brush (*Chrysothamnus*). Additional small plants are winterfat (*Eurotia lanata*), gray molly (*Kochia vestita*), pickleweed (*Allenrolfea occidentalis*), glasswort (*Salicornia*), and desert teas (*Ephedra*). There are also numerous other species of *Artemisia* and *Atriplex*. Grassland steppes include fescue–wheatgrass (*Festuca–Agropyron*) mixes and wheatgrass–bluegrass (*Agropyron–Poa*) mixes.

This is a cold desert, which exists in the rain shadow of the Pacific-coastal mountains: the Sierra Nevadas and the Cascades (Fig. 3.31). The seasonal precipitation is distributed relatively evenly, but the summer amounts are somewhat less than during the rest of the year (Fig. 3.30). The annual totals are generally less than 250 mm. There are extended periods of subfreezing temperatures and considerable snow.

The Sonoran Desert

This is the next smaller of the North American deserts. It is generally a lowland desert, with most of its area being below 600 m above sea level. However, some peaks in northern Baja California exceed 3000 m elevation, while the Salton Sink in southern California dips to a minimum of 72 m below sea level. The surface consists of gravelly and sandy plains, drainage basins, extensive sand-dune fields, bajadas, and north–south-oriented mountain ranges and isolated mountains. There are numerous lava flows and volcanic hills. Some sand dunes are up to 100 m high. The Gran Desierto del Altar is North America's only large erg.

The Sonoran Desert is said to be the ecological twin of the Monte Desert in northern Argentina, with the naked eye not being able to distinguish the difference. Vegetation is abundant and diverse; this is one of the most complex desert biomes in the world. Some of the more abundant vegetation types are noted here. Trees and shrubs include paloverde (*Cercidium microphyllum, C. floridum*), smoke tree (*Dalea spinosa*), ironwood (*Olneya tesota*), mesquite (*Prosopis juliflora*), catclaw (*Acacia greggii*), elephant tree or torchwood (*Bursera microphylla* and *B. hindsiana*), ashy jatropha (*Jatropha cinerea*), creosote bush (*Larrea tridentata*), burro bush (*Franseria dumosa*), ocotillo (*Fouquieria splendens*), white bursage (*Ambrosia deltoidia*), and big galleta (*Hilaria rigida*). The northern limit of creosote bush distribution approximately marks the boundary of the hot deserts of North America with the cold Great Basin Desert to the north. Cactus are especially plentiful: teddy bear cholla (*O. bigelovii*), chain-fruited cholla (*O. fulgida*), buckhorn cholla (*O. acanthocarpa*), pencil cholla (*O. arbuscula*), organ pipe cactus (*Stenocereus thurberi*), senita (*Lophocereus schottii*), fish-hook barrel cactus (*Ferocactus wislizenii*), cardón (*Pachycereus pringlei*), hecho (*P. pecten-aboriginum*), and saguaro (*Carnegiea gigantea*). The height of the saguaro can exceed 15 m and that of the cardón 20 m, with the latter cactus possibly being the largest in the world. Other important genera include *Agave, Yucca*, and *Dudleya*. Common grass genera are *Bouteloua, Hilaria*, and *Sporobolus*.

This is often considered to be the hottest of the North American deserts. Even though the Mojave has the highest recorded temperature in the Americas at a location below sea level in Death Valley, its winters are colder than those of the Sonoran Desert. In the Sonoran Desert, daily summer temperature maxima can frequently exceed 120 °F over large areas. Subfreezing temperatures rarely last for more than a day. There are well-defined rainy seasons in both winter and summer, with this desert receiving more rain than the other two hot deserts of North America. In the western desert, the California climate of dry summers and wet winters prevails (e.g. San Diego, Fig. 3.30), whereas in the eastern and southern desert, summer rainfall dominates (e.g. Guaymas, Fig. 3.30) as a result of the monsoon and generalized convection. Annual totals in the east are 250–400 mm, where moisture can reach from the Gulf of Mexico. In the west, in the rain shadow of the Pacific Coast Ranges, annual totals are as small as 50 mm. Sonoran Desert weather is also influenced by the heat low that is located in this area in the summer (Douglas and Li 1996). The Baja California Peninsula is a special situation in that it is a cool foggy desert.

The Mojave Desert

The smallest and driest of the North American deserts is the Mojave. Elevations vary from 85 m below sea level, the lowest point in the Americas, to about 1500 m above sea level, with most of the desert at 600–1500 m above sea level. There are many small sand-dune fields, with exceptional dunes in Eureka Valley that rise over 200 m above the desert floor. Otherwise, the surface is composed of the typical mix of mountains with interior basins.

Bursage (also burro bush, *Franseria dumosa*) and creosote bush (*Larrea tridentata*) constitute 70% of the Mojave vegetation. The oldest living plant known is a creosote bush in the Mojave, radiocarbon-dated at 9400 years. The curious Joshua tree (a tree-sized yucca, *Yucca brevifolia*) occurs at higher elevations. Other species are saltbush (*Atriplex*), blackbrush (*Coleogyne ramosissima*), the Mojave yucca (*Yucca schidigera*), single-leafed piñon (*Pinus monophylla*), desert holly (*Atriplex hymenelytra*), cattle spinach (*A. polycarpa*), brittlebush (*Encelia farinosa*), and galleta grass (*Hilaria rigida*).

The weather is often considered to be transitional between the cold Great Basin Desert and the hot Sonoran Desert. The highest recorded screen-level air temperature in the western hemisphere, 56.7 °C (134 °F), was recorded in the Mojave at Furnace Creek in Death Valley. Ground temperatures in this area have been measured at 88 °C (190 °F). Winters are colder than in the Sonoran Desert, with cold-air drainage into valleys producing nocturnal temperatures of below − 15 °C (5 °F) in the winter. The Mojave is in the rain shadow of the southern Sierra Nevadas, the Tehachapi, the San Gabriel, the San Bernardino, and the Little San Bernardino Mountains. Precipitation is extremely low, 25–50 mm in the east to 125 mm in the west, occurs mostly in winter, and is of Pacific-storm origin.

A hot and dry wind that originates in the Mojave Desert is the Santa Ana wind. When high pressure is established over the Great Basin Desert, the wind

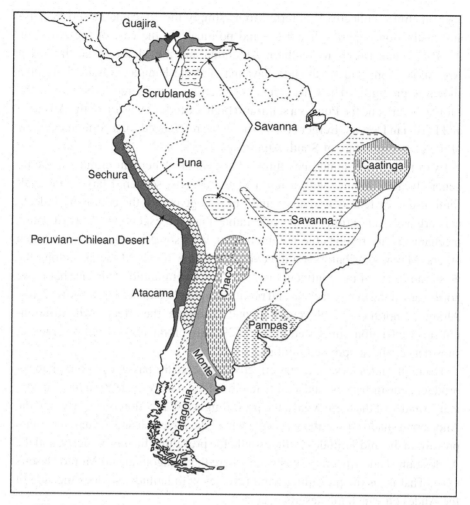

Fig. 3.35 The deserts and biomes of South America. (Adapted from Mares 1999.)

over the Mojave is easterly or northeasterly. As the air drains downward from the desert plateau into coastal southern California, the wind speeds are high and the air temperature rises as a result of adiabatic warming. The air temperature is thus even higher than it was over the Mojave, and the relative humidity is even lower. The strong winds and low humidity can last for days, desiccating the surface and causing devastating fires.

South American deserts

The Atacama and Peruvian Deserts are narrow, extremely arid areas extending along the South American coast to the west of the Andes, from north of Chiclayo in northern Peru to almost Santiago in central Chile (Fig. 3.35). They are sometimes collectively referred to as simply the Atacama Desert because any distinction is purely historical and political. To the east of the Andes is the Patagonian Desert in southern Argentina and the Monte in northwestern Argentina. Semi-arid scrublands, savanna, and the Pampas grassland are also shown in the figure. The Chaco area is sometimes considered to be part of the Monte Desert, and the Puna is a semi-arid high-altitude extension of the Atacama and Peruvian Deserts. Berger (1997) should be referenced for a complete review of the geomorphology of South American deserts.

Even though we sometimes think of the South American climate as being generally quite humid, with a few isolated deserts, scattered large and small semi-arid to extremely arid areas make up a large part of the continent. Indeed, it is arguable that aridity is the rule rather than the exception. At least a few mechanisms are responsible for the aridity, with some better understood than others. Major contributors are the existence of the Andes Mountains along the western margin of the continent, and the Peru, or Humboldt, Current that flows northward along the west coast and produces upwelling of cold water (Fig. 2.14). Along the north coasts of Venezuela and Colombia, the effect of the differential surface friction, illustrated in Fig. 2.29, may produce low-level divergence, subsidence, and suppressed rainfall.

The high Andes serve as a very effective barrier to the passage of precipitation-producing disturbances, and a rain-shadow effect is very evident. In the subtropical latitudes of the trade winds, the prevailing wind direction is easterly, and the Andes rain shadow is on the west side where the Atacama and Peruvian Deserts prevail. In the mid-latitudes farther south, the prevailing flow is westerly, and the Andes rain shadow is on the east side where the Patagonian and Monte Deserts exist. That is, as the prevailing wind reverses with latitude, so does the side of the Andes on which the deserts occur.

A significant climate-controlling contrast between the deserts of South America and those of North America and Asia is that South American deserts lack any large poleward area of continental interior. Thus, wind flow from the poleward direction imports relatively mild maritime–polar rather than continental–polar airmasses. Thus, there are no outbreaks of extremely cold air that penetrate into the mid-latitude or the subtropical deserts.

The Atacama and Peruvian Deserts

Good summaries of the geography of these deserts may be found in Bowman (1924), Cornett (1985), and Meigs (1966). They exist in a narrow strip of land, about 3700 km long, between the Andes and the Pacific Ocean: the longest desert and semi-desert of the world. In Peru, the driest area is the lowland desert

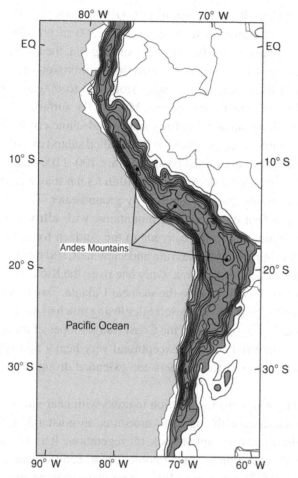

Fig. 3.36 Terrain elevation for the region of the Peruvian and Atacama Deserts. The contour interval is 500 m, and heavy contours are used for every 1000 m interval. Elevations greater than 1000 m are shaded.

along the coast, which is widest in the north between Piura and Chiclayo, where it is covered with migrating sand dunes and is known as the Sechura Desert (Fig. 3.35). South of Chiclayo, to Pisco, the Andean slopes reach down to sea level. In a typical year, 10–20 rivers have sufficient volume to maintain their flows through the desert until they reach the sea. More than that number are consumed by the desert on their way from the mountains. In the Atacama, three parallel, north–south-oriented zones are definable: the coastal mountain range to the west, the western slopes of the Andes to the east, and a depression with salt flats in the middle. Fig. 3.36 illustrates the terrain. The total width of the desert varies from a few kilometers to 150 km. The geomorphology of the area is quite complex. There is generally no coastal plain, with coastal mountains or plateaus dropping off rapidly to the sea from 600–900 m elevations. In one location, 175 km north of Antofagasta, a summit of 2300 m prevails within 5 km of the coast. Along the coast to the south of Antofagasta, the coastal range rises abruptly from the sea to 1800 m in a stretch that is unbroken for 150 km. Inland, the elevation of the desert basin ranges from 300 to 900 m, with the lower elevations on the north and south margins. Much of the surface consists of stony and pebbly areas, large areas of shifting sands, and saline crusts. North–south-oriented chains of nitrate and sulfate salt flats (called salars) are 40–80 km wide and lie in depressions or undrained basins that are 700–1100 m above sea level. The largest is the great Salar de Atacama. Within a salar may be a marshy area with salty shallow ponds that are supplied by groundwater. The salt pans tend to be at the eastern foot of the coastal mountains, with **alluvial fans** of gravel and sand sloping upward to the canyons in the Andean foothills to the east. The groundwater originates as Andes rain and snowmelt, and moves slowly westward in the aquifer under the Atacama. Only one river, the Rio Loa, reaches through the basin and the coastal range to the sea near Calama. Two other rivers entering the Atacama disappear into the desert after losing much of their water to irrigation. Numerous dry arroyos exist on the Coast Range slopes of the western desert, but they carry water only during exceptional very heavy rains that are said to occur once or twice per lifetime, between extended droughts of 10–50 years' duration.

Even though there is some scrub vegetation in areas with near-surface water that is not saline, or in areas with fog-bank moisture, an outstanding feature of these deserts is the almost complete lack of vegetation. It is so dry that decomposition of dead vegetation is almost non-existent; dead vegetation may be thousands of years old. There is even little or no vegetation on the eastern Atacama margins, up to 2500 m ASL on the western slopes of the Andes. Near 3000 m ASL, some grasses and shrubs appear. In the northern Atacama where wet valleys dissect the desert, there are plant communities composed of trees (e.g. *Prosopis, Salix, Schinus, Acacia*) and shrubby and herbaceous plants. In the southern Atacama, where it is a little wetter, there are numerous annuals that appear, grow, flower, fruit, and drop seeds over a few days after a rain, with the seeds sometimes waiting years for the next rain before the cycle repeats.

The Peruvian and Atacama Deserts have the cold upwind Humboldt (Peru) Current. Lowest water temperatures of typically 15.6 °C (60 °F) are found within a mile of the coast, north to Punta Pariñas, Peru, within five degrees latitude of the equator. Because of this current, the coastal air temperatures are remarkably similar and moderate along the entire extent of the deserts. Over the 1200 km distance from Callao, near Lima in mid Peru, to Antofagasta, well south into

Table 3.3 *Monthly average temperatures for the coolest and warmest months, and annual mean temperature, for four coastal locations in the Peruvian and Atacama Deserts*

Station	Latitude (°S)	Warmest month mean temperature (°F–°C)	Coolest month mean temperature (°F–°C)	Annual mean temperature (°F–°C)
Callao, Peru	12	71.0–21.7	62.5–16.9	66.5–19.2
Arica, Chile	18	71.0–21.7	62.0–16.7	66.0–18.9
Iquique, Chile	20	70.5–21.4	61.5–16.4	64.8–18.2
Antofagasta, Chile	23	71.0–21.7	62.0–16.7	65.5–18.6

Source: From Miller (1961).

the Chilean Atacama, Table 3.3 shows that the coolest and warmest months at four selected coastal locations are within 0.5 K (1.0 °F) of each other. The mean annual temperatures for the four locations differ by only 1.0 K (1.7 °F). These temperatures are lower than for any other desert area in the world of similar latitude.

Along much of the coast, stratus clouds exist over the sea in a layer about 600–1500 m above sea level. The clouds advect inland, and intersect the Coastal Range of mountains to produce fog and sometimes drizzle (see the cloud in Fig. 2.28). Often the base of the cloud layer descends to the surface at night, with the base ascending and the cloud virtually evaporating during the day. Sometimes, especially in the winter when the water is the coldest, the cloud remains at the surface as fog over the water and coast, where the fog is known in Peru by the name *garua*. This fog, with a distinct upper limit at typically 500–700 m AGL, can advect inland through the Andean valleys a distance of 30–40 km. Fog droplets collecting on tree leaves and flowing down the stems to the ground can provide a substantial amount of water annually. Between the coastal mountains and the coast, cloudy and humid, but rainless, conditions prevail. To the east of the mountains in the large expanse of desert, the sky is clear. There is much seasonal and geographic variability to the characteristics of the cloud and fog. The Austral summer from November to April is generally cloudless, whereas the winter is cloudy. The average relative humidity on the coast rarely falls below 50%, and in some areas averages about 75%, in spite of the extreme aridity (Mossman 1910).

It is claimed that the desert of northern Chile is the driest location on Earth, although it is difficult to compare locations where there has been virtually no

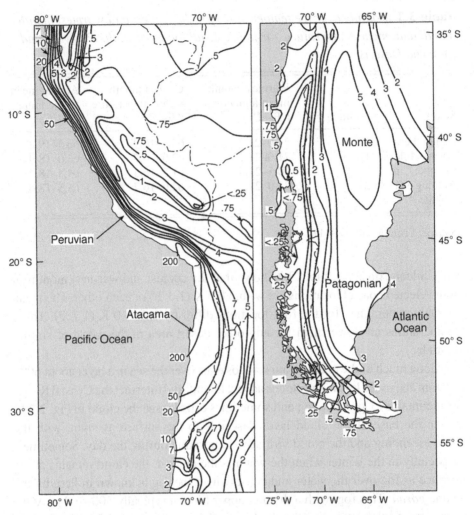

Fig. 3.37 For the Peruvian, Atacama, Patagonian and Monte Deserts, the distribution of aridity based on the Budyko ratio. This is the depth of water that could be evaporated by the observed total annual net radiation, divided by the depth of the observed annual rainfall. (From Henning and Flohn 1977.)

recorded rainfall or where rainfall is less than 1 mm per year. For example, at Arica, Chile, out of 39 years, only four had annual rainfall totals greater than 4 mm. The annual-average precipitation is provided in Plate 5, and the Budyko index is shown in Fig. 3.37. Summer rains are limited to the desert's eastern margins (Fig. 3.38) where the terrain lifts the air to 3000 m before cloud, and rain or snow, result. Over the Andes' western slopes below 3000 m, some locations receive neither summer nor winter precipitation. See Box 3.3 for insight into farming the Atacama with minimal rainfall.

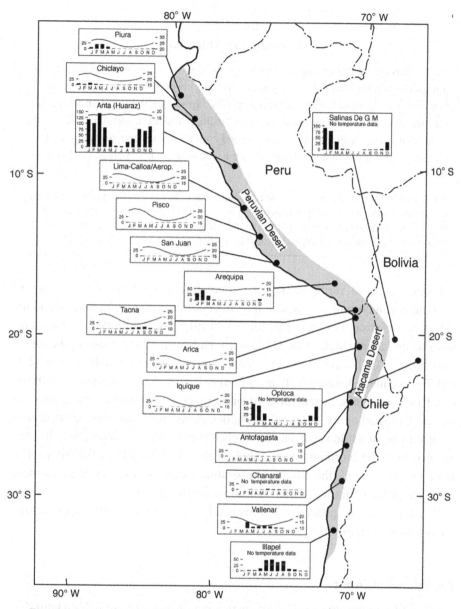

Fig. 3.38 Monthly distributions of rainfall (mm) and temperature (°C) for selected locations (dots) in and around the Peruvian and Atacama Deserts. The area of these deserts is shaded.

Box 3.3
Farming the Atacama

Indigenous desert farmers have learned how to eke out a crop with only a meager amount of precipitation, especially in desert margins and along humid coastlines. For example, Darwin (1860) remarks upon the farmers on the southern fringe of the Atacama Desert taking advantage of a "trifling amount of moisture": tilling the ground after one brief shower, planting seed corn after a second, and reaping a harvest after a third. In a similar reference to the Atacama, Mossman (1910) states that

for areas where the aridity is not absolute, a single heavy shower benefits pastures and fields and brightens the outlook of hundreds of people. Two showers bring a year of plenty, and three or more showers make the year memorable, if indeed they do not bring floods and greater disaster than several years of drought.

Large east–west variations in the temperature result from the existence of the cold ocean current, the radiational effects of the coastal cloudiness, and the elevation variation. For example, Iquique, Chile (20° S latitude), on the coast, has a mean January high of 28 °C (82.4 °F). However, 70 km inland to the east at Canchones (1050 m elevation, in a depression) the mean monthly high is 32 °C (90 °F). The mean January daily minimum is 13 °C (55.4 °F) at Iquique and 0 °C (32 °F) at Canchones. In general, cold air drainage at night into the numerous salt basins, and shading by surrounding higher elevations, cause the temperatures locally to be much lower than over higher surrounding terrain. The salt lakes are frozen in the morning during much of the year, in spite of the subtropical latitudes. In general, the annual temperature range is twice as large in these basins as in locations at a similar latitude along the coast.

Coastal winds are generally southwesterly, and are frequently strong during the afternoon and evening.

The Patagonian Desert

This desert is unique to world deserts in that it is on an east coast in high latitudes. It is the largest desert of the Americas, a cold-winter, low-altitude, high-latitude desert covering most of southern Argentina. The desert is larger than Spain and Portugal combined, with the Atlantic Ocean to its east, the Andes to the west, the Rio Colorado to the north, and the Straits of Magellan to the south. Coastal terraces produce sharp cliffs along the Atlantic coastline, and flat tablelands gently ascend to the foothills of the Andes in the west, which form the 1600 km long western border of the desert. A few rivers with roots in the Andes cut deep canyons through the plateaus or terraces, and descend from about 1500 m at the foot of the mountains. Much of the surface is covered by sometimes deep layers of gravel that is cemented with clays into pavement, with also some granite and volcanic basalt surfaces. Depressions often contain shallow lakes with high concentrations of salts. Substantial erosion exists because of the common high winds; manifestations of this are exposed compacted soil and gravel pavement, and rapidly moving sand tongues that are oriented parallel to the wind direction. Some dunes exist near the sea.

These tablelands or steppes are almost devoid of trees, supporting only scrub and dry grass. Steppe shrubs dominate in the east, and grasses prevail in the higher-altitude and more humid west. Riparian vegetation exists in valley bottoms. The approximate percentages of the different types of surfaces are 45% shrub desert, 30% shrub–grass semi-desert, 20% grass steppe, and 5% water surface and other categories. The grasses include species of *Poa, Festuca, Agrostis, Deschampsia, Carex*, and *Juncus*, and some of the shrubs are *Berberis, Azorella, Verbena, Nardophyllum, Mulinum, Senecio*, and *Adesmia*.

At these high latitudes (37°–50° S) much of the precipitation, including snow, is normally in the winter and associated with storms in the westerlies. As such, the high southern Andes (Fig. 3.39) upwind of the Patagonian Plateau very effectively block the storms, and create an elongated rain shadow to the leeward side. In addition, there is the effect of the cold Falkland Current off the east coast. When the air flows from the east, it is cold, stable, and has little moisture. Thus, flow from this direction, even though it is upslope, does not produce precipitation.

Pronounced temperature contrasts are lacking in spite of the range of latitudes and elevations. The variability in the east–west direction is small because the warmer climate normally associated with the lower elevations near the coast is compensated for by the proximity to the cold Falkland Current. The north–south variability in mean annual temperature is also small, partly because the narrowing of the continent to the south means that the effects of continentality and latitude on the winter climate are compensating. Much of the precipitation occurs as snow, sometimes heavy, during the winter months of June, July and August (Fig. 3.40). Temperature extremes can be large. Winter temperatures in the south are as low as –29 °C (–20 °F) while summer temperatures can be well over 37 °C (100 °F). The number of frost-free days per year is generally less than 100. Fig. 3.37 shows the Budyko dryness index and Plate 6 shows the mean annual precipitation.

Winds are relatively constant and strong, with the surface flow mainly being from the west and southwest. As the northern desert is approached, the dominant surface flow becomes more northerly because of a trough on the lee side of the

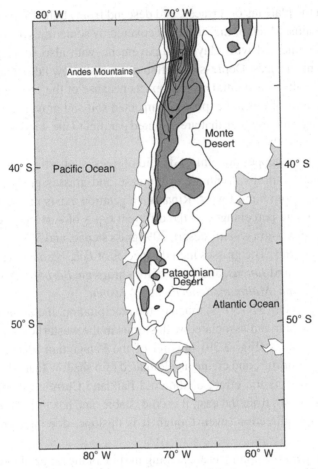

Fig. 3.39 Terrain elevation for the region of the Patagonian and Monte Deserts. The contour interval is 500 m, and heavy contours are used for 1000 and 2000 m. Elevations greater than 1000 m are shaded.

Andes. The dryness of the Patagonian summer is not so much a result of the high temperature but rather of the high evaporation by the strong winds.

The Monte Desert

The Monte's topography is more irregular than that of the Patagonian Desert, with extensive mountains, valleys, plateaus, sandy plains, and saline depressions. Substrates range from rock to clay, depending on location, elevation, and slope. It is floristically rich, with much of the surface covered by a relatively homogeneous growth of thorn scrub, and annual grasses appearing after rain. If a single plant characterizes the area, it is the genus *Larrea*, which forms very large

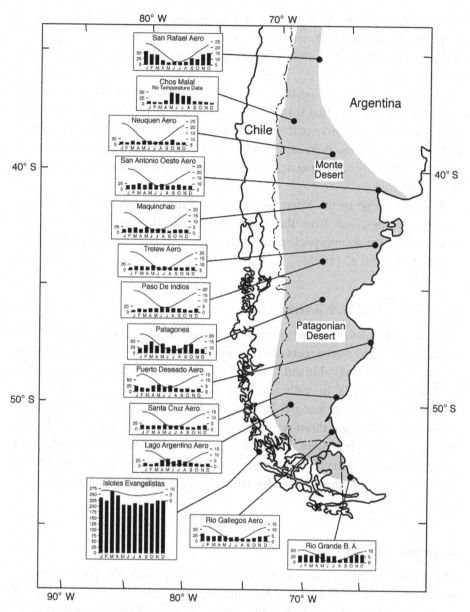

Fig. 3.40 Monthly distributions of rainfall (mm) and temperature (°C) for selected locations (dots) in and around the Patagonian and Monte Deserts. The data for the location along the west coast are shown to illustrate the large spatial contrast produced by the rain shadow of the Andes Mountains. The overall arid area is shaded.

almost single-species shrub growths. The dominant genera of the higher plants of the Monte are *Acacia, Bougainvillea, Cassia, Larrea, Opuntia, Prosopis*, and *Trichocereus* (Mares *et al.* 1985). Large cardon cacti grow in the northern desert on rocky slopes.

The Monte climate is similar to that of the Patagonian in that its aridity is also related to the fact that it is in the rain shadow of the Andes. In the more subtropical latitudes, more of the rainfall occurs in the summer months (e.g. San Rafael, Fig. 3.40), but there is a trend toward more winter precipitation in the south. It is the only Argentine desert with more precipitation in the warm season than in the cold.

A notable wind of the Monte is the hot and dry katabatic flow, called the *zonada*, that descends from the Andes and causes dust storms. A complete discussion of the geomorphology, climate, and vegetation of the Monte can be found in Mares *et al.* (1985).

Other arid areas of South America

There are numerous other arid to semi-arid regions in South America that are not normally considered to be parts of the major deserts (Fig. 3.35). For example, the Guajira Desert of Colombia and Venezuela is an extremely arid region located on the Guajira Peninsula. In northeastern Brazil, the semi-arid Caatinga scrubland experiences periodic prolonged droughts. In Bolivia, Paraguay, and Argentina is the semi-arid Chaco thorn-scrub area. The Pampas of Argentina is a broad, flat, generally treeless plain that appears similar to the Great Plains of North America. Even though precipitation is relatively high here, so is the evaporation. The Puna is a semi-desert dry plateau of the high Andes, at medium to low latitudes.

Australian deserts

The Australian deserts are the Great Sandy, the Gibson, the Great Victoria, the Tanami, and the Simpson, which includes the Sturt Stony (Fig. 3.41). Of all the continents, Australia has the largest fraction of its area occupied by deserts: about 70% that is classified as arid or semi-arid (Fig 3.1). There are no expansive mountain ranges in the arid area (Fig. 3.42), with the Hamersley Ranges, the Macdonnell Ranges, the Musgrave Ranges and the Isa Highlands having elevations not exceeding 1500 m ASL. The substrate consists of bare rock, pebbles, red soils, and wind-blown sand; Fig. 3.9 shows their distribution. The central part of the continent has many very shallow saline lakes and basins, with external drainage being limited to areas near the coast. A large fraction of the

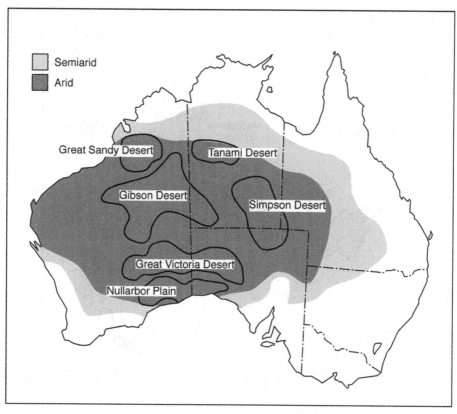

Fig. 3.41 The deserts of Australia, with the semi-arid area in light shading and the arid area in dark shading. (From Mares 1999.)

Great Sandy is composed of fixed sand dunes, with an east–west alignment that is roughly parallel to the direction of the trade winds. During rainy periods such as the late 1970s, interdune areas contain temporary lakes and more abundant vegetation (Williams and Calaby 1985). It has been claimed that 38% of the world's dune fields are found in Australia, and that the dune fields represent 40% of the continent's land surface (Croke 1997). Where the Great Sandy meets the coast in a flat sandy plain, the coastline is one of the most desolate in the world. The Simpson has the character of a rocky hammada in the north, transitions to shifting sand dunes in the central part, and is a stony reg in the south. The Gibson is a broad rocky hammada plain that transitions southward into fixed dunes in the Great Victoria. Croke (1997) should be consulted for further information on the geomorphology of Australian deserts.

The xerophytic desert vegetation is dominated by species of *Acacia* throughout. Mulga (*Acacia aneura*) is the primary species in the Victoria, Gibson, and Great Sandy Deserts, and brigalow (*A. harpophylla*) is dominant in the Simpson

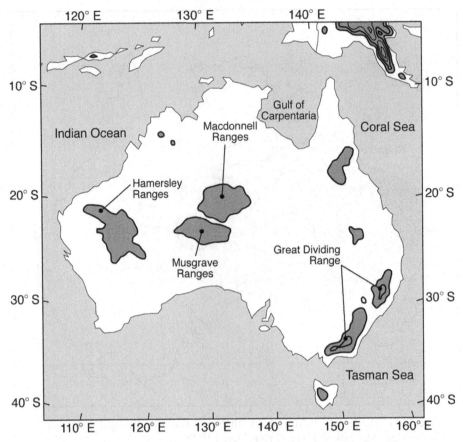

Fig. 3.42 Terrain elevation for the region of the Australian Deserts. The contour interval is 500 m, and heavy contours are used for 500 and 1000 m. Elevations greater than 500 m are shaded.

Desert. Species of eucalypt are also scattered through the arid areas, in addition to smaller shrubby plants and grasses such as needlewood (*Hakea leucoptera*), species of *Grevillea*, and tussocks of spiny grass called spinifex or porcupine grass (*Triodia* spp.). Williams and Calaby (1985) summarize the arid vegetation communities as follows: arid hummock (*Triodia*) grasslands, 1.6 million km^2; arid tussock (*Astrebla*) grasslands, 0.5 million km^2; *Acacia* shrublands, 1.6 million km^2; and low shrublands (Chenopodiaceae), 0.4 million km^2. There are also small areas of *Eucalyptus* shrublands and of low woodlands with *Casuarina* and *Heterodendrum* spp. Moore and Perry (1970) provide maps of the vegetation of the arid areas of Australia.

The dominant reason for the aridity of the continent is its location within the heart of the subtropical high-pressure belt. Good summaries of the prevailing

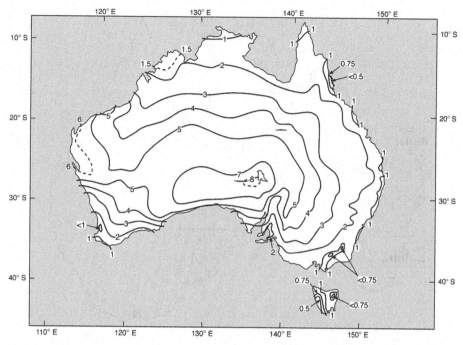

Fig. 3.43 For the deserts of Australia, the distribution of aridity based on the Budyko ratio. This is the depth of water that could be evaporated by the observed total annual net radiation, divided by the depth of the observed annual rainfall. (Adapted from Henning and Flohn 1977.)

meteorological processes may be found in Gentilli (1971), Williams and Calaby (1985), and Gibbs (1969). In addition to the effect of the subtropical high, there is also the fact that the elevated terrain of the Great Dividing Range, on the southeast margin of the continent, blocks the inland transport of moisture by the southeasterly trade winds, and a rain-shadow effect likely results. In general, the aridity increases toward the interior, from the more humid coasts (Fig. 3.43). Figs. 3.6 and 3.44 illustrate the seasonality of the rainfall, and Plate 7 shows the annual average. Toward the tropics, the rainfall is concentrated in the summer, whereas toward mid-latitudes the climate is more Mediterranean with somewhat dry summers and wet winters.

As we know, the weather regimes tend to follow the sun's migration across latitudes. Here, the subtropical high-pressure belt is located near the southern coast of the continent in the summer, at 35–40° S, and migrates northward with the sun to about 27–30° S in the winter. In the summer, heat lows form over the western edge of the continent and over the land to the south of the Gulf of Carpentaria.

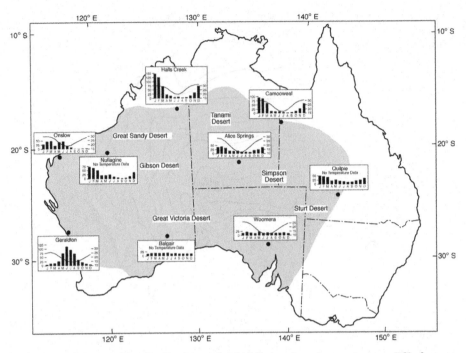

Fig. 3.44 Monthly distributions of rainfall (mm) and temperature (°C) for selected locations (dots) in and around the deserts of Australia. The arid area is shaded.

There are four types of weather system that produce rainfall (Williams and Calaby 1985). In the north, monsoon rains and tropical cyclones can penetrate into the desert, with the former producing thunderstorms with locally heavy rain, and the latter sometimes causing heavy, prolonged, and widespread rain. In the southern deserts, widespread winter rainfall can result from mid-latitude fronts as well as upper-atmospheric low-pressure systems. In exceptional circumstances, the southern and northern systems can affect the entire desert. Dew and fog are unimportant sources of moisture in the Australian Deserts. The hottest area is in the northwest Great Sandy Desert, where the temperature maximum exceeds 38 °C (100 °F) on over half of the days of the year.

Asian deserts

Asia has deserts that are within the subtropical high-pressure belt, as well as high-latitude, cold deserts. In the subtropics, there is a relatively continuous series of deserts extending from the Arabian Desert, which abuts the Sahara, to the Thar Desert of Pakistan and India. To the north, there is a belt of cold deserts, from the Turkestan Deserts near the Aral Sea, eastward to the Gobi Desert of

Mongolia and China. Asia contains most of the cold deserts of the world because it has more land area within the 35°–50° latitude belt than any other continent north or south of the equator. The aridity of the subtropical deserts has the same basic cause as the aridity of the Sahara: the subsidence associated with the large-scale global circulation. A major contributing cause of the cold deserts' aridity is the great distance from major moisture sources. They are at least 1600 km from the nearest ocean, and their interiors are even farther inland. Because of the large number of small and large mountain ranges that punctuate the surface in these latitudes, rain-shadow effects likely contribute to some of the local variability in aridity.

The major named deserts of Asia are described below. Many smaller desert areas exist, but are not shown on the maps. Mares (1999) should be consulted for more information. A review of the geomorphology of Asian deserts is found in Derbyshire and Goudie (1997).

East Asia: the Gobi and Taklamakan Deserts

In the Mongolian language, the word *gobi* means "waterless place;" in Chinese it means "gravelly pebbly plain." Thus, the term gobi is used in Asia to refer, in general, to any area of relatively flat, rocky or gravelly desert plains. In contrast, the word *shamo* is used to refer to sandy deserts. The name Taklamakan is derived from a Turkic word that means labyrinth (or, once you enter, you never return). The approximate boundaries of the Gobi and Taklamakan Deserts are shown in Fig. 3.45, as is the larger, semi-arid and arid area in which they are embedded. There are a number of other named stony and sandy deserts within this large area of aridity, such as the Bei Shan, Dzungaria, Tsaidam, Ala Shan and Ordos Deserts, but the Gobi is the largest desert in Asia. The terrain elevations in this area are depicted in Fig. 3.46. Elevations in these deserts are generally from 900–1500 m. The high elevations in the south are the northern edge of the Tibetan Plateau. The Taklamakan Desert is in a large basin of interior drainage, the Tarim Depression, that is virtually surrounded by mountains. To the north are the Tien Shan and the east–west-flowing, landlocked Tarim River. To the west are the Pamir Mountains and to the south the Kunlun Shan. The Tarim Depression opens eastward into the Lop Nor Depression. To the north of the Tien Shan Mountains are the Altai Mountains, with the break between the two known as the Dzungarian Gates; an historical east–west passage as well as a link that makes the aridity continuous between eastern and western Asia.

The Gobi Desert surface consists of mountains that merge with stony slopes and plains. The substrate is fine gravel and sand. There is little water, except for some surface ponds. Dunes cover about 5% of the total area, with maximum

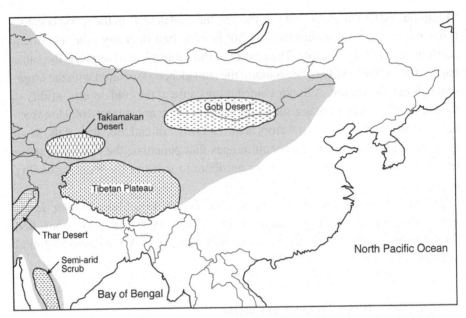

Fig. 3.45 The approximate boundaries of the Gobi and Taklamakan Deserts of East Asia. Many smaller named deserts are not shown, within the large area of aridity outlined in gray shading. (Adapted from Mares 1999.)

heights of about 200 m. One mountain gorge in the Gobi has permanent ice, a small glacier in the desert (Man 1999). The Taklamakan is one of the world's largest sandy deserts, and the largest in Asia, with dunes that are commonly 100 m tall, and sometimes reach 300 m. There are some gravel plains and internal mountains, and the water table can be within 2 m of the surface.

Vegetation is sparse to non-existent (generally less than 10% coverage) in the desolate waste of the western Gobi, but in the eastern section the condition gradually changes into a rolling plain that is covered with "Gobi sagebrush" and short bunchgrass. Except near water sources, vegetation is limited to occasional very sparse grasses, and low shrubs and semi-shrubs. In some protected mountain valleys, stands of scrub trees such as tamarisks and poplars (*Populus diversifolia*) survive. Vegetation varies greatly with the soil characteristics. In the mostly barren west, major plants are *Nitraria sphaerocarpa* and *Reaumoria soongarica*, with many others. In the north, the desert steppes or semi-desert support *Stipa* and *Allium* grasses. On gravelly sites, especially, *Anabasis brevifolia* is common. In sandy areas, saxaul trees (*Haloxylon ammodendron*) are characteristic. Vegetation cover is greater where the water table is high.

In the Taklamakan, significant stands of shrubs and trees are found only in riparian valleys and in moist, dune-valley depressions with water 1–2 m below the

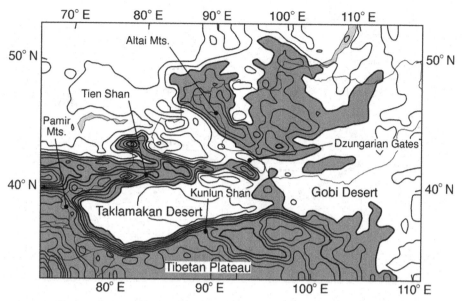

Fig. 3.46 Terrain elevation for the region of the Taklamakan and Gobi Deserts. The contour interval is 500 m, and heavy contours are used for 1500 and 3000 m. Elevations greater than 1500 m are shaded.

surface. Scattered dry-steppe vegetation, small shrubs, and grasses are found on low mountains and hills. The stabilized dunes support some grasses and trees. In stoney areas are xerophytic shrubs such as *Ephedra przewalskii, Zygophyllum xanthoxylon, Z. kaschgaricum, Gymnocarpos przewalskii, Anabasis aphylla,* and localized areas of *Calligonum roboroski* and *Nitraria sphaerocarpa.* The psammophytic herbaceous vegetation types in the sandy parts of the desert are *Aristida pennata, Agriophyllum arenarium,* and *Corispermum* species. Woody vegetation on the edge of the sand deserts is primarily *Tamarix ramosissima.* Many halophytic species are found in lake depressions.

Figure 3.47 shows the monthly variation in temperature and precipitation. The Gobi Desert is the coolest of the Asian deserts, and is noted for its winter northerly surges of very cold continental–polar air. The seasonal variation in temperature is clearly large, at least partly because of the extreme continentality and the relatively high latitude. For example, the annual range in monthly average temperature is greater than 40 °C (72 °F) at some of the stations in Mongolia and northern China for which data are plotted. Summer temperature extremes can reach 38 °C (100 °F), and, in winter, daily-average temperatures are commonly −10 °C (16 °F). Gobi rainfall is predominantly in the summer months because of the rain-suppressing influence of the vast winter continental anticyclone (Fig 2.5a). Annual-average precipitation varies greatly with location, depending

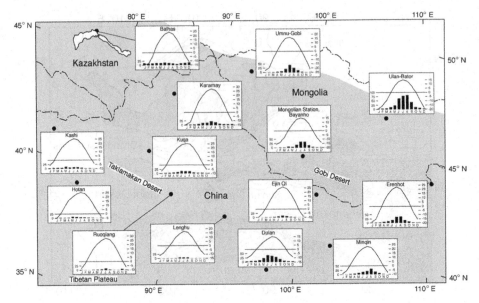

Fig. 3.47 Monthly distributions of rainfall (mm) and temperature (°C) for selected locations (dots) in and around the Taklamakan and Gobi Deserts, with the arid area shaded.

on many factors, but can range from less than 1 cm up to 10 cm. The meager Taklamakan rainfall is also in the summer, almost exclusively, with about 30% from local convective showers. One contributor to the dryness of this area is the fact that the Tibetan Plateau and the other mountains surrounding the Tarim Depression block the entry of rain-producing disturbances and moisture from the Indian Ocean (e.g. monsoon-related rainfall) and the westerlies. Plate 8 illustrates the annual-average rainfall. The paucity of data in this region means that this analysis should be interpreted cautiously, especially in the sense that it does not represent mesoscale features in the rainfall pattern.

Figure 3.48 shows the Budyko ratio for the area. The greatest aridity is in China, in the Taklamakan area. In addition to the above-noted effect of the mountain-blocking of moisture and storms, the aridity of this area of central and eastern Asia is also a result of the fact that it is distant from both the cyclonic storms that originate over the western part of the continent and the warm, moist Pacific air that is carried inland by East Asia's summer monsoon. In the figure can be seen the tongue of higher aridity than extends to the east of Lake Balkash through the Dzungarian Gates, connecting the deserts of eastern and western Asia. This can also be seen in the annual-average rainfall plots in Plates 8 and 9.

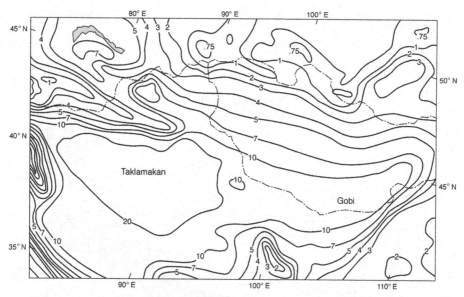

Fig. 3.48 For the Taklamakan and Gobi Deserts, the distribution of aridity based on the Budyko ratio. This is the depth of water that could be evaporated by the observed total annual net radiation, divided by the depth of the observed annual rainfall. (From Henning and Flohn 1977.)

The Turkestan-area deserts

The western edge of this arid area begins at the shores of the Caspian Sea, and extends eastward toward the Taklamakan. In the south, it is bordered by the mountains of Iran and Afghanistan (Fig. 3.49), and in the north it merges through Kazakhstan with the plains of the Russian steppes. Fig. 3.50 shows the locations of some of the named deserts of the area. Adjacent to the arid eastern shore of the Caspian Sea, which is 28 m below sea level, is the gravelly Ustyurt Desert. To the south are the Turanian Plains whose hills are clayey and stony. To the southeast and east are two vast sandy deserts, the Kyzylkum (meaning red sand) Desert and the Karakumy (meaning black sand) Desert, primarily in Uzbekistan and Turkmenistan, respectively. Their surfaces consist of rock, clay, sand, and saline depressions. To the northeast in Kazakhstan is the large stony plateau of the Betpak–Dala Desert. Yet farther north are the generally barren and rocky hills of the arid and semi-arid Kazakh Upland. Toward the Taklamakan Desert is the Pamir Plateau, a very high and cold desert where elevations reach 7495 m. There are two large rivers in the area. The Amu Darya and the Syr Darya originate in the mountains to the southeast of the Kyzylkum, and flow northwestward toward the Aral Sea. The Ili River originates in the Tien Shan Mountains, and flows northward into Lake Balkhash.

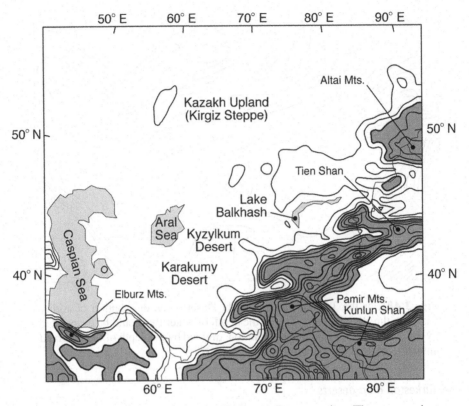

Fig. 3.49 Terrain elevation for the Turkestan Desert region. The contour interval is 500 m, and heavy contours are used for 1500 and 2500 m. Elevations greater than 1500 m are shaded.

Vegetation is sparse over most of this area. In the north, from the Caspian Sea to Lake Balkhash, steppe grasslands predominate. In the southern part of this belt there are some shrubs. Where the soils are only slightly saline, *Artemisia maritima* is common, but on the more saline soils *A. pauciflora* prevails. Another very common low shrub is the salt bush (*Atriplex canum*). Mixed with the shrubs are various steppe grasses, especially the fescues (*Festuca sulcata, F. valesiaca*), koeleria (*Koeleria gracilis*), and feather grasses such as *Stipa capillata*. In sandy areas, *Aristida pennata* is common. Taller vegetation includes the Siberian salt tree (*Halinodendron argenteum*), white saxaul (*Ammodendron conollyi*), and other species of saxaul. The flood plains of major rivers are often meadowland with tall reeds and small deciduous trees such as Euphrates poplar (*Populus diversifolia*), bloomy poplar (*P. primosa*), willow (*Salix*), Russian olive (*Eleagnus angustifolia*), and several tamarisk species (*Tamarix* spp.).

In the Kyzylkum and Karakumy Deserts, a precipitation minimum occurs in the summer months (Fig. 3.51), with a maximum in the spring. This transitions to a summer precipitation maximum in the steppe grasslands of northern

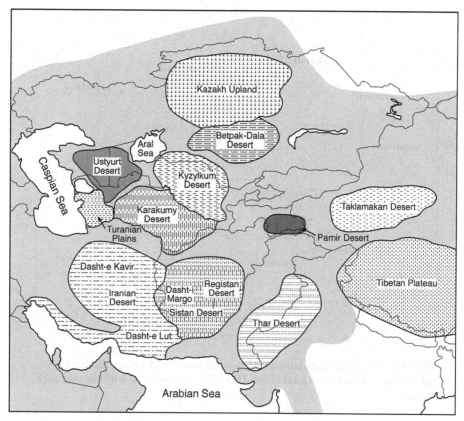

Fig. 3.50 The Turkestan area and surrounding deserts. There are many other smaller named deserts that are not shown, within the large area of aridity indicated with gray shading. (Adapted from Mares 1999.)

Kazakhstan, and in the southeast toward the desert of the Pamir Plateau. This summer precipitation is of convective origin. The general lack of significant winter precipitation anywhere in the area results from the fact that the storm tracks lie to the south in that season. In the summer, the storm tracks lie to the north. The spring and/or fall precipitation maxima for some stations in the Kyzylkum and Karakumy Deserts result from the fact that the storm track is migrating through the area at those times.

As is the case to the east toward the Gobi, annual ranges of monthly average temperatures are very large, for the same basic reasons, exceeding 43 °C (78 °F) in the north. Here, up to half the months have averages at or below freezing. Annual ranges decrease somewhat toward the south, but they are still large. This decrease is partly attributable to the fact that the cyclonic disturbances in the winter here cause more cloudiness, which reduces the daytime maximum and increases the nighttime minimum temperatures. Hot, exceptionally dry, winds that sometimes cause dust storms, are called *sukhovei* (sukhovey, singular), and

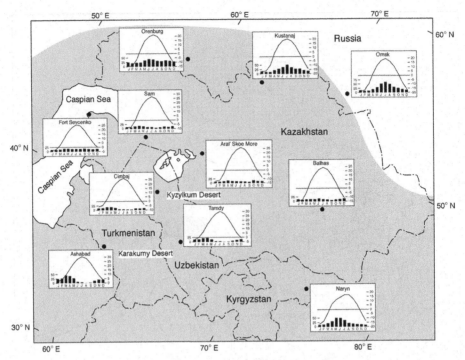

Fig. 3.51 Monthly distributions of rainfall (mm) and temperature (°C) for selected locations in and around the Turkestan-area deserts, with the arid area shaded.

have a frequency maximum over a wide area to east of the Caspian Sea (Lydolph 1964).

The Budyko index (Fig. 3.52) shows the greatest aridity to be in an area that spreads eastward from the Caspian Sea, through the Ustyurt Desert and the Turanian Plains, and into the Kyzlkum and Karakumy Deserts. These high values of the index in the Kyzlkum and Karakumy are partly attributable to the fact that the mean cloudiness there is very low, at less than 5%, when the greatest potential radiation is available in the summer. There is also an aridity maximum around Lake Balkhash in eastern Kazakhstan. The aridity index becomes even greater in China to the southeast. The annual-average precipitation for the area is shown in Plate 9.

The Thar Desert

The Thar Desert spans the border between India and Pakistan (Fig. 3.50), with the two countries sometimes applying local names to their parts of the Thar. In Pakistan, the southern part is known as the Sind Desert and the northern part is the

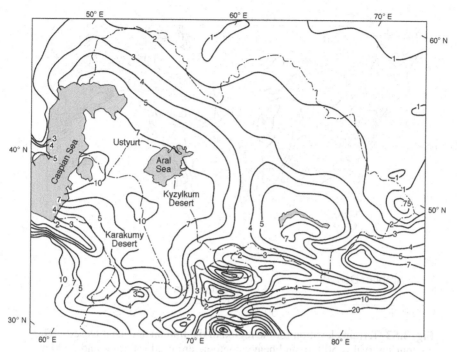

Fig. 3.52 For the Turkestan-area deserts, the distribution of aridity based on the Budyko ratio. This is the depth of water that could be evaporated by the observed total annual net radiation, divided by the depth of the observed annual rainfall. (From Henning and Flohn 1977.)

Cholistan Desert. A small area to the northwest of the Thar in Pakistan is known as the Thal Desert. In India, a larger area of aridity that includes the Thar Desert is known as the Rajasthan or Rajputana Desert, or the Great Indian Desert. The eastern desert boundary in India is the Aravalli Mountain Range (Fig. 3.53). To the west of this range is a vast sea of sand on a plateau that is interrupted by low rock hills. In general, there are both stable and unstable dunes, but much of the area of the Thar is level to gently sloping rock plains. The nature of the vegetation is influenced strongly by whether the local substrate is sand, gravel, or rock, with thorny vegetation generally being dominant. In very arid areas, *Haloxylon* and *Calligonum* are common. In the subdesert are *Prosopis*, *Tamarix*, and *Acacia*, and in the more moist areas are *Acacia*, *Tamarix*, *Anogeissus*, *Capparis*, *Cassia*, and *Salvadora*. For additional information about Thar Desert vegetation, see Gupta (1986).

This desert is in a transition zone between precipitation-producing mechanisms. In the east and central desert, more than 90% of the rainfall is produced by the Southwest Asian Monsoon, from mid-June to mid-September (Fig. 3.54),

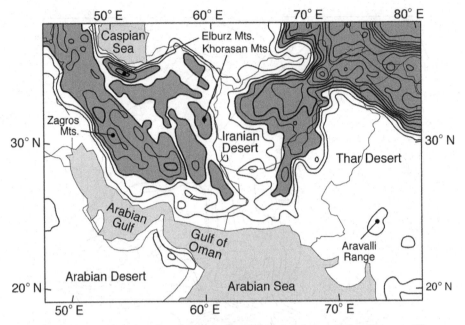

Fig. 3.53 Terrain elevation for the region of the Iranian and Thar Deserts. The contour interval is 500 m, and heavy contours are used for 1500 and 2500 m. Elevations greater than 1500 m are shaded.

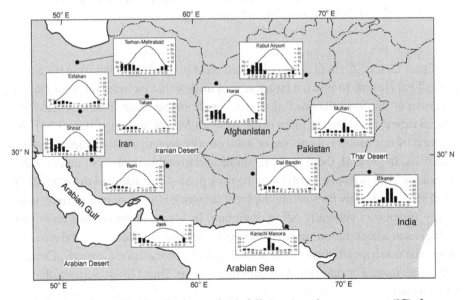

Fig. 3.54 Monthly distributions of rainfall (mm) and temperature (°C) for selected locations (dots) in and around the Iranian and Thar Deserts, with the overall arid area shaded.

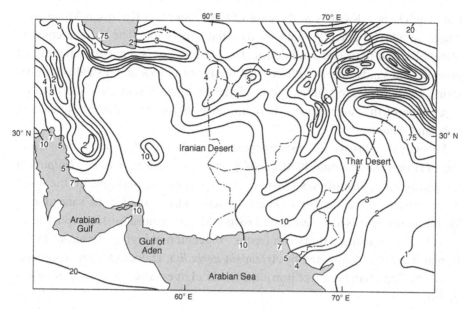

Fig. 3.55 For the area of the Iranian and Thar Deserts, the distribution of aridity based on the Budyko ratio. This is the depth of water that could be evaporated by the observed total annual net radiation, divided by the depth of the observed annual rainfall. (From Henning and Flohn 1977.)

with annual amounts ranging from 100 to 300 mm. In the northern and western portions, mid-latitude cyclones of Mediterranean origin produce moderate winter precipitation. The interannual rainfall variability is high in locations that rely on the monsoon because the Thar is on the western edge of the monsoon's influence, and inter-annual shifts in the monsoon's position have great consequences. Periods of several years without rainfall have been recorded. The annual range of temperature is less than in the higher-latitude cold deserts discussed above. In Plate 10, the tongue of low rainfall on the right side of the figure (green color) represents the minimum in this area of the Thar Desert. Figure 3.55 shows that the greatest aridity is on the Pakistan side of the border.

The deserts of Iran and Afghanistan

Immediately to the south of the complex of cold deserts that extends to the east from the Caspian Sea are the deserts of Iran and Afghanistan (Fig. 3.50). The term Iranian Desert is usually used to refer to the entire contiguous arid area within Iran, western Pakistan, and Afghanistan. The Iranian Desert is bounded on the west by the Zagros Mountains (Fig. 3.53). To the east of the mountains, within the broad area of the Iranian Desert, are the deserts of Dasht-e Kavir and Dasht-e

Lut. The Dasht-e Kavir is at the base of the Elburz and Khorasan Mountains, where an alluvial gravelly plain forms a flat depression. The surface consists of pebbles, salt flats, and sand dunes. The Dasht-e Lut is a mostly unvegetated gravelly and sandy desert with internal mountain ranges and a sand sea in the south. Farther east, in southwestern Afghanistan, the Sistan Desert exists in a depression of the same name. There are areas covered by moving sands, salt flats, and swamps.

Vegetation consists of psammophytic shrubs (*Calligonum, Haloxylon, Salsola*) on sands, annual herbaceous plants (*Carex, Poa, Gagea, Tulipa*) on the plains, and **gypsophytes** (*Salsola, Anabasis*) on the plateaus. To the north of the Sistan is the Registan Desert, which consists mainly of sand dunes. To the north also is the desert of Dasht-i Margo, whose surface consists of gravel, sandy basins, and dry salt basins. Vegetation at higher elevations in the Iranian Desert is dominated by *Artemisia herbalba*. Halophytes are also common. The Great Salt Desert of Iran is the world's largest absolutely vegetationless desert.

Precipitation in the deserts of Iran and Afghanistan is almost exclusively in the winter and spring seasons (Fig. 3.54), associated with cyclones that develop in the Mediterranean Basin, with summers being very dry. Rarely, the Southwest Asian Monsoon can extend into the region, with resulting very heavy rainfall that has led to the loss of life and property (Ramaswamy 1965). The amplitude of the annual temperature wave is, as expected, less in the south and near the coast than in more continental areas to the north.

The Arabian Desert and surroundings

The Arabian Desert, as defined here, extends across the Arabian Peninsula from the Arabian Sea to the Mediterranean Sea. Because there are semi-arid areas to the northwest in Turkey, they will also be included in this discussion. Meteorologically, these areas are simply part of the broad expanse of contiguous subtropical aridity that includes the Sahara, the East African deserts, and the Iranian and Thar Deserts (Fig. 3.1). Fig. 3.56 shows the locations of some of the named deserts of the region.

The variation in terrain elevation is significant across this area (Fig. 3.57). On the western side of the peninsula is an elongated plateau that runs parallel to the Red Sea coast, with elevations of 1000–1500 m. The plateau is relatively narrow in the north, but broadens southward to cover as much as one-half the width of the peninsula in some places. The northern and southern parts of this plateau are the Hejaz and Asir Mountains, respectively, near the Red Sea coast. A connecting, even higher plateau exists in the southwest corner, with elevations

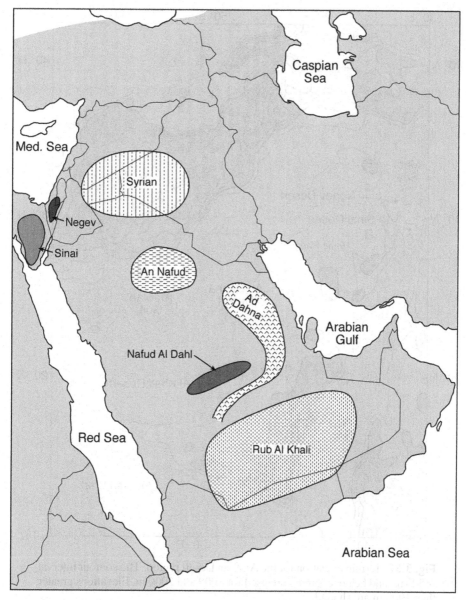

Fig. 3.56 Some of the named deserts of the Arabian Peninsula, within the large area of general aridity shown by the gray shading.

generally between 2000 and 3000 m. These two plateaus represent an unbroken escarpment for 1000 km along the Red Sea coast. From these plateaus, the terrain slopes steeply down on the west to a coastal plain on the Red Sea, and gradually downward to the east. In the east is a vast desert plain with elevations of generally less than 200 m, where this transitions in the extreme east to the Oman Plateau

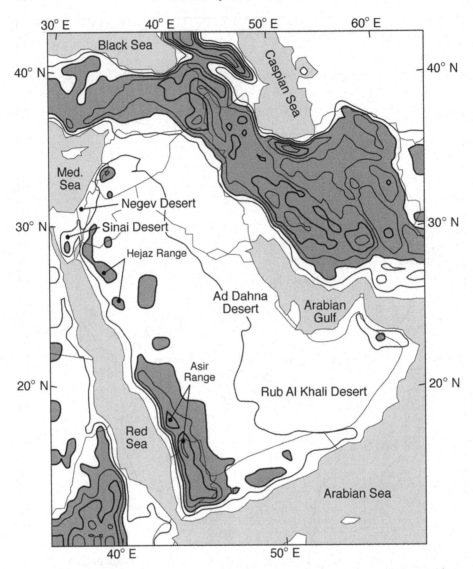

Fig. 3.57 Terrain elevation for the Arabian Desert region. The contour interval is 500 m, and heavy contours are used for 1000 and 2000 m. Elevations greater than 1000 m are shaded.

with elevations of 500 to 1500 m. The Jebel Akdhar mountains on the plateau rise to 3300 m along the coast of the Gulf of Oman. To the northwest of the Arabian Gulf in the Tigris–Euphrates Valley of Iraq are elevations of less than 100 m. Mountains also prevail to the east of the Mediterranean and in semi-arid areas of Turkey to the north. This arid area is bordered on the east by the Zagros Mountains in western Iran. Near the Mediterranean, in the Jordan Valley, is the Dead Sea depression with the world's lowest elevation, 395 m below sea level.

The lower peninsula has no important rivers; only wadis that are dry except after a rain. To the north, perennial rivers are more numerous, such as the Tigris, the Euphrates, and the Jordan. The Tigris River in Iraq empties into the Arabian Gulf; the Euphrates River begins far to the north in Turkey before merging with the Tigris. Within the Tigris–Euphrates Valley in Iraq are numerous large lakes and significant areas of marshland. Marshlands are also common on the Red Sea coast. Subsurface water is abundant, with its source as subsurface flow from the mountains. Wells, springs, and natural waterholes make life and travel possible, even in much of the interior. Extensive groundwater is also mined for agricultural irrigation, especially in Iraq.

There are a number of smaller, individual deserts within and near the greater Arabian Desert, such as the An Nafud in the north; the Ad Dahna in the east central part, the Rub Al Khali in the south, the Sinai Desert on the Sinai Peninsula, and the Negev Desert to its north. The Rub Al Khali is the largest uninterrupted sand desert in the world, with dunes that reach heights of 250 m. In the north is the Syrian Desert. The surface is largely sand, gravel and rock. In the east, sand dunes occur in an arc through Israel, Jordan, northern Saudi Arabia, eastern Saudi Arabia, and south to the Rub Al Khali. Another arc of rocky and barren desert mountains, and gravel plains cut by wadis, exists on the west side of the Arabian Desert, through western Iraq, Jordan, Israel, western Saudi Arabia, Yemen, and southern Saudi Arabia. In the east, especially near the Arabian Gulf, are extensive salt flats with surfaces that are sometimes salt-encrusted. Throughout the region near shores and inland depressions are soils in which salts are naturally present. Salt-laden soils that have resulted from irrigation cover large areas of alluvial plains in eastern Iraq. The remainder of the surface consists largely of mountains, rocky escarpments, regs, and wadis. In northern areas, there are vast plains of desert pavement. Allison (1997) and Holm (1960) should be consulted for a complete discussion of the geomorphology of the Arabian Peninsula.

The vegetation in this area is similar to that of the Sahara. Indeed, McGinnies *et al.* (1968) point out that the same general types of arid-land vegetation prevail in the Sahara, the Arabian Peninsula, and across Asia to India. Extensive natural vegetation over the Arabian Peninsula is largely limited to areas where surface or near-surface water is available, and to the higher elevations of mountains where rainfall is more abundant. The mountain climates support juniper (*Juniperus*) and acacia (*Acacia*) trees. Also, the rocky plains will temporarily respond to winter and early spring rains by producing ephemeral grasses and other plants. Some vegetation is found even in the driest places, such as the Rub Al Khali.

Along the wadis, tall phreatophytic trees with extensive root systems, such as tamarinds and Christ's thorn, draw on subsurface water. Inland marshy areas support lush vegetation, and salt-tolerant vegetation inhabits sea-coast marshes. Along the foggy southeastern coast there are dense stands of shrubs, and short and tall grasses. Refer to Abd El Rahman (1986) for more information about Arabian Desert vegetation.

In the southern part of the peninsula, the Southwest Asian Monsoon has a major influence on the low-level wind field. As seen in Fig. 2.5, the January sea-level pressure pattern over the Indian Ocean and Arabian Sea is a result of a deep thermal low-pressure area over northern Australia and a high-pressure region over Siberia. The result is a northeasterly low-level wind in the Arabian Sea and western Indian Ocean. In the Northern Hemisphere summer when the continents warm, a low-pressure center develops over southern Asia and the wind reverses to a southwesterly direction. In July, a monsoon trough and the ITCZ are positioned along a line that crosses the Strait of Hormuz and follows the northern borders of Oman and Yemen across the peninsula. The area to the south of this line is considered to have a monsoon climate.

The highly variable orography and the complex coastlines that surround the area produce much detailed fine structure to the climate within the broad classifications. In many areas, sea-breeze circulations modulate the local climate in summer when the coastal temperature contrast is the greatest. These sea breezes can penetrate inland and influence the desert boundary-layer temperature and winds for hundreds of kilometers. For example, Steedman and Ashour (1976) show that the coastal breeze on the eastern shore of the Red Sea penetrates over 200 km inland over the desert on some summer days. Lieman and Alpert (1992) show the significant influence of the Mediterranean Sea breeze and orography on the boundary-layer depth over Israel. Warner and Sheu (2000) use a model to simulate the 100 km penetration into the desert of the Arabian Gulf coastal breeze on the eastern side of the Arabian Peninsula. There are also interesting coastal microclimates along the Arabian Sea where seasonal cloud and fog prevail near the Somali Current (Price *et al.* 1985).

There are two rainfall climates: a Mediterranean climate in the north with dry summers, and a more monsoon-like climate in the south with drier winters. Rainfall amounts are larger in the north (Plate 11), where Mediterranean cyclones cause the maximum in winter (Fig. 3.58) in northern Saudi Arabia, Iraq, the western Arabian Gulf, Turkey, and Syria. A southern storm track can, however, allow the cyclones to penetrate into the deserts of the southern peninsula. For example, Membery (1997) describes a complex winter storm that caused a prolonged period of record rainfall. Spring rainfall is also significant because of persisting cyclonic activity and because increased solar heating causes the

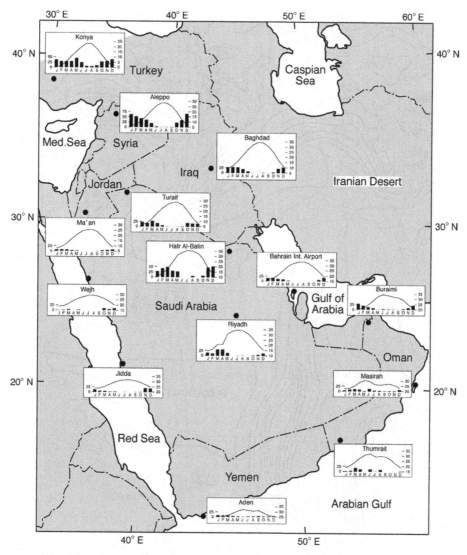

Fig. 3.58 Monthly distributions of rainfall (mm) and temperature (°C) for selected locations (dots) in and around the Arabian Desert, with the arid area shaded.

development of coastal and mountain-valley circulations with associated moist convection. A general summary of winter and spring weather in this area can be found in Pedgley (1974). By summer, the primary synoptic-scale feature is a monsoon trough that becomes established over the peninsula, with conditionally unstable southwesterly flow over the southern part. Fig. 3.12 shows that the monsoon flow from the Gulf of Guinea reaches this area, and is responsible for substantial rain in the mountains of Yemen. But the stabilizing influence of cold

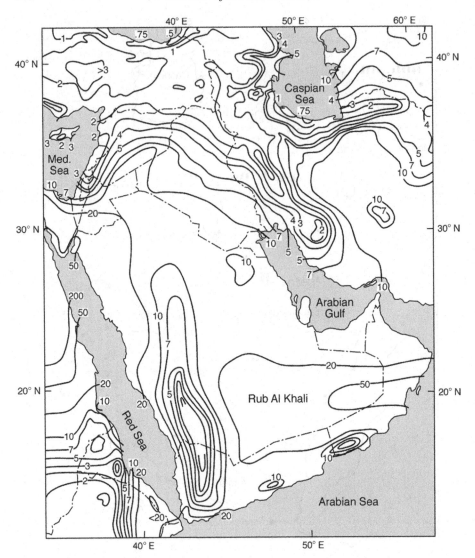

Fig. 3.59 For the Arabian Desert, the distribution of aridity based on the Budyko ratio. This is the depth of water that could be evaporated by the observed total annual net radiation, divided by the depth of the observed annual rainfall. (From Henning and Flohn 1977.)

upwelling on the south coast in the Arabian Sea causes the summer rainfall there to be minimal (Fig. 3.58). The greatest aridity is found in the Rub Al Khali in the southeast, and along the Red Sea coast to the northwest (Fig 3.59).

Summer temperatures in the Arabian Desert have reached 54 °C (129 °F), but in spite of the low latitudes winter temperatures are sometimes below freezing. The annual variation in the monthly temperatures is less than 10 K (18 °F) at some locations along the coast, such as Aden (Fig. 3.58), but at inland places with more continental climates, such as Baghdad, the annual range is over 30 K (54 °F).

As noted in Chapter 2, a heat low forms over the Arabian Peninsula in the summer and significantly affects the local weather, including the influence of the Southwest Asian Monsoon. Further discussion of this heat low can be found in Ackerman and Cox (1982); Blake *et al.* (1983); Smith (1986a,b); Bitan and Sa'aroni (1992); and Mohalfi *et al.* (1998).

Winds in many areas of the Arabian Desert "blow constantly and fiercely" (Petrov 1976). Some of the named ones are as follows.

- *kaus* – a strong southerly or southwesterly wind over the eastern peninsula and the Arabian Gulf, which develops in advance of a surface low-pressure center. This is the name used in Arabia for the sirocco of North Africa, the shlour of Syria and Lebanon, and the sharqi of Iraq.
- *shamal* (Arabic for north) – a northwesterly wind, especially over the eastern Arabian Peninsula. The winter shamal is intermittent and strongest, and follows cold-frontal passages. The summer shamal is weaker and steadier, and is forced by the pressure gradient of the Indian and Arabian thermal lows (Fig. 2.5). Perrone (1979) and Stranz (1974) describe the winter shamal.
- *suhaili* – southwest winds that prevail after the passage of a pressure trough. They are sometimes of sufficient strength to be a danger to vessels in the Arabian Gulf.
- *aziab* – a hot and dry southerly wind that develops over the peninsula in March ahead of a low-pressure center.

European arid areas

The Iberian Peninsula is the only area in Europe containing semi-arid lands. Even though the most arid part of the region is in the southeast of Spain, in the depression of the Ebro River where the annual rainfall is less than 40 cm, much of the peninsula is considered to be semi-arid. Most of the precipitation occurs in the spring and fall seasons, with the intervening months being largely rainless. Temperature maxima range from 35–40 °C (95–104 °F), and minima are below 4 °C (39 °F). Natural vegetation is generally limited to shrubs such as lavender (*Lavandula latifolia*), saltwort (*Salsola vermiculata*), white sage (*Artemisia herba-alba*), thyme (*Thymus vulgaris*), rosemary (*Rosmarinus officinalis*), and grasses (*Stipa parviflora* and *Brachypodium retusum*). There are areas with salt flats that support only halophytic vegetation. Much of the area has been historically exploited for agriculture, and overgrazing has created serious erosion problems. A description of the arid areas of Spain can be found in Faus-Pujol and Higueras-Arnal (1998).

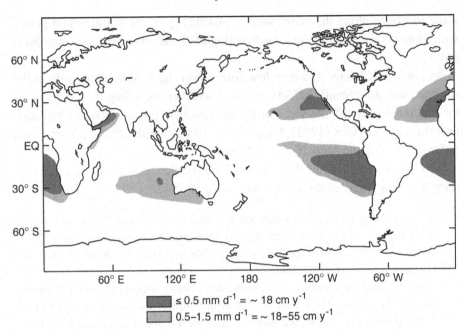

60° N

30° N

EQ

30° S

60° S

60° E   120° E   180   120° W   60° W

≤ 0.5 mm d⁻¹ = ~ 18 cm y⁻¹

0.5–1.5 mm d⁻¹ = ~ 18–55 cm y⁻¹

Fig. 3.60 Precipitation over the oceans with average annual totals of less than 55 cm yr^{-1} (1.5 mm d^{-1}). This analysis is produced by merging information from five kinds of satellite, from global-model-based precipitation re-analyses, and from precipitation gauges (Xie and Arkin 1996, 1997). The resolution is 2.5° latitude by 2.5° longitude. (Courtesy of United States National Aeronautics and Space Administration.)

Ocean deserts

Figure 2.14 includes a sketch of the vast area of the world's oceans that is estimated to be arid. Because of a lack of data with good spatial density, the boundaries shown are only estimates. An attempt at a more quantitative estimate of the boundaries of ocean deserts is seen in Fig. 3.60. For visual simplicity, only rainfall over oceans is shown, as well as only areas with precipitation of less than 55 cm per year.

In general, large areas of ocean to the west of the major land masses have arid climates. The aridity in these subtropical maritime regions can only be associated with the global-scale circulations discussed in the last chapter, the north–south Hadley circulation and east–west Walker-type circulations. From South America, aridity spreads westward over 8000 km to a point west of the Marquesas Islands along the equator. Similarly, the aridity of the Namib and Kalahari Deserts extends westward almost to the coast of South America, and the dryness of Australia reaches across the Indian Ocean most of the way to Africa. To the north, the dryness of the Sahara prevails through the Sargasso

Sea, almost to Bermuda and the West Indies, and the North American Deserts reach almost to the longitudes of the Hawaiian Islands. The boundary-layer structure in these regions of maritime subsidence has not been thoroughly studied, but it is reasonable that the low relative humidity of the stable, subsiding air will be substantially increased by the surface evaporation, at least in a shallow layer.

Suggested general references for further reading

A few of the references suggested below do not deal directly with desert weather, climate, or surface conditions. Nevertheless they are recommended because their contents provide descriptions about life in that desert, and subjective impressions of the weather and surface conditions there.

General references

Amiran, D. H. K., and A. W. Wilson (Eds.), 1973: *Coastal Deserts: Their Natural and Human Environments* – description of the climate, geomorphology, hydrology, and other aspects of all the world's coastal deserts.

Arakawa, H. (Ed.), 1969: *Climates of Northern and Eastern Asia* – climate descriptions include the areas of the deserts of China.

Bender, G. L. (Ed.), 1982: *Reference Handbook on the Deserts of North America* – general discussion of North American deserts, without much technical treatment of the meteorology. A summary is provided of historical and current desertification.

Bryson, R. A., and F. K. Hare (Eds.), 1974: *Climates of North America* – climate descriptions include all North American deserts.

Evans, D. D., and J. L. Thames (Eds.), 1981: *Water in Desert Ecosystems* – good treatments of desert climate, soils, geology, and hydrology, with an emphasis on the latter.

Evenari, M., 1985a: *The desert environment* – describes the overall climate, geomorphology, geological history, soils, and the hydrologic cycle of deserts.

Evenari, M., 1985b: *Adaptations of plants and animals to the desert environment* – briefly summarizes the various adaptations of vegetation to arid environments.

Ferrari, M., 1996: *Deserts* – pictorial presentation of the surface conditions of many of the world's deserts.

Griffiths, J. F. (Ed.), 1972c: *Climates of Africa* – climate descriptions include the areas of the African deserts.

Howe, G. H., *et al.* 1968: *Classification of World Desert Areas* – summary of surface conditions, vegetation, and climate of the world's deserts.

Jaeger, E. C., 1957: *The North American Deserts* – reviews the climates, surface conditions, flora, and fauna of the North American deserts.

Lydolph, P. E., 1977: *Climates of the Soviet Union* – climate descriptions include the Turkestan-area deserts.

MacMahon, J. A., 1985: *Deserts* – summarizes the vegetation of the North American deserts.

MacMahon, J. A., and F. H. Wagner, 1985: *The Mojave, Sonoran and Chihuahuan Deserts of North America* – describes the climate, geomorphology, soils, and vegetation.

Mares, M. A., 1999: *Encyclopedia of Deserts* – a complete non-technically oriented encyclopedia of desert biology, geomorphology, and climate, with a bit lighter emphasis on the latter.

McGinnies, W. G., *et al.,* 1968: *Deserts of the World* – appraisals of research about, and summaries of, the vegetation, surface materials, geomorphology, hydrology, and weather and climate of desert environments. This contains an especially good review of the literature up to the date of publication.

McKee, E. D., (Ed.), 1979: *A Study of Global Sand Seas* – a thorough review of the sand seas of the world, including wind field climatologies and a satellite analysis of dune structure.

Meigs, P., 1966: *Geography of coastal deserts* – very good summary of all aspects of coastal deserts, worldwide.

Noy-Meir, I., 1985: *Desert ecosystem structure and function* – general discussion of desert ecosystems.

Orshan, G., 1986: *The deserts of the Middle East* – describes the overall climate, geomorphology, soils, and vegetation of the deserts of the Middle East.

Rumney, G. R., 1968: *Climatology and the World's Climates* – a detailed description of the world's climates, including those of arid areas.

Schwerdtfeger, W. (Ed.), 1976: *Climates of Central and South America* – climate descriptions include all South American deserts.

Shmida, A., 1985: *Biogeography of the desert flora* – describes desert ecosystems, vegetation types, and causes of aridity.

Takahashi, K., and H. Arakawa (Eds.), 1981: *Climates of Southern and Western Asia* – climate descriptions include the areas of the deserts of the Indian Subcontinent and the Near East.

Trewartha, G. T., 1961: *The Earth's Problem Climates* – treats the climates of most arid (and non-arid) areas.

The Sahara Desert

Ayyad, M. A., and S. I. Ghabbour, 1986: *Hot deserts of Egypt and the Sudan* – reviews the climate, geomorphology, and vegetation of Egypt and the Sudan.

Cloudsley-Thompson, J. L., 1984: *Sahara Desert* – covers the climate, microclimates, geology, soils, and plant ecology of the Sahara.

Dubief, J., 1959: *Le Climat du Sahara. Vol. 1* and Dubief, J., 1963: *Le Climat du Sahara. Vol. 2* – these are the best references available about the climate of the Sahara (in French).

Gautier, E.-F., 1935: *Sahara: The Great Desert* – describes the surface conditions and climate of the Sahara.

Le Houérou, H. N., 1986: *The desert and arid zone of northern Africa* – reviews the climate, geomorphology, soils, hydrology, plants, and desertification of the northern Sahara.

Monod, T., 1986: *The Sahel zone north of the Equator* – describes the climate, changes in climate, geomorphology, soils, and vegetation of the Sahel.

The Namibian Desert

Walter, H., 1986: *The Namib Desert* – reviews the climate, soils, vegetation, and ecosystems.

The Kalahari Desert

Owens, M., and D. Owens, 1984: *Cry of the Kalahari* – descriptions of the Kalahari by researchers who lived in it for years.

Thomas, D. S. G., and P. A. Shaw, 1991: *The Kalahari Environment* – the primary focus is the surface landforms of the Kalahari, but there is some material on environmental change and a brief summary of the climate.

The Karoo Desert

Werger, M. J. A., 1986: *The Karoo and southern Kalahari* – describes the climate, geomorphology, soils, and ecology.

The Sonoran Desert

Clements, T., *et al.*, 1957: *A study of desert surface conditions* – a description of the distribution of playas, desert flats, bedrock fields, alluvial fans, dunes, mountains, and volcanic fields in the Sonoran and Mojave Deserts in the United States.

Clements, T., *et al.*, 1963: *A Study of Windborne Sand and Dust in Desert Areas* – reviews the causes of dust storms and their properties, primarily for the northern Sonoran and Mojave Deserts.

Gehlbach, F. R., 1993: *Mountain Islands and Desert Seas* – a natural history of the desert borderlands of the United States and Mexico.

Hartmann, W. K., 1989: *Desert Heart: Chronicles of the Sonoran Desert* – a natural and human history of the Sonoran Desert.

Hastings, J. R., and R. M. Turner, 1965: *The Changing Mile* – describes microclimates and microenvironments within an area of the northern Sonoran Desert, and uses co-located photographs to illustrate how the vegetation changed during roughly the first half of the twentieth century.

Krutch, J. W., 1951: *The Desert Year* – description of the Sonoran Desert environment.

The Great Basin Desert

West, N. E., 1983a: *Overview of North American temperate deserts and semi-deserts* – provides a summary of the climate, the regional subdivisions, and ecological subdivisions of the Great Basin Desert, with some coverage of the Mojave and northern Sonoran Deserts.

West, N. E., 1983b: *Great Basin-Colorado Plateau sagebrush semi-desert* – describes the climate, soils, geology, vegetation, ecosystem, and the human impacts for the central and southern Great Basin Desert.

West, N. E., 1983c: *Western intermountain sagebrush steppe* – discusses the climate, geology, vegetation, and human impacts for the northern Great Basin Desert.

West, N. E., 1983d: *Intermountain salt-desert shrubland* – describes the climate, geology, soils, vegetation, ecology, human impacts, and erosion for the salt deserts that are primarily within the Great Basin Desert.

The Mojave Desert

Clements, T., *et al.*, 1957: *A study of desert surface conditions* – a description of the
    distribution of playas, desert flats, bedrock fields, alluvial fans, dunes, mountains, and
    volcanic fields in the Sonoran and Mojave Deserts in the United States.
Clements, T., *et al.*, 1963: *A Study of Windborne Sand and Dust in Desert Areas* – reviews
    the causes of dust storms and their properties, primarily for the northern Sonoran and
    Mojave Deserts.
Darlington, D., 1996: *The Mojave: A Portrait of the Definitive American Desert* – a
    layman's description of the weather and surface conditions of the Mojave, with an
    emphasis on the great impact of human activities.
West, N. E., 1983e: *Colorado Plateau-Mohavian blackbrush semi-desert* – summarizes the
    ecosystem that overlaps the Mojave Desert in which blackbrush (*Coleogyne
    ramosissima*) dominates.

The Chihuahuan Desert

Tweit, S. J., 1995: *Barren, Wild and Worthless: Living in the Chihuahuan Desert* – personal
    accounts of exploring the natural history of the northern Chihuahuan Desert.
Gehlbach, F. R., 1993: *Mountain Islands and Desert Seas* – a natural history of the desert
    borderlands of the United States and Mexico.

The Atacama Desert

Bowman, I., 1924: *Desert Trails of Atacama* – describes early explorations of the Atacama,
    with descriptions of the surface conditions and the weather.
Rauh, W., 1985: *The Peruvian-Chilean Deserts* – reviews the climate, vegetation, and soils
    of the Atacama Desert.

The Peruvian Desert

Rauh, W., 1985: *The Peruvian-Chilean Deserts* – reviews the climate, vegetation, and soils
    of the Peruvian Desert.

The Patagonian and Monte Deserts (and the Puna and Chaco)

Mares, M. A., *et al.*, 1985: *The Monte Desert and other subtropical semi-arid biomes of
    Argentina, with comments on their relation to North American arid areas* – discusses
    the climate and vegetation of the Monte Desert, and compares the Monte with the
    Sonoran Desert.
Soriano, A., 1983: *Deserts and semi-deserts of Patagonia* – discusses the climates,
    paleoclimates, vegetation, geology, and desertification of the Patagonian Desert.

The Australian Deserts

Cogger, H. G., and E. E. Cameron, (Eds.), 1984: *Arid Australia* – contains a number of
    short papers on various aspects of the Australian deserts, with an emphasis on the
    biological setting.

Gentilli, J., 1971: *Climates of Australia and New Zealand* – describes the general
  circulation of the Southern Hemisphere, factors that control the Australian climate, the
  dynamics of the Australian troposphere, and climate fluctuations.
Shephard, M., 1995: *The Great Victoria Desert* – summary of geomorphic and vegetation
  conditions in the Great Victoria Desert, but with an emphasis on the history, flora,
  fauna, and human inhabitants.
Williams, O. B., and J. H. Calaby, 1985: *The hot deserts of Australia* – summarizes the
  climate, geomorphology, soils, and vegetation.

The Gobi Desert

Andrews, R. C., 1921: *Across Mongolian Plains* – a description of travels in the Gobi.
Man, J., 1999: *Gobi: Tracking the Desert* – a description of travels in the Gobi.
Walker, A. S., 1982: *Deserts of China* – a review of the geography and surface conditions
  of China's deserts, with some discussion of desertification.
Walter, H., and E. O. Box, 1983f: *The deserts of central Asia* – describes the climate,
  vegetation, and ecology of the Gobi Desert and nearby desert areas.

The Taklamakan Desert

Blackmore, C., 1995: *The Worst Desert on Earth: Crossing the Taklamakan* – a description
  of a recent expedition to cross the Taklamakan in the east–west direction (the long
  way). It provides a good personal account of the weather and surface
  conditions.
Walker, A. S., 1982: *Deserts of China* – a review of the geography and surface conditions
  of China's deserts, with some discussion of desertification.
Walter, H., and E. O. Box, 1983f: *The deserts of central Asia* – describes the climate,
  vegetation, and ecology of the Taklamakan Desert and nearby desert areas.
Walter, H., and E. O. Box, 1983g: *The Pamir – An ecologically well-studied high-mountain
  desert biome* – describes the climate, vegetation, and water budget of the Pamir
  Mountains to the west of the Taklamakan Desert.

The Turkestan-area Deserts

Walter, H., and E. O. Box, 1983a: *Caspian lowland biome* – describes the climate,
  geomorphology, soils, ecosystems, and human impacts of the semi-desert lowlands to
  the north of the Caspian Sea.
Walter, H., and E. O. Box, 1983b: *Semi-deserts and deserts of central Kazakhstan* –
  describes the climate, geomorphology, soils, vegetation, ecosystems, and human
  impacts in Kazakhstan, from the Caspian Sea to east of Lake Balkhash.
Walter, H., and E. O. Box, 1983c: *Middle Asian deserts* – discusses the climate, soils,
  vegetation, and ecosystems of the lowlands in southern Kazakhstan, including the
  Karakumy and Kyzylkum Deserts.
Walter, H., and E. O. Box, 1983d: *The Karakum Desert, an example of a well-studied
  eu-biome* – discusses the Karakumy Desert as an ecological unit, including the soils,
  geology, river system, climate, microclimates, water balance, vegetation, and human
  impacts.

Walter, H., and E. O. Box, 1983e: *The orobiomes of middle Asia* – describes the vegetation of the high-altitude deserts of southern and southeastern Kazakhstan.

The Thar Desert

Chowdhury, A., *et al.*, 1980: *Meteorological aspects of arid and semi-arid regions of India with special reference to the Thar desert* – brief discussions of the rainfall, evaporation, temperature, droughts, floods, severe weather, and hydrology of the Thar.
Gupta, R. K., 1986: *The Thar Desert* – a review of the climate, geomorphology, soils, plants, and animals of the Thar.

The Iranian Desert

Breckle, S. W., 1983: *The temperate deserts and semi-deserts of Afghanistan and Iran* – discusses the climate, hydrology, physiography, vegetation, ecology, and human impacts of the deserts of Afghanistan and Iran.

The Arabian Desert

Abd El Rahman, A. A., 1986: *The deserts of the Arabian Peninsula* – reviews the climate, geology, soils, vegetation, and groundwater.
Al-Shalash, A. H., 1966: *The Climate of Iraq* – includes factors controlling the climate, temperature characteristics, and precipitation.
Brody, L. R., 1977: *Meteorological Phenomena of the Arabian Sea* – contains a discussion of the climatology of, especially, the monsoonal wind regimes along the coast of the Arabian Peninsula and Somalia.
Perrone, T. J., 1979: *Winter Shamal in the Persian Gulf* – a summary of the weather conditions associated with the shamal in the Arabian Gulf and the desert of the eastern Arabian Peninsula.
Sanger, R. H., 1954: *The Arabian Peninsula* – mostly a discussion of civilization in the area, but there are observations of the weather and surface conditions.
Satchell, J. E., 1978: *Ecology and environment in the United Arab Emirates* – reviews the climate, physiography, soils, water, agriculture, forestry, and indigenous vegetation of the United Arab Emirates.
Walters, K. R., and W. F. Sjoberg, 1988: *The Persian Gulf Region: A Climatological Study* – a complete review of the different large-scale and regional weather regimes that influence the Arabian Peninsula.

The Negev Desert

Evenari, M., *et al.*, 1971: *The Negev: The Challenge of a Desert* – contains descriptions of the landforms and geology, water availability, and vegetation.

The arid region of Madagascar

Rauh, W., 1986: *The arid region of Madagascar* – a review of the climate and vegetation.

Questions for review

(1) For what deserts is summer precipitation clearly on the equatorial side and winter precipitation clearly on the poleward side?

(2) List the cold deserts of the world. Why do we call them cold deserts?

(3) List the foggy deserts of the world, and the nearby ocean currents.

(4) What monsoon systems are responsible for providing summer rainfall to deserts, and which deserts are affected?

(5) Why does the South American cold desert experience less severe cold-weather outbreaks than does the North American cold desert?

(6) What areas of the world experience aridity that is at least partly attributable to continentality?

(7) What areas of the world experience aridity that is at least partly attributable to mountain-produced rain shadows?

(8) List areas within arid zones where mountains lessen the degree of aridity.

(9) Cite examples of how the seasonal north–south migration of subtropical anticyclones is reflected in the monthly variations in precipitation, as reported in the figures of this chapter.

(10) Explain how sea-breeze circulations affect the regional climate of coastal areas.

(11) Describe the different types of substrate that are typically found in deserts.

Problems and exercises

(1) Fig. 2.9 contrasts the different indices of aridity for the Arabian Peninsula. Using the original references and whatever other sources that you can locate, perform a similar comparison for other arid areas. By how much do they differ in terms of defining the areas of greatest aridity?

(2) Speculate about why the arid maritime areas shown in Fig. 3.60 are adjacent to the western coasts of deserts.

(3) Look over the various maps of the monthly distributions of precipitation, and identify locations and months where the onset of a summer monsoon is apparent.

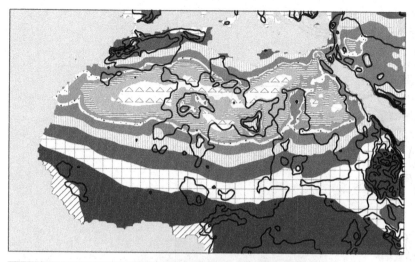

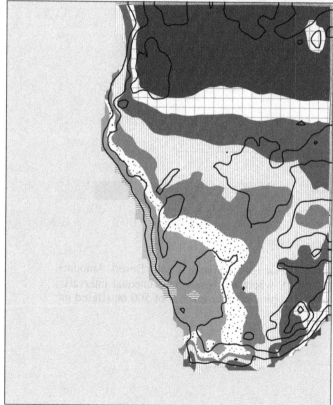

Plate 1 (above) Average annual precipitation for the Sahara Desert. Amounts are shown on the grey bar (mm), where the bands have unequal intervals. Terrain elevation is also plotted, with a contour interval of 500 m. (Based on data described in New *et al.* 2000.)

Plate 2 Average annual precipitation for the Namib and Kalahari Deserts. Amounts are shown on the grey bar (mm), where the bands have unequal intervals. Terrain elevation is also plotted, with a contour interval of 500 m. (Based on data described in New *et al.* 2000.)

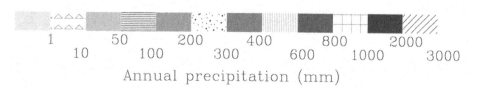

1   50   200   400   800   2000
  10   100   300   600   1000   3000

Annual precipitation (mm)

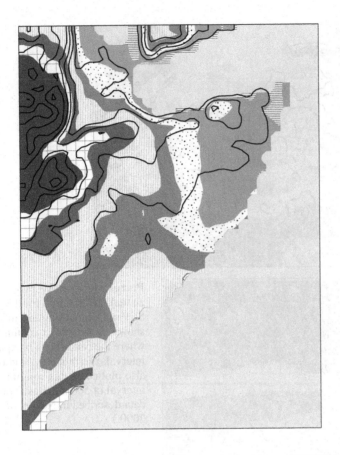

Annual precipitation (mm)

Plate 3 Average annual precipitation for the Somali–Chalbi Desert. Amounts are shown on the grey bar (mm), where the bands have unequal intervals. Terrain elevation is also plotted, with a contour interval of 500 m. (Based on data described in New *et al.* 2000.)

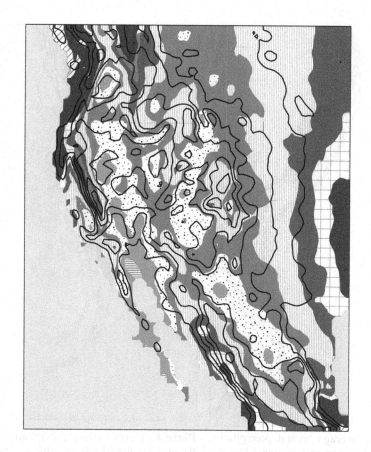

1	50	200	400	800	2000
10	100	300	600	1000	3000

Annual precipitation (mm)

Plate 4 Average annual precipitation for the North American deserts. Amounts are shown on the grey bar (mm), where the bands have unequal intervals. Terrain elevation is also plotted, with a contour interval of 500 m. (Based on data described in New *et al*. 2000.)

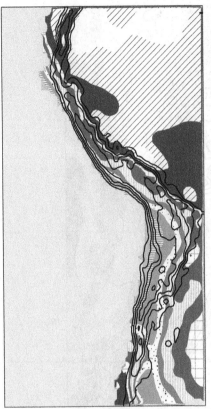

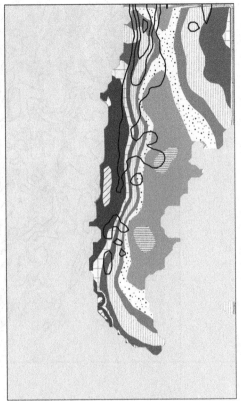

Plate 5 Average annual precipitation for the Peruvian and Atacama Deserts. Amounts are shown on the grey bar (mm), where the bands have unequal intervals. Terrain elevation is also plotted, with a contour interval of 500 m. (Based on data described in New *et al.* 2000.)

Plate 6 Average annual precipitation for the Patagonian and Monte Deserts. Amounts are shown on the grey bar (mm), where the bands have unequal intervals. Terrain elevation is also plotted, with a contour interval of 500 m. (Based on data described in New *et al.* 2000.)

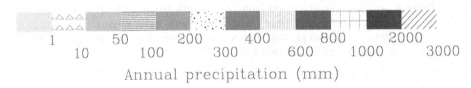

1   10   50   100   200   300   400   600   800   1000   2000   3000

Annual precipitation (mm)

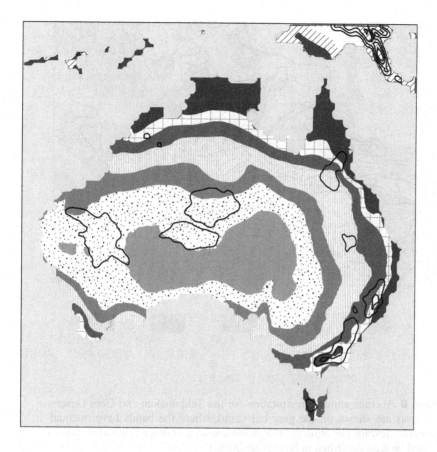

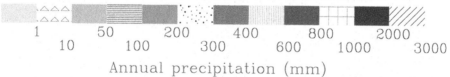

Annual precipitation (mm)

Plate 7 Average annual precipitation for the Australian Deserts. Amounts are shown on the grey bar (mm), where the bands have unequal intervals. Terrain elevation is also plotted, with a contour interval of 500 m. (Based on data described in New *et al*. 2000.)

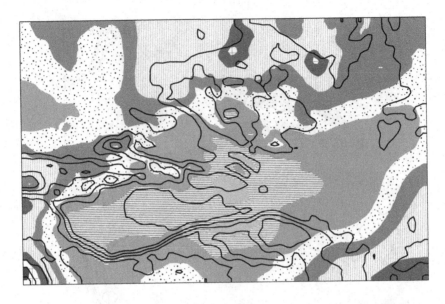

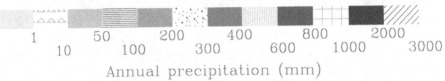

Annual precipitation (mm)

Plate 8 Average annual precipitation for the Taklamakan and Gobi Deserts. Amounts are shown on the grey bar (mm), where the bands have unequal intervals. Terrain elevation is also plotted, with a contour interval of 500 m. (Based on data described in New *et al.* 2000.)

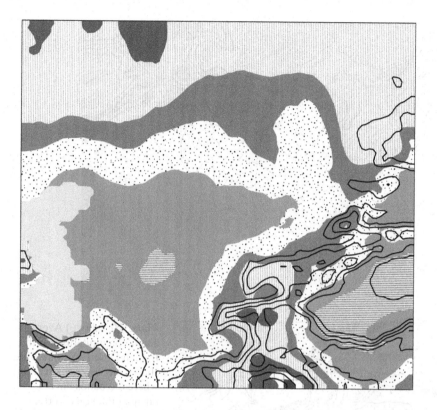

| | | | | | | | | | | |
|1| |50| |200| |400| |800| |2000|
| |10| |100| |300| |600| |1000| |3000|

Annual precipitation (mm)

Plate 9 Average annual precipitation for the Turkestan Desert. Amounts are shown on the grey bar (mm), where the bands have unequal intervals. Terrain elevation is also plotted, with a contour interval of 500 m. (Based on data described in New *et al.* 2000.)

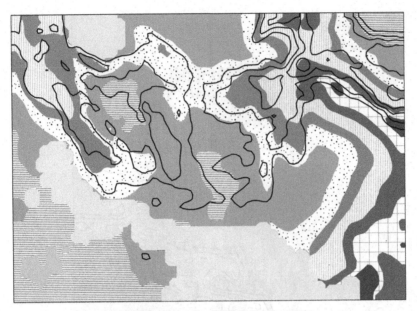

Plate 10 (above) Average annual precipitation for the Iranian Desert. Amounts are shown on the grey bar (mm), where the bands have unequal intervals. Terrain elevation is also plotted, with a contour interval of 500 m. (Based on data described in New *et al.* 2000.)

Plate 11 Average annual precipitation for the Arabian Desert. Amounts are shown on the grey bar (mm), where the bands have unequal intervals. Terrain elevation is also plotted, with a contour interval of 500 m. (Based on data described in New *et al.* 2000.)

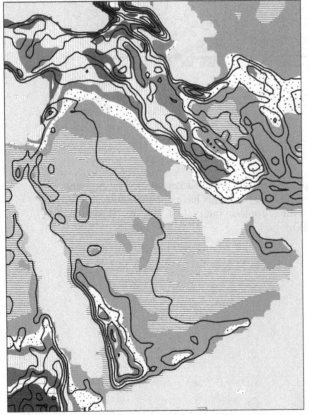

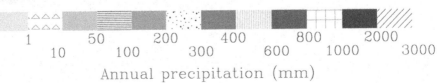

Annual precipitation (mm)

4

Atmospheric and surface energy budgets of deserts

When finally they climbed atop it, the ereg [erg, sand sea] *with its sea of motionless waves lay all about them. They did not stop to look: absolute silence is too powerful once one has trusted oneself to it for an instant, its spell too difficult to break.*

Paul Bowles, American/Moroccan writer and composer
The Sheltering Sky (1949)

No man can live this life and emerge unchanged. He will carry, however faint, the imprint of the desert, the brand which marks the nomad; and he will have within him the yearning to return, weak or insistent according to his nature. For this cruel land can cast a spell which no temperate clime can match.

For this was the real desert where differences of race and colour, of wealth and social standing, are almost meaningless; where coverings of pretence are stripped away and basic truths emerge. It was a place where men live close together. Here, to be alone was to feel at once the weight of fear, for the nakedness of this land was more terrifying than the darkest forest at dead of night.

Wilfred Thesiger, British explorer
Arabian Sands (1964)

One way of better understanding desert climates and microclimates is through their atmospheric and surface energy budgets. In this chapter, a review will first be provided of the overall concepts of atmospheric and surface energy budgets, and then these budgets will be contrasted for non-arid and arid climates. In the subsequent three chapters, additional material will be presented about the effects of desert land-surface properties, such as vegetation and substrates, on the interaction of the atmosphere and the desert surface.

Components of the atmospheric and surface energy budgets

Some basics of solar and terrestrial radiation

Because electromagnetic radiation is an important mechanism for energy transport in the Earth–atmosphere system, its characteristics and some governing laws will be briefly reviewed. All radiation travels at a speed of 3×10^8 m s^{-1}, and all substances with a temperature greater than absolute zero (0 K = -273.2 °C) emit radiation. The rate at which energy is radiated at a given temperature, summed over all wavelengths, is proportional to the fourth power of the absolute temperature, which is the basis for the Stefan–Boltzmann law. With an added dimensionless coefficient, ε, the law states that

$$\text{Energy emitted} = \varepsilon \sigma T^4, \tag{4.1}$$

where σ is the Stefan–Boltzmann constant, equal to 5.67×10^{-8} W m^{-2} K^{-4}. If the radiating body emits the maximum possible radiation per unit area per unit time, the **emissivity**, ε, is equal to unity, and the emitter is called a **black body**. Less efficient radiators have emissivities between zero and unity. See Chapter 8 for the emissivities of desert surfaces. The intensity of the radiation emitted by a black body at different wavelengths is a function of temperature, and is prescribed by Planck's Law. This intensity distribution with wavelength has a very similar shape for emitters of any temperature (see the examples of Fig. 4.1), and has a single maximum. The particular wavelength composition of the emitted energy depends on the temperature of the emitter, such that a temperature increase not only increases the total amount of energy emitted (Eqn. 4.1), but it also increases the fraction of the shorter wavelengths. That is, the curve in Fig. 4.1 shifts to the left as temperature increases, and the wavelength of the peak emission, λ_{max}, moves accordingly, such that

$$\lambda_{\text{max}} = 2.88 \times 10^{-3}/T, \tag{4.2}$$

where λ_{max} is in meters and T is in degrees Kelvin.

In most atmospheric applications, we are only concerned with wavelengths in the ultraviolet, visible, and infrared portions of the full **electromagnetic spectrum** shown in Fig. 4.2. Most of the sun's energy is in the visible part of the spectrum, from 0.36 μm (violet) to 0.75 μm (red), as shown in Fig. 4.1, which depicts the energy intensity for the different wavelengths emitted by the sun, whose surface temperature is about 6000 K. There is also significant solar energy emitted in the wavelength bands that are shorter than the violet (the ultraviolet) and longer than the red (the infrared), with the total solar ultraviolet–visible–infrared band extending from about 0.15 μm to about 3.0 μm. The infrared energy in this solar spectrum is referred to as the solar infrared, to contrast it with the longer-wavelength infrared that is emitted by the

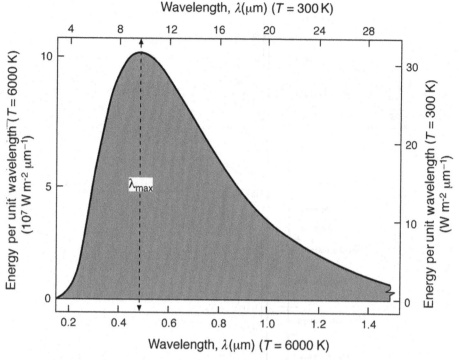

Fig. 4.1 The spectral distribution of radiant energy emitted from a black body at 6000 K (left vertical and lower horizontal axes) and 300 K (right vertical and upper horizontal axes). The λ_{max} is the wavelength at which the energy output per unit wavelength is a maximum. Approximately 10% of the energy is emitted at wavelengths longer than those shown. (1 μm $= 10^{-6}$ m.) (Adapted from Monteith and Unsworth 1990.)

cooler Earth and its atmosphere. Fig. 4.1 also shows the spectrum of the energy emitted by the Earth and its atmosphere, which have a temperature of roughly 300 K (26.8 °C, 80.2 °F). The wavelength band is all within the infrared, and extends from about 3 μm to 100 μm. Thus, the **solar spectrum** from 0.15 μm to 3.0 μm is referred to as **short-wave radiation**, while Earth's spectrum in the range 3.0–100 μm is **long-wave radiation**. Note that the maximum intensity of the sun's radiation is about a factor of 10^7 greater than that for Earth. The peak intensity of the solar spectrum is at 0.48 μm (λ_{max} is green, in the middle of the **visible spectrum**), whereas for the Earth–atmosphere system it is at about 10 μm.

The Earth–sun astronomical relationship, and atmospheric diurnal and annual cycles

The sole source of energy for the Earth–atmosphere system is, of course, the sun (except for small geothermal sources). The daily rotation of Earth on its axis

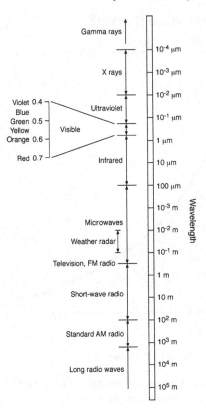

Fig. 4.2 The electromagnetic spectrum. The wavelengths of the radiation are shown on the right, and the names of the different bands are on the left.

is responsible for the diurnal cycle of temperature and other weather, and the annual revolution of Earth around the sun determines the annual cycle of the seasons. The instantaneous solar input at any point on the surface, with a clear atmosphere above, is determined by the sun angle, and the total daily energy input is related to the evolution of the sun angle during the daylight hours. Fig. 4.3 depicts the relationship between the sun angle and the flux of direct solar energy on a horizontal surface, where the surface flux has dimensions of energy per unit area per unit time. The circular area illuminated by a beam of light on a surface that is normal to the beam is related to the elliptical area illuminated on a horizontal surface by the same beam, through the cosine of the **zenith angle** (Z). Because the total **radiant energy** that illuminates the normal and horizontal areas is the same, the intensities of the radiation, that is the energy flux, on the two surfaces is also related through the cosine of the zenith angle. In addition to this direct solar energy received from the disk of the sun, the sun also illuminates the surface indirectly from all points in the sky "dome" through

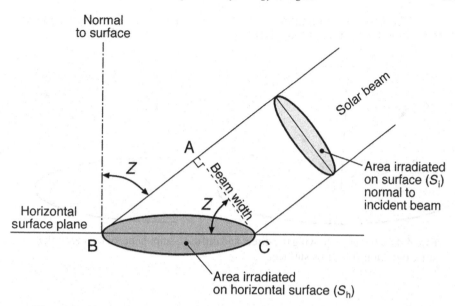

Fig. 4.3 The relationship of the intensity of the radiation on a surface, S_i, that is normal to an incident beam of light, to the intensity of the radiation on a horizontal surface S_h. The circular area irradiated on the surface normal to the incident beam is $\pi (0.5 \text{ AC})^2$, and the larger elliptical area irradiated on the horizontal surface has area $\pi (0.5 \text{ AC})^2/\cos Z$. The intensity of the light (the energy flux) on the horizontal surface S_h is less than the intensity on the normal surface S_i, with the intensities being related by the ratios of the areas ($\cos^{-1} Z$). That is, as the zenith angle approaches 90° the intensity on the horizontal surface goes to zero, and as the zenith angle approaches 0° the intensity on the horizontal surface approaches the intensity on the normal surface.

scattering by (1) air molecules (responsible for the blue color), (2) natural and anthropogenic particulates, and (3) clouds. Fig. 4.4 illustrates the geometry of the **direct** and **indirect** (or **diffuse**) **solar radiation**. Of course most substrate surfaces are not horizontal, so the slope and aspect (compass orientation of the slope) of the surface must be accounted for when calculating the intensity of the solar radiation on it.

In order to understand the seasonal variations in the radiation and energy budgets, the annual evolution of the Earth–sun geometry must be considered. Fig. 4.5 shows that, throughout the annual cycle, Earth's axis of rotation is tilted 23.5 ° with respect to the plane in which Earth revolves around the sun (the ecliptic plane). During the summer, the sun is at a relatively high angle (zenith angle, Z, small) during the day, and the period of daylight is relatively long. During the winter, the sun is at a lower angle and the period of daylight is shorter.

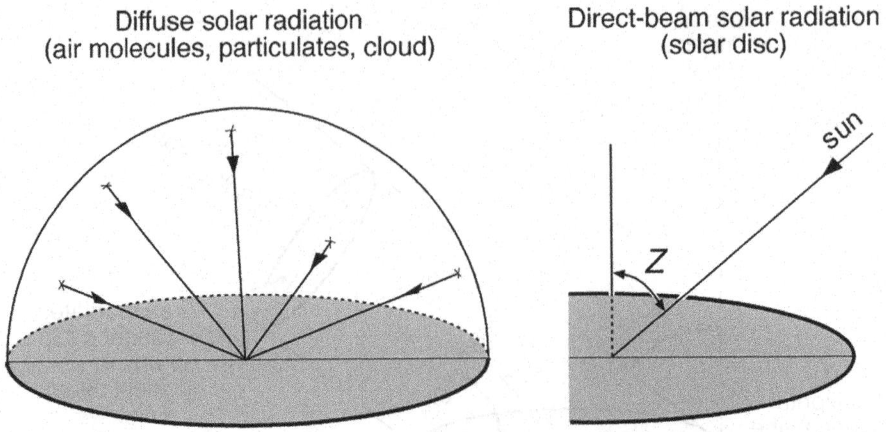

Diffuse solar radiation
(air molecules, particulates, cloud)

Direct-beam solar radiation
(solar disc)

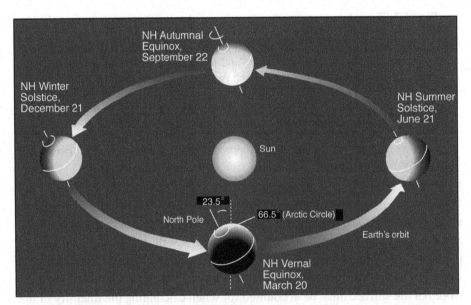

Fig. 4.4 Schematic showing the direct and diffuse components of the incoming solar radiation at Earth's surface.

Fig. 4.5 The geometric relationship between Earth and the sun during the evolution of the seasons. The figure of Earth at the Northern Hemisphere Vernal Equinox shows the 23.5° tilt of Earth's rotational axis (solid line through Poles) with respect to the plane on which it revolves around the sun (dotted line is normal to this plane). The designations of the solstices and equinoxes are for the Northern Hemisphere (NH). The dates are approximate.

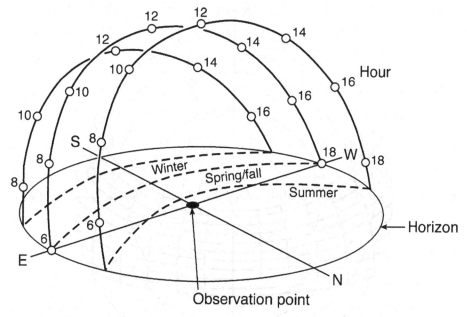

Fig. 4.6 The path of the sun across the sky during the solstices and the equinoxes for a mid-latitude Northern Hemisphere location. The numbers on the sun-path curves represent times of day, and the dashed lines are the projections of the sun paths on the horizontal plane. To apply the figure to the Southern Hemisphere, the compass points as well as the times on the sun-path curves must be reversed.

Fig. 4.6 illustrates the paths of the sun during the **Summer** and **Winter Solstices** and the **equinoxes**, for mid-latitudes in the Northern Hemisphere. Not only is the period of daylight greater by about 6 h at the Summer Solstice, resulting in more possible energy received, but the mid-day solar zenith angle is smaller and the intensity on a horizontal surface is therefore greater (see Fig. 4.3). The dashed lines represent the projection of the sun's position on the horizontal plane, where such paths are shown for 20 times of the year in the sun-track diagram of Fig. 4.7. The latter diagram applies for 30° north or south latitude, the subtropics where many of the world's deserts exist. On the Winter Solstice, the sun rises and sets well to the south of the east and west compass directions, and its elevation above the horizon at noon is between 30° and 40° (90°−latitude−23.5°). On the Summer Solstice, the sun almost reaches the zenith at noon, and the day length is about 4 h greater than on the Winter Solstice. Similar diagrams for other latitudes can be found in List (1966), and mathematical expressions for the sun's position are in Sellers (1965). Fig. 4.8 illustrates the seasonal and latitudinal variation of the possible daily-total (without atmospheric attenuation) solar radiation receivable

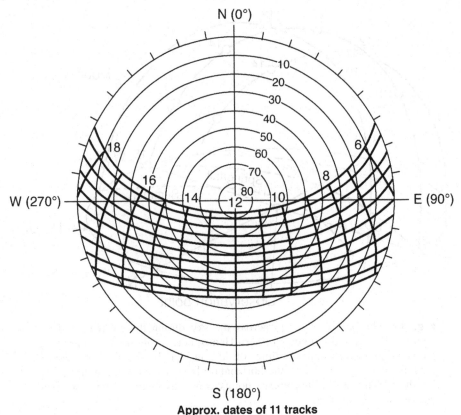

Approx. dates of 11 tracks

22 June (most northerly track)
21 May, 24 July
1 May, 12 August
16 April, 28 August
3 April, 10 Sept.
21 March, 23 Sept.
8 March, 6 Oct.
23 Feb., 20 Oct.
9 Feb., 3 Nov.
21 Jan., 22 Nov.
22 Dec. (most southerly track)

Fig. 4.7 Sun-path diagram for 20 dates at 30° latitude. The long heavy lines represent the sun paths, and the shorter, approximately orthogonal, lines indicate the local time (labeled every 2 h) for points along the path. The azimuth (compass direction) of the sun at any time is defined by the azimuth scale on the outer circle (degrees), and the angle of the sun above the horizon is indicated by the radial coordinate (degrees) that is labeled along the top half of the north–south axis. Sun-path diagrams for other latitudes can be found in List (1966).

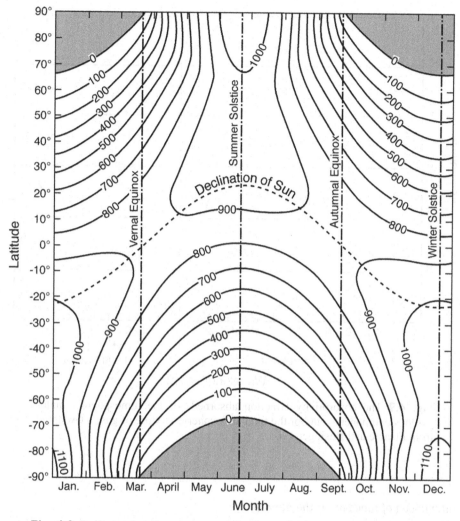

Fig. 4.8 Daily-total radiant energy received on a horizontal surface at the top of the atmosphere, by time of year and latitude. The isopleths are labeled in cal cm^{-2}. (From List 1966.)

on a horizontal surface. The values represent the time integral of the unattenuated direct solar energy flux during daylight hours, which is based on the intensity of the sun's radiation, and the Earth–sun geometry that controls the daily evolution of the sun angle. No meteorological effects are accounted for (reflection by clouds, scattering, etc.), so this is the maximum possible energy receivable. For high-latitude, cold deserts (50° latitude), the daily radiation available on the Summer Solstice is approximately five times that available on the Winter Solstice. In mid-latitudes (40°), and in the subtropical latitudes (20°–30°) of many of the world's deserts, the difference is closer to factors of three and two, respectively.

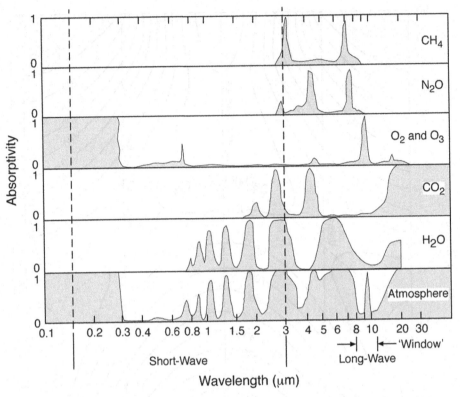

Fig. 4.9 The absorptivity (fraction absorbed) of the atmosphere's major gaseous components, and of the total atmospheric mixture of gases. (Adapted from Fleagle and Businger 1963).

Attenuation of radiation by the atmosphere

Because of the existence of clouds, dust, and **optically active** (absorbing and emitting) gases, Earth's atmosphere is far from transparent to the passage of the radiant energy emitted by Earth and the sun. Each of the constituents has its own unique effect on each individual wavelength band of the radiation. In terms of the bulk effect of the atmosphere, part of the radiation is reflected or scattered, part is absorbed, and the remainder is transmitted. The fraction of the incident radiation that is absorbed (**absorptivity**) individually by the atmosphere's major gaseous components, and by the total atmospheric mixture of gases, is shown in Fig. 4.9. It is clear that the gaseous medium totally absorbs some wavelength bands of radiant energy, while it is relatively transparent to others. Within the visible wavelengths of the solar spectrum, from 0.36 to 0.75 μm, where the greatest

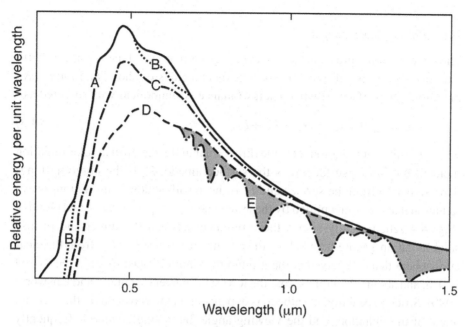

Fig. 4.10 Attenuation of the extraterrestrial solar spectrum (A) by ozone absorption (B), molecular scattering (C), aerosol scattering (D), and water vapor and oxygen absorption (E). Curve E is terrestrial sunlight. (From Henderson 1977.)

energy is represented (Fig. 4.1), not much absorption occurs from gases. At ultraviolet wavelengths shorter than 0.30 μm, ozone effectively absorbs virtually all of the energy, and water vapor becomes an important absorber at wavelengths longer than 0.80 μm. There is a "window" in the long-wave absorption at about 8–11 μm, which encompasses the wavelength of the peak emission at 10 μm from the Earth–atmosphere system that emits at about 300 K (Fig. 4.1). Fig. 4.10 shows the effects of both scattering and absorption on the attenuation of the solar beam as it penetrates the cloud-free atmosphere. The outer curve, A, is the solar spectrum at the top of the atmosphere, which differs from the smooth black-body spectrum of a 6000 K radiator shown in Fig. 4.1. Here, the degree to which the emissivity departs from unity depends on wavelength. The remainder of the curves show the progressive modification of the top-of-the-atmosphere spectrum by ozone absorption, **molecular scattering**, **aerosol scattering**, and water vapor and oxygen absorption. The lowest curve (E) shows the energy that survives the downward transit of the solar beam through the atmospheric medium.

The surface radiation budget

The net radiation represents the rate of energy gain or loss at the surface of Earth after accounting for all the various sources and sinks of short- and long-wave radiation. The surface radiative energy balance is symbolically represented as

$$R = (Q + q)(1 - \alpha) - I\!\uparrow + I\!\downarrow, \tag{4.3}$$

where R is the net radiation, Q is the direct solar and q the diffuse solar radiation incident on Earth's surface, α is the surface albedo, $I\!\uparrow$ is the outgoing long-wave radiation from the surface and $I\!\downarrow$ is the absorbed downwelling long-wave radiation that has been emitted by the atmosphere (gas, particulates, and clouds). (Fig. 4.4 schematically shows the difference between the direct and diffuse solar radiation.) More than half of the downward long-wave flux from the clear atmosphere that is received at the ground is emitted by gases within the lowest 100 m, and about 90% comes from the lowest kilometer (Monteith and Unsworth 1990). Strictly speaking, the albedo is dependent on the wavelength, the incident angle of the radiation, and the viewing angle, but a single value is frequently used as an approximation. The quantities R, Q, q, and I are energy fluxes. The sign of R depends on whether a summation of the sources and sinks of radiant energy on the right corresponds to a net loss (R negative) or gain (R positive) of energy by the surface. The term $I\!\uparrow = \varepsilon \sigma T_g^4$ and $I\!\downarrow = \varepsilon I$ (incident), where I (incident) is the downwelling long-wave radiation that is incident at the surface and T_g is the absolute temperature of Earth's surface. Again, see Chapter 8 for examples of desert albedoes and emissivities. The ε coefficient of I (incident) represents the fraction of the atmospheric long-wave radiation that is absorbed by the surface. Incident radiation can either be absorbed, transmitted, or reflected by a medium. An energy-conservation statement would be that the total of the radiation in the three categories must equal the incident radiation. If we defined the **transmissivity** (ψ_λ), the **reflectivity** (α_λ), and the **absorptivity** (ζ_λ) in terms of a fraction of the incident radiation, we have

$$\psi_\lambda + \alpha_\lambda + \zeta_\lambda = 1, \tag{4.4}$$

where the subscript indicates that this relationship applies, strictly speaking, only for individual wavelengths. In practice, we often use it for fairly broad bands of radiation. For example, the albedo in Eqn. 4.3 is a broad-band reflectivity for all of the solar spectrum. Now, Kirchhoff's Law states that good absorbers are good emitters, for the same temperature and wavelength. That is, the absorptivity equals the emissivity:

$$\zeta_\lambda = \varepsilon_\lambda. \tag{4.5}$$

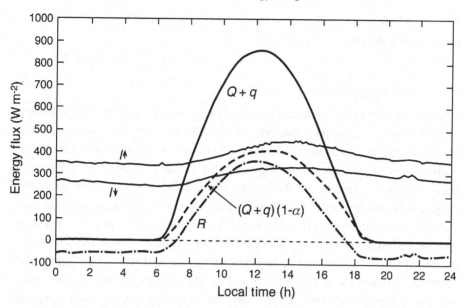

Fig. 4.11 Components of the radiation budget measured on a clear day (7 September 1987) at a dry location in the Great Basin Desert in North America. See text for the meaning of the symbols. (From McCurdy 1989.)

For an opaque surface such as the land, $\psi_\lambda = 0$, and therefore, from Eqns. 4.4 and 4.5,

$$\varepsilon_\lambda = 1 - \alpha_\lambda. \tag{4.6}$$

This is why the ε coefficient of I (incident) represents the fraction of the incident long-wave radiation that is absorbed by the surface, namely $1 - \alpha_\lambda$. Because long-wave emissivities for natural surfaces are typically greater than 0.9, this long-wave albedo is often considered to be zero, and ignored.

An example of the diurnal variation of these components of the radiation budget in Eqn. 4.3 is shown in Fig. 4.11. These data are for a cloud-free September day in a semi-arid area in the northern Great Basin Desert of North America, but the relations among the components would be similar for most places in summer. The curve for direct and diffuse solar radiation shows the expected sinusoidal shape, with a peak at local noon. The outgoing long-wave radiation ($I\!\uparrow = \varepsilon\sigma T_g^4$) increases during the daylight hours, with a peak in the afternoon, and is an expression of the diurnal change in the ground temperature. The downwelling long-wave flux increases during warmer daylight hours as well, because the atmosphere also emits according to the fourth power of its absolute temperature. The reflected solar energy (($Q + q)(1 - \alpha)$) is proportional to $Q + q$ itself.

The net radiation (R) is, of course, the sum of the gains and losses at the surface, expressed by the other components.

The surface energy budget

We now turn from the surface radiation budget to the surface energy budget, of which the net radiation is one component. The other major elements include (1) the energy loss and gain associated with the evaporation, condensation, and transpiration (by vegetation) of water; (2) the vertical conductive exchange of heat between the surface and the interior of the substrate; and (3) the exchange of heat between the substrate and the atmosphere through conduction and convection. Less important, often neglected, processes are (1) vertical convection and **advection**, and horizontal advection, of air within the interstitial air spaces in the substrate; (2) melting or freezing of surface and subsurface water; and (3) those associated with the energetics of the photosynthesis process. One often sees the components written in the form of an energy-conservation equation, such as

$$R = LE + H + G, \tag{4.7}$$

where L is the latent heat of evaporation, E is the evaporation or condensation rate, H is the **sensible-heat** exchange between the ground and the atmosphere, and G is the sensible-heat exchange (conduction) between the surface and the substrate. (The term "sensible" refers to heat that can be sensed, in contrast to heat that is "latent", in the form of water vapor, that is released or consumed only upon condensation and evaporation, respectively.) The quantity LE is the **latent-heat** flux, and H and G are heat fluxes also, all with the same dimensions as R. This equation applies to the Earth–atmosphere interface, and simply states that the net radiative flux at the massless interface must equal the sum of the other three fluxes at the interface. If the net radiative flux is positive, for example during a sunny day, conservation of energy requires that this radiative energy gain at the surface be partitioned among the sensible-heat flux into the substrate, the sensible-heat flux to the atmosphere, and the latent-heat flux associated with the evaporation or transpiration of water into the atmosphere. When averaged over an annual cycle, the heat gained by the ground during the warm months and the heat lost by the ground during the cold months approximately cancel, and G is zero. Thus, as an annual average

$$R = LE + H. \tag{4.8}$$

Alternatively, another energy-conservation equation can be written for a unit mass or unit area of the surface that is experiencing gains or losses of energy

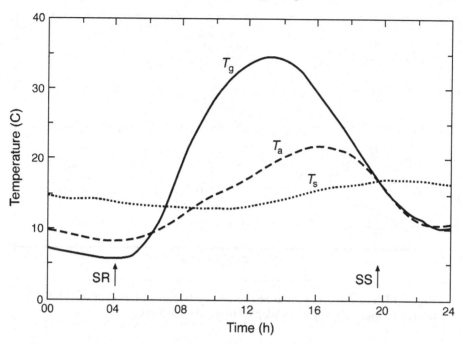

Fig. 4.12 The diurnal variation of surface temperature (T_g), atmospheric temperature at 1.2 m AGL (T_a), and soil temperature at 0.2 m below the surface (T_s), corresponding to the surface energy-budget components for the non-desert location shown in Fig. 4.16b. The surface was moist bare soil, the date was 30 May, and the latitude was 49° N. The times of sunrise (SR) and sunset (SS) are indicated. (From Novak and Black 1985.)

from the same four processes represented in Eqn. 4.7. For example, the first law of thermodynamics states that, at a particular time and place,

$$C\frac{\partial T_g}{\partial t} = R - LE - H - G, \tag{4.9}$$

where T_g is the temperature of the surface layer of substrate and C is its **thermal capacity** per unit area (energy per unit horizontal area per degree). The thermal capacity is the amount of energy required to raise the temperature of a unit area by one degree. Fig. 4.12 shows the diurnal surface temperature (T_g) trace corresponding to the surface energy-budget components for the non-desert location shown later in Fig. 4.16b. The surface temperature increases until 1300 LT in this example. During the period of increase, the sum of the terms on the right side of Eqn. 4.9 is positive: that is, the energy-source term, the net radiation, exceeds the energy-loss terms associated with evaporation and heat fluxes from the surface to the deeper substrate and to the atmosphere. After 1300 LT, the net radiation decreases sufficiently (see Fig. 4.16b) so that the energy losses exceed

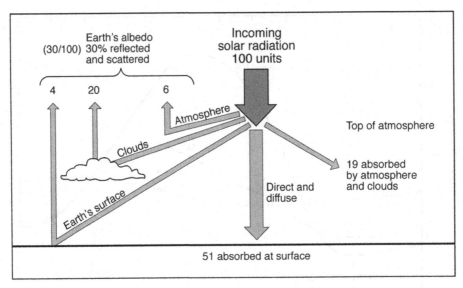

Fig. 4.13 The estimated atmospheric energy budget in terms of the global annual average components. (Adapted from Ahrens 2000.)

the energy gains, and the temperature decreases. The air temperature at 1.2 m (T_a) reaches a maximum late in the afternoon near 1600 LT, a few hours after the surface temperature maximum. Throughout most of the afternoon, $T_g > T_a$, and thus the sensible heat flux is upward from the surface to the atmosphere, and T_a continues to increase. The 0.2 m (below the surface) soil-temperature (T_s) maximum is even later, near midnight.

The complete atmospheric and surface energy budget

Figure 4.13 shows the disposition of the solar energy that enters the Earth–atmosphere system, in terms of global-average values (continental and maritime areas, and arid and non-arid areas combined). Of 100 units of radiation entering the system annually from the sun, 30 units are both reflected and scattered back to space. This includes 4 units that are reflected from Earth's surface, 20 units that are reflected and scattered from clouds, and 6 units that are reflected and scattered from molecules and dust in the atmosphere. Nineteen units are absorbed by the atmosphere and clouds. The remaining 51 of the 100 units are direct and diffuse solar radiation that are absorbed by Earth's surface. Fig. 4.14, again based on a global annual average, describes what happens to the 70 units of energy that are shown in Fig. 4.13 to be absorbed by both Earth's surface and atmosphere. Energy gains at the surface, totaling 147 units, are 51 units from

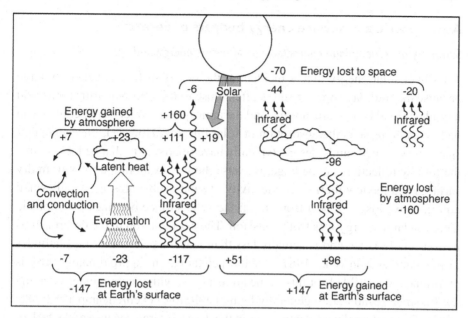

Fig. 4.14 The elements of the energy exchange between the surface and the atmosphere, based on a global annual average. (Adapted from Ahrens 2000.)

direct and diffuse solar radiation $((Q + q)(1 - \alpha)$ and 96 units of infrared from the clouds and gas of the atmosphere $(I\downarrow)$. Surface energy losses, which also must total 147 units, include 117 units of infrared emitted to space and the atmosphere $(I\uparrow)$, 23 units lost by evaporation (LE), and 7 units lost to the atmosphere through sensible heat fluxes (H). The net radiation, R, is the radiative gain of 147 units minus the radiative loss of 117 units. Also, the 160 units of energy gains by the atmosphere from the various sources on the left must equal the losses from the processes on the right. These estimates of the various components of the budgets are being updated frequently, and there are some important differences among the estimates (see, for example Mitchell 1989; Kiehl and Trenberth 1997). It is important to remember that the magnitudes of these components are global averages, and thus they represent a large contribution from conditions over the oceans. Some energy-budget diagrams of the sort in Figs. 4.13 and 4.14 show an actual energy flux at the top of the atmosphere, rather than an arbitrary base value of 100 units. The global-average, top-of-the-atmosphere energy flux on a horizontal surface is about 342 W m^{-2} (see problem 3). The individual components of the budget thus represent partitions of this base value.

Atmospheric and surface energy budgets of deserts

Summary of some unique characteristics of desert energy budgets

The direct solar energy that is incident at the desert surface is generally large because the mid-day summer sun is often close to the zenith at arid subtropical latitudes, cloud is typically absent, and the atmospheric water-vapor content can be low and thus so is the absorption of solar infrared. However, unvegetated or sparsely vegetated sandy deserts typically have relatively high albedoes, so that the loss by reflection can be large. Because the low evaporation rates from dry surfaces and the low thermal conductivity of sandy substrates allow only small non-radiative losses of heat from the surface, the surface becomes very hot and there is a large energy loss from emission. The net radiation in the desert is thus not especially large, and is actually less than that for bare soil in mid-latitudes in the same season (Oke 1987). Of the net radiation, approximately 90% is partitioned to sensible heat flux to the atmosphere, with the remainder warming the substrate (evaporation generally being negligible) (Vehrencamp 1953; Oke 1987). The nocturnal net radiative loss in the desert is large because of the lack of cloud, the dry atmosphere, and the initially high surface temperature. Note that, with the assumption of no evaporation as a first approximation, the equivalent of Eqn. 4.8 for extreme deserts is

$$R = H, \tag{4.10}$$

as an annual average. That is, integrating over a year, the net radiation is approximately equal to the sensible heat flux to the atmosphere.

In summary, the components of the energy and radiation budgets for the desert differ from the global averages in Figs. 4.13 and 4.14 in the following respects.

- Higher surface temperatures cause the intensity of infrared emission from the substrate to be larger (Eqn. 4.1).
- Lack of cloud and low water-vapor content cause atmospheric absorption of Earth's emitted infrared to be smaller.
- Higher surface temperatures cause greater sensible heating of the atmosphere through convection.
- Low surface-water availability (dry substrates) cause less surface heat loss through evaporation. Fig. 4.14 shows that the greatest amount of net heat transfer between the surface and the atmosphere, for the global average, takes place through evaporation at the surface and condensation in the atmosphere. The net transfer through infrared radiation is smaller because of roughly equal amounts transferred in each direction, and the sensible-heat transfer is also smaller. For the desert, this important mechanism of latent-heat transfer can be negligible.

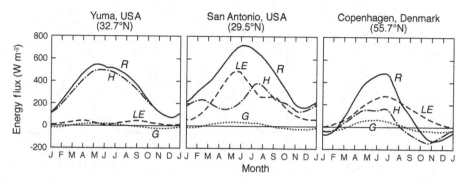

Fig. 4.15 Seasonal variation in the components of the surface energy balance for Yuma and San Antonio, United States and Copenhagen, Denmark. Curves are labeled as follows: R, net radiation; H, sensible heat flux to the atmosphere; LE, latent heat flux to the atmosphere; G, heat flux into the ground. (From Sellers 1965.)

- Low water-vapor content of the atmosphere and weak mechanisms for producing cloud cause less heat gain by the atmosphere through the latent heat of condensation (cloud and fog formation).

- Low water-vapor content of the atmosphere causes weaker infrared emission of the atmosphere to the surface and space.

- Higher albedoes of sandy desert substrates cause greater reflection of solar energy by the surface.

- Less cloud contributes to less atmospheric reflection. Note that, for the global average shown in Fig. 4.13, more than half the planetary albedo results from the existence of clouds. This contribution over deserts is generally small.

- Sparse vegetation causes less surface heat loss through transpiration, and the albedo is higher without vegetation.

- Potentially higher dust concentrations in the atmosphere greatly modify many components of the energy budget. For example, there is increased atmospheric reflection of incoming solar radiation, more diffuse radiation, greater absorption of incoming solar radiation, and increased absorption and re-emission of terrestrial infrared by the atmosphere. A more detailed discussion of the effects of aerosols on the atmospheric energy budget is found later in this chapter.

Examples of desert and non-desert energy budgets

To illustrate quantitative differences between the surface energy balance components for desert and non-desert land surfaces, some general examples will be provided of the seasonal and diurnal variability. Even though these examples may apply for specific days and surface conditions, they are good representatives for desert and non-desert conditions. Fig. 4.15 shows the average *seasonal*

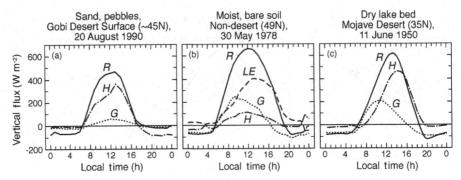

Fig. 4.16 Diurnal variation in the components of the surface energy balance for three locations: (a) the Gobi Desert, (b) a moist bare-soil surface in western Canada, and (c) a dry lake in the Mojave Desert. Curves are labeled as follows: *R*, net radiation; *H*, sensible heat flux to the atmosphere; *LE*, latent heat flux to the atmosphere; *G*, heat flux into the ground. The latent heat flux over the dry Mojave lake surface was too small to measure. (Adapted from Wang and Mitsuta (1992) (a), Novak and Black (1985) (b), and Vehrencamp (1953) (c).)

variation, for three Northern Hemisphere locations, of net radiation, latent and sensible heat fluxes from the surface to the atmosphere, and the sensible heat flux into the substrate. Yuma is extremely arid with approximately 6.6 cm (2.6 in) of rainfall per year; San Antonio is at about the same latitude but receives about ten times the annual rainfall of Yuma; and Copenhagen is non-arid, and at a much higher latitude. The summer net radiation (*R*) is less for Yuma than for San Antonio, for the reasons mentioned above: over desert surfaces there are often large losses from reflection due to the high albedo, and large losses from emission because of the high surface temperature that results from the poor thermal conductivity and low moisture content of the substrate. At Yuma, approximately 90% of the net radiation is partitioned to sensible heating (*H*) of the boundary layer, and less than 10% contributes to evaporation (*LE*). The evaporation rate approaches zero in the summer months. At the non-desert locations, the sensible heat fluxes to the atmosphere are considerably less than at Yuma. This, of course, is a consequence of the greater moisture content of the substrate at locations with greater precipitation. At all three locations, there is a gain of heat by the ground (*G*) during the summer months, and a loss during winter. Budyko (1958) contains additional interesting comparisons of the annual variation of the surface energy-balance components for desert and non-desert locations.

Examples of the *diurnal* variation in the energy-budget components are shown in Fig. 4.16 for two desert locations, one in the Gobi Desert with a very sparsely vegetated sandy and pebbly surface and one in the western Mojave Desert that is a dry lake bed. Data are also shown for a non-desert location in western

Canada, with moist, bare soil. The dates are all in late spring or summer. As in Fig. 4.15 for the annual cycle, the diurnal maximum in net radiation (R) is larger for the non-desert location in spite of the fact that it is the highest in latitude and receives less intense solar input. Also similar to the annual-cycle energy balance is the large fraction of the net radiation that is used for sensible heating of the atmosphere (H) at the desert locations. At the non-desert location with a moist surface, much more of the available net radiation is used to evaporate surface water (LE) than for sensible heating of the atmosphere. The asymmetric shape (with respect to local noon) of some of the curves for the daytime sensible heating of the ground (G) and the atmosphere (H) is likely related to the fact that the latent and sensible heating of the atmosphere become more efficient during late morning and afternoon as the atmosphere becomes more unstable and the surface wind speed increases. During this time there is less energy available for heating of the substrate. Additional studies that report on measurements of the diurnal variation in the surface energy budget for other desert locations include Stearns (1967) for the Peruvian Desert; Braud *et al.* (1993) for an arid area of Spain; Menenti (1984) for a playa in Saudi Arabia; Aizenshtat (1960) for the Kyzylkum Desert; Smith (1986a) for the Empty Quarter of Saudi Arabia; Unland *et al.* (1996) for the Sonoran Desert; and Terjung *et al.* (1970), Albertson *et al.* (1995), and Vehrencamp (1951) for the Mojave Desert.

In contrast with wetter environments, the surface energy budget in extreme deserts is sometimes quite temporally consistent in terms of the percent of the available net radiation that is partitioned to the various components: heating of the soil, sensible heating of the atmosphere, and evaporation. Pike (1970) calculated daily averages of the various components for two days that are separated by three months, 16 February 1969 and 14 May 1969, for the same location in a playa in Saudi Arabia that had a relatively shallow and constant water-table depth. Table 4.1 shows, remarkably, that the percentages of the net radiation are, within 1%, almost exactly the same for the two days. Even though the large fraction of the net radiation that is partitioned to soil heating and evaporation reflects the fact that this study area is a playa, with higher thermal conductivity and water availability than would be normal for a dry sand-desert environment, this temporal consistency may be typical of deserts because the infrequent rainfall does not cause the soil moisture to change significantly. This rhythmic and diurnally repetitive nature of the desert energy budget is shown graphically in Fig. 4.17 for a location in the dry Rub Al Khali Desert of the Arabian Peninsula. For this seven-day period, which was typical of the whole month, there is very little day-to-day change in the amplitudes and shapes of the plots for each of the components. This regularity is typical of very arid areas.

Table 4.1 *Values of the surface energy-balance terms on 16 February and on 14 May 1969 for a playa in Saudi Arabia.*

Values are millimeters per day of equivalent water depth. Also shown are the relative percentages of the net radiation for the different terms.

	16 February 1969		14 May 1969	
Term	Value	% of net radiation	Value	% of net radiation
Net radiation	+9.40	—	+15.62	—
Soil heat flux	−3.00	32	−5.00	32
Sensible-heat flux	−5.25	56	−8.84	57
Latent-heat flux	−1.15	12	−1.78	11

Source: From Pike (1970).

Because of the large near-surface heat-flux convergence[4.1] in the substrate during the day, and the large divergence at night, substrate surface temperatures[4.2] have a large diurnal variation. Garratt (1992b) discusses observed and theoretical land-surface temperature maxima, some for desert environments. In one study (Peel 1974) over the central Sahara Desert, described by Oke (1987), the diurnal substrate surface temperature ranged over 37 K (67 °F), from 28 °C (82 °F) to 65 °C (149 °F). However, in the sand only 0.3 m below the surface, the substrate temperature was virtually constant at 38 °C (100 °F) for the diurnal period. Campbell (1997) reports that recorded substrate surface temperatures in the North American Desert have reached 88 °C (190 °F), while 84 °C (183 °F) has been observed elsewhere (Cloudsley-Thompson 1977). Oke (1987) claims that diurnal surface-temperature amplitudes can approach 80 K (144 °F), and screen-level (1.5 m AGL) temperature amplitudes have reached 56 K (101 °F) at Tucson in the United States. Similarly large surface-temperature amplitudes of 78 K (140 °F) have been observed in high-latitude sandy deserts in Asia (Walton 1969). Associated with the large diurnal amplitudes are rapid surface-temperature changes. Blake *et al.* (1983) report a temperature rise from 24 °C (75.2 °F) to 48 °C (118.4 °F) in the 45 min after sunrise in the Rub Al Khali Desert: about 0.5 K min^{-1} (1.0 °F min^{-1}).

[4.1] A heat-flux convergence means that more heat is flowing toward a point in a given direction than away from it in the same direction. In the context of the substrate near the surface during the day, the rate at which energy is being gained by solar input from above is greater than the rate of loss by conduction to the substrate below. This heat-flux convergence causes the near-surface temperature to increase. At night, the surface substrate's heat loss by infrared emission to space is occurring at a greater rate than the gain of heat from the substrate below, leading to a flux divergence and a decrease in temperature.

[4.2] The surface temperature of a substrate can be the temperature as measured by a thermometer that is placed in contact with the surface, or it can be the **radiative temperature** that is estimated through observation of the emitted long-wave radiation (see, for example, Eqns. 4.1 and 4.2).

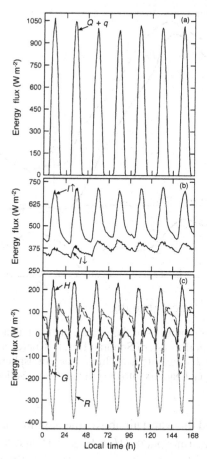

Fig. 4.17 The surface energy- and radiation-budget components observed during a seven-day period in June 1981 in the Rub Al Khali Desert of the Arabian Peninsula. Shown is the total solar direct and diffuse radiation ($Q + q$), the upward and downward infrared radiation (I), the sensible heat flux to the atmosphere (H), the net radiation (R), and the heat flux into the substrate (G). (Adapted from Smith 1986a).

The "oasis effect" on the energy budget

An anomalous desert surface energy budget can be found near oases, and represents what is known as the **oasis effect**. The oasis has a surface that is moist from irrigation, and contains transpiring vegetation. Because of the dry air that flows across the oasis from the surrounding desert, there is a large evaporation rate with an associated rapid loss of heat from the surface. This maintains a low surface temperature over the oasis. For example, the mid-day surface temperature of unshaded soil in an irrigated cotton field in an Asian desert was about 34 °C (93.2 °F), whereas the arid surrounding surface was 60 °C (140 °F) (Aizenshtat

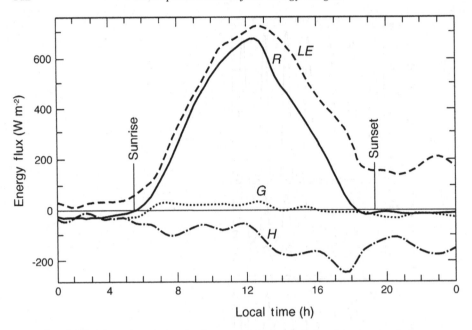

Fig. 4.18 Diurnal variation in the components of the surface energy balance for a stand of irrigated Sudan grass in the eastern Sonoran Desert. Curves are labeled as follows: R, net radiation; H, sensible heat flux to the atmosphere; LE latent heat flux to the atmosphere; G, heat flux into the ground. (From Sellers 1965.)

1960). Because of the cool surface and the warm air above, the daytime sensible heat flux between the surface and the atmosphere (H in Figs. 4.15 and 4.16) is *downward*. If the sensible heat flux into the substrate is ignored to simplify the argument, the surface in the daytime is cooled by evaporation and warmed by both the net radiation and the downward sensible heat flux from the atmosphere. Thus, the evaporative cooling rate can exceed the heating rate from the net radiation. It is typical for evaporation to continue throughout the night. An example of the vertical fluxes over an oasis is shown in Fig. 4.18 for a stand of irrigated Sudan grass in the eastern Sonoran Desert. The evaporation rate remained especially high after sunset in this case because the wind speed increased. A larger-scale example of this effect, described by Slatyer and McIlroy (1961), is when the hot dry sirocco winds from the deserts of North Africa invade semi-humid areas along the coastal Mediterranean. Here, also, the rate of heat loss by evaporation can exceed the rate of heat gain by the net radiation. Similar situations must inevitably also prevail in many other locations where hot dry desert air advects over moist surfaces of less desert-like surroundings. There is a scale limit to the anomalous surface heat budget, however, in that the lower atmosphere's temperature and humidity will be progressively modified the longer that it is in contact with the moist and cool surface of the oasis. Thus, the oasis effect will be largest on the upwind side where the air is driest and hottest. This and other aspects of the oasis effect are discussed by Gay and Bernhofer (1991), Prueger *et al.* (1996), Lang *et al.* (1974), de Vries (1959), Yinqiao and Youxi (1996), Holmes (1969), Slatyer and Mabbutt (1965), Sokolik (1961), Tsukamoto *et al.* (1992), and Philip (1960).

Another unusual characteristic of surface energy budgets that is occasionally observed surrounding oases in deserts is a downward flux of water vapor near the surface. Outside of arid areas, the surface is generally a source of water vapor during the day and the flux is positive (upward). In arid areas, the vapor flux may be small because of the relatively dry surface, but it is still generally upward. However, when oases with moist surfaces, or nearby water bodies, create a humid boundary layer that moves over typically dry surfaces downwind, the low-level water-vapor flux has been measured to be downward. This situation has been observed by Wang and Mitsuta (1990, 1992) for a location in the Gobi Desert where oases are located at a distance of about 2 km downslope to the north. During the daytime heating of the surface, thermally forced northerly winds move upslope and advect moist boundary-layer air from the oasis to the observation site. The vertical flux measurements at 2.5 m AGL for this site indicate a downward transport of latent heat that exceeds 25 W m^{-2}. Tsukamoto (1993) shows that this flux estimate is about 20% too large, but the direction is still downward. In the night, the wind becomes downslope toward the oasis, and the surface water-vapor flux returns to near-zero values. Such downward fluxes have also been observed in another desert area of China (Harazono *et al.* 1992). Exactly what is happening in these situations is yet to be determined.

The effect of mineral aerosols on the desert radiation budget

Atmospheric mineral aerosols, or dust, have multiple physical effects on the radiation and energy budgets: there is increased atmospheric reflection, absorption, and scattering of incoming solar radiation, and increased absorption and re-emission of outgoing terrestrial infrared radiation. The net effect of the dust should be placed in the context of the radiation balance in general, and that of deserts in particular. Charney (1975) points out the general rule that the annual sum of the incoming solar radiation at the top of the atmosphere in low latitudes is greater than the annual sum of the outgoing infrared radiation, leading to a net gain from radiation. At high latitudes, there is a negative balance with more outgoing than incoming top-of-the-atmosphere radiation. The oceans and atmosphere transfer the excess low-latitude energy poleward to balance the deficit there. But Charney (1975) also shows a specific example of the satellite-estimated net radiation at the top of the desert atmosphere, where the outgoing exceeds the incoming (a deficit) over the Sahara and Arabian Deserts. The expected excess prevailed at all the other non-desert longitudes in this latitude belt. Because there are no significant non-radiative, local sources of heat in the desert, such as latent heating from condensation, this radiative deficit implies that heat must have been transported into the desert laterally through advection in order for an energy balance to exist.[4.3] The radiative deficit observed over the desert was presumably related to the strong surface emission of infrared that resulted from the high surface temperature, the large fraction of the solar radiation reflected from the high-albedo surface, and a relatively transparent atmosphere.

In contrast to the above observation of a net radiative cooling over the desert, there is evidence that the presence of dust in the desert boundary layer changes the balance such that there is a net radiative warming. The dust, generally confined to the mixed layer below 600–800 mbar, has been observed to double the short-wave absorption (Ackerman and Cox 1982). This heating by the dust overcomes the local radiative deficit of the clear desert atmosphere observed by Charney (1975), producing either a balance of incoming versus outgoing radiative energy, or a slight excess (Smith *et al.* 1986b). With respect to the vertical distribution of the dust effects on the energy budget, Tegen *et al.* (1996) show a net cooling at the surface with heating above in the dust layer. This would stabilize the temperature profile. When the dust layer was elevated above the surface over the Arabian Peninsula, Mohalfi *et al.* (1998) showed the existence of a 50–150 m deep, surface-based temperature inversion below the dust layer, but neutral to unstable conditions within the dust layer.

The particle-size distribution, the vertical distribution of the particles, and the mineralogical composition of the dust have been shown to be fundamental to the nature of the radiative effects (Tegen and Fung 1994; Sokolik and Golitsyn 1993; Sokolik and Toon 1996). Fig. 4.19 shows a model-based estimate of the vertical distribution of the effect of dust on the radiative heating rate in the lower atmosphere. The dust layer extended from the surface to 3 km, the solar zenith angle was small, and the characteristics of both Afghanistan and Saharan dust were employed. The solar and infrared radiative effects are shown separately and in combination. The effect of the dust on the solar was greater heating, and the effect on the infrared was greater cooling. The mineral aerosols produced

---

[4.3] An energy balance must exist over an annual cycle unless there is some interannual cooling or warming trend.

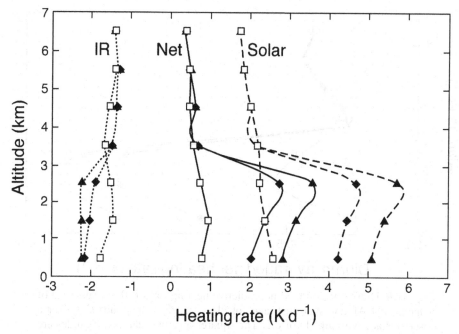

Fig. 4.19 Solar (long-dashed lines), infrared (short-dashed lines), and net (solid lines) radiative heating rates of Saharan and Afghanistan dust, and for clear (no dust or cloud) sky, over a desert surface with albedo of 0.3 and a high sun angle. The squares are for clear sky, triangles are for Saharan dust, and diamonds are for Afghanistan dust. The dust, with an optical thickness of 0.5 at 0.5 μm, extends from the surface to 3 km. (From Quijano *et al.* 2000.)

a net heating in the layer where they were located, with the Saharan dust producing greater net heating than the Afghan dust because of its lower albedo. Given the greater net heating near the top of the dust layer, the effect of the dust is to stabilize the temperature lapse rate near the surface. When the net heating rate is computed for an entire diurnal cycle, Fig. 4.20 shows that the dust effect is still one of net warming in the dust layer, relative to a clear atmosphere. The model simulations also showed that the existence of the dust caused a more positive, or less negative, top-of-the-atmosphere radiative forcing, and that the mineralogical composition of dust could determine the sign of the radiative forcing. That is, the more strongly heated Saharan dust could lead to a positive radiative forcing, whereas Afghan dust might not be sufficient to change the forcing from negative to positive. Additional discussions of the effects of mineral aerosols on the desert radiation budget can be found in Ackerman and Chung (1992), Sokolik *et al.* 2001, Carlson and Benjamin (1980), and Chou *et al.* (1992).

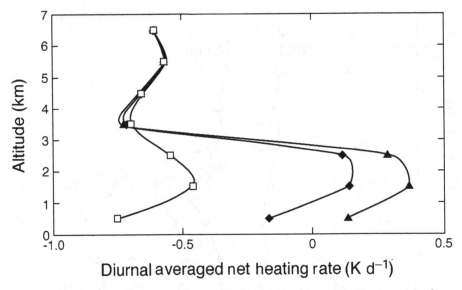

Fig. 4.20 Diurnally averaged net radiative heating rates (30° N, summer) of Saharan and Afghanistan dust, and for clear (no dust or cloud) sky, over a desert surface with albedo of 0.2. The squares are for clear sky, triangles are for Saharan dust and diamonds are for Afghan dust. The dust, with an optical thickness of 0.5 at 0.5 μm, extends from the surface to 3 km. (From Quijano *et al.* 2000.)

Suggested general references for further reading

Brutsaert, W. H., 1982: *Evaporation Into the Atmosphere: Theory, History and Applications* – even though, as the title suggests, the emphasis here is on moisture fluxes, there is a good general treatment of the surface–atmosphere energy exchange.

Budyko, M. I., 1958: *The Heat Balance of the Earth's Surface* – a classic reference on the subject of the radiation and surface energy budgets.

Monteith, J. L., and M. H. Unsworth, 1990: *Principles of Environmental Physics* – the focus is the micrometeorology of plant and animal life, but the introductory chapters thoroughly deal with energy transport by radiative and other processes.

Oke, T. R., 1987: *Boundary Layer Climates* – a good summary is provided of surface energy and mass exchanges. A relatively qualitative approach is employed, even though many laws of radiation physics are explained.

Stull, R. B., 1988: *An Introduction to Boundary Layer Meteorology* – a general reference on boundary-layer physics, which contains a treatment of the surface energy and radiation budgets.

Questions for review

(1) Explain the concept of net radiation.
(2) What is the difference between the energy conservation equations 4.7 and 4.9?
(3) Describe the typical energy balance within, and downwind of, oases within arid areas.

(4) Summarize the effects of the following on surface and atmospheric energy budgets.
- clouds
- atmospheric water vapor
- surface albedo
- surface moisture
- vegetation
- atmospheric dust

(5) Explain the ways in which the radiation and energy budgets differ for arid and non-arid areas.

(6) Explain why the sensible heating of the ground in two of the plots in Fig. 4.16 has a maximum in the morning.

Problems and exercises

(1) The largest net loss of energy from Earth's surface to the atmosphere is through evaporation, based on the global-average conditions shown in Fig. 4.14. If Earth's surface was all desert landscape, with no evaporation, make a rough estimate of how much warmer the surface would be. What assumptions did you have to make?

(2) Apply Eqn. 4.2 to confirm the wavelengths of maximum energy emission by Earth at 300 K and the sun at 6000 K.

(3) It is noted in this chapter that the global-average heat flux on a horizontal surface at the top of the atmosphere is 342 W m^{-2}. Show how this is obtained. Assume that the solar constant[4.4] is 1368 W m^{-2}.

(4) At what times of the day on the solstices is the intensity of the direct solar radiation changing the most rapidly?

(5) In simplifying Eqn. 4.7 to obtain the annual-average energy-conservation equation for deserts, it is assumed that LE is very small compared to the other terms, and that the annual-average G is near zero. Are these generally reasonable assumptions? When might they not be reasonable?

(6) Figure 4.8 shows that the summer solar-energy supply on a horizontal surface (one that is parallel to Earth's surface) at the top of the atmosphere is very similar for the latitudes of subtropical deserts and polar regions. Why is the desert so much warmer?

(7) Assume that the sun is 30° above the southern horizon (zenith angle of 60°), and that the flux on a surface perpendicular to the incident beam of light is S (W m^{-2}). What is the intensity on a horizontal surface? Assume that a sand dune has an east–west orientation, and north and south slopes of 20°. What is the intensity of the sun's energy on the north and south slopes?

(8) Speculate on why summer temperature maxima for mid-latitude cold deserts are as high as for lower-latitude hot deserts, but winter temperatures are much lower for cold deserts.

[4.4] The solar constant is the solar flux on a surface that is perpendicular to the sun's radiation at the top of the atmosphere.

5

Surface physics of the unvegetated sandy desert landscape

It was as if we were on a wholly lifeless planet. If one strayed behind a moment and let the caravan out of one's sight, a loneliness could be felt in the boundless expanse such as brought fear even in the stoutest heart. And the deeper we penetrated into the sand ocean, the stronger this feeling was. If the wind or a storm be a sign of life, the lack of it, however annoying it had been, had an almost crushing effect. Nothing but sand and sky; . . . all else is dead.

Gerhard Rohlfs, German explorer
Briefe aus der Libyschen Wüste (Reports From the Libyan Wasteland)
(Karl Alfred von Zittel 1875)

I was standing there wholly alone, my body the center of a vast empty disk. The horizon around me formed one uninterrupted circle. The sky was a glaring colorless brightness, with not a cloud to be seen. Aside from me and the ground on which I stood, there was nothing but the brilliantly white, shimmering disk of the sun . . . An indescribable sense of loneliness and forsakenness overpowered me. Suddenly I no longer seemed to have any sense of spatial or temporal dimensions. Here, where there were no visible standards of measurement, I felt as if I had lost all the inner standards that gave me an awareness of time and place. An infinite distance appeared to separate me from the nearest living being, and it would take me forever, it seemed, to cover the distance.

Uwe George, German naturalist and desert explorer
In the Deserts of This Earth (1977)

Processes that involve the movement of heat and water within the plant canopy and the ground beneath it are important to the study of regional meteorology because the land surface has an intimate interaction with the atmosphere. Through the movement of heat and water across the land–atmosphere interface, properties of the land surface such as temperature and wetness are felt by the atmospheric boundary layer and the free atmosphere above. The study of this heat and water exchange within the substrate and the plant canopy is the province of land-surface physics. This chapter first provides an introduction to the principles

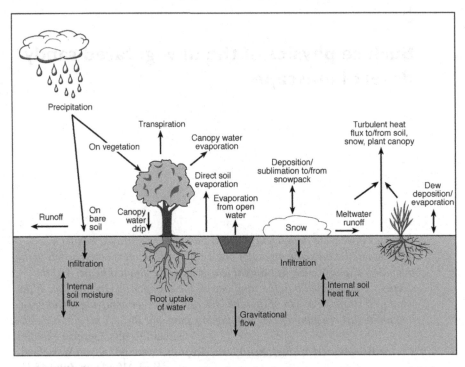

Fig. 5.1 Schematic showing land-surface physical processes that are associated with the movement of heat and water within the substrate and near the surface. (Adapted from Chen and Dudhia 2001.)

of land-surface physics, and then describes the processes by which heat and moisture are transferred within, and at the surface of, unvegetated, sandy desert substrate: classic conditions of an extreme desert. Extensions to these concepts, which are required by the existence in the desert of vegetation and different substrate conditions, will then be discussed in the following two chapters. The effect of the landscape on the frictional stress felt by the air moving over it is more the subject of boundary-layer meteorology than of land-surface physics, so most of the discussion of this topic will be deferred until Chapter 10.

Introduction to the concepts of land-surface physics

Perhaps the best approach to developing an appreciation for the breadth of the topic of land-surface physics is through the diagram in Fig. 5.1 of the various physical processes that must be dealt with. The processes involve heat transfers and the movement and transformation of water in its various forms. Within the substrate, there are the following processes.

Box 5.1
Capillary and surface-tension effects in soils

Most of us have experienced capillary effects without perhaps having been aware of the name. For example, if we immerse a soda straw in a liquid, the liquid will rise a short distance in the straw. If we have a smaller-diameter tube of glass, often called a capillary tube, the liquid will rise even farther. This water movement is a consequence of surface-tension effects that result from imbalances in molecular forces near the surface of a liquid. Now, imagine soil with small pore spaces filled with air. Liquid water will be drawn into these narrow spaces through the same forces that draw it up the capillary tube against the force of gravity. And, as with the straw and the capillary tube, the water will rise farther through small soil pores than through large ones. For example, the capillary rise of water from a shallow water table will be inhibited by shallow layers of sand because the sand has large pore spaces.

- Liquid water is transported downward through gravity drainage and in all directions through **capillary** effects (Box 5.1). Water also can rise and fall through changes in the water table.
- Water vapor moves vertically through the air spaces by convection and molecular diffusion.
- The roots of vegetation draw water from the substrate within the root zone.
- The substrate water freezes and thaws, with the release and consumption of the **latent heat of fusion**.
- Evaporation and condensation occur, with the consumption and release of the latent heat of condensation.
- Heat is conducted.

At the interface between the substrate surface and the atmosphere, there are the following exchanges.

- Rainwater, snowmelt water, irrigation water, and dew enter the substrate.
- Water from the substrate evaporates and **sublimates** into the atmosphere.
- Heat is exchanged between the atmosphere and the substrate.
- Liquid water passes from the underground roots to the above-ground stems and leaves.

Immediately above the interface, the following processes are important.

- Rain falls on the bare ground and vegetation.
- Water drips from vegetation onto the bare ground or onto other vegetation.
- Snow accumulates on the bare ground and vegetation.
- Snow and frost melt and sublimate, consuming heat.

- Dew and frost form on the bare ground and vegetation, releasing latent heat.
- Fog deposits on the bare ground and vegetation.
- Water evaporates from the leaf surfaces of vegetation, and transpires from vegetation, with the consumption of heat.

The following sections will review some of the physical principles that control the above processes.

Vertical heat transport within the substrate

Vertical heat transport within substrates is mostly through conduction (i.e. molecular diffusion), even though convective and advective movement of air can transport heat when the porosity (percent air space) is high. Because the substrate is a medium potentially consisting of solid, liquid, and gaseous phases, the thermal conductivity will depend on the proportions and characteristics of these components. The direction of the conductive heat transfer is from higher temperature to lower temperature, and the magnitude of the heat flux is proportional to the temperature gradient. Mathematically,

$$Q_s = -k_s \frac{\partial T_s}{\partial z}, \tag{5.1}$$

where Q_s is the heat flux in the soil (positive upward), k_s is the soil *thermal conductivity* (W m^{-1} K^{-1}), z is distance on the vertical axis (positive upward), and T_s is the soil temperature. That is, the heat flux is proportional to the temperature gradient multiplied by a factor that reflects the ability of the substance to transfer heat. This is called a flux-gradient form of equation. The negative sign indicates that the flux is in the direction of lower temperature. The thermal conductivity is formally defined as the quantity of heat that flows through a unit cross-sectional area per unit time, when there exists a temperature gradient of one degree per meter perpendicular to the cross section. Chapter 8 lists the thermal properties of different desert substrates, and Table 5.1 contrasts values for a variety of different natural materials. In general, a soil consists of solid particles, water, and air spaces, and the relative contributions from the conductivities of these three components determines the total soil conductivity. The existence of water in the soil dramatically increases the conductivity, not only because the conductivity of water is high, but because the water displaces the air, which has an especially low conductivity (i.e. air is a good thermal insulator). Specifically, the conductivity of air is about two orders of magnitude lower than that of wet soil or rock (Table 5.1). Thus, tabulated values of conductivity should specify the **soil-moisture content** and the porosity of the soil. Fig. 5.2 shows qualitatively how the conductivity and other thermal properties of soil depend on the soil-moisture content.

Table 5.1 *Thermal properties of natural materials*

Material	Remarks	Heat capacity, C ($J m^{-3} K^{-1} \times 10^{6}$)	Thermal conductivity, k ($J s^{-1} m^{-1} K^{-1}$)	Thermal diffusivity, K ($m^2 s^{-1} \times 10^{-6}$)	Thermal admittance, μ ($J m^{-2} s^{-1/2} K^{-1}$)
Sandy soil	Dry	1.28	0.30	0.24	620
(40% pore space)	Saturated	2.96	2.20	0.74	2550
Clay soil	Dry	1.42	0.25	0.18	600
(40% pore space)	Saturated	3.10	1.58	0.51	2210
Peat soil	Dry	0.58	0.06	0.10	190
(80% pore space)	Saturated	4.02	0.50	0.12	1420
Rock	Granite	2.13	2.71	1.27	2402
Snow	Fresh	0.21	0.08	0.10	130
	Old	0.84	0.42	0.40	595
Ice[a]	0 °C, pure	1.93	2.24	1.16	2080
Air[a]	10 °C, still	0.0012	0.025	21.50	5

[a] Property depends on temperature.
Source: From List (1966).

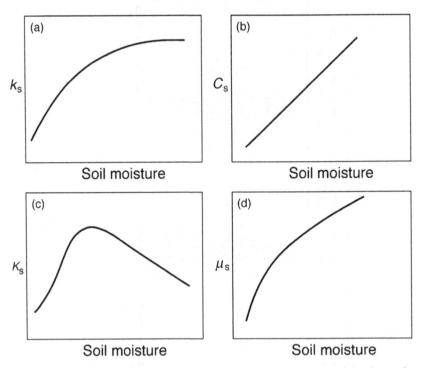

Fig. 5.2 General relation between soil-moisture content and (a) thermal conductivity, (b) heat capacity, (c) thermal diffusivity, and (d) thermal admittance for most soils. (From Oke 1987.)

Even though tabulated values of thermal conductivity do not generally reflect its dependence on temperature, this should be accounted for in desert soils because of the large near-surface diurnal temperature variation. Sepaskhah and Boersma (1979) measured soil thermal conductivity at 25 °C (77 °F) and 45 °C (113 °F), a typical range for desert conditions, and showed the conductivity to be three times larger at the higher temperature when the moisture content was moderate.

Another important physical property of a substrate is the **thermal**, or **heat**, **capacity** $(C, \mathrm{J\,m^{-3}\,K^{-1}})$, that describes how much heat (J) is required to raise the temperature of a unit volume $(\mathrm{m^{-3}})$ by one degree (K). As with the conductivity, the value of the soil heat capacity (C_s) depends on the fractions of the soil solids, water, and air. Air has a low heat capacity, so displacing air with water, as soil is moistened, raises the heat capacity of the air–water–solid mixture (Fig. 5.2b). Thus, more heat is required to raise the temperature of a moist soil than of a dry soil. A related quantity is the specific heat $(c, \mathrm{J\,kg^{-1}\,K^{-1}})$, which is the amount of heat required to raise the temperature of a unit mass (kg) by one degree (K). Thus, it is equal to the heat capacity divided by the density of the substrate

($c = C/\rho$). The heat capacity and the specific heat are sometimes referred to as the soil's "thermal sensitivity."

A quantity that is related to the heat capacity and the thermal conductivity is the thermal diffusivity ($K_{Hs} = k_s/C_s$). We will see that it determines the speed with which a temperature change propagates through a medium such as soil. Imagine the surface of the substrate heating up during the daytime heating cycle, which creates an upward temperature gradient immediately below the surface. Eqn. 5.1 predicts a downward heat flux that is directly proportional to the soil's thermal conductivity. Because the temperature gradient and therefore the heat flux are still small some distance below the surface, there would be a heat-flux convergence between the surface and that level. That is, the downward heat flux into the layer from the top is greater than the downward heat flux out of the layer at the bottom. This will raise the temperature of the soil in the layer in inverse proportion to its heat capacity, as shown in Eqn. 5.2.

$$\frac{\partial T}{\partial t} = -\frac{1}{C_s}\frac{\partial Q_s}{\partial z}. \tag{5.2}$$

In the above example, Q_s just below the surface has a large negative value, and is less negative with greater depth. The vertical derivative is thus negative, providing for a temperature increase. Combining Eqns. 5.1 and 5.2 gives

$$\frac{\partial T}{\partial t} = \frac{1}{C_s}\frac{\partial}{\partial z}\left(k_s\frac{\partial T_s}{\partial z}\right) = \frac{k_s}{C_s}\frac{\partial^2 T_s}{\partial z^2} = K_{Hs}\frac{\partial^2 T_s}{\partial z^2}, \tag{5.3}$$

where the simplification has been made that k_s does not change with depth. Here we see that the rate of temperature change is proportional to the diffusivity and the second derivative of temperature with respect to depth. Fig. 5.3 shows a schematic of the idealized temperature distribution immediately above and below the air–ground interface, for both nighttime and daytime conditions. During the day, the curvature of the temperature profile below the surface is such that the second derivative in Eqn. 5.3 is positive and the temperature increases. When the curvature reverses at night, the substrate cools. Note that a constant rate of temperature change with depth (a straight, sloping line in Fig. 5.3) would produce the same flux everywhere (Eqn. 5.1), and no temperature change (Eqn. 5.2).

Figure 5.2 shows that diffusivity is directly proportional to soil moisture, for low soil moisture, because the conductivity increases faster with increasing soil moisture than does the heat capacity. When the conductivity curve develops less slope than does the heat capacity curve at higher soil moistures, the diffusivity begins to decrease.

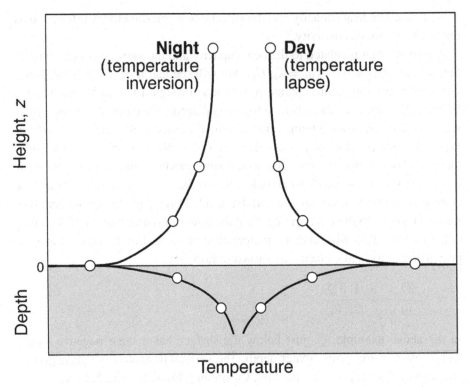

Fig. 5.3 Idealized vertical profiles of temperature in the soil and the atmosphere near the surface. (Adapted from Oke 1987.)

Microclimate conditions are less extreme for high soil diffusivities. Where the heat gained during the day at the surface can penetrate rapidly and deeply, the surface in contact with the atmosphere is lower in temperature and the atmosphere is heated less by it. At night, the heat stored in the deep substrate reservoir can analogously force an upward-moving temperature wave that maintains the surface temperature at relatively high values. Thus, high soil diffusivities allow for smaller day and night temperature extrema near the surface, in both the atmosphere and the substrate. See Table 5.1 for examples of thermal diffusivities.

Another way of understanding how the diurnal temperature wave at and below the surface is related to substrate properties is through the concept of thermal admittance, which is a property of the interface between two media (e.g. the substrate and the atmosphere). It is defined as $\mu_s = (K_s C_s)^{1/2}$, and is a measure of the ability of a surface to accept or release heat. Consider your feet in contact with a tile floor that has a temperature that is lower than your skin temperature. The temperature gradient at the skin–tile interface causes a flux from you to the tile, and a critical factor that determines whether the surface "feels" cold to you is whether the tile surface warms very rapidly to your skin temperature, or whether it remains colder. In the case of the tile, the conductivity is large so that the heat it gains from your body is conducted rapidly from the surface to the tile material below. Thus, the surface temperature does not rise rapidly, as it would if you were standing on a wood floor having a lower conductivity (i.e. wood is a better thermal insulator). Analogously, if the material has a high heat capacity, its temperature is not going to rise rapidly in response to heat input from your skin. Thus, high conductivity and high heat capacity (i.e. large admittance) contribute to sustaining a large heat transfer across an interface because the temperature contrast is maintained. Of course, the temperature response of the second medium is equally important, so in the case of the surface–air interface, the admittance of the air must also be considered.

Thus, when the surface in contact with the atmosphere has high admittance (i.e. high conductivity and/or high heat capacity), the temperature of the surface does not increase as much during the daytime as it would for a low-admittance surface. This has implications for the daytime surface energy budget in terms of smaller sensible heat fluxes to the atmosphere (a cooler boundary layer) and weaker long-wave emission from the ground. Miller (1981) provides an illustration of how contrasting substrate admittances influence surface fluxes. When a cold air mass ($-17\,°C$) moved over an area near Leningrad, the surface flux was 45 W m^{-2} over frozen bare soil with an admittance of 1000 J m^{-2} K^{-1} s$^{-1/2}$, but was only 15 W m^{-2} over an adjacent snow-covered surface with an admittance of about 330 J m^{-2} K^{-1} s$^{-1/2}$. The high admittance of the bare soil caused the substrate surface to remain much warmer than the air, and this air–surface temperature difference maintained a larger heat flux than prevailed over the snow surface with the smaller admittance. Again, see Table 5.1 for examples of thermal admittances.

The amplitude and time lag of the temperature wave that propagates downward into the substrate during the day also depend on the conductivity and the specific heat. The amplitude of the daily temperature oscillation is greater for large conductivities and for small heat capacities. That is, the amplitude is proportional to the diffusivity, $K_{Hs} = k_s/C_s$, such that

$$(\Delta T_s)_z = (\Delta T)_0 \exp(-z(\pi/K_{Hs}P)^{1/2}), \tag{5.4}$$

where z is the depth below the surface, exp is the base of natural logarithms, P is the wave period (24 h is the dominant period in this discussion of the diurnal temperature wave), $(\Delta T)_0$ is the amplitude of the temperature wave at the surface ($z = 0$), and $(\Delta T_s)_z$ is the amplitude at depth z. That is, the amplitude of the diurnal temperature oscillation decreases exponentially with depth. Fig. 5.4a

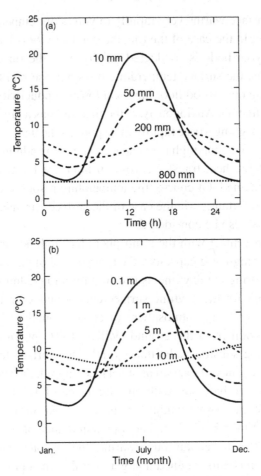

Fig. 5.4 Idealized (a) diurnal and (b) annual cycles of temperature at different depths within the substrate. (From Oke 1987.)

shows idealized diurnal temperature variations at different depths on a cloudless day. The time lag with depth shown in the figure is defined by

$$(t_2 - t_1) = \frac{(z_2 - z_1)}{2}(P/\pi K_{Hs})^{1/2}, \tag{5.5}$$

where t_1 and t_2 are the times that a temperature-wave maximum or minimum reaches depths z_1 and z_2, respectively. In other words, the difference is the time required for the temperature wave to pass from one depth to the other. Other variables have the same definition as before. The temperature wave travels faster in substrates with higher thermal conductivity and with lower heat capacity. The time lag means that the near-surface substrate can be cooling while warming continues a short distance below. At some depth, the curves are out of phase so that the time of the temperature maximum at that depth corresponds to the temperature minimum at the surface. Eqns. 5.4 and 5.5 apply also to the annual cycle, where the period corresponds to that of the seasonal rather than the diurnal cycle (Fig. 5.4b). Eqn. 5.4 shows that, with a period of 365 days, the depth to which the thermal wave penetrates is about 14 times greater than for the diurnal period. If the diurnal thermal wave penetrates to 0.5 m with a particular amplitude, the annual wave will penetrate to 7 m with the same amplitude.

Vertical water transport within the substrate

There are two general ways for liquid water to enter the substrate between the water table and the surface. Water can move upward from the water table through capillary action (or the water table itself can rise). And water can enter the substrate from the surface by the process called **infiltration**. The efficiency of infiltration depends on a number of factors such as rainfall intensity, total rainfall amount during a storm, the physical composition of the substrate, and antecedent moisture. This rate of water movement into the substrate is important because it determines the potential for runoff and flooding, the amount of water near the surface that is available for evaporation, the availability of water for plants, and the extent of groundwater replenishment.

The upward and downward transport of water within soils can take place through five physical mechanisms that will be described in the following paragraphs: three apply to liquid water and two to water vapor. For liquid water, there are two forces that operate. One is gravity and the other is related to the surface tension between the soil particles and the water. It perhaps seems unusual to refer to surface tension as a force, but molecules at the surface of a liquid experience molecular forces that are not symmetrical and, therefore, not balanced like those experienced by molecules in the interior away from the surface. Consider that a volume of soil (or, more familiarly, a sponge) will not drain completely dry after being wetted. Eventually, the drainage rate will approach zero because surface-tension forces, which promote retention of water within the soil (or sponge), balance the gravity force. Water in the vapor phase can be transported through buoyancy forces when the vertical lapse rate of temperature becomes unstable. This is an effect of gravity again, because gravity is the force behind buoyancy. Also, molecular motion causes water vapor to diffuse from areas of higher to lower concentration, without the need for any movement of air on scales larger than the molecular.

In the first mechanism, liquid water can move vertically as a result of a change in the pressure head, which causes the water table to change. That is, a water surface must be in dynamic equilibrium with its surrounding fluid, so local water excesses or deficits compared to surroundings are reconciled through water movement that equalizes the pressure. A simple illustration of this effect is found in desert environments where a hydrologically closed basin, perhaps containing a salt flat, is surrounded by mountains. Rainfall over the mountains refreshes the water table below, which increases the water pressure and causes the ground water to move laterally (from high to low pressure) into the central basin until equilibrium is attained with a higher water table there. Analogously, extraction of water at a well site decreases the pressure there and causes inflow from the surroundings, lowering the water table over a wider area. This movement of water in response to pressure differences is forced by gravity; that is, the force of gravity is responsible for static pressure within a fluid. (Static pressure is related to the weight of the fluid above a point, whereas dynamic pressure is a result of fluid movement.)

Secondly, liquid water can move through soils by capillary action. This movement results from surface-tension effects between the water and soil particles. For example, water can move from the water table into the dry layer above through capillary effects. The capillary-rise layer has a lower bound at the water table and an upper bound that depends on soil properties. In general, capillary effects contribute to the spread of water from wetter to dryer soil. These surface-tension forces that bind the water to the soil particles are determined by the soil porosity and the soil moisture itself. The more porous the soil and the more dry the soil, the weaker are the surface-tension effects. For example, capillary movement of water can be blocked by a layer of open-textured soil (e.g. coarse sand or gravel) or by a dry layer of soil.

Thirdly, downward liquid-water transport between the substrate particles is forced by gravity, where this water is supplied through infiltration: the entry at the surface of water from rain, snow melt, or irrigation. The infiltration rate is limited by the rate of soil water movement below, called **percolation**, with any excess running off laterally or ponding at the surface. The efficiency of this type of soil-water movement is important to groundwater recharge; the evaporation rate; runoff, erosion and flood production; the availability of water for plant uptake; and chemical changes such as salinization. The rate of this downward water movement, defined as the hydraulic conductivity, is controlled by the surface-tension effects between the soil particles and the water. The force of gravity draws the water downward, but surface-tension forces between the water and soil particles promote retention of the water. This latter effect of water retention is quantified in terms of the soil moisture potential, which can be visualized as the amount of energy necessary to extract water from the soil matrix. Tight soils such as clay have a high potential, or water-retention capacity, compared with sand. Also, dry soils have a higher potential than wet soils, i.e. it takes less energy to extract a unit of moisture from a wet soil than from the same soil after it has become drier. Thus, the hydraulic conductivity is greater when the soil is wet and porous.

The amount of liquid water in soils is generally defined in terms of the **soil-moisture content**, which is the percent of the volume of a soil that is occupied by water. The upper limit of the soil-moisture content is determined by the porosity. Coarse-textured soils, such as sand, tend to be less porous than fine-textured soils, even though the mean pore size is greater in the former. The temporal change in the moisture content at any point in a soil can be represented by an equation that is similar to Eqn. 5.2, which expresses the local temperature change in terms of the difference in the vertical heat fluxes into and out of a layer (as represented for a point by the vertical derivative of the flux). Analogously, for soil moisture the vertical derivative of liquid-water fluxes must be represented. If the soil-moisture flux toward a point is greater than the flux away from it, the soil moisture increases, and visa versa. The following equation expresses local changes with time in soil-moisture content (Θ) as a result of vertical variations in the flux (q). The term $D_\Theta \partial\Theta/\partial z$ is associated with capillary (i.e. surface tension) water movement and K_Θ represents gravity-forced water movement. These are the second and third mechanisms described above, respectively.

$$\frac{\partial\Theta}{\partial t} = \frac{\partial q}{\partial z} = \frac{\partial}{\partial z}\left(K_\Theta + D_\Theta \frac{\partial\Theta}{\partial z}\right) = \frac{\partial K_\Theta}{\partial z} + \frac{\partial}{\partial z}\left(D_\Theta \frac{\partial\Theta}{\partial z}\right). \quad (5.6)$$

Here, Θ (dimensionless) is the volumetric soil water content, q is a vertical volume flux of liquid water (volume per unit area per unit time (m s^{-1}, positive upward), K_Θ (m s^{-1}) is hydraulic conductivity and D_Θ (m^2 s^{-1}) is soil-water diffusivity. The subscripts on K and D imply their dependence on Θ. The terms conductivity and diffusivity have been borrowed from the equations for the molecular diffusion and conduction of heat, which have terms similar in form to those above. Unfortunately, this terminology does not reflect the actual physical processes that are represented in the equation. The hydraulic conductivity and soil-water diffusivity can be expressed as

$$K_\Theta = K_{\Theta_s}(\Theta/\Theta_s)^{2b+3}, \quad \text{and} \quad (5.7)$$

$$D_\Theta = -(bK_{\Theta_s}\Psi_s/\Theta)(\Theta/\Theta_s)^{b+3}, \quad (5.8)$$

respectively, where K_{Θ_s} is the saturation hydraulic conductivity, Θ_s is the saturation volumetric soil moisture content, Ψ_s is the saturation soil moisture potential (a negative number), and b is an empirically defined coefficient (Ek and Cuenca 1994). All of these quantities are a function of the soil type.

The above mechanisms apply to liquid-water movement. However, water vapor can also move vertically above the water table through porous, dry soil, by the two mechanisms noted above: convection and vapor diffusion. Convection requires that the temperature of the soil, and that of the air within the soil, decrease upward in the soil more rapidly than the dry adiabatic lapse rate that is required for the triggering of buoyant motion. The flux of water by vapor diffusion is proportional to the gradient of the water-vapor content of the air within the soil.

Liquid-water transport within vegetation, and transpiration

Vegetation is important to the moisture budget because the roots access shallow and deep moisture that is not otherwise directly available for evaporation at the surface. This moisture is transferred through the xylem up the stems to the leaves, where it evaporates within the intercellular spaces of the leaves and is released through the stomata into the atmosphere. The latent heat consumed in this process is provided by the foliage, in contrast to evaporation from bare ground where the latent heat comes from the substrate. In either case, the energy loss is part of the surface energy budget.

The rate of transpiration from vegetation depends on many factors, including vegetation type and density, atmospheric humidity, time of day, season, and the degree of heat and water stress to which the vegetation has been subjected. There has been considerable historical controversy about the dependence of the transpiration rate on soil moisture. A wilting-point value of soil-moisture content has been defined as the limit below which the vegetation permanently wilts and transpiration ceases. This is a convenient concept, but ignores the fact that the moisture content within the root zone is not uniform, and that different co-existing vegetation types have greatly different tolerances for soil dryness. Field capacity is another threshold on the soil-moisture scale with implications for vegetation. It is defined as the moisture value below which internal drainage ceases. That is, for a soil-moisture content that is less than the field capacity, the soil would retain the moisture and none would drain downward. This is another concept that has gained popularity because of its simplicity rather than its strict accuracy. Some have used the assumption that water is equally available to vegetation for any moisture value above the wilting point. Others have assumed that the vegetation is under stress for wetnesses between the wilting point and field capacity, with a transpiration rate that is dependent on soil moisture, and that only for wetnesses above field capacity is there no longer a stress.

Heat and water-vapor exchange between the surface and the atmosphere

It was mentioned previously that the vertical transfer of sensible heat at the substrate–atmosphere interface occurs through conduction. This takes place within a very shallow layer of atmosphere, called the laminar (non-turbulent) sublayer, having a depth ranging from a few molecules to, at most, a few millimeters. Above this layer, the transfer is through turbulent eddies of air. This turbulence does not contribute to the flux at the surface because the eddies cannot exist there, where the velocity normal to the surface must be zero.

Because all non-radiative transfer of heat at the surface is through conduction, the heat flux can be represented by the same sort of flux-gradient relation employed to represent heat transport by conduction within the substrate (Eq. 5.1). A similar expression is used for the vapor flux. That is

$$H = -C_a k_{Ha} \frac{\partial T}{\partial z}\bigg|_0 = -\rho c_p k_{Ha} \frac{\partial T}{\partial z}\bigg|_0 \quad \text{and} \tag{5.9}$$

$$LE = -\rho L k_{Wa} \frac{\partial q}{\partial z}\bigg|_0 = -\rho L k_{Wa} \frac{\partial q}{\partial z}\bigg|_0, \tag{5.10}$$

where C_a is the heat capacity of the atmosphere, c_p is the specific heat at constant pressure of the atmosphere, k_{Wa} and k_{Ha} are the diffusivities of water vapor and heat in the air, respectively, q is specific humidity, and the vertical derivatives are evaluated within the laminar sublayer near the surface. These near-surface fluxes are virtually the same as the fluxes a meter or so above the surface where the turbulence is responsible for the transfer. If it is assumed that the same types of equation can be applied to turbulent transfer as are used for molecular transfer, the above equations can be rewritten with the molecular diffusivities replaced with eddy diffusivities (e.g. K_{Ha} instead of k_{Ha}):

$$H = -C_a K_{Ha} \frac{\partial T}{\partial z}\bigg|_0 = -\rho c_p K_{Ha} \frac{\partial T}{\partial z}\bigg|_0 \quad \text{and} \tag{5.11}$$

$$LE = -\rho L K_{Wa} \frac{\partial q}{\partial z}\bigg|_0 = -\rho L K_{Wa} \frac{\partial q}{\partial z}\bigg|_0. \tag{5.12}$$

A challenge to applying these equations is that the exchange coefficients are functions of distance from the surface and static stability, varying by over three orders of magnitude from day to night. Alternative expressions for H and LE can be obtained if we vertically integrate these equations with the assumption that the fluxes do not vary much with height within the first couple of meters. The resulting expressions are

$$H = \rho c_p D_H (T_s - T_a) \quad \text{and} \tag{5.13}$$

$$LE = \rho L D_W (q_{s,sat}(T_s) - q_a), \tag{5.14}$$

where D_H and D_W are transfer coefficients that are integral functions of K_{Ha} and K_{Wa}, T_s is the temperature of the surface, and T_a and q_a are the temperature and specific humidity, respectively, of the air at a specified height near the surface. The value of the specific humidity at the surface, q_s, is equal to the saturation value, $q_{s,sat}$, at the temperature T_s of any surface at which evaporation is occurring: water bodies, damp soil, leaf stomata.

Thus, the sensible- and latent-heat fluxes between the substrate and the atmosphere can be represented in terms of the differences between the temperature and humidity at and immediately above the surface. The direction of the fluxes depends on the sign of the difference, and the magnitude of the fluxes depends on the degree of the contrast between the conditions at the surface and above. The transfer coefficients are functions of factors that affect the intensity of the turbulence, such as the roughness of the surface, the vertical shear of the horizontal wind from which turbulent energy can be derived, and the vertical lapse rate of atmospheric temperature that determines whether turbulent energy is available from buoyancy. For example, evaporation rates (LE) are high when the atmosphere is dry (small q_a), the surface is warm (large $q_{s,sat}$,), and the near-surface wind speed is high (producing a large shear, and thus a large transfer coefficient, D_W).

Another way of visualizing the controls on the surface heat flux is through the concept of thermal admittance. Most of the earlier discussion about admittance was in the context of the substrate properties; however, it was pointed out that the admittance of the atmosphere on the other side of the interface is equally important in determining the heat fluxes. This atmospheric admittance is defined as $\mu_a = (k_a C_a)^{1/2} = (C_a K_{Ha})^{1/2}$, where K_{Ha} is the eddy diffusivity of Eqn. 5.11. For example, suppose that the surface is receiving solar radiation during the day. This heating of a thin layer of substrate and air at the interface will produce a temperature gradient within both the air and the substrate (see Fig. 5.3). The energy not lost by long-wave emission and evaporation will be partitioned between sensible heat fluxes into the atmosphere and the substrate in proportion to the relative admittances of the two media. Say, for example, that the substrate has a very low admittance because of poor thermal conductivity, but that the boundary layer has a large admittance because the turbulence is well developed and the eddy diffusivity is large. The heat flux into the soil (Eqn. 5.1) will thus be small in spite of the large $\partial T_s / \partial z$, but the heat flux into the atmosphere (Eqn. 5.11) will be large because the eddy diffusivity, K_{Ha}, is large. Thus, more of the radiant-energy input to the surface will be partitioned to the sensible heat flux to the atmosphere rather than to the substrate. Alternatively, at night the radiative cooling of the surface draws heat from the air and the substrate in proportion to their admittances. Because calm, near-surface winds and a stable profile of temperature with height mean that turbulence is weak and the eddy diffusivity is small, the atmospheric admittance is small at night and most of the surface heat lost to space by radiation is provided by the substrate rather than by the atmosphere. Methods of estimating atmospheric and substrate admittances are discussed in Novak (1986).

Land-surface physics of unvegetated sandy desert

The above basic concepts of land-surface physics apply to all climates, including arid ones. However, there are some idiosyncrasies associated with the surface physics of deserts, and these will be the focus of the remainder of this chapter, and the next two chapters.

Infiltration of rainfall, and substrate liquid-water movement

One of the factors than controls the infiltration rate, and the subsequent movement of the water within a sandy desert substrate, is the existence of fine particles in the size distribution of the sand. Finer particles inhibit the water movement by lowering the hydraulic conductivity. For example, infiltration rates were determined to be about three time larger on a sand dune in eastern Oman than in the bottom of a nearby dry sandy wadi (Agnew 1988). Compared with the dune sands, the wadi sands contained more fine particles, which were produced by abrasion when the water course flowed. Interestingly, the infiltration rate even varied by over 50% for different locations on the same dune, such as at the summit, the base, and the upwind and downwind faces. It is possible that the wind drifting graded the particle sizes across the dune's cross-section. The infiltration rates of $10\text{--}15$ mm min^{-1} measured for this dune are large, but are not greatly different than those found for sands in other locations such as the Sahara. Agnew and Anderson (1992) suggest that very high sustained rainfall intensities in excess of 100 mm h^{-1} would be required in order to initiate surface runoff in these conditions. There are numerous examples of lower values of infiltration rate for different sandy substrates. In a study of the northern Sonoran Desert in a regime of summer monsoon convection (Schreiber and Kincaid 1967), only about 7 mm of rain were required before runoff began. It was also determined that the average runoff depended most on rainfall quantity and relatively little on the existence of a vegetation canopy or antecedent soil moisture. Even though many desert substrates have infiltration rates that are much less than these for sandy conditions, it is estimated that only 10% of the rain events in arid areas produce any runoff (Slatyer and Mabbutt 1965). That is, over 90% infiltrates or evaporates.

Of the water that infiltrates a desert substrate, it is useful to consider the fraction that percolates to the water table to replenish it. Recall that the ability of the substrate to retain water is reflected in the field capacity. This was found to be quite low in the Oman sands, compared with most typical soils (3.6% of the water was retained by the sand), largely because of the coarse texture and the lack of organic matter. Agnew (1988) suggests that this low field capacity would allow the water to penetrate 8–11 m during a wet year with 300–400 mm of rain. If the sands were wet to start with, the rain might refresh the water table at 25–30 m. Dincer *et al.* (1974) measured the moisture content of the sands in the Dahna dune region in the Arabian Peninsula (annual rainfall of about 80 mm). After the spring rains, the top 20 cm was relatively dry with a moisture content of about 1.2% by mass. Below that, the water content was fairly uniform at 2.5–4.5% down to a depth of more than 3 m. By late summer, the drier layer had increased to 50 cm in depth. They state that, in dune regions, surface runoff can be neglected. As an example, near Riyadh, Saudi Arabia, no runoff was observed when 30 mm of rain fell on dune sand in about 15 min. For coarse dune sand with a mean particle diameter of 0.3 mm, 50 mm of rainwater penetrated down to 100 cm in 24 h, sufficiently far to escape evaporation and possibly recharge the water table. With finer sand particles having a higher field capacity, it was observed that water was retained by the sand in a shallower layer, and lost to evaporation.

The fact that the hydraulic diffusivity is dependent on soil moisture leads to some interesting properties of the process by which moist, unvegetated soils dry out. Assume that a layer of soil in contact with the atmosphere has an initially uniform wetness. Evaporation from the surface will begin to dry a shallow layer near the top. As this layer dries, its hydraulic diffusivity decreases, which reduces the upward transport of water from below through the development of a "vapor barrier" in the soil. This accelerates the rate at which the surface layer desiccates and the rate at which the hydraulic conductivity declines. Capehart and Carlson (1997) employ a model to show that this soil-moisture decoupling effect is greatest for sandy soils, and that the decoupling of the near-surface soil moisture from the moisture below occurs quickly, below a mid-range, soil-moisture threshold that depends on soil type. This has implications related to the surface energy and moisture budgets of barren, sandy deserts. The desiccated upper soil layer that is decoupled from the reservoir of moisture in the substrate below will have a low latent-heat flux to the atmosphere, and the low thermal conductivity of the dry upper layer will allow only a small penetration of heat into the substrate during the day. These factors will cause the surface temperature to be higher than if the surface substrate had access to the deeper moisture, and this higher temperature will cause the sensible heat flux to the

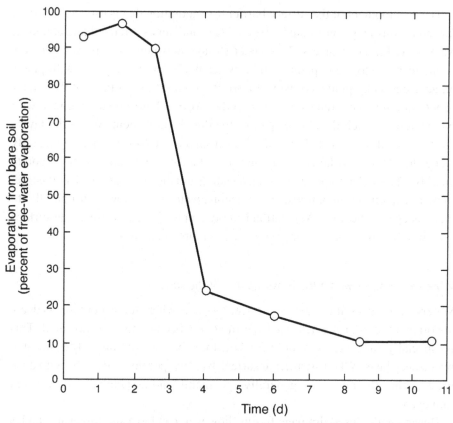

Fig. 5.5 Observed evaporation from bare soil in Alice Springs, Australia. (Adapted from Slatyer and Mabbutt 1965.)

atmosphere as well as the long-wave emission to be large. This rapid decrease in evaporation when the surface layer of the substrate becomes decoupled from the moisture below has been observed in deserts. Fig. 5.5 illustrates this for arid Alice Springs, Australia, where the evaporation rate remained approximately uniform for two days, and then dropped by almost 70% in little more than a day. This precipitous decrease resulted from the rapid decoupling of the surface from the subsurface soil moisture when a certain moisture (and hydraulic diffusivity) threshold was reached. The residual surface vapor flux after this time was likely a result of vapor transport through the dry layer from the evaporation zone below. The existence of any vegetation will enable water to be transported through the root system to the surface, bypassing the dry surface layer with low hydraulic conductivity and leading to a more rapid drying of the deeper layers within the root zone.

The development of the surface barrier to evaporation that is especially strong for sandy soils may contribute to the fact that sand dunes often contain abundant water stored in deep layers. This water storage between the rare rain showers is one reason why dune plants can grow relatively large compared with plants on adjacent rocky plains (Bowers 1986). Bagnold (1954) points out that water penetration into sand dunes depends on the slope of hard-crusted laminae with fine material, which develop as part of the dune's movement. When the layers are horizontal, the crusted surface is firm underfoot and water runs laterally along the compacted layer, remaining near the surface where it can evaporate quickly. The sand in these areas is unvegetated and the deep sand lacks moisture. In contrast, where the laminae slope steeply, water flows downward parallel to them, deep into the dune. Vegetation is more common in these areas, the surface sand is soft underfoot, and deep moisture is more abundant.

Water-vapor fluxes within the substrate and at the surface

Water-vapor fluxes at the desert surface, for areas where the water table is deep, are often difficult to characterize experimentally because they are so small. This is especially true for extreme deserts where the substrate is completely desiccated over a deep layer. When significant surface water-vapor fluxes do exist with deep water tables, the fluxes are generally ephemeral, and associated with a recent rain event.

Deserts with closed drainage basins, however, can have shallow water tables with dry porous (e.g. sandy) desert substrates above. Here, evaporation does not occur at the surface, but rather at some distance below it. If there is capillary rise of water with its source at the water table, the evaporation is at the upper boundary of this moist layer, at most 40 cm above the water table in sandy soils (Bagnold 1954). If there is no capillary rise, the evaporation is directly from the water table. The water-vapor flux is virtually always upward between this evaporation depth and the surface, and takes place through the interstitial spaces in the substrate matrix through two processes: vapor diffusion and convection. This vapor flux is naturally most significant when there is a high permeability to air flow, i.e. in coarse soils that have low water content, a condition often found in deserts. Under such conditions, the intensity of the in-soil vapor flux, and hence the intensity of the surface flux, is strongly dependent on the soil vertical temperature gradient (related to convection: vertical buoyancy-driven motion) and the soil vertical vapor-pressure gradient (related to vapor diffusion) between the evaporation depth and depths nearer the surface. Convection of air within the soil generally only occurs at night or in the early morning after the upper soil layers have cooled and the temperature lapse rate becomes unstable. This process is especially important in deserts, given the large diurnal swings in the vertical temperature gradients within the upper layer of soil (Fig. 5.3). Vapor-diffusion effects are proportional to the vapor-pressure gradient between the evaporation depth and the upper soil layers. An additional mechanism can be responsible for vertical water-vapor transport in the sandy layer near the surface that experiences significant diurnal temperature fluctuations. As the humid interstitial air in the near-surface sand warms during the day, the air expands, and water vapor is expelled into the desert atmosphere. At night when the sand near the surface cools, and the air within it contracts, dry desert air will be drawn into this layer. This diurnal breathing and drying of the layer of heated sand only occurs in a shallow layer, of depth 10–20 cm perhaps, because of the poor thermal conductivity of sand. Surface-pressure fluctuations associated with the migration of synoptic-scale high- and low-pressure systems can also cause expansion and contraction of air in the substrate, and contribute to upward surface vapor fluxes.

Model-based sensitivity studies have shown that a realistic treatment of soil vapor fluxes is essential for simulating the overall energy and moisture budget of the desert (Niu *et al.* 1997; Braud *et al.* 1993). In calculations based on field-experiment data with non-desert soil, Cahill and Parlange (1998) showed that vapor transport was responsible for 10–30% of the total moisture flux. Similar values were obtained by Westcot and Wierenga (1974). Numerous data from other field experiments, with soils and large diurnal temperature oscillations that are typical of deserts, substantiate the importance of the vapor flux (Barnes *et al.* 1989; Rose 1968). Ambroggi (1966) states that evaporation from depths of as much as 20 m beneath Sahara sands has a significant effect on surface vapor fluxes, and that such subsurface evaporation, which is between one thousandth and one ten-thousandth of the potential surface value, could represent a loss of as much as 3000 m^3 of water per year per square kilometer of surface.

The following discussion of field-program data illustrates some of the complexities of subsurface vapor transport. For a multi-year period, hourly surface-evaporation data were obtained by using sand tanks (lysimeters) in the Namib Desert (Hellwig 1973a,b, 1978). The sand porosity was 33.5%. As summarized in Table 5.2, the water-table depths employed ranged from the surface of the sand to 60 cm below the surface. When the water table was at the surface of the sand, the surface was characterized as wet; when it was at 10 cm and 20 cm, the surface was characterized as moist; and when it was at 30 cm and greater depth, the surface was dry. Over 70 weeks of data were used, after periods affected by rain were eliminated from the data set. Fig. 5.6 summarizes the diurnal

Table 5.2 *Water-table depths maintained in seven lysimeters used to estimate evaporation from sandy soils in arid climates*

Tank number	1	2	3	4	5	6	7
Depth of water table below sand surface (cm)	0	10	20	30	40	50	60
Surface condition	wet	moist	moist	dry	dry	dry	dry

Source: From Hellwig (1973a).

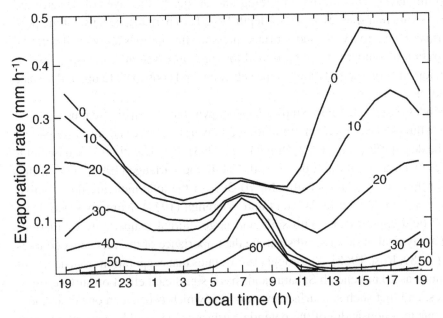

Fig. 5.6 Diurnal variation in the hourly surface evaporation, averaged over all seasons for a multi-year period during which data were obtained using sand tanks (lysimeters) in the Namib Desert. Seven lysimeters were employed, each with a different water-table depth that was maintained at a constant value throughout the study period. The curves are labeled in terms of the water-table depth (cm). (From Hellwig 1978.)

variation of the hourly surface evaporation from the system, averaged over all seasons for the entire period. Two maxima are revealed: one is in the early morning and the other is in the afternoon and evening. For deeper water tables, the sunrise maximum is the greater perturbation, whereas for shallower water tables the late-day maximum is larger. It can be seen that a significant water-table-dependent phase shift exists for both maxima. For the 60 cm water table, compared with the surface water table, the early-morning maximum is about 3 h later and the late-day maximum is about 8 h later.

The causes of the late-day maximum will be discussed first. For the shallow subsurface water tables (10 cm and 20 cm), the surface is moist as a result of capillary rise, and thus the evaporation takes place at the surface. The amplitude and phase of the evaporation curves are thus potentially a function of (1) the diurnal variation of the radiant energy input that determines the surface temperature, (2) the diurnal variation of the relative humidity of the air above the surface, (3) the diurnal variation of the wind speed (no data provided), and (4) the resistance of the sand to capillary rise (i.e. how efficiently evaporated surface water is replenished from below). In this case, the higher late-afternoon surface temperature and the possibly higher afternoon surface wind speeds (associated with convection) contribute to this maximum. For the deeper water tables (30–60 cm), where the surface is dry, capillary rise of water does not reach the surface. Here, convection or vapor diffusion must transport water vapor from the evaporation depth, at the upper limit of the capillary rise, to the surface. Because subsurface temperature lapse rates are generally stable (decreasing temperature with depth) at this time, vapor diffusion must be responsible for the transport. At least two factors are responsible for the decreasing amplitude of the evaporation curve with increasing water-table depth: the effective resistance to the vapor diffusion is greater for greater transport distances; and the temperature at the evaporation depth is lower for deeper water tables, and thus the evaporation rate will be lower also. The fact that the evaporation-rate maximum is later in the day for deeper water tables is a result of the different diffusive transport times and the fact that the temperature maximum at the evaporation depth occurs later for greater water-table depths (Fig. 5.4a). The water-table-depth diurnal temperature profiles shown in Fig. 5.7 exhibit maxima that very closely coincide with the times of the surface-moisture-flux maxima in Fig. 5.6.

Regarding the surface vapor-flux maximum near sunrise, there are at least two potential contributing processes. First, for shallow water tables, the small evaporative loss from the cool surface at night allows the capillary rise to replenish the surface water at a rate faster than it is lost. Thus, the surface should be more moist near sunrise than at any other time of day, and the evaporation rate should increase rapidly after sunrise when solar energy becomes available. However, the data show a rise in evaporation rate well *before* sunrise for near-surface water tables, making this argument suspect. Secondly, for deeper water tables where the capillary rise does not reach the surface and fill the pores with liquid water, convective transport of vapor within the sand will be most intense when the downward temperature gradient is largest, and the stability is least between the surface and the evaporation depth; this period of strongest convective transport

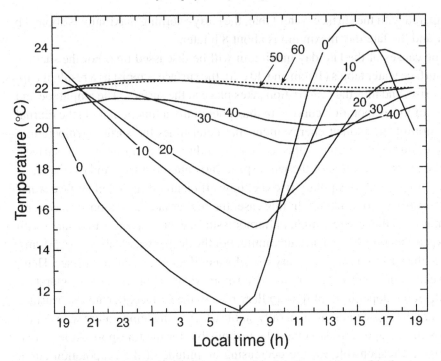

Fig. 5.7 Diurnal variation in the water-table-level temperature for the seven lysimeters for which the evaporation rates are plotted in Fig. 5.6. The curves are labeled in terms of the water-table depth (cm). (From Hellwig 1978.)

should be near sunrise. The cause of the morning maximum for the very shallow and surface water tables, with wet surfaces, is more problematic.

It is revealing that Hellwig's data show that, for the water table at the surface, the moist sand is warmer than the air only at night between about 0100 and 0700 LT, with the maximum difference of 2 K at 0500 LT. Such a situation is opposite to the normal one in which the surface is warmer than the air during the day, and colder than the air at night (Fig. 5.3). This means that, based on Eqn. 5.13, this is the only period with a positive (upward) sensible heat flux (H). Such positive H can lead to convective mixing, a larger eddy diffusivity, and increased evaporation (Eqn. 5.12), which could explain the stronger observed evaporation between 0300 and 0900. This non-natural air–surface temperature difference is likely an unavoidable consequence of the experimental conditions employed in that the air flowing over the moist lysimeter sand has been in thermal equilibrium with the surface of the surrounding dry desert. The lysimeter sand may be warmer at night than the surrounding desert sand because its greater moisture content and higher resulting thermal conductivity allowed storage of

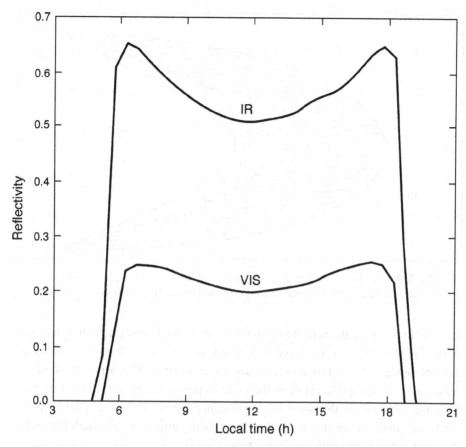

Fig. 5.8 Variation in the June-average reflectivity of a sandy surface in the Rub Al Khali Desert as a function of time, for visible and solar-infrared wavelengths. (Adapted from Smith 1986a.)

more heat from the previous day. This is an aspect of the oasis effect discussed earlier.

Thermal processes within the substrate and at the surface

It has been mentioned that the albedo of sandy desert substrates is typically larger than those of other surfaces in temperate latitudes. This contributes to a smaller net radiation. Two other properties of the albedoes of desert substrates are illustrated in Fig. 5.8, which depicts the June-average reflectivity of a sandy surface in the Rub Al Khali Desert. The albedo is shown as a function of time of day, for both visible and solar-infrared wavelengths. First, it is clear that the reflectivity is dependent upon the angle between the incident radiation and the

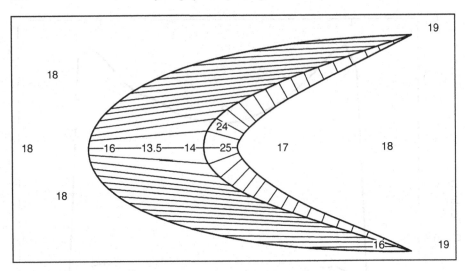

Fig. 5.9 Reflectivity (%) of a sandy surface at different locations on and surrounding a barchan sand dune in the Peruvian Desert. (From Bryson 1978.)

surface. In addition, the reflectivity in the solar infrared is much greater than that in the visible. This is typical for non-desert surfaces as well, but the factor of 2.5 difference in the maxima here is larger than normal. The dependence of the reflectivity on sun angle between the twilight periods is greater in the infrared. Deering *et al.* (1990) also show the significant dependence of the reflectivity on the zenith angle of the sun, as well as the viewing angle, for an alkali flat and a flat sand dune in the northern Chihuahuan Desert.

It is tempting to assume that the albedo of an expansive, sandy desert landscape is uniform, but this is not the case. Sand grains of different sizes and density may be made up of predominantly different minerals having different reflectivities, so that if the wind sorts the grains by size, areas of different albedo result. Fig. 5.9 illustrates this for a barchan sand dune in the Peruvian Desert. The wind is oriented from left to right in the figure. The lighter-weight and more reflective (in this case) grains are blown over the crest onto the right slip face of the dune. The denser and darker grains remain on the upwind dune face, explaining the significantly lower albedo on the upwind side. The area around the dune with the low albedo is a rocky surface with a light coating of sand.

To understand the dependency of the subsurface heat fluxes on the mixture of solids, liquid water, and vapor in a sandy substrate, let it be reasonably assumed that the grains are composed of quartz. The relative thermal conductivities of quartz, water, and air are $333 : 23 : 1$, respectively (Jury *et al.* 1991). The importance of water to the bulk thermal conductivity of a granular medium is two-fold. First, the water displaces the air, with the conductivity of water

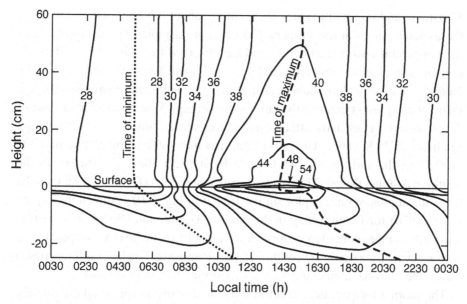

Fig. 5.10 The temperature (°C) just above and below the surface of sparsely vegetated sandy plains in the Sonoran Desert, averaged for July and August 1956. The dashed and dotted lines define the time of the temperature maximum and minimum at each height or depth. Note the non-uniform isotherm interval. (Adapted from Dodd and McPhilimy 1959.)

being over 20 times greater. Also, even a small amount of water can provide a high-conductivity contact medium among the highly conductive quartz grains. In extreme deserts with long periods between rain events, the dryness of the sandy medium will mean that the high conductivity of the grains will be mitigated by the poor conductivity of the interstitial air spaces. The importance of this effect will depend on the pore space as well as the size of the sand grains, with both factors affecting the contact among the grains. The thermal conductivity of wet sand has been measured to be from eight to twelve times greater than that for the same sand when dry (see Chapter 8). As the grain size decreases, the increased pore space and reduced contact among the grains is reflected in reduced conductivity. For this reason, the relative thermal conductivity of different soils has the order sand > loam > clay. Table 5.1 shows that the thermal diffusivity is larger for a sand substrate than for the other soil types listed. Eqn. 5.4 thus indicates that the diurnal and seasonal temperature wave in sand will be of higher amplitude than for other soils.

Figure 5.10 shows an example of the diurnal temperature wave in a sandy substrate, and the temperature just above the surface, for a sparsely vegetated sandy plain in the Sonoran Desert. The diurnal amplitude at the surface exceeded 50 K (90 °F), while at 20 cm below the surface it was about 10 K (18 °F).

Maximum and minimum values at 20 cm are about 6 h later than the times of the extreme values at the surface. The observed damping of the amplitude of the temperature wave with depth, and the observed time lag, are qualitatively consistent with the values predicted by Eqns. 5.4 and 5.5 (see problem 3 below). Note that the vertical gradient near the surface is sufficiently large so that it is difficult to represent graphically. Substitution in Eqn. 5.9 of a reasonable value of H for a desert surface in the afternoon produces vertical atmospheric temperature gradients of 30 K mm^{-1}. Griffiths (1966) has measured a temperature of 77 °C (160 °F) at the surface, and 49 °C (120 °F) at 5 cm above the surface in the southern Arabian Desert. This is a lapse rate that is 550 000 times larger than the dry adiabatic lapse rate. A plot similar to that in Fig. 5.10 is shown in Terjung *et al.* (1970) for Death Valley in the Mojave Desert, and Dincer *et al.* (1974) shows the temperature–depth plots for different times in the daytime part of the heating cycle for a sandy area of the Arabian Peninsula. The latter reference also shows the seasonal temperature wave penetrating to 6 m in the sand.

The diurnal temperature variation illustrated above is for an approximately horizontal surface. However, because the surface flux of solar energy depends greatly on the angle between the solar beam and the surface (Fig. 4.3), the slope and orientation of bare or sparsely vegetated surfaces represent a strong control on the energy budget of the substrate and the lower atmosphere. For example, for seasons and latitudes for which the solar zenith angle is large, equatorward-facing slopes will receive considerably more daily total solar energy than will poleward-facing slopes. With the greater energy available for evaporation, the equatorward-facing slopes will thus also be drier. Fig. 5.11 illustrates this effect in terms of the surface temperatures in the area between the roughly north–south-oriented crests of two sand dunes spaced about 50 m apart in the Karakumy Desert of Asia during the summer. The leeward-facing east slope had an angle of about 30° to the horizontal, with a lesser angle on the windward west side. At night, there was very little spatial variability in the surface temperature. But soon after the eastward-facing slope became exposed to the rising sun, the temperature of its sand surface increased to a value that was over 23 K (41 °F) higher than the temperatures between the dunes and on the west-facing slope. Later, during the afternoon hours, the westward-facing slope began to receive more direct solar radiation, and it warmed while the eastward-facing slope cooled. Wind speeds near the surface were 1–3 m s^{-1} most of the time. For surface temperature variations on this small horizontal scale, air parcels are exposed to surface temperature anomalies for an insufficient time to be greatly modified, so the 2 m air temperatures measured at the different sites generally varied by less than 1 K.

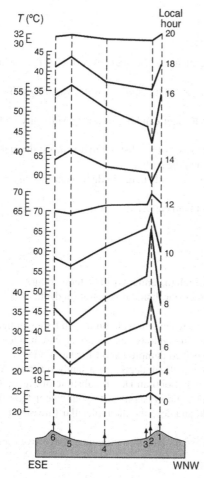

Fig. 5.11 The diurnal temperature variation of the sand surface in the Karakumy Desert in the summer at six locations (noted in the terrain cross-section) between sand-dune crests that are separated by *c.* 50 m. (Adapted from Aizenshtat 1960.)

Suggested general references for further reading

Hillel, D., 1998: *Environmental Soil Physics* – contains a thorough treatment of soil physics, including the water and energy budgets of the soil. There are also chapters on water-use efficiency in irrigation, soil erosion, and soil-moisture uptake by plants.

Oke, T. R., 1987: *Boundary Layer Climates* – a discussion is provided of physical processes at the surface and in the boundary layer. This is a good reference for students from a variety of backgrounds.

van Wijk, W. R., (Ed.) 1963: *Physics of Plant Environment* – primarily treats the thermal conditions of the soil and the atmosphere near the surface.

Questions for review

(1) Explain the concept of admittance.
(2) List the various phase transitions of water that occur at and below the surface, which consume and release latent heat.
(3) Explain the mechanisms involved in the vertical transport of water vapor by diffusion and convection in a substrate.
(4) Summarize the five mechanisms by which water can be transported vertically within a substrate.
(5) Describe how the rapid drying near the surface of a substrate that is moist through a deep layer can inhibit further evaporation at the surface.
(6) Diagram the various components of the water and energy budgets near the surface, in the substrate and the atmosphere.
(7) Explain the concepts of wilting point and field capacity.

Problems and exercises

(1) Show mathematically that the depth to which the annual cycle's temperature wave penetrates into the soil is about 14 times greater than the penetration depth of the diurnal cycle's temperature wave.
(2) Explain why dry sand feels so much hotter when you walk on it with bare feet than does a macadam road surface, which has a much lower albedo, or a cement sidewalk.
(3) Use Eqns. 5.4 and 5.5 to compute the expected damping and time lag of the temperature wave between the surface and a 20 cm depth, for a sandy substrate. What must the thermal diffusivity be for the sandy Sonoran Desert substrate of Fig. 5.10, in order that the observed damping and time lag correspond with your calculated values?
(4) Refer to a physics or chemistry textbook, and describe the molecular forces that produce surface-tension effects.
(5) How does the subsurface temperature-wave propagation influence convective transport of water vapor in the substrate?

6

Vegetation effects on desert surface physics

The difficulty in surviving the severe weather and climate of the desert sometimes stimulates development of a strict desert social code of honor, solidarity, and hospitality. For example

Theodore Lascaris, an emissary of Napoleon, and his Syrian dragoman, Fathallah Sayigh, were given shelter by a poor old widow of the Sardiyya Bedouin in Jordan, who slaughtered her one and only sheep in their honour. "Grandmother," they said to her, "why such waste?" To which she replied: "If you entered the dwelling of a living person and did not find hospitality there, it would be as though you had paid a visit to the dead."[6.1]

Joseph Chelhod
Islands of Welcome in a Sea of Sand (1990)

The mesquite is God's best thought in all this desertness. It grows in the open, is thorny, stocky, close grown, and iron-rooted. Long winds move in the draughty valleys, blown sand fills and fills about the lower branches, piling pyramidal dunes, from the top of which the mesquite twigs flourish greenly. Fifteen or twenty feet under the drift, where it seems no rain could penetrate, the main trunk grows, attaining often a yard's thickness, resistant as oak.

Mary Austin, American naturalist and writer
The Land of Little Rain (1903)

Even though the expression "desert vegetation" may seem contradictory, nothing could be further from the truth in some places. Deserts, in fact, may have a richness of flora and fauna that is unparalleled in more humid areas such as woodlands. Such desert vegetation – primarily shrubs, herbs, grasses, succulents, and small trees – has a number of influences on the land-surface physics and, hence, the atmosphere through modification of the near-surface fluxes of heat, water vapor and momentum. The vegetation influences the land-surface and atmospheric processes through the same mechanisms that would prevail without

---

[6.1] Originally in *Memoirs* by Fathallah Sayigh, an Arabic dialect manuscript purchased by the French poet Alphonse de Lamartine in Syria in 1832, and preserved in the Bibliothèque Nationale, Paris.

the aridity, but sometimes the effects are unique. The types of effect are summarized as follows, with more detail provided in the later sections of this chapter.

- albedo – Vegetation generally has a different albedo than does a bare surface with the substrate exposed. Because bare desert substrates (e.g. sandy soils) often have a relatively high albedo, the effect of vegetation is to lower the aggregate value in proportion to the vegetation density.
- interception of rainfall – Foliage intercepts rain and snow, with multiple effects. The fact that the kinetic energy of raindrops is absorbed by the foliage instead of by the bare surface results in less sealing and erosion of the substrate. Also, the water retained on the surfaces of the vegetation evaporates after the rain event, and is not available to infiltrate and moisten the substrate.
- soil permeability – Vegetation, through a number of mechanisms, often increases the permeability of the soil to rainfall, thus increasing the ratio of infiltration to runoff, and reducing the flooding potential. There are, however, mechanisms with the opposite effect.
- transpiration – Because the roots of vegetation can tap into subsurface soil moisture, the result is generally an increase in the loss of soil moisture to the atmosphere when such root-zone moisture is available.
- reduced near-surface winds – The effect of the vegetation reduces the wind speed very near the surface. This reduces the rate of erosion by the wind. The lower wind speed also reduces the surface sensible heat flux and the evaporation rate.
- soil crusting and binding – Shallow roots bind the soil and inhibit wind and water erosion. Cryptogamic crust, a crusty black ground covering made up of lichens, mosses, and blue-green algae, also reduces erosion. Microscopic plants, called microphytes, can form crusts that influence the water infiltration.
- interception of solar and terrestrial radiation – Vegetation has an important effect on the disposition of solar radiation that is independent of the fact that its albedo is different from that of the substrate. The solar energy that is intercepted and absorbed by leaves and stems, in contrast to that absorbed by the substrate, is more effectively transferred to the atmosphere through sensible heating. This is because the vegetation has very little mass for heat storage, and the air is in contact with a relatively large foliage surface area. Vegetation surfaces also intercept upwelling infrared energy emitted by the substrate.
- shading of the bare substrate – The shading of the substrate by vegetation limits the solar energy input. Less energy is thus available for evaporation from, and for sensible heating of, the substrate. The shading also reduces the amplitude of the diurnal temperature variation in the substrate near the surface. This will reduce the loss of substrate water vapor that results from the diurnal surface exchange of air (caused by the expansion and contraction of substrate air), and from convection in the substrate.
- induced terrain-elevation variation – The existence of larger vegetation structures such as trees and shrubs in environments that are prone to wind erosion can influence the erosion pattern, and hence the terrain profile.

Because water availability and substrate properties are often marginal for vegetation survival in deserts, and may vary considerably over small distances, the place-to-place contrasts in natural vegetation can be much greater than is typical in other environments. Vegetation can be relatively lush in riparian areas or where the water table is high, but a short distance away, solid rock or highly alkaline soil or water may support no life, or perhaps a totally different plant community that is adapted to the drastically different conditions. Because of the effects of terrain slope and aspect on extremes of temperature and soil moisture, one will sometimes find east–west-running hills with one slope clothed in vegetation and the other bare, or with totally different vegetation types on the opposite slopes that are perhaps just hundreds of meters apart. Aerial photography reveals many curious patterns in desert vegetation density and distribution that must be related to the substrate type and water content. For example, Goudie and Wilkinson (1977) report on the existence of regular bands of vegetation in Somalia that resemble the black stripes on a tiger's skin, and thus are termed *brousse tigrée* by the French. Mountain ridges that run parallel to foggy coastlines often separate lushly vegetated ecosystems from barren surfaces. Wilfred Thesiger (1964) recounts the vegetation contrast that he observed along the southeast coast of the Arabian Peninsula.

As we climbed the mountain-side I noticed paradise fly-catchers, rufous and black, with long white streamers in their tails, and brilliant butterflies. They were in keeping with the jungles which surrounded us, and as unexpected in Arabia. Then we came out on the downs and camped near the top of the mountain. I walked to the watershed, anxious to see what lay beyond, and found myself standing between two worlds. To the south were green meadows where cattle grazed, thickets, and spreading trees, whereas a stone's throw to the north was empty desert – sand, rocks, and a few wisps of withered grass. The transition was as abrupt as it is between the irrigated fields and the desert in the Nile valley. Here the dividing line followed the crest of the mountains.

Hastings and Turner (1965) give an especially thorough account of the observed vegetation changes with altitude, latitude, longitude, and local edaphic and microclimatic conditions in a small area of the Sonoran Desert.

The particular areas in arid climates where vegetation can be best supported have been termed "niches." Even though there are many factors involved, and regional idiosyncrasies make generalization difficult, the niches are listed below in order of increasing wetness and potential for vegetation survival (Goudie and Wilkinson 1977):

- slopes with impermeable soils and high rates of runoff
- flood terraces with compacted surface soils
- clay soils on level flats with high evaporation rates

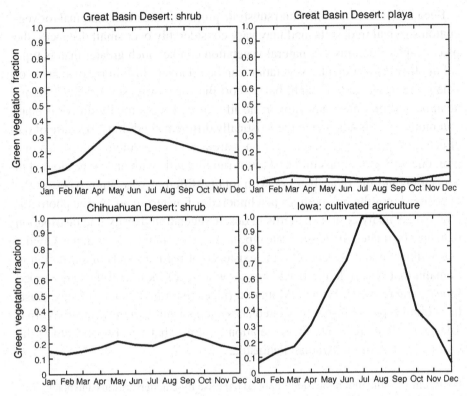

Fig. 6.1 The seasonal variability of green vegetation fraction for various desert surfaces and a non-desert surface, for a typical year. (Courtesy of Fei Chen, National Center for Atmospheric Research.)

- sand flats that are easily penetrated by rain
- stable sand dunes
- talus (scree) slopes and stony plains with low evaporation rates
- fissured rocky outcrops that can hold water easily
- bottoms of slopes with inflows of seepage
- erosion channels carrying water temporarily and storing water in the underlying soil; and
- dry valleys containing a continuously flowing stream of water at no great depth below the surface.

As in temperate climates, desert vegetation undergoes seasonal variations. The seasonality of the greenness can depend on both temperature as well as the availability of moisture from rainfall. In the latter case, some annual types of plant can emerge and flower and produce seed within a matter of days after rain. Likewise, perennials can greatly increase the amount of green vegetation when intermittent seasonal moisture becomes available. And, unlike natural vegetated surfaces in non-arid areas, the amount of vegetation can differ markedly over very short distances. Fig. 6.1 shows the satellite-derived green vegetation fraction for three desert surfaces and, for comparison, for a non-desert surface. The green vegetation fraction is the fraction of the surface area that is masked by vegetation, and can be thought of as the horizontal density of the vegetation cover. Of the three desert locations, the Great Basin shrubland has the largest seasonal variation, with the greenness responding to spring-season rainfall. The northern Chihuahuan Desert location experiences some spring-season rainfall, as well as rain beginning in July from the North American monsoon; hence the two modest peaks in the green vegetation fraction. The Great Basin playa, not far from the shrubland, has a very low fraction, that does, nevertheless, show a slight seasonality.

In addition to such *intra*-annual changes in vegetation amount, the *inter*annual variability can also be great in arid areas. This is because the average rainfall is sometimes not much greater than what is necessary for vegetation to thrive, or even survive, and thus in years for which the rain is much less than the average, the amount of vegetation is less. This can have important implications in terms of the year-to-year variation in the impact of vegetation on the land-surface physics. Fig. 6.2 illustrates the interannual variability in vegetation for an area of the Sahel on the southern margin of the Sahara, and for an area of the central Sahara. The metric for the vegetation amount is the Normalized Difference Vegetation Index (NDVI), which is based on satellite measurements in the red and solar-infrared spectral region. The spectral response in the red region is inversely related to the chlorophyll density (see the minimum for live vegetation in Fig. 6.5), and in the solar infrared the response is related to scattering by leaves. The figure shows the existence of considerable interannual variation in the vegetation, primarily as a result of rainfall variability. For example, 1984 was one of the driest years of the century in that area.

Even though a number of examples have been shown already of surface energy budgets for desert areas, most of the areas have been relatively unvegetated. Thus, before going into detail about particular effects of vegetation on the desert surface and atmosphere, an example will be provided of the budget for an area of the Great Basin Desert where the vegetation cover varies from 40 to 75%. Fig. 6.3 shows the monthly total energy, for a complete annual cycle, associated with net radiation, latent heat flux, sensible heat flux, and soil heat flux. Table 6.1 is based on the data in this figure, but describes the fraction of the net radiation partitioned to the sensible and latent heating of the atmosphere, and to the soil heating. The annual-average fraction partitioned to atmospheric sensible heating (85.3%) is not as large as the 90% rule of thumb noted before for non-vegetated sandy deserts; however, it is still large compared with non-desert environments. The largest latent-heat fluxes occurred in April and May after the vegetation

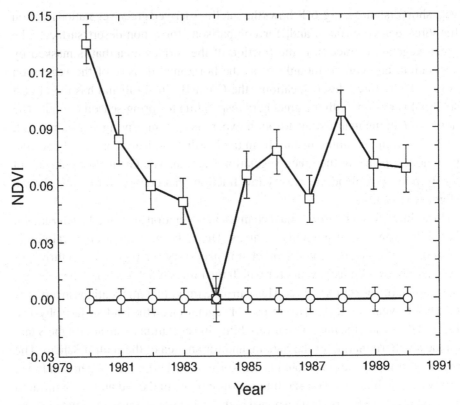

Fig. 6.2 Satellite-derived Normalized Difference Vegetation Index (NDVI) for a zone (the Sahel) with long-term average rainfall between 200 and 400 mm y^{-1} along the southern boundary of the Sahara Desert (squares), and for the Sahara Desert itself (circles) for the period 1980–1990. (From Tucker *et al.* 1991.)

was fully leafed and while the spring rain showers were still active. In June and July, the evapotranspiration decreased significantly, and the latent-heating fraction of the net radiation dropped to its lowest monthly value (8.4%). This is consistent with the decrease in rainfall during the summer at Salt Lake City and Winnemucca in the Great Basin Desert (Fig. 3.30).

The effects of vegetation on the desert surface energy budget

Because of the high reflectivity of a sandy desert substrate, live vegetation with lower albedo will cause the aggregate albedo of a mixture of vegetation elements and exposed substrate to be lower than that of an unvegetated surface. Thus, the vegetation albedo will have a first-order effect of contributing to a larger value of net radiation (Eqn. 4.3). This effect is complicated by the fact

Table 6.1 *Monthly fraction of the net radiation partitioned to sensible, latent, and soil heating for a location in the Great Basin Desert of North America*

Month	H/R	LE/R	G/R
Oct	0.786	0.226	−0.012
Nov	0.842	0.178	−0.020
Dec	0.797	0.213	−0.010
Jan	0.854	0.164	−0.018
Feb	0.863	0.141	−0.005
Mar	0.842	0.155	0.003
Apr	0.791	0.205	0.004
May	0.780	0.213	0.007
Jun	0.905	0.088	0.007
Jul	0.908	0.084	0.008
Aug	0.874	0.123	0.003
Sep	0.896	0.104	−0.001
Annual	0.853	0.146	0.001

Source: Adapted from Malek and Bingham (1997).

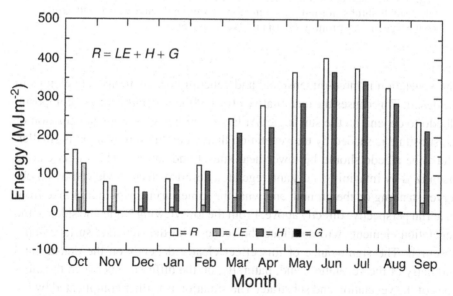

Fig. 6.3 Monthly total energy for a complete annual cycle, associated with net radiation (R), latent heat flux (LE), sensible heat flux (H), and soil heat flux (G), for a location in the Great Basin Desert of North America. (Adapted from Malek and Bingham 1997.)

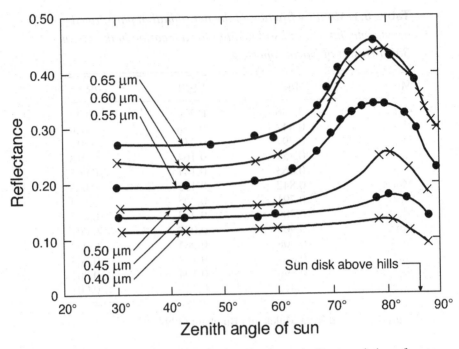

Fig. 6.4 Measured spectral reflectance of a desert surface consisting of scattered small shrubs in sand, as a function of sun angle and wavelength in the visible spectrum. (From Ashburn and Weldon 1956.)

that vegetation is three-dimensional and generally occurs in discrete clumps in the desert. This causes the quantitative effect of the vegetation on the composite albedo to depend on the sun angle. At large zenith angles, a greater fraction of the substrate is shaded by the vegetation than when the zenith angle is smaller. Thus, the albedo should be lower near sunset and sunrise. This effect can be visualized by imagining a distant vegetated desert surface as viewed by an observer standing on the ground, and what the same surface would look like from an aerial perspective directly above. From the low viewing angle, the sides of the vegetation elements will dominate the scene, with the substrate surface being obscured. The significance of this effect clearly depends on the density and geometry of the vegetation, and, naturally, on the difference between the albedoes of the vegetation and substrate. The situation is further complicated by the fact that substrates themselves can have albedoes that depend on the angle of the incident light. For example, Ashburn and Weldon (1956) measured the spectral (wavelength-dependent) reflectance of a desert surface consisting of scattered small shrubs in sand, and showed a significant dependence of the reflectance on sun angle and wavelength. Fig. 6.4 shows that, for longer wavelengths in the visible spectrum (e.g. 0.60–0.65 μm), the reflectance varies by almost 0.2 for sun angles that are more than 60° from the zenith. The albedo response between 80° and 90° is the aforementioned effect of the vegetation shading the sand; the dependence at smaller zenith angles is related to the fact noted above that the sand is not behaving as a perfect diffuse reflector, i.e. the reflectivity is larger at lower sun angles (larger zenith angles). Bowker and Davis (1992) and Cess and Vulis (1989) further discuss this effect for desert surfaces. It is reasonable that such temporal changes in albedo near sunrise and sunset (large zenith angles) could have a significant meteorological impact because boundary-layer transitions between nocturnal and daytime-convective regimes are large during these times (Fig. 2.3a).

But it is not just *live* vegetation that is important to the desert albedo. In particular, it is common for the space between the live vegetation to be covered with dead plant debris rather than bare soil because the highly irregular rainfall in deserts causes periodic mortality of perennial plants, because of the vestiges of previous years' crops of annual plants, and because the low humidity inhibits decay. Otterman (1977a,b, 1981) points out that this debris can actually have a greater impact on the area-average albedo than do the live plants, especially when the area covered by live vegetation is small. Fig. 6.5 illustrates this with spectral reflectivity measurements from satellite and hand-held radiometers for the Negev and Sinai Deserts. Clearly, over a wide band of solar wavelengths, the plant debris has considerably lower reflectivity than does the unvegetated soil. The slopes of the reflectivity curves shown in this figure for the Negev and Sinai, as well as the relative displacements of the curves for the bare surface, the live vegetation, and the dead vegetation, are very similar to observations described for the Sonoran Desert (Whitlock and LeCroy 1987).

A complicating factor in determining the effect of vegetation on desert albedo is that some desert plants modify the reflectivity of their surfaces depending on the moisture stress. When moisture is plentiful and photosynthesis is taking place, the leaves are deep green and the reflectivity is low. However, under dry conditions the plant can avoid being strongly heated and consequently desiccated by making the leaves more reflective and lowering the net radiation. This is accomplished by the leaf color changing from green to gray or silvery, or by the growth of very fine, light hairs on the leaf surface. The reflectivity in the visible wavelengths of some species can vary by 30–40% through this adaptation to water availability (Bowers 1986).

In addition to albedo effects on the net radiation, vegetation impacts the surface energy budget through its influence on the partitioning of the net radiation among the sensible heat flux to the atmosphere, heat storage by the absorbing medium (substrate or vegetation), and evapotranspiration. Specifically, because

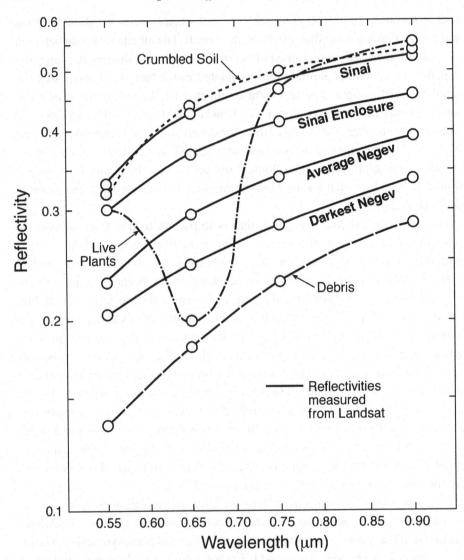

Fig. 6.5 Spectral reflectivities of the overgrazed Sinai Desert (Sinai), an en-closure in the Sinai in which human activities had been excluded for three years (Sinai Enclosure), and of the Negev Desert, which is less affected by people, based on Landsat satellite data. Data for two areas of the Negev Desert are pro-vided: one is for normally vegetated Negev conditions (Average Negev), and one is for a more heavily vegetated area (Darkest Negev). The sun elevation was 51° (zenith angle = 39°). A hand-held field radiometer was used to provide additional spectral reflectivities for live vegetation (Live Plants), the dead plant material (Debris), and crumbled soil that is devoid of live or dead vegetation. (Adapted from Otterman 1981.)

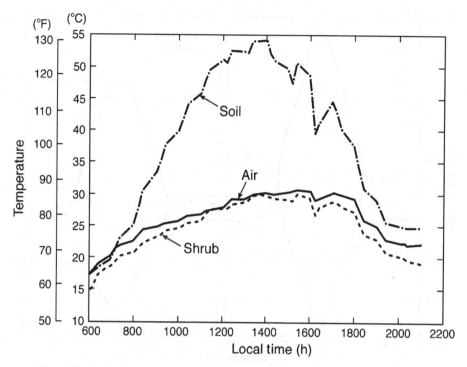

Fig. 6.6 The air temperature at 2 m above the ground (Air), the temperature of the stems and leaves of scattered sunlit vegetation (Shrub), and the surface temperature of the sunlit substrate (Soil) for a location in the Sonoran Desert in summer. (Adapted from Humes *et al.* 1994.)

solar-energy-absorbing leaves and stems are much better ventilated by moving air than is the absorbing bare substrate surface, and because the small mass of the leaves and stems does not permit significant heat storage, a greater fraction of the solar energy is partitioned to sensible heating of the atmosphere by the vegetation than by the bare substrate. This effect increases in importance with greater vegetation density and greater zenith angle. Also, the vegetation absorbs upwelling long-wave radiation from the ground that would otherwise have been emitted to space. This absorbed energy contributes to the sensible heating of the lower atmosphere by the vegetation. The importance of these effects is demonstrated by the fact that vegetation has a daytime temperature that is very similar to that of the air, and one that is lower than that of bare soil. This lower temperature of the vegetation than of the soil surface exists in spite of the fact that the vegetation absorbs a larger fraction of the incident solar energy than does the substrate (because of its lower albedo). Fig. 6.6 shows the temperature of air at 2 m above the surface, and the temperatures of shrub vegetation and nearby bare ground for a location in the Sonoran Desert in summer. Clearly the vegetation

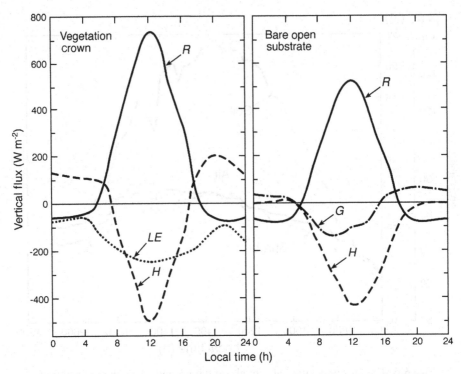

Fig. 6.7 The net radiation (R), the sensible heat flux to the atmosphere (H), the latent-heat flux (LE), and the heat flux into the substrate (G) for two nearby locations in the Kyzylkum Desert. The left panel is for the crown of dense saxaul vegetation and the right panel is for a nearby area of bare, open substrate. (Adapted from Aizenshtat 1960.)

is in thermal equilibrium with the air because the sensible heat transfer is so efficient. The cooler vegetation is also radiating infrared with much less intensity than is the hotter bare ground (Eqn. 4.1). This contributes to a larger value of net radiation for the vegetation. This effect of desert vegetation on the net radiation and sensible heat flux to the atmosphere has been well documented. Terjung *et al.* (1970) compute a net radiation for mesquite and desert holly that is 30–50% higher than for the nearby bare substrate in Death Valley in the Mojave Desert, and Otterman *et al.* (1993) estimate a sensible heat flux to the atmosphere that is almost a factor of two larger for a vegetated area of the Sinai Desert than for a nearby devegetated area, for a zenith angle of 60°. An example of the diurnal variation of net radiation and sensible heat flux to the atmosphere for adjacent bare substrate and dense saxaul vegetation is illustrated in Fig. 6.7 for the Kyzylkum Desert. Within the crown of the saxaul plant, the large net radiation is divided between the latent and sensible heat fluxes to the atmosphere, with no significant storage term for the vegetation. Over the bare and dry substrate, the smaller net radiation is partitioned between the sensible heat fluxes into the atmosphere and the substrate. The condition illustrated in Fig. 6.6, in which the vegetation and air temperatures are very similar, also was documented here.

The above estimates of the net radiation and sensible heat fluxes over bare substrate (see, for example, Fig. 6.7) apply to situations where large areas of bare surface exist far from the vegetation. However, when sunlit bare ground patches exist between vegetation elements, the vegetation has an impact on the energy budget of the bare ground. One of the mechanisms is related to the fact that the existence of the vegetation reduces the wind speed near the surface, and this reduces the efficiency of the turbulent sensible heat transport from the sunlit substrate, between the vegetation elements, to the atmosphere. The result of this effect in the above study in the Kyzylkum Desert was a mid-day substrate surface temperature between the saxaul plants that was 4–6 K (8–11 °F) higher than the substrate surface temperature in a nearby large clearing. Thus, the following surfaces are listed in order of increasing daytime sensible heat fluxes to the atmosphere: the sunlit bare substrate between vegetation elements, the bare substrate in open areas, and the vegetation surface. In contrast to the situation during mid-day hours, the surface of the bare substrate between the vegetation elements is cooler in the evening and morning than the substrate in the open area because the vegetation shades the surface. That is, the temperature contrast reverses.

Studies of the type described above are sometimes hampered by the fact that the contrast between the naturally vegetated and unvegetated areas exists primarily because of different substrate properties, so differences in fluxes are not due solely to vegetation. Exceptions are studies that compare conditions for a vegetated desert ecosystem with a nearby area that has had the vegetation removed, for example by overgrazing. In such a study, Otterman (1989) arrived at conclusions similar to those above. The sensible heat flux to the atmosphere was 1.5–2.0 times greater over a naturally vegetated area of the Sinai Desert compared to a nearby area that had been devegetated by grazing.

The effects of vegetation on the desert surface water budget

Vegetation can affect the desert's surface moisture budget through a variety of mechanisms. It shades the substrate and reduces the solar energy available for evaporation. It reduces the ventilation of the surface by the wind, and thus reduces the evaporation rate. And, it draws subsurface water through the roots and transpires it into the atmosphere. The importance of these processes to the

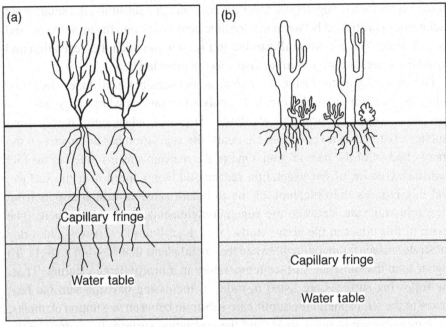

Fig. 6.8 A cross-section showing the relationship between the root structure and the water table for two types of desert vegetation. Phreatophytes (a) send roots deep, and directly tap the groundwater. Xerophytes (b), with shallow roots, conserve water through various mechanisms during periods between rain when the near-surface soil layers dry out. (From Goudie and Wilkinson 1977.)

water budget depends on the type and abundance of the vegetation, the health of the vegetation, the recentness of rain, and the season.

Compared to evaporation from the surface, transpiration from vegetation can have an especially large relative impact on the water and energy budgets in the desert, compared with non-arid areas. That is, except after a rain or over playas with high water tables, evaporation from the surface of the substrate is negligible, especially because some substrates have a high resistance to water movement. However, the deep-rooted nature of some desert vegetation and the high water table in some areas allow transpired subsurface moisture to enter the atmosphere in spite of dry and sometimes relatively impervious surfaces. Fig. 6.8 illustrates typical root structures of phreatophytes that send roots down to the water table or to the capillary-rise zone above it, and of xerophytes having shallow and sometimes horizontally extensive root systems. Because xerophytes survive arid conditions through conservation of water through various mechanisms, lesser amounts of water are discharged into the atmosphere when the soil is dry. An example of a phreatophyte is the mesquite (*Prosopis*) tree, the roots of which have been measured to depths of 80 m. The consumption of ground water by phreatophytes, and the consequent release of it into the atmosphere, can be enormous. Some desert streams have been observed to stop flowing at the surface during the daylight hours because this non-water-conserving vegetation draws down the water table when it is transpiring (Childs 2000). For the same reason, larger gauged desert streams can show a diurnal variation in discharge, partly because of the diurnal cycle of transpiration (Slatyer and Mabbutt 1965). Zones (1961) calculated that, for the dry Winnemucca Lake Valley in the Great Basin Desert, the sparse phreatophytes (greasewood, saltgrass, and saltbush), which cover an area of about 2100 ha, transpire about 10 cm of ground water annually beneath that area. Similarly, Taylor (1934) estimates that, in the western Mojave Desert, in an area where the water table is high, the transpiration by native phreatophyte vegetation annually consumes a depth of 2 m of ground water. On a larger scale, more than six million hectares containing phreatophytes in 17 western states of the United States transpire about 40–50 cm of water depth annually (Slatyer and Mabbutt 1965; Robinson 1958; Bouwer 1975).

The first-order effect of the shading of the substrate by vegetation is to reduce the evaporation from the surface, if any moisture is available. The reduction by the vegetation of the near-surface wind speed also contributes to a lower evaporation rate. However, such a lower evaporation rate may delay or avoid the rapid desiccation of the surface layer of the substrate that would create a barrier to additional upward water movement (Fig. 5.5). Thus, even though the rate of evaporation from a vegetation-shaded, moist substrate may be lower than if no vegetation were present, the evaporation may be sustained for a longer period and the total water loss may be greater. Fig. 6.9 illustrates this effect for two nearby locations near arid Alice Springs, Australia. One area consisted of a grove of acacia (*Acacia aneura*) shrubs, and the other area between the groves had about one-third the vegetation density. When the soil was moist, the evaporation in the less shaded area was relatively high because of the greater solar energy available at the unshaded surface. Where the denser vegetation shaded the surface, the evaporation rate was considerable less. But, as the substrate dried, at about 2.2 cm of available soil water the evaporation rate plummeted for the less shaded surface where the earlier rate had been the greater, probably because the vapor barrier formed at the desiccated surface and reduced the hydraulic diffusivity. For the lower soil-moisture values, the more vegetated substrate maintained a higher evaporation rate.

Numerous other comparisons have been made of evapotranspiration at vegetated and unvegetated desert sites. For example, in the northern Sonoran

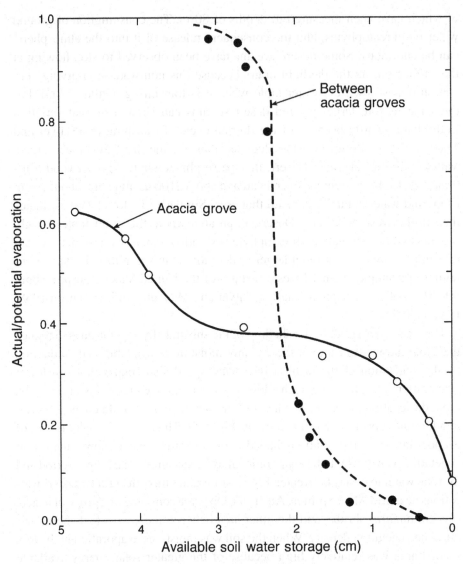

Fig. 6.9 Evaporation rate, near arid Alice Springs, Australia, from the shaded substrate in a grove of acacia shrubs and from the sunlit substrate in a less vegetated nearby area between acacia groves, as a function of the integrated soil moisture during a drying period. The evaporation rate is expressed in terms of the fraction of the potential evaporation. (From Slatyer and Mabbutt 1965.)

Desert, evapotranspiration rates were determined for bare soil and for locations with three types of vegetation: desert hackberry, mesquite and native perennial grasses. During the late winter and spring when the soil water content was relatively high (5–10%) and the vegetation was in full foliage, the 2–4-week average evapotranspiration was sometimes 2–4 times higher from the vegetated surface than from the bare one, but during the rest of the year the evapotranspiration was very similar. The multi-week-average evapotranspiration values from the mesquite and grasses were strongly related to the precipitation and to the season in which the precipitation fell. Typical winter values for mesquite were 1–2 mm d^{-1}, with summer rates ranging from 2 to over 5 mm d^{-1}. The evapotranspiration from the desert grasses peaked in the summer, and was more responsive to rainfall during that season, with values approaching 10 mm d^{-1}. In contrast, Evans *et al.* (1981) report on the results of a field study in which evapotranspiration was estimated for different types of desert vegetation and for nearby bare soil at another Sonoran desert location, and found no significant differences in the long-term average evapotranspiration associated with vegetated areas and bare-soil areas. Sammis and Gay (1979) estimated that transpiration made up 7% of the water loss from a northern Sonoran Desert site with creosote bush (*Larrea tridentata*), as estimated by using a lysimeter.

The salt concentration in water in desert environments can be high, and temporally and spatially variable. Unfortunately, this means that estimating transpiration from deep-rooted plants in deserts is difficult because the transpiration can depend strongly on the ion concentration in the water supply. Experimental results from a six-year study with salt cedar showed that, for a decrease in the electrical conductivity of the water by a factor four, over an environmentally realistic range of values, the transpiration was observed to increase by a factor of six (van Hylckama 1970). This sensitivity can be important because the groundwater mineralization can vary greatly with location, from 5000 to 100 000 ppm dissolved solids (Motts 1965). Such dependence of transpiration on ion concentration is independent of the additional factor that different species of phreatophtye, with different transpiration characteristics, prevail in regions with different levels of groundwater mineralization.

Another effect of vegetation is on the porosity of the surrounding surface, which influences the retention of scarce rainwater through infiltration. For most desert conditions, the vegetation increases the porosity of the substrate, and thus the infiltration rate, by providing shelter for burrowing animals; by intercepting rainfall which, when heavy, can compact the surface and wash away sand, leaving stone cover; and by adding organic matter to the soil (Abrahams and Parsons 1991). After the vegetation dies and decays, the substrate in the vicinity may still remain more porous than originally unvegetated surroundings because of

Table 6.2 *Infiltration rates measured at three distances from plant stems*

The measurements near creosote bushes were made in two areas.

	Infiltration rate (cm min^{-1})		
Plant species	Near	Intermediate	Far
Paloverde	0.644	0.357	0.260
Creosote bush (1)	0.401	0.212	0.094
Both species	0.522	0.285	0.177
Creosote bush (2)	0.520	0.480	0.197

Source: From Lyford and Qashu (1969).

the disturbed soil in the root zone. Table 6.2 illustrates these effects based on measurements of the infiltration rate at three distances from creosote bushes and paloverde trees. The "near" locations were very close to the stem or trunk of the plant, the "far" locations were 1 m beyond the edge of the canopy of the plant, and the "intermediate" locations were one-half the "far" distance. The infiltration rate is roughly a factor of three greater near the vegetation, compared to more-distant locations.

A contrasting situation occurs in playas, where the existence of vegetation can lead to a *decrease* in the surface porosity. When there is no vegetation, and therefore no groundwater sink from transpiration, the water often rises to the surface through capillary flow from a subsurface water table or from direct discharge from a surface water table. These processes lead to the development of a rough, "puffy," porous character to the surface, which also often develops contraction cracks after drying. However, where phreatophyte roots draw upon the subsurface water, there may be no surface discharge to create the above effects, and the surface is hard and smooth (Motts 1965). Another mechanism by which some vegetation can inhibit infiltration results when a type of oil is exuded onto the sand grains beneath, which makes the surface hydrophobic, inhibiting the entry of water. Agnew (1988) describes how *Prosopis* vegetation (mesquite) in Oman (and that is ubiquitous elsewhere) has this effect. It is reasonable that such an organic surface coating would also inhibit evaporation. A similar very firm crust of 20–30 mm thickness was observed under a sausage tree (*Kigelia africana*) in Tanzania, where the soil outside the canopy was soft (Perrolf and Sandstrom 1995). There is much dispute about whether microphytic crusts increase or decrease water infiltration and evapotranspiration, with Dunkerley and Brown (1997) citing many references on both sides of the argument. Further discussion of the effects of vegetation on infiltration is found in Kironchi *et al.* (1992).

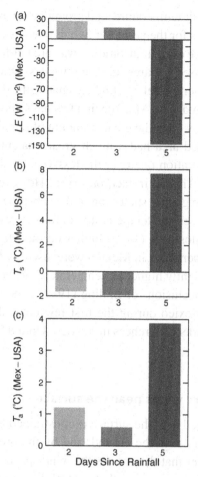

Fig. 6.10 Difference in daily-average surface evaporation rate (a), surface temperature (b), and air temperature (c) between two nearby paired areas on opposite sides of the Mexico–United States border, on days two, three, and five after a rain event. Because of grazing practices, the vegetation amount was greater in the area in the United States. (Adapted from Bryant *et al.* 1990.)

Differences in vegetation density across boundary fences offer opportunities for evaluating the effect of vegetation on the surface moisture and energy budgets. For example, in the Sonoran Desert, long-term overgrazing in Mexico has resulted in more bare soil and less plentiful grasses compared with the conditions on the United States side of the border. Bahre and Bradbury (1978) estimate that, in that area of the Sonoran Desert, Mexico has a 5% higher albedo and 29% more bare soil than adjacent land in the United States. For paired sites on opposite sides of the border, Bryant *et al.* (1990) have measured the effect of the vegetation differences on the water and energy budget after a rain event. Fig. 6.10 shows the average-daily surface evaporation rate, air temperature, and ground temperature differences between the two nearby sites for the period of two to five days after a rain event. With less vegetation, surface evapotranspiration was greater for the first few days after the rain event because the barer substrate was less shielded from sunlight, and the wind speed at ground level was higher. That is, the drying occurred more rapidly in the less vegetated area. After the first few days, the transpiration from the denser vegetation drew upon the greater amount of remaining subsurface rain water, and the vegetated area had a much greater latent-heat flux. A result of the greater early evaporation rate over the less-vegetated area should be lower surface temperatures. This is, in fact, observed. However, by day five the greater transpiration and the greater sheltering of the substrate from solar radiation in the United States caused the temperature contrast to reverse, with the surface temperature being almost 8 K (14 °F) higher on the Mexican side of the border. By day five, air temperatures in Mexico were also 4 K (7 °F) higher than in the United States. The slightly higher air temperatures in Mexico on days two and three contrast with conclusions of Balling (1989), who noted that air temperatures were cooler in Mexico during the first few days after a rain. Additional discussion of cross-border differences in Mexico–United States climate is found in Balling (1988).

The effects of vegetation on desert winds near the surface

The profile with height of the wind speed near the surface is constrained by the fact that the speed must go to zero very near the ground, and by the pressure gradient at the top of the boundary layer that determines the wind speed there. When the surface structures, such as bushes or trees, that control the roughness are not too close together, the wind speed increases approximately logarithmically with increasing height above the surface (see Chapter 10). In general, the larger the roughness elements, such as vegetation, that block the flow of the wind, the thicker will be the layer of affected air and the more gradual will be the increase of wind speed with height (Fig. 6.11). That is, for smooth surfaces the wind speed increases rapidly with distance above the ground. When shrubs or trees are very close together, however, the top of the canopy acts itself like a surface, so the base of the logarithmic wind profile, with a zero speed, may exist there. If the vegetation consists of shrubs or trees, where the lower stem or trunk has no foliage and there is space for air to flow, the wind speed may be greater near the ground than above in the vegetation canopy. The effect of desert-shrub spacing on the wind field is shown schematically in Fig. 6.12 in terms of a wake zone within which the wind field is modified by the obstacle to the flow. The regularity in the spacing of the shrubs in the figure is consistent with that condition

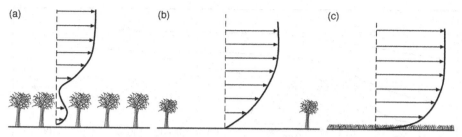

Fig. 6.11 Schematic of the vertical profile of the horizontal wind speed for (a) an area that is densely vegetated with trees or shrubs, (b) an area with similar but more scattered vegetation, and (c) an area with a grassy surface. (Adapted from Stull 1988.)

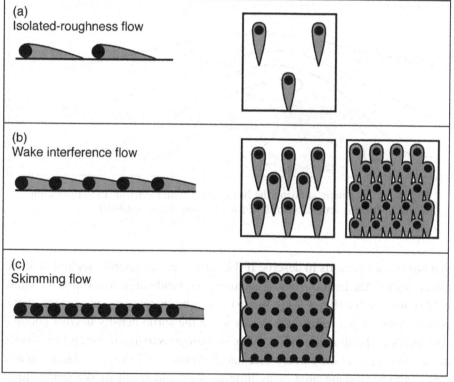

Fig. 6.12 Schematic of the area affected by isolated shrubs, or roughness elements. In (a), the shrubs are widely spaced. The wakes do not interfere with each other, and undisturbed areas exist. In (b) the shrubs are more closely spaced, and there is some wake interference. In (c) the shrubs are so densely packed that the flow skims over the top. (From Wolfe and Nickling 1993.)

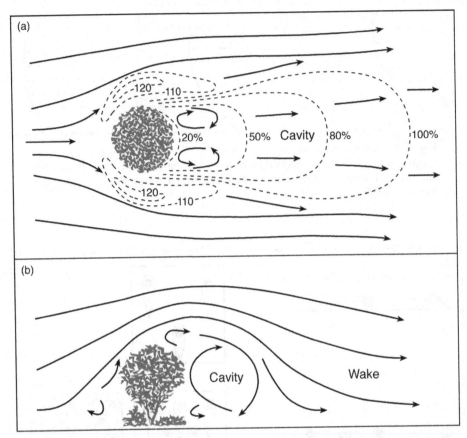

Fig. 6.13 Schematic showing the horizontal (a) and vertical (b) perturbation to the air flowing around a shrub. (From Ash and Wasson 1983.)

that sometimes prevails in deserts. If the shrubs are sufficiently separated, their effect on the wind field is isolated and there is considerable area where the wind field is undisturbed (Fig. 6.12a). When the shrubs are more closely spaced, there is some interference among the wakes when the shrub density is great enough (Fig. 6.12b). The situation in Fig. 6.12c is analogous to that in Fig. 6.11a, where the flow skims over the surface. Lee and Soliman (1977) suggest that a ground cover with a masking area of as little as 40% can result in skimming flow. Fig. 6.13 provides more detail about the nature of the flow around an isolated shrub, when there is no interference from other obstacles. Fig. 6.13a illustrates the horizontal perturbation to the wind field. There are relatively low speeds on the lee side, with counter-rotating vortices. Where the flow is channeled around the sides of the shrub, the speeds are greater than the undisturbed values.

The main feature in the side view of the flow in Fig. 6.13b is, again, the vortex on the lee side, and the general suggestion of turbulence near the obstacle.

As noted above, desert shrubs can represent an unusual collection of roughness elements because they are sometimes very regularly spaced, as a result of the fact that they compete with each other for substrate water. However, field studies over natural surfaces with such scattered and regular roughness elements have not been plentiful, especially those that consider the roughness-element spacing as a control variable. The spatial density, or concentration (C), of roughness elements is commonly quantified in terms of a dimensionless parameter defined as $C = hw/d^2$, where h and w are the height and width of the roughness elements (hw being the silhouette area), respectively, and d is the mean spacing of the elements. As noted earlier, for closely spaced vegetation roughness elements, the flow skims the surface of the top of the vegetation, with little interaction with eddies that exist between the vegetation elements, and the effect of the vegetation on the wind speed above the canopy is small. Conversely, for very widely spaced vegetation roughness elements, the roughness of the unvegetated substrate between the vegetation elements is most important in the interaction of the flow with the surface, and thus the effect of the vegetation on the roughness is also small. Thus, the roughness effect of the vegetation on the wind profile should be a maximum for some intermediate C. A wind-tunnel study of flow over regularly spaced geometric shapes (Raupach *et al.* 1980) shows that the roughness is a maximum for values of C between 0.1 and 0.2. However, Marshall (1970) estimates that, for shrubs in semi-arid southern Australia, values of C greater than 0.0253 correspond to no contribution of shrubs to the roughness. Lee (1991a,b) estimated the roughness based on wind-profile measurements at seven locations in the Mojave and Sonoran Deserts having a variety of vegetation roughness-element types with different spacing. The values of C for the measurement sites range from about 0.035 to 0.09, where the wind-profile-estimated roughnesses are larger at the sites with larger values of C. Using the above formulae for C, and assuming a roughness maximum at $C = 0.1$, roughness elements consisting of creosote bush with heights and widths of 2 m would produce a maximum roughness if their spacing were about 6 m. The vertical structure of the vegetation roughness elements is also important. Shrubs or trees with a low spatial density are rougher when they have a high center of gravity (most of the mass near the top), and those with a high spatial density are rougher when the center of gravity is low (Shaw and Pereira 1982).

An additional way in which vegetation can influence the roughness of the desert surface, and hence the low-level wind profile, is through its modification of the orography. There are two possible mechanisms (Motts 1965; Bullard 1997;

Table 6.3 *The response of various characteristics of the atmosphere and surface to an increase in the vegetative cover of an area with size greater than about 100 km^2*

The question marks indicate a greater degree of uncertainty in the sign of the response.

Increase	Decrease
Albedo in near-infrared wavelengths	Albedo in visible wavelengths
Absorption of solar energy	Infrared emission
Roughness	Wind speed near the surface
Turbulence intensity	Runoff
Evapotranspiration	Wind and water erosion
Net radiation	Maximum temperature
Substrate moisture retention	Substrate heat storage
Relative humidity	
Vapor pressure	
Minimum temperature	
Clouds (?)	
Rainfall (?)	
Mean upward motion (?)	
Sensible heat flux	
Rainfall infiltration	

Source: This is based on discussions in this chapter, as well as Anthes (1984), Charney (1975), Charney *et al.* (1975), Sagan *et al.* (1979), Henderson-Sellers (1981), and Dickinson (1983).

Bagnold 1954). One is related to the fact that phreatophytes in playas where the water table is high discharge large amounts of water through transpiration, which lowers the water table around the plants and causes subsidence of the surface. More spectacularly, vegetation captures wind-blown sand and silt which forms small hills known as *phreatophyte mounds*, or nebkas, or rhebda, which are surmounted by a single plant. These are commonly 1 m high, but can reach heights of more than 6 m and diameters of 25 m. For example, in areas of the Sahara, cones of sand rise as high as 15 m out of the flat desert plain. They are surmounted by a few branches of tamarisk that lean to leeward, struggling to survive (Bagnold 1935). A lee dune (also called shadow dune, or rehba) is similar to a phreatophyte mound except that the sand forms an elongated dune on the lee side of the plant rather than a symmetric mound.

As a summary of the previous discussion, Table 6.3 lists the differences in surface and atmospheric properties that are associated with bare and vegetated surfaces. The caveat noted in the table regarding the surface area being greater than 100 km^2 is associated with the fact that larger areas would be required to influence cloud and rainfall.

Suggested general references for further reading

Chapman, V. J., 1960: *Salt Marshes and Salt Deserts of the World* – describes the physiography and development of alkali deserts, the soil properties, and the physiology of halophytic vegetation. Specific information is also provided about some of the salt deserts of the world.

Hastings, J. R., and R. M. Turner, 1965: *The Changing Mile* – a thorough discussion of the vegetation variability with elevation and microclimate in the Sonoran Desert.

Hillel, D., 1998: *Environmental Soil Physics* – contains a chapter on soil–plant–water relations.

Oke, T. R., 1987: *Boundary Layer Climates* – discusses the climates of vegetated surfaces.

Shmida, A., 1985: *Biogeography of the desert flora* – a good summary of desert vegetation types, and how they relate to the geomorphology.

Questions for review

(1) What factors cause the net radiation to be larger over vegetated desert surfaces than over bare surfaces?

(2) Summarize the mechanisms by which vegetation influences soil permeability to water.

(3) How does the surface energy budget differ for a desert substrate with dead-vegetation debris relative to one with no debris? Does dead vegetation have a greater impact than live vegetation? Might there be desert substrates for which the existence of live or dead vegetation increases the aggregate albedo?

(4) Summarize the ways in which xerophytes adapt to reduced availability of soil moisture.

(5) Why are sensible heat fluxes to the atmosphere greater over vegetated areas than unvegetated areas?

(6) Review Table 6.3 and explain the different responses of the desert atmosphere and surface to the existence of vegetation.

Problems and exercises

(1) It is sometimes proposed that the natural vegetation cover in arid lands be removed in order to increase the discharge in rivers and streams for human use. Aside from the environmental sanity of such a practice, list the mechanisms by which the discharge would increase.

(2) It has been proposed (see, for example, Anthes 1984) that vegetating large areas of semi-arid lands might enhance the rainfall and reduce the aridity. Based on your knowledge of how the physics of the land surface and the atmosphere are affected by vegetation, speculate on the validity of this proposition.

(3) The discussion in Chapter 5 of thermal admittance was in the context of a heated substrate in contact with the atmosphere. How can this admittance concept be adapted to explain the large sensible heat flux to the atmosphere that results when vegetation is heated by solar energy?

(4) Under what conditions might the evapotranspiration over an area of unvegetated substrate be similar to that over a nearby vegetated area?

7

Substrate effects on desert surface physics

Deserts have been the setting for many expressions of the literary and visual arts because of the romanticism of the barren landscape. The following discusses a French film of army life in the Sahara in terms of the psychological impact of the desert.

The originality of the film lies in its treatment of the character's difficult physical, and above all, psychological adaptation to the desert, where people used to city life must come to terms with a scale of values that has been turned upside down. They discover the futility of trying to master time and space in a constantly changing environment. The sand erodes everything except the memory of passions. Even speech becomes meaningless when the characters realize that the desert is a world where life is measured out in silence.

> Mouny Berrah, Algerian sociologist and journalist
> *Screenplays in the desert* (1994)

The author describes areas of desert sand dunes, or ergs.

Instead of finding chaos and disorder, the observer never fails to be amazed at the simplicity of the form, and exactitude of repetition, and a genetic order unknown in nature on a scale larger than that of a crystalline structure. In places vast accumulations of sand weighing millions of tons move inexorably, in regular formation, over the surface of the country, growing, retaining their shape, even breeding, in a manner which, by its grotesque imitation of life, is vaguely disturbing to an imaginative mind.

> Ralph A. Bagnold, British scientist and adventurer
> *The Physics of Blown Sand and Desert Dunes* (1954)

Desert landforms, rather than the vegetation, represent our mental image of the highly arid landscape: bare-surfaced mountains from horizon to horizon; rock plateaus with deep gashes cut by now-dry rivers; perfectly flat and featureless vast stoney plains; and sand seas, some with dunes the size of small mountains.

The substrates[7.1] of which these features are built can have a wide range of mineralogical and hydrological properties that exert strong controls on the surface heat and water budgets of the desert. These physical properties include albedo, thermal conductivity, heat capacity, porosity, and water content. Chapter 5 treats the land-surface properties of the sandy desert substrate, and this chapter will extend the discussion to other substrates. The desert microclimate variability, which is the focus of Chapter 11, is a manifestation of, among other things, these substrate contrasts.

The desert substrates are largely composed of inorganic detritus having a range of particle sizes, where particle-size categories are commonly defined as follows.

- clay – very fine particles of hydrous aluminum silicate and other minerals, with diameters less than 0.002 mm
- silt – fine particles of rock with diameters greater than 0.002 mm and less than about 0.05 mm
- sand – fine particles of rock with diameters of greater than 0.05 mm and less than about 2.0 mm
- gravel – small rock particles that are larger than sand
- rock – consolidated solid mineral matter

Other definitions of particles in terms of size are shown in Jury *et al.* (1991). Because actual substrates are generally characterized by a range of particle sizes, descriptors are used such as loam (a mixture of clay, silt, and sand), silty clay, silty loam, sandy loam, and gravelly sand. Loess is a loamy deposit that has been chiefly deposited by the wind.

Desert substrates are distinctive in a number of respects. They are low in organic matter, often being composed primarily of sand, gravel or rock. The surface is often hard, with no protective vegetation canopy or layer of vegetation litter. The high substrate temperature near the surface contributes to a high evaporation rate when water is available, which leads to surfaces that are almost always dry. Soil porosity varies widely, from near-zero values for rock and clay to very large values for coarse sand. The paucity of vegetation roots contributes to low porosity, as does the compaction of the surface by large thunderstorm raindrops with high terminal velocity. There are many unique surface conditions that have little parallel in other climates. For example, desert pavement is a sheet of pebbles or gravel that is cemented to the substrate below. At higher desert elevations, the substrate may be dominated by solid rock. Below that, an

---

[7.1] The terms *substrate* and *soil* will be used somewhat interchangeably in this text, even though the biological use of the term "soil" has implications related to an ability to sustain vegetation. The geological use of the term soil is more similar to the term substrate; both refer to any type of surface and subsurface material, including bedrock.

alluvial fan contains particles whose sizes decrease in the downslope direction, graded by the water-erosion process, with a progression from gravelly sand, to sandy loam, to loam. Farther below, particles are generally small, and the water table is occasionally near the surface in bare-soil depressions or flat basin floors. Here, there is generally a significant salt content, with surface crusting common.

The following section describes in more detail the various types of surface that are common in deserts. These surface types and landforms are similar in most deserts because of the prevalence of the same water and wind erosional processes that have configured the surfaces over millennia. The next two sections describe how the substrate types influence the surface energy and water budgets.

Types of desert substrate

Salt flats, playas, salars, sabkhas, salt pans, etc.

In the lower portions of arid basins with internal drainage, flat and generally barren areas exist. They periodically flood and accumulate sediment, and sometimes have water tables near the surface. Salt crusts of various depths are common. In most English-speaking places they are called *playas* (Spanish for beach or shore). But their commonness throughout the arid world is reflected in the fact that they go by a variety of names: *salar* (Chile and other Spanish-speaking places); *sabkha* or *sebkha, chott* (Arabia, northern Africa); *qa* (Jordan); *pans, salt pans* (southern Africa, Australia); *takyr* (Russia); and *kavir* (Iran). The term *dry lake* is used informally sometimes, referring to the fact that these features were often non-ephemeral lakes during wetter climates, but this is paradoxical because lakes are wet most of the time and playas are dry most of the time (Neal 1975).

On the small end of the size spectrum, playas may be only a few tens of square meters in area, but large ones may exceed $10\,000$ km^2. Most of the larger playas were lakes during previous pluvial periods (e.g. the Pleistocene), as sometimes reflected by remnants of old shorelines. The smaller playas originated as a result of regional erosion patterns. There are over $50\,000$ playas on Earth (Rosen 1994), but perhaps fewer than 1000 cover more than 100 km^2 (Neal 1975). Using North America as an example, there are approximately 300 playas with areas of more than 5 km^2, virtually all being in the desert environments of the West (Neal 1965), with $30\,000$–$40\,000$ smaller playas in the semi-arid grasslands of Texas and New Mexico (Steiert 1995; Shaw and Thomas 1997).

Playas can have a variety of different characteristics in spite of the fact that they share the common name, as discussed in Neal (1975), Menenti (1984),

Motts (1965), Rosen (1994), and Lines (1979). For example, of two types of hard-crusted playa, one can be composed of almost 90% clay with a salt content of a few percent, while the other can be almost pure salt. When the water table is nearer the surface, some playas have a soft, sticky, wet surface. Other types with a near-surface water table have a soft dry surface.

Depending on the local hydrology and rainfall climatology, the existence of playa salt crusts can vary greatly from day to day or with season. Dry meteorological conditions or a high water table associated with a positive hydrostatic head can be conducive to crust formation and deepening (Motts 1970), whereas surface flooding from precipitation or runoff inhibit it.

Sand surfaces

The origin of most desert sand is Pleistocene (see Chapter 15) streams, now dry, which formed and deposited the sand during more humid times. Some of the desert sand, however, originates along present ocean shorelines. It is estimated that local production of sand by abrasion of coarser rock is greatest for annual precipitation amounts of about 30 cm. Much more precipitation and there would be sufficient vegetation to stabilize the substrate to erosion; much less and there would be too little runoff to grind the rock into sand. The present locations and formations of the sand are a result of its movement from its sources by the wind. This wind-transported sand is called aeolian sand. Aeolian sand can exist in the form of relatively flat sheets, plains, or fields where sand flow-through dominates over sand accumulation. Where sand accumulates, dune fields (ergs) form. The combined areas of flat sand sheets and dunes are called sand seas; dunes typically average 60% of the sand-sea area. Dunes, which take on various geometrical forms, can be either active (moving) or stable (fixed). Sand fields and dunes cover 20–50% of the surface area of many hot deserts, but the percentage is quite variable. For example, sands cover less than 1% of the deserts of the Americas (Lancaster 1995), but almost half the area of the Australian Deserts (Mabbutt 1977).

The size distribution of ergs is fairly peaked toward the larger sizes (Cooke and Warren 1973). Isolated dunes of small to medium size are fairly rare. It has been estimated that 99.8% of aeolian sand is found in ergs that are greater than 125 km^2 in area, and 85% is in ergs of greater than 32 000 km^2 in extent. The modal size is about 188 000 km^2 (Cooke and Warren 1973), with the largest being the Rub Al Khali sand sea in Arabia whose area is over 560 000 km^2.

Dunes can be classified in terms of their relationship to the formative winds. Linear or longitudinal dunes have long axes that are roughly parallel to the wind and the resultant sand drift direction. Transverse or crescentic dunes have their

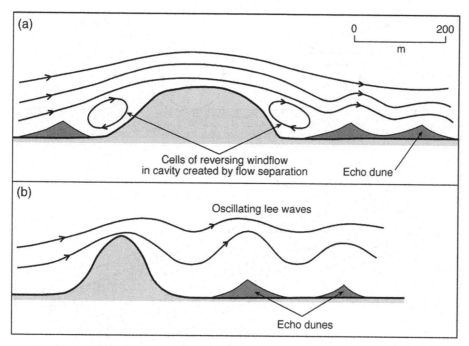

(a)

0   200

m

Cells of reversing windflow
in cavity created by flow separation   Echo dune

(b)

Oscillating lee waves

Echo dunes

Fig. 7.1 The airflow over elevated terrain in the sandy desert, and how it influences nearby dune formations. (Adapted from Thomas 1997b.)

major axes perpendicular to the drift direction. Crescent-shaped, barchan dunes are in this category. Star dunes have at least three arms radiating from a central peak, and result from complex and changing wind regimes. Thomas (1997a) contains a good discussion and depiction of the dune types.

Because local orography can influence the wind flow, the orography's signature can also often be found in the locations and shapes of dunes. Fig. 7.1 is a schematic that shows two wind-flow patterns over a terrain-elevation maximum. In the first case (Fig 7.1a), there are no orographically produced waves in the airflow on the lee side, but vortices are created at the base of the terrain slope where the large-scale flow separates from the ground. The scouring and shaping of the sand by the vortices can create dunes at the base of the slopes. The airflow over the first lee-side dune is perturbed, and produces an echo dune. When the large-scale airflow develops lee waves (Fig. 7.1b), these can also be seen in the downwind dune pattern.

One process by which linear dune fields (called **siefs**) form reveals something about prevailing local atmospheric winds. These dunes have been observed to have lines of shallow cumulus clouds, called "cloud streets," above the dune crests, with cloud-free areas between the dunes. It is speculated (Goudie and Wilkinson 1977) that these dunes might be formed by the development of

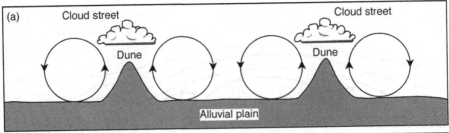

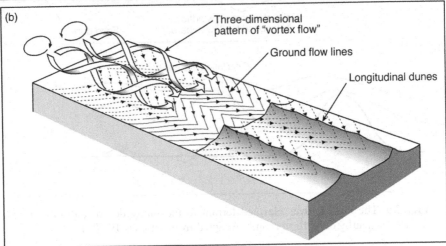

Fig. 7.2 (a) Schematic cross-section of circulations associated with roll vortices over a sandy desert surface. The roll vortices may be responsible for the development of longitudinal dunes and the clouds that appear to be aligned with the dune ridges. (b) Three-dimensional structure of the roll vortices. (Adapted from Goudie and Wilkinson 1977.)

helical-roll vortices within a steady prevailing mean wind field (Fig. 7.2). With sufficient water vapor available, clouds are produced in the upward-motion field within the vortices. The upward motion over the dune crests might be reinforced by the surface heating through the mechanism illustrated in Fig. 2.6f. Additional discussion of the wind flow around dune forms can be found in Mainguet (1991).

Pavements, crusts, and varnishes

A hard, smooth, stony surface in the desert, formed by a few different mechanisms, is called desert pavement or stone pavement. This is one of the most-common surface types, with other local names including *gibber plains* or *stony mantles* in Australia, *gobi* or *säi* in central Asia, and *reg* in northern Africa. One formation process involves prolonged wind erosion and drought (Hayden 1998). During drought periods, plants die and constant wind blows away the

Box 7.1
Desert pavement as a landing strip

Antoine de Saint-Exupéry (1939) was a legendary French aviator and writer, and flew for the first airmail service. He describes his luck at encountering desert pavement, although he does not use the term, when crash landing his aircraft in 1935 in the Libyan Desert

We owed our lives to the fact that this desert was surfaced with round black pebbles which had rolled over and over like ball-bearings beneath us. . . . The ground was sand, covered over with a single layer of shining black pebbles. They gleamed like metal scales and all the domes (dunes) about us shone like coats of mail.

surface layer of light substrate. This raises to a common level the stones within the soil, which form a tight surface mosaic (see Box 7.1). The washing away of finer material by water could also result in this type of surface. Saline clay, from the weathering of the rock, forms just below the layer of stones, enhancing the tightness of the mosaic and preventing the regrowth of vegetation during succeeding wet periods. Also, repeated freezing and thawing of moist stony soils in colder deserts forces the embedded stones to the surface (Campbell 1997). Similarly, expansion and contraction of clays, in the wetting and drying process, can force stones to the surface. Many desert pavements are essentially permanent, and represent a "climax" state. The pavement may be various shades of gray, black, or brown, depending on the thickness of the desert varnish coating (discussed below).

Another type of crust that is formed by chemical action can be much deeper than the desert pavement. This cement-like layer of **hardpan** is formed by the accumulation of various salts at or near the surface. Gypsum crusts (gypcrete); calcium, or lime, crusts (calcrete, called *caliche* in North American deserts); silica crusts (silcrete); halite crusts (salcrete); and iron-rich crusts (laterites) are either currently forming because there is insufficient precipitation to completely leach the salts from the substrate, or they formed during previous periods. When the hardpan forms as a result of drying in a substrate where rainwater develops a high dissolved mineral content, the depth of the layer depends on how deeply rains penetrate. In southwestern North America, the caliche forms where the rainfall is between 10 and 45 cm per year (Reeves 1976). It characteristically forms in well-drained gravelly substrates, which are especially common in regions of alluvial fans. Within the same drainage basin, the differing mineralogical composition of the fans causes different degrees of calichification (Lattman 1983). Gypsum crusts are common in parts of the Namib Desert and the Australian Deserts, and may be 30–100 cm thick. Calcrete layers can be 50 m thick. Crusts can also develop as a result of upward movement of groundwater and salts to the surface, where surface or near-surface evaporation leaves the salt to accumulate in a crust. Dunkerley and Brown (1997) contains a detailed discussion of these types of crust.

Rocky desert substrates sometimes undergo weathering processes, which is relevant to this discussion because the weathering can significantly affect the albedo and thus the surface energy budget. A shiny red or blue-black coating, called "desert varnish," is created by bacterial action and water-borne deposition of manganese and iron oxides. Even though a varnish can sometimes form in as little as 25 years, the normal rate is probably much slower (Goudie and Wilkinson 1977). El-Baz (1986) states that sand is darkened by this process in the North American, Namibian, Australian (Sturt), Arabian, and eastern Saharan Deserts. See Dunkerley and Brown (1997) for further discussion.

Rock surfaces

Rock substrates can consist of large expanses of relatively solid bedrock, or they can be the crumbled rocky material that remains after any finer-grained material has been carried away by the wind. Such surfaces can be quite varied, and include rugged fields of large rocks or boulders (*hammadas, hamadas*) on plateaus, or large plains at lower elevations that consist of gravel or pebbles that have been deposited by ancient rivers or alluvial fans (*serir* in the central and eastern Sahara, *reg* in the western Sahara). The cemented crust of small stones, described above, is also a type of rock surface. At least 40–50% of the area of hot deserts is typically made up of rocky surfaces (Evenari 1985a), and Monod (1973) estimates the percentage to be as large as 80% for the Sahara. In addition to the above types of rocky plains, large fractions of desert surfaces can consist of mountains that are virtually all rock. Large, isolated and sometimes bizarrely shaped rock features protruding from relatively flat surroundings are common because of differential erosion of contrasting types of rock; the features can be called *inselbergs* or *caprocks*. When a mountain erodes, the rock fragments are carried downslope by water, and alluvial fans form in the plains at the ends of the water channels. When these fans merge laterally, they form a laterally contiguous debris field that is called a **bajada**. The characteristic conical profile at the sides of distant isolated terrain features is evidence of bajada formation.

Dry river beds

Dry river beds (*wadi* in northern Africa, *arroyo seco* in Spanish-speaking areas) are another ubiquitous feature of hot deserts. They occasionally fill with water and debris after heavy rains, but these are often rare occurrences. The dry river

Table 7.1 *Comparison of the percent of the total area associated with different surface types for a few deserts*

Each column adds up to 100%.

Group	Southwest USA	Sahara Desert	Libyan Desert	Australia	Arabia
Playas	1.1	1	1	1	1
Bedrock fields (including hammadas)	0.7	10	6	14	1
Desert flats	20.5	10	18	18	16
Regions bordering through-flowing rivers	1.2	1	3	13	1
Fans and bajadas	31.4	1	1	—	4
Dunes	0.6	28	22	38	26
Dry washes	3.6	1	1	—	1
Badlands	2.6	2	8	—	1
Volcanic cones and fields	0.2	3	1	—	2
Mountains	38.1	43	39	16	47

Source: Adapted from Clements *et al.* (1957)

beds are sometimes very broad, with steep side walls, because the floods often erode more laterally than vertically.

Loess

Loess is made up of silt and clay particles, with some sand, and is composed mainly of quartz and calcium carbonate. Silt-sized particles are produced by mechanical grinding in glacial environments, and salt weathering and aeolian abrasion in deserts. The loess of glacial origin is generally transported by the wind from its region of formation. Loess is thus aeolian silt. There are large areas of loess in arid areas, including the Negev Desert, the Tunisian Sahara, and the arid areas of China, India, and Pakistan. Loess deposits occupy twice the area on Earth of aeolian sand.

An example of the spatial distribution of substrate types

An example of the distribution and variability of desert substrates is shown in Fig. 3.9 for the Australian Desert. There is, however, much microstructure to the landscape that cannot be described in large-scale analyses such as this. Table 7.1 lists the fractional coverage of surface types for a few different desert

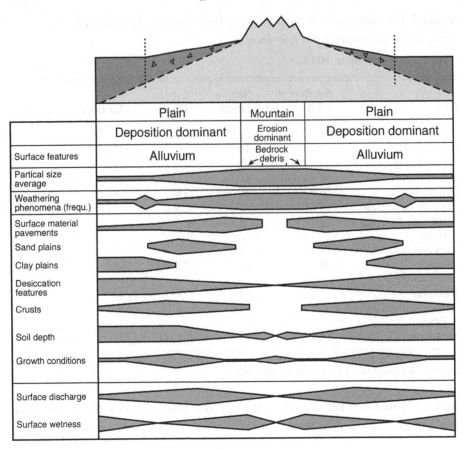

Fig. 7.3 A cross-section of a typical mountain, alluvial slope, and valley terrain profile, and the associated surface conditions. The width of each bar is proportional to the existence of the property or process. (Adapted from Cooke and Warren 1973.)

areas. There is a lack of uniformity in the use of various scientific and local terms that describe types of desert substrates, and this makes it difficult to compare statistics. For example, from Table 7.1 it is difficult to estimate the percent of the surface covered by sand, because the "dune" is the only sand category listed explicitly.

Because erosion processes play such a strong role in the distribution of desert substrates, it is informative to employ a schematic, such as that in Fig. 7.3, that relates substrate properties to the process by which mountains erode to form alluvial slopes and valley sediments. Where the solid rock substrate of the mountain is exposed, erosion is dominant. Surrounding this area, erosion products are washed downslope, with the finest particles reaching the lowest elevations to form fine sediment in the inter-mountain plains. Pavements require gravel-sized rock, and would thus be found relatively high upslope near the source of the rock. Sand plains are farther downslope where average particle sizes are smaller. Even farther downslope are clay plains in the inter-mountain valleys, which perhaps contain playas. Precipitation, which falls over the mountains, flows downslope on the surface within channels, and infiltrates the alluvium to flow below the surface. The infiltrated water can surface and produce discharges that form oases along the slopes of the alluvium. The overall wetness of the surface would be potentially greatest at higher elevations where the rainfall is greater and where larger rocks would impede sheet runoff and enhance local water retention. The surface wetness would also be high at the bottom of the alluvial slope where the water table is high, for example below a playa surface.

Box 7.2
Singing, booming, roaring, thundering, musical sands

These words have been used by various desert travelers to describe the sounds that emanate from sand flowing down the face of a high dune. Ralph Bagnold (1954) was camped in the Sahara in the 1930s.

Suddenly . . . a vibrant booming so loud that I had to shout to be heard by my companion. Soon other sources, set going by the disturbance, joined their music to the first, with so close a note that a slow beat was clearly recognized. This weird chorus went on for more than five minutes continuously before silence returned and the ground ceased to tremble.

Similarly, Mildred Cable (Cable and Francesca 1942) was exploring the Gobi Desert in about the same period: "Once at midnight we were awakened by a sound like a roll of drums . . . " Cable also describes other travelers' report of the sounds of a ship's siren coming from sand hills in the Arabian Desert. Marco Polo referred to sounds like distant thunder coming from sand hills. Baron von Humboldt believed the sounds heard so often at night from Peruvian sands were from water that moved underground as the temperature changed. Over 1000 years ago, a Chinese writer referred to the "hill of the sounding sand." Native legends have attributed the sounds to the bells from an underground monastery long ago engulfed by the sands, or to sirens who try to lure travelers to a parched doom. For reasons that are not yet determined, most accounts of the sounds refer to them occurring late in the day or at night. It is possible that the sands must be dry, because the sounds are sometimes only reported in the dry season, and not after a rain (Yarham 1958). The "barking" sound that sand sometimes makes when walked upon may be from the same cause.

Effects of desert substrate types on the surface energy budget

The discussion of land-surface physics in Chapter 5 serves as an introduction to the influence of substrates on the surface and subsurface energy budget. This section will expand upon that general discussion by considering the land-surface physics of different desert substrates. For example, albedo variations among the different substrates can have important effects on the energy budget. In particular, playa albedoes are especially complex. There can be large seasonal variations in the albedo of playas with thin salt crusts. In the summer, the salt crust can develop to a few millimeters in depth and have a relatively high albedo. However, in the winter, or after heavy rain in the summer, the salt can dissolve and the albedo decreases significantly. McCurdy (1989) reports that when a Great Basin Desert playa salt crust was moistened from rain its albedo was 0.24, but when it was completely dry it had a late afternoon albedo that exceeded 0.75. Even significant *diurnal* variations in albedo result from the fact that, when the water table is high, the crust can rehydrate at night, and then become progressively drier and more reflective during the day (McCurdy 1989). Also, wind speed and direction can influence the intra-playa distribution of albedo. When, as is sometimes the case, 5–10 cm of water covers the playa, wind can shift the location of the shallow pond of water from one location to another, and thus the albedo distribution shifts as the changes in wind speed and direction dry and moisten different areas of the playa.

Desert varnish, whose formation was discussed earlier, can significantly affect the albedo. Reeves *et al.* (1975) showed that weathered volcanic rock has an albedo that is 0.05–0.15 lower than that of unweathered rock, depending on the wavelength. Sand albedoes also show the impact of weathering. For example, in the Namib Desert where dunes migrate from west to east away from the source of the sand, the older sands to the east, 30–120 km inland, are of brick red color while the sands to the west near the coast are a brighter white-yellow (Logan 1960). Similarly, in Australia, dune reddening is observed with increasing distance from the sand source (El-Baz 1986). In the Sahara, "blackened rock and soil" are common (Arritt 1993). Such patterns are seen clearly in photographs from space. Thus, in defining the albedo of a rocky or sandy desert substrate for energy budget calculations, one has to be aware of the degree of local weathering, and not simply use a standard tabulated value.

Figure 7.4 summarizes the reflectance, at various wavelengths within the visible spectrum, for different desert substrates. The one surface with some vegetation was, nevertheless, primarily sand covered. The basaltic lava has an extremely low albedo, while the salt-covered playa surface has a very high albedo. Such contrasts can sometimes occur on quite small scales in deserts.

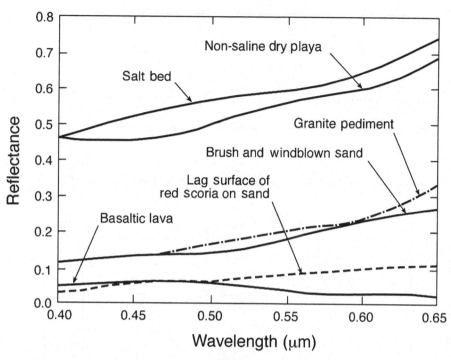

Fig. 7.4 The reflectance of different desert substrates as a function of wavelength within the visible spectrum. (From Ashburn and Weldon 1956.)

For example, the white gypsum sand in the northern Chihuahuan Desert has an albedo that is even higher than that of the salt bed in the figure, and these sands are within a few tens of kilometers from basaltic lava flows with an albedo of less than 10%.

Thermal conductivities and heat capacities also vary greatly among the substrates, and influence the surface heat budget. For example, Fig. 5.10 illustrates the diurnal variation in temperature above and below the surface for a sandy substrate in the Sonoran Desert, averaged over two summer months. Similar data were also obtained from two nearby locations: one had a surface of desert pavement and the other was at a higher elevation with a deep, rock substrate. The albedo of the sand was measured to be 35% and that of the desert pavement was 15%. Fig. 7.5 shows the difference between the August diurnal temperatures above and below the surface, for the sand and desert-pavement substrates (temperatures for desert pavement minus temperatures for the sand). The greatest difference between the two locations is at 2.5 cm below the surface, where the desert pavement substrate is 7.8 K (14 °F) warmer than the sandy substrate. The surface temperature of the sand is 2.2 K (4 °F) warmer than that of the desert pavement in mid-afternoon, and in the early morning the sand surface is 3.9 K

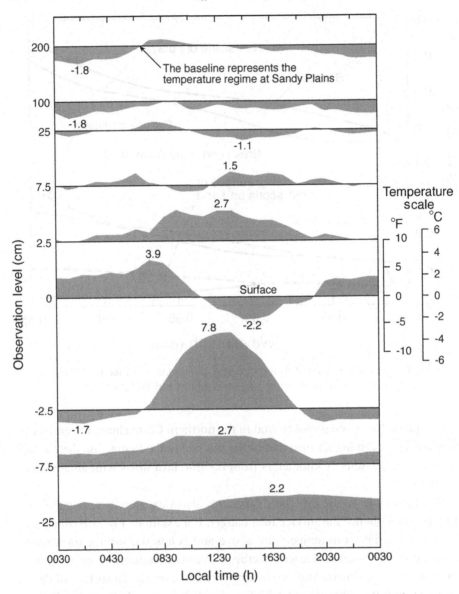

Fig. 7.5 The temperature difference between a sandy substrate and a substrate consisting of desert pavement (temperatures for desert pavement minus temperatures for sand) for different heights and depths. The temperatures for the sandy substrate are shown in Fig. 5.10. The numbers on the curves indicate the maxima or minima (°C), (From Dodd and McPhilimy 1959.)

(7 °F) cooler. The thermal diffusivity of the dry sand is smaller than that of the stone pavement (see Table 5.1), less of the heat gained during the day penetrates into the sandy substrate, and the afternoon surface temperature is higher and the subsurface temperature is lower for the sand. The night surface temperature is higher for the stone pavement because of the greater amount of heat stored by the substrate. An analogous plot in Fig. 7.6 for the rock surface shows a qualitatively similar relation to the conditions for the sandy substrate. That is, compared to the sandy substrate, night surface, subsurface, and atmospheric temperatures are 3–4 K (5–7 °F) higher, and the afternoon surface temperature is 5 K (9 °F) cooler. These departures from the diurnal temperature cycle for sand, shown in Fig. 5.10, are complicated by the fact that, for all the substrates, there is likely a vertical gradient in the substrate composition, and hence in the conductivity and heat capacity as well. For further examples of substrate effects on the heat budget, see Whalley *et al.* (1984), who describe diurnal surface temperature variations for a variety of different types of rock surfaces in the Karakoram Mountains to the southwest of the Taklamakan Desert.

Playa substrates also strongly contrast with their surroundings in terms of the thermal diffusivity. The often moist playa substrate and its compactness relative to sand have a profound effect on the thermal conductivity and the heat capacity. For example, Tapper (1991) estimates the thermal conductivity for salt-lake sediments and nearby non-playa sand to be 2.32 W m^{-1} K^{-1} and 0.33 W m^{-1} K^{-1}, respectively.

Effects of desert substrate types on the surface water budget

The rate at which water infiltrates desert surfaces varies widely depending on the substrate. At the one extreme are some sandy surfaces. For example, Agnew and Anderson (1992) report infiltration rates of over 500 mm h^{-1} for sands in Niger and Oman. Such high permeability to water is not characteristic of many desert surfaces, which often have only thin layers of soil, have significant slope, have surface crusts, or have sheets of rock at or near the surface. This localized or large-scale low water permeability is reflected in the fact that more people perish in the desert from an excess of surface water, i.e. drowning in floods, than from a lack of it, i.e. thirst (Nir 1974). Childs (2000) notes

This place is stained with such ironies, a tension set between the need to find water and the need to get away from it.

Hoogmoed (1986) suggests that typical desert infiltration rates are 30 mm h^{-1}, and that the percentage of annual rainfall that is partitioned to surface runoff

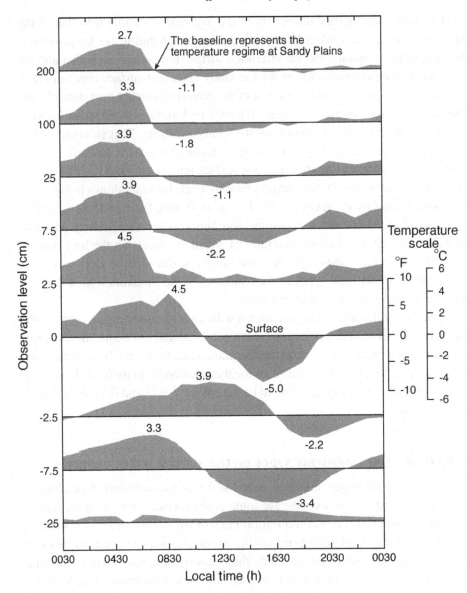

Fig. 7.6 The temperature difference between a sandy substrate and a substrate consisting of solid rock (temperatures for rock minus temperatures for sand) for different heights and depths. The temperatures for the sandy substrate are shown in Fig. 5.10. The numbers on the curves indicate the maxima or minima (°C), (From Dodd and McPhilimy 1959.)

at various desert sites in India and Niger ranges from 5 to 15%. Wells (1983) states that, for the North American deserts, of the generally less than 250 mm of annual rainfall, less than 25 mm (about 10%) is partitioned to runoff. It is clear that this partitioning will vary greatly among different desert substrates, and will depend on the characteristic intensity of the rainfall from convective or stable processes. The pervasive evidence of often severe water erosion in most desert environments certainly attests to the existence of sheet flow of surface water during at least the more severe rainfall events.

A process that tends to seal the surface and reduce infiltration is the pounding of the soil by raindrops that have not been intercepted by vegetation. The fine clay that is common to desert substrates is dispersed as a colloid in the surface water, and the colloid swells and seals the surface to penetration by air and water. In the Negev Desert, Peel (1975) reports that the infiltration rate for undisturbed soil with such a sealed surface was 2 mm h^{-1}, but when the surface layer was disturbed, the infiltration rate was 10 mm h^{-1}. When the soil was covered and protected by stones, the physical effects of the rain did not form the compacted layer, and the infiltration rate was high. If the stones were removed, after a few rain events the surface became more impermeable and the runoff greatly increased. Yair (1990) reports similar results, also for the Negev Desert. For fine sandy loam, McIntyre (1958) found the permeability of the soil beneath a surface rainfall-impact seal to be about 200 times that of the lower part of the seal and 2000 times that of the upper part of the seal. See Dunkerley and Brown (1997) for further discussion of such inorganic crusts or seals.

The hydrologic implications of hardpans, for example caliche layers, are obvious. Impervious layers such as this represent a barrier to upward or downward movement of liquid water or vapor, and, depending on their depth and thickness, inhibit or prevent the growth of many plant species whose existence strongly influences the surface energy and moisture budgets. The hardpans also affect flooding. For example, Lattman (1983) reports that flooding of alluvial slopes (see, for example, Fig. 7.3) in an area of the Great Basin Desert is much greater in magnitude and frequency for calichified than for non-calichified substrates. For the same geographic area, Cooley *et al.* (1973) describe the effect of caliche on infiltration and state that a major factor in determining infiltration, runoff, and flooding is the depth below the surface of the caliche layer.

When the alluvial fan around eroding mountainous slopes does not contain caliche, there is often an elevation dependence to the permeability of the surface to water. The rocky debris high on the slope will slow down the sheet runoff from heavy rainfall, which will aid in penetration. And, the coarser sand and gravel that are less high on the slope will allow the runoff from brief storms to penetrate more quickly than will the finer soil particles even farther downslope.

A manifestation of the resulting gradient in soil moisture can sometimes be seen in contrasts in vegetation density as a function of elevation.

It should be emphasized that, for the above reasons, *lateral* water movement in deserts is thought to be primarily through overland (Hortonian) flow rather than through subsurface flow, although Agnew and Anderson (1992) review arguments that this is an oversimplification. Nevertheless, the common occurrence in arid lands of shallow soils, rocky substrates, hardpan layers, and heavy rainfall of short duration points toward overland-flow dominance.

Regarding playas, the texture and chemical composition of the near-surface soil determine the degree to which subsurface water is available for evaporation at the surface and the rate of infiltration of rainwater. For the Bonneville Salt Flats in the Great Basin Desert, Lines (1979) estimates that the infiltration of precipitation and wind-driven surface brine by thick perennial salt crusts is almost five times greater than that by more silty playa surfaces. In addition, evaporation rates from thick salt crust was two to over six times greater than from the clay–silt type of playa surface. This is partly a result of the fact that **osmotic** forces exist when there is a salt gradient, causing water to move toward the greater salt concentration (which is generally at the surface). For some playas, significant upward latent-heat fluxes in the atmosphere are only observed in the early morning after the surface has been moistened through a night of water rise from capillary and osmotic effects. After the surface evaporation rate has been elevated during the morning, the playa surface dries and the latent-heat flux to the atmosphere dramatically decreases. For other types of playa surfaces, the rise of water more rapidly replenishes the water at the surface that is lost by evaporation, and the latent-heat flux is more important in the surface energy budget throughout the day. A more-complicated pattern was described by Pike (1970), for a playa in Saudi Arabia, where the evaporation rate decreased during the morning, with a rapid increase in the sensible heat flux, but the evaporation increased again during the afternoon. This pattern is similar to that described earlier for the lysimeter study of Hellwig (1973a,b; 1978). Even though very coarse-grained playa sediments allow more rapid direct penetration of water from a high water table, the capillary discharge of water at the surface is not as great as it would be with finer-grained sediments. "Sand lenses" embedded within a fine-grained sediment can effectively block all capillary rise because of their coarse-grained nature, and render some areas of a playa as non-discharging and dry at the surface (Motts 1965).

The existence of sometimes deep, polygonal contraction cracks in playa surfaces can make the estimation of evaporation difficult because the cracks expose the water table more directly to the atmosphere. It has been estimated by Menenti (1984) that the upward water-vapor flux in the cracks could be as much as two orders of magnitude larger than what would be expected from a diffusion-like process in the substrate.

As an illustration of the large contribution of playa evaporation to desert hydrology, Menenti (1984) calculated that the groundwater loss by evaporation from the approximately 20 large playas in southern California represented 6% of the total groundwater exploitation (1979 usage) by agriculture for the entire state. Pike (1970) also states, based on measurements, that groundwater evaporation from playas in Saudi Arabia is large, and important to the local water budget. His measurements showed losses of 50–84 mm per month, with an annual loss of 0.73 m. Note that estimating long-term evaporative losses from playas where rainfall is significant can be difficult, unless long measurement periods are employed, because water tables can sometimes change quickly after rainfall events.

When estimating evaporation from damp surfaces and standing water in playas, and from lakes in arid areas, the salinity of the water must be accounted for because the dissolved salt affects the saturation, or equilibrium, vapor pressure over the liquid. The effect is a reduction of the evaporation rate. The salinity (percent salts, by mass) of playa water is generally high, and it is also often quite high for desert lakes. Whereas sea water has a salinity of 3–4%, it is 18% for the Great Salt Lake (Great Basin Desert), 24% for the Dead Sea (Middle East), and 33% for Lake Van (Turkey). Allison and Barnes (1985) speculate that this salinity effect accounts for the fact that the annual evaporation from an almost saturated sediment in a playa in Australia was only about 5% of what would be expected from a surface of pure water. Water salinity also has a significant impact on the hydraulic conductivity of soils and the infiltration rate of water into soil (Agassi *et al.* 1996).

The effects of desert pavements on the hydrology are unclear. Contrary to the intuitive conclusion that the pavement would increase runoff, Thames and Evans (1981) speculate that the roughness of the surface would impede it, and that a vesicular layer that sometimes exists below the pavement might actually enhance infiltration.

Suggested general references for further reading

Cooke, R. U., and A. Warren, 1973: *Geomorphology in Deserts* – includes a discussion of desert hydrology, surface conditions, fluvial landscapes, and aeolian processes.

Cooke, R., *et al.*, 1993: *Desert Geomorphology* – includes discussions of desert surface conditions, fluvial domains in deserts, wind in desert geomorphology, and the evolution of desert landforms.

Issar, A. S., and S. D. Resnick, 1996: *Runoff, infiltration and subsurface flow of water in arid and semi-arid regions* – includes hydrologic balance studies of two desert watersheds, a discussion of the effects of the physical and chemical properties of arid soils on infiltration rates, a description of the quantitative prediction of runoff events in the Negev Desert, and a description of soil sealing, infiltration and runoff.

McKee, E. D., *et al.*, 1977: *Desert sand seas* – includes satellite images of various desert sand seas.

Neal, J. T. (Ed.), 1975: *Playas and Dried Lakes* – discusses the climatological causes of playas, their hydrology, the variations in playa types, and their changes and surface features.

Perrolf, K., and K. Sandstrom, 1995: *Correlating landscape characteristics and infiltration – A study of surface sealing and subsoil conditions in semi-arid Botswana and Tanzania* – a summary of the effects of surface seals on evaporation in semi-arid areas.

Thomas, D. S. G., 1997a: *Arid Zone Geomorphology: Process, Form and Change in Drylands* – a review of dune types, and extensive general reference on surface properties and processes in deserts.

Questions for review

(1) Summarize the different particle-size categories for substrates.

(2) Explain why a moist playa substrate would remain cooler during the day and warmer at night, relative to a nearby dry, sandy substrate.

(3) Describe two mechanisms by which playas form.

(4) How much of an effect can desert varnish have on the albedo of sand or rock?

(5) Explain why the afternoon surface temperature for the deep rock substrate was less than that for the sandy substrate in Fig. 7.6.

(6) Explain the different processes that can lead to the formation of desert pavement.

(7) Over what range of values can desert substrate albedoes vary? How do you think that this variability compares with that for non-desert substrates or for vegetated non-desert areas?

(8) Review the different types of substrate in the region of alluvial slopes, and how the infiltration rate depends on location.

Problems and exercises

(1) If it is assumed that the "dust" that is transported great distances in dust storms is fine silt with a diameter of 0.002 mm, and that the sand particles that move shorter distances within sand seas have a diameter of 1 mm, calculate the ratio of the masses of the two types of particle. For simplicity, assume that the densities are the same.

(2) Using tabulated values from earlier chapters or other sources, estimate the thermal diffusivity ratio of a moist playa substrate and the surrounding dry sand.

(3) In the schematic of Fig. 7.3, speculate about where natural oases would exist.

(4) The term "patterned ground" is used to refer to any desert surface in which the occasional drying of moist substrates causes the development of regularly shaped polygonal cracks. Where would surfaces be found that develop such patterns?

(5) Find a good reference on the types of desert sand dune, and describe them.

(6) Using Eqn. 5.4 and tabulated values of the heat capacity and conductivity of different desert substrates, calculate the depth at which the annual temperature wave is 10% of the surface value.

8

Desert-surface physical properties

Charles Darwin, after his visit to the Patagonian Desert of South America in the last century, was puzzled by his degree of fascination with it.

I find that the plains of Patagonia frequently cross before my eyes . . . They can be described only by negative possessions; without habitations, without water, without trees . . . they support merely a few dwarf plants. Why then . . . have these arid wastes taken so firm possession of my mind.

Charles Darwin, English naturalist
Journal of Researches into the Natural History and Geology of the Countries Visited During the Voyage of the H.M.S. Beagle Round the World (1860)

The following text summarizes some of the historical mis-characterizations of desert environments.

The desert is evil. It is deadly and barren and lonely and foreboding and oppressive and godforsaken. Its silence and emptiness breed madness. Its plant forms are strangely twisted, as are its citizens, who live there because they cannot get along anywhere else. Charlie Manson. Brigham Young. Bugsy Siegel. Carlos Castañeda and Don Juan's demons. Moses. In history and myth, the desert is the barrier to the Promised Land – our realm of trial and exile, the place where people go to be punished, seeking wisdom born of misery. Since modern transportation has compromised this relationship, reducing the desert's danger by abbreviating the time required to cross it, about the best thing you can say about the place is that it is boring. Aside from random deposits of minerals, it is valuable mainly as a zone in which to shed the shackles of civilization, race cars on infinitely straight highways, bask naked in unending sun, shoot guns at rocks and road signs, careen across sand hills on dirt bikes and dune buggies. In the desert, one is liberated from all restraint because nobody's around to notice. That's why we explode atom bombs there. Or dump things that create problems elsewhere. They can't do any damage in the deserts; there's nothing out there to damage.

David Darlington, American writer
The Mojave: A Portrait of the Definitive American Desert (1996)

Our ability to accurately estimate desert substrate properties is important because they affect the land-surface physics (as seen in the previous three chapters). And the land-surface physics determines desert microclimates (Chapter 11) and the mesoscale circulations (Chapter 12) that result from horizontal contrasts in surface heat fluxes. The numerical-model simulations described in the next chapter also require an accurate specification of the substrate properties. Thus, because of their importance this chapter will more completely review estimates of these physical properties.

Albedo

Recall that the albedo of a surface represents the fraction of the incident light that is reflected, and generally pertains to the entire spectrum of incident radiation. In contrast, reflectivity applies to a particular wavelength band. Albedo values exhibit a wide variability in deserts. Dry gypsum sand may have a very high albedo, equivalent to that of fresh snow (0.7), whereas nearby lava flows may have an albedo of 0.1. Table 8.1 summarizes some of the numerous estimates of desert albedo that have been reported in the literature. There are clearly place-to-place variations in the albedo of a particular surface type (e.g. dry sand), depending on the exact chemical composition, the degree of weathering, etc. Pielke (1984) provides a tabulation of albedoes for a wide range of natural and artificial surfaces, and Hovis (1966) shows plots of spectral reflectance for desert surfaces such as gypsum sand, silica sand, dry salt, dry-lake soil, beach sand, a salt pool, and Chilean nitrate soil.

Thermal properties

These properties include the thermal conductivity and heat capacity (or specific heat), and parameters that represent combinations of them such as thermal diffusivity and admittance. Table 8.2 shows values for some typical desert substrate types. These quantities depend strongly on the substrate's pore space and moisture content, so considerable variability should be expected within a particular substrate type. Estimates of these properties for other desert and non-desert locations can be found in Miller (1981) and Pielke (1984).

Aerodynamic roughness

The roughness length is a surface property that is related to the amount of drag experienced by wind flowing over it. Its correct specification in numerical models is thus important to accurate simulation of the wind profile and turbulence

Table 8.1 *Estimates of desert albedoes*

Method of estimation: s, satellite; g, ground; a, aircraft; u, unknown.

Surface type	Value	Method of estimation	Location	Source
salt playa, dry	0.46—0.74, from 0.40 to 0.65 μm	a	Mojave	Ashburn and Weldon (1956)
non-saline playa, dry	0.46—0.69, from 0.40 to 0.65 μm	g	Mojave	Ashburn and Weldon (1956)
bare sand, dry	0.30	s	Sahara	Vonder Haar (1980)
	0.28	s	Sahara	Kondratyev et al. (1981)
	0.342	s	Sahara	Carlson and Wendling (1977)
	0.297	s	Sahara	Carlson and Wendling (1977)
	0.30	u	Sahara	Menenti (1984)
	0.18	u	Kalahari	Menenti (1984)
	0.25	a	Mojave	List (1966)
	0.24—0.28	a	Mojave	List (1966)
sand, wet	0.18	g	unknown	List (1966)
clay, dry	0.09	g	unknown	List (1966)
clay, wet	0.23	u	unknown	Chudnovskii (1966)
silt loam, wet	0.16	u	unknown	Chudnovskii (1966)
	0.10—0.60 for various wavelengths and soil water	u	unknown	Bowers and Hanks (1965)
volcanic rock, weathered	0.07—0.11, from 0.45 to 0.70 μm	u	unknown	Reeves et al. (1975)
volcanic rock, unweathered	0.10—0.25, from 0.45 to 0.70 μm	u	unknown	Reeves et al. (1975)
granite bedrock	0.11—0.33, from 0.40 to 0.65 μm	g	Mojave	Ashburn and Weldon (1956)
brush and windblown sand	0.11—0.27, from 0.40 to 0.65 μm	g	Mojave	Ashburn and Weldon (1956)
cindery lava on sand	0.03—0.11, from 0.40 to 0.65 μm	g	Mojave	Ashburn and Weldon (1956)
basaltic lava	0.05—0.02, from 0.40 to 0.65 μm	g	Mojave	Ashburn and Weldon (1956)
composite surface	0.32, visible	s	Sinai area 1	Otterman and Fraser (1976)
	0.42, visible		Sinai area 2	
	0.34, visible		Sahel	
	0.34, visible		Thar	
	0.31, visible		Afghanistan	

Table 8.2 *Thermal conductivity, specific heat, and thermal diffusivity of desert substrates*

Material	Thermal conductivity (W m^{-1} K^{-1})	Thermal diffusivity (m^2 s^{-1} × 10^{-6})	Specific heat (J kg^{-1} K^{-1})	Source
quartz sand, medium fine, dry	0.26	0.20	795	List (1966)
sand, dry	0.23	0.35	—	Geiger (1966)
quartz sand, 8.3% moisture	0.59	0.33	1004	List (1966)
sand, 23% saturation	1.67	0.70	—	Geiger (1966)
sandy clay, 15% moisture	0.92	0.37	1381	List (1966)
sandstone	2.59	1.13	879	List (1966)
salt lake sediment	2.34	—	—	Tapper (1991)
sand	0.33	—	—	Tapper (1991)

Table 8.3 *Estimates of desert aerodynamic roughness lengths*

u, Unknown.

Surface type	Value (cm)	Location	Reference
bare sand	0.001	u	Bagnold (1954)
(1 mm diameter)	0.018	Peruvian Desert	Stearns (1967)
	0.03	u	Oke (1987)
	0.07	u	Stull (1988)
moving sand	0.3	u	Rasmussen *et al.* (1985)
bare, smooth playa	0.01	u	Sullivan (1987)
sand with small desert shrubs	0.173	u	Greeley and Iverson (1985)
very rough lava	1.0	Mojave Desert	Greeley and Iverson (1985)
sand with small desert shrubs	0.105	Sonoran Desert	Lee (1991a)
smooth desert	0.03	u	Deacon (1953)
dry lake bed	0.003	Mojave Desert	Vehrencamp (1953)
smooth mud flats	0.001	u	Deacon (1953)
stones (7 cm diameter)	0.028	Peruvian Desert	Stearns (1967)

Table 8.4 *Estimates of emissivities for arid-land surfaces*
u, Unknown.

Surface type	Value	Location	Reference
bare sandy soil	0.93	North American Desert	Hipps (1989)
shrub (*Artemisia tridentata*)	0.97	North American Desert	Hipps (1989)
bare loamy sand	0.914	Botswana	van de Griend *et al.* (1991)
shrub	0.986	Botswana	van de Griend *et al.* (1991)
stony area	0.959	France	Labed and Stoll (1991)
dry sand	0.95	u	Lee (1978)
wet sand	0.98	u	Lee (1978)
dry sand	0.914	u	Paltridge and Platt (1976)
wet sand	0.936	u	Paltridge and Platt (1976)
desert	0.84–0.91	u	Oke (1987)
dry playa	0.927	Great Basin Desert	McCurdy (1989)
saturated playa	0.967	Great Basin Desert	McCurdy (1989)

Table 8.5 *Estimates of saturation hydraulic conductivity and porosity for arid-land substrates*

Surface type	Saturation hydraulic conductivity (cm min^{-1})	Porosity (%)
sand	1.0	25–50
sandy loam	0.2	—
sandy clay loam	0.04	—
sandy clay	0.01	30–60
clay	0.00001	40–70
gravel	100	25–40
silt	0.001	35–50

intensity above the desert surface. Table 8.3 shows typical roughness lengths for different types of surface. The concept of roughness length will be more completely described conceptually and mathematically in Chapter 10.

Emissivity

Emissivity is defined in Chapter 4 as the ratio of the actual radiation emission and the maximum possible radiation emission (the black body emission). Emissivities vary between zero and unity. Accurate calculation of the radiative cooling of substrates and vegetation, for example in numerical models of the atmosphere, requires the use of good values of emissivity. Table 8.4 shows emissivities of desert substrates. Most are between 0.95 and 0.99.

Hydraulic properties

Saturation hydraulic conductivity is one of the more important substrate properties that relates to vertical water movement. Estimates are provided in Table 8.5, for various soil types that might be found in arid lands. In general, sandy soils have higher values, and those with clay have lower values. Typical values for porosity (percent air space) are also shown.

Suggested general references for further reading

Hillel, D., 1998: *Environmental Soil Physics* – contains numerous estimates of substrate physical properties.

Jury, W. A. *et al.*, 1991: *Soil Physics* – a summary of substrate physical properties.

Menenti, M., and J. C., Ritchie, 1994: *Estimation of effective aerodynamic roughness of Walnut Gulch watershed with laser altimeter measurements* – describes the estimation of aerodynamic roughness length in the northern Sonoran Desert using an airborne laser altimeter.

Oke, T. R., 1987: *Boundary Layer Climates* – contains tabulated values of many substrate physical properties.

Pielke, R. A., 1984: *Mesoscale Meteorological Modeling* – contains tables of substrate physical properties.

Stull, R. B., 1988: *An Introduction to Boundary Layer Meteorology* – explains the concept of aerodynamic roughness length.

Questions for review

(1) What desert substrates have the highest and lowest albedoes?
(2) What types of desert substrate have the highest thermal diffusivity?
(3) Rank different desert substrates in terms of porosity.

Problems and exercises

(1) Speculate on why wet sand has a lower albedo than does dry sand.

9

Numerical modeling of desert atmospheres

There is an unaccountable solace that fierce landscapes offer to the soul. They heal as well as mirror, the brokenness we find within. Moving apprehensively into the desert's emptiness, up the mountain's height, you discover in wild terrain a metaphor of your deepest fears. If the danger is sufficient, you experience a loss of competence, a crisis of knowing that brings you to the end of yourself, to the only true place where God is met . . . Some people die in the desert. Others flee as quickly as possible before it can effect them in any serious way. Only a few remain long enough to discover a hard-headed, unromanticized compassion, stripped of the sentimentalism that too often substitutes for love. They are the ones who manage to sustain the terrible and the good, without compromising either one.

Belden C. Lane, American educator and writer
The Solace of Fierce Landscapes (1998)

In reference to Chile's harsh Atacama Desert:
It had not rained for half a century there . . . The bare earth, plantless, waterless, is an immense puzzle. In the forests or beside rivers everything speaks to humans. The desert does not speak. I could not comprehend its tongue: its silence . . .

Pablo Neruda, Chilean poet
Confieso que he vivido (I Confess That I Have Lived) (1974)

Numerical models of the atmosphere are used in a variety of applications. They can be employed for operational weather forecasting, with forecast periods of as long as ten days. Other versions of the models can be used for longer-range forecasts of interseasonal and interannual climate change. Also, the models can be used for weather and climate research. Here, the model simulations generate gridded data sets that can be analyzed to develop a better understanding of weather and climate processes. This improved understanding of physical processes can then enable the further development of more accurate forecast models.

Examples of what has been learned about arid-land atmospheric processes through the use of such models are reported throughout the other chapters. This chapter will briefly review the basic concepts of numerical weather prediction, and show how the models are being used for arid-lands forecasting and research (see Box 9.2).

General concept of numerical weather prediction

The atmospheric models are based on the equations of fluid dynamics and thermodynamics, which have been adapted so that they apply to Earth's atmosphere. This set of time-dependent, non-linear, non-homogeneous, partial differential equations must be solved by numerical techniques. To integrate the equations forward in time, initial values and boundary values must be provided for all of the meteorological variables. The specific variable set depends on the formulation of the equations, but a typical one would include the three orthogonal components (east–west, north–south, vertical) of the wind-velocity vector, and the atmospheric temperature, humidity, pressure, and density. The initial values of the variables, which apply at the beginning of the integration of the equations with respect to time, are based on an analysis of observations obtained from surface-based sensors, radiosondes, satellites, etc. Initial values for the models are typically defined at midnight and noon Greenwich time (0000 UTC and 1200 UTC), when radiosondes are launched to define the conditions in the atmosphere above the surface. Models whose computational area spans the entire planet have boundary conditions that are naturally periodic, so there are no "edges" to the computational area for which boundary values need to be specified. But for limited-area, or regional, models, the values of the variables at the lateral boundaries need to be specified from some external data source during the period of the forecast or simulation[9.1]. In an operational forecast setting, these boundary values would be obtained from the forecast of a global model. In a research setting where a historical period is being simulated, the values are based on an analysis of observations.

The model equations represent all of the physical processes that are important to atmospheric dynamics and thermodynamics. These include the interaction of solar and terrestrial radiation with the atmosphere and the surface; the effects of atmospheric turbulence on the transport of heat, water vapor, and momentum within the planetary boundary layer; and all the processes that are associated with the atmospheric component of the hydrologic cycle, including cloud and precipitation physics. The close physical interaction between atmospheric and

---

[9.1] A model *forecast* is produced when the equations are integrated into the future, whereas a *simulation* is produced when the model equations are applied to an historical period.

land-surface processes means that the atmospheric models also need to represent the land-surface processes that determine the surface temperature and water availability. In general, the processes depicted in Fig. 5.1 need to be represented.

The numerical structure of a model determines the scales of atmospheric processes that can be represented. For example, consider the model equations applied on a three-dimensional grid, or matrix, of points. The spacing of the grid points determines the model's ability to resolve different phenomena. A general rule is that about ten grid points are required to properly represent, or resolve, a wave or a feature in the atmosphere. Thus, if a desert valley exists between two mountain ridges that are 100 km apart, and it is necessary to model the subsidence in the valley and the upward motion over the ridges during the daytime heating cycle (see Fig. 2.6f), at least ten grid points that are spaced about 10 km apart would be required over the valley. The spacing of the grid points in the vertical direction also needs to be sufficient to resolve the vertical variation in the atmospheric variables. Box 9.1 describes the relation between the model's horizontal resolution and the time required to produce a simulation or forecast.

Some examples of atmospheric model applications in arid areas

In other chapters there are numerous examples of how models have been used to provide a better understanding of atmospheric processes in arid environments. The models can be used for this purpose in at least two ways. First, they can use available observations as input, and the model-simulated meteorological variables defined on the regularly spaced model grid points can be analyzed to provide insights into processes that cannot be obtained from the data themselves. Secondly, sensitivity experiments can be performed in which some aspect of the physical system is represented differently in the model in order to determine its importance. For example, Figs 2.22 and 2.23 illustrate the difference between the rainfall produced in model simulations that were performed with the observed mountains as the lower boundary of the atmosphere, and with an artificial condition in which there were no mountains. Comparison of the rainfall in the two cases, and the storm characteristics, allowed a judgement to be made about the effects of the mountains on the distribution of aridity. An additional application of these models that has not been illustrated thus far is their use for operational weather forecasting in arid areas. In the remainder of this chapter will be provided a few case studies of how models have been applied to arid regions for varied purposes.

Box 9.1
The trade-off between forecast speed and resolution of atmospheric models

One of the practical constraints on numerical weather models is that they must be able to produce forecasts, or simulations of historical periods, in a reasonable amount of time. In the former case, for forecasts to be useful they need to be available relatively quickly. That is, the numerical evolution of the atmosphere in the computer needs to progress more rapidly into the future than does the real atmosphere. A 24 h forecast that required 24 h to produce would not be of any value. Even though there is no such constraint with research simulations of historical periods, there is still a practical limitation that a simulation should be completable within perhaps a few weeks. Given these requirements, what controls how fast the computer can integrate the model equations forward in time? One basic factor is the number of model grid points in the three-dimensional matrix of points on which the equations must be solved. The time required is linearly related to the number of these points. It is also linearly related to the number of time steps required in the integration of the equations for a particular forecast period. For each time step, the equations need to be solved at each of the grid points. The size of the time step is controlled by numerical criteria that must be met in order for the integration of the equations to be numerically stable. One of the more-stringent criteria is $U\Delta t/\Delta x < 1$, where U is the speed of the fastest meteorological wave, Δx is the horizontal spacing of the grid points, and Δt is the time step used in the time integration. So, what would be the effect on the computation time required to produce a forecast of a given length if we want to double the horizontal resolution over the area of the grid? This will, of course, increase the number of grid points by a factor of four. However, halving Δx requires that we also halve Δt, and twice as many time steps will be needed to integrate the equations for a forecast of a given length. The doubling of the horizontal resolution will therefore increase the computational requirements by a factor of eight. This is a heavy penalty, and is the reason why models used for operational forecasting sometimes have relatively coarse resolutions.

Operational weather prediction for the United Arab Emirates

In support of a research project in the United Arab Emirates (UAE), an operational model-based forecasting system was established. The objective was to produce forecasts for the country that represented both synoptic-scale and mesoscale meteorological processes. Mesoscale effects included the thermally forced sea-breeze circulation that strongly controls the daytime temperature and humidity near the coast. That is, during the day the sea-breeze circulation advects cooler, more humid, maritime air inland, and the timing of the movement of the leading edge of the maritime air, and its inland penetration distance, determines the maximum temperature at inland locations.

The computing hardware that is available generally determines how many computational grid points can be employed to produce an operational forecast. For example, a 24 h forecast needs to be completed within a few hours of elapsed time in order for the products to be operationally useful. The more computing power available, the more grid points for which computations can be performed during the few-hour period. Given a geographic area that must be spanned by the model, this computational limitation determines the spacing of the grid points in the horizontal and the vertical. In this case, the computing hardware permitted about a 3 km grid spacing to be used over the UAE, and 32 computational levels to resolve the vertical structure of the atmosphere.

Lateral boundary conditions for these forecasts were obtained from larger-scale forecasts that had already been completed by a global model of the United States National Weather Service. In situations such as this, where only a relatively small area must be spanned by a high-resolution model grid, it is advantageous to use a nested system of interacting model grids that have progressively higher horizontal resolution as they "telescope" down to the small focus area of the forecasts. Fig. 9.1 shows the configuration of the grid nest for the UAE forecast model. The grids become progressively higher in resolution toward the smallest center grid, with the grid spacings being 30 km, 10 km and 3.3 km.

Figure 9.2 shows an example of the daytime sea breeze and the nighttime land breeze circulation, as forecast by the model. The forecast winds and temperature are for 40 m AGL, the lowest computation level of the model. The initial (start) time of the forecasts was 1200 UTC 30 June 2002 (local time is Greenwich time + 4 h). Fig 9.2a shows the forecast land-breeze circulation at 0700 LT, based on a 15 h forecast. The weak flow from the land to the sea dominates half the area of the UAE, and extends well into the Arabian Gulf. For the sea-breeze circulation during the day, Fig. 9.2b shows a much stronger sea-to-land circulation dominating a large area at 1600 LT (a 24 h forecast). The inland advection of the cooler maritime air is clearly seen in terms of the temperature distribution. Objective skill statistics show that the error in the forecast of the sea breeze's arrival at Abu Dhabi is typically 1–2 h.

A study of small-scale winds in the Great Basin Desert

This example is of a research study whose objective was to determine the causes of the locally forced, mesoscale, low-level winds in an area of the Great Basin Desert. That is, the synoptic-scale weather determines the general wind pattern,

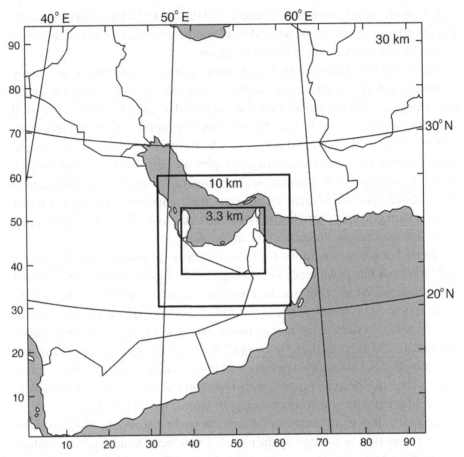

Fig. 9.1 Model grid configuration for the United Arab Emirates forecast system. The grid spacings of the different model grids are indicated. The left and bottom scales are labelled in terms of the grid points on the outer model grid. (Courtesy of Roelof Bruintjes and Tara Jensen, National Center for Atmospheric Research.)

but mesoscale perturbations to the large-scale flow are caused by contrasting local rates of surface heating and cooling during the diurnal cycle. The sea breeze discussed above is one example of such a local wind, which is forced by the different heating and cooling rates of the land and sea. In this study of the Great Basin Desert, the focus was on the development of local boundary-layer circulations by different surface heating rates around playas, lakes, and mountains in the region. The numerical experiments that were performed, how they were interpreted, and what was learned from them, are described below.

Figure 9.3 shows the nested system of four computational grids that were employed, and the grid-point spacing for each of the grids. The model solutions on

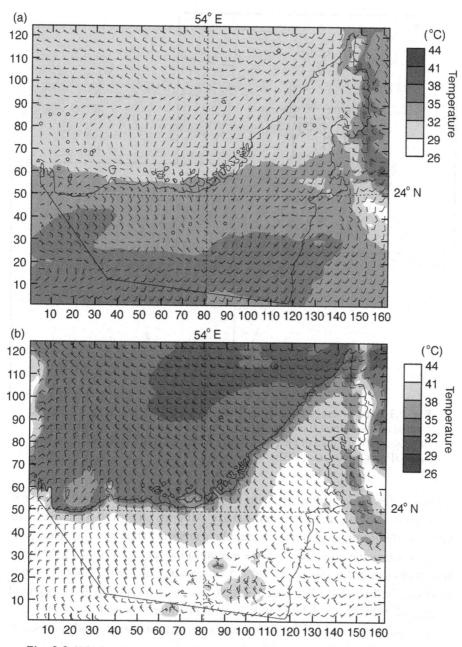

Fig. 9.2 Wind and temperature forecasts for 40 m AGL over the United Arab Emirates on the eastern Arabian Peninsula. (a) A 15 h forecast of the land breeze at 0700 LT. (b) A 24 h forecast of the sea breeze at 1600 LT. The temperatures are shown by the gray shades, with band widths of 3 K. On the wind symbols, long barbs represent 5 m s^{-1} and short barbs represent 2.5 m s^{-1}. The left and bottom scales show grid-point numbers. (Courtesy of Roelof Bruintjes and Tara Jensen, National Center for Atmospheric Research.)

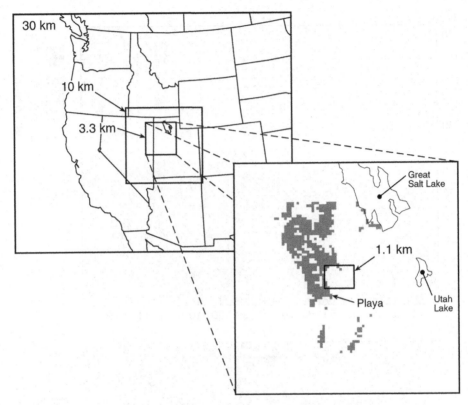

Fig. 9.3 Area coverage for the four computational grids, with the inner grids located over the Great Basin Desert. The grid spacings for each of the grids are indicated, as is the area of the playa (gray shade) and the boundaries of the two lakes. (From Rife *et al.* 2002.)

the two highest-resolution grids, with 1.1 km and 3.3 km grid spacings (shown in the expanded view), were used in the study. All grids employed 31 levels in the vertical. The locations of the lakes and playa areas are shown in this figure, and Fig. 9.4 shows the complexity of the terrain in the region. The playa substrate was damp and had a salt crust, and the surrounding area was generally sandy with scattered shrubs. The model simulations of 24 h duration were begun at 1200 UTC (0500 LT) 14 July 1998. Lateral boundary conditions for the outer model grid were provided by global analyses of near-surface and radiosonde data.

   This case study is discussed here in order to illustrate how models can be used to isolate certain physical processes from the many that can influence the atmosphere at any particular time and place. In this case; mountain-valley circulations exist, with near-surface upslope flow (Fig. 2.6f) during the day and downslope drainage at night; lake breezes occur near the Great Salt Lake and

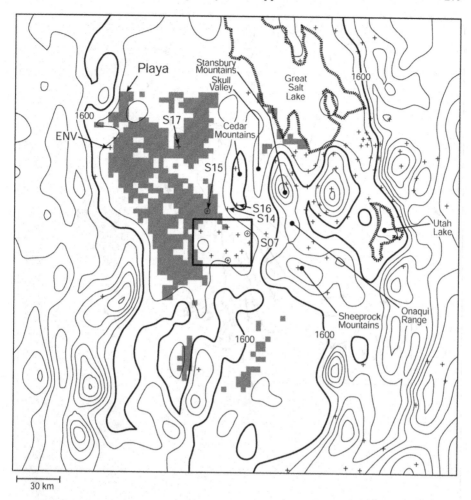

Fig. 9.4 The topography for the model grid with a grid spacing of 3.3 km, shown in Fig. 9.3. The contour interval is 200 m, and the bold line indicates the 1600 m elevation. The area of the playa is shaded, and shores of the two lakes are hatched. The plus signs indicate surface observations and the circles identify radiosonde observations used as input to the model. The inner box denotes the boundaries of model grid 4. (From Rife *et al.* 2002.)

Utah Lake, directed at low levels from water to land during the day (Fig. 2.6e) and from land to water at night; and the playa breeze flows near the surface from the playa to the surrounding sandy surface during the day, reversing at night (see discussion about Fig 12.2). Because of these multiple causes of local winds, it is difficult without a model to isolate the relative contributions to the small-scale wind pattern at any particular time and location. For example, Figs. 9.5 and 9.6 show observations of the 10 m AGL winds at 1400 LT 14 July and 0500 LT 15 July (during the model-simulation period), respectively. There is clearly much

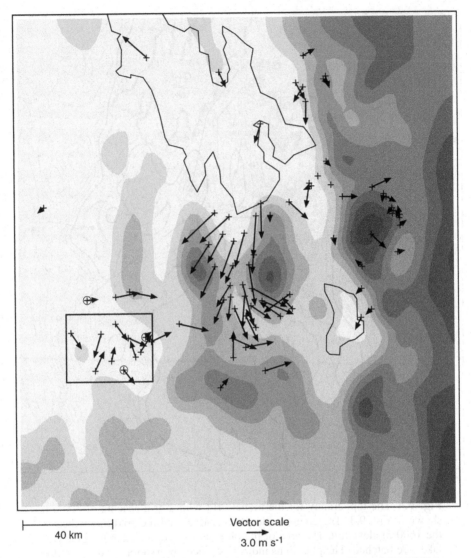

Vector scale
3.0 m s⁻¹

40 km

Fig. 9.5 Observations of the 10 m AGL wind (see vector scale) at 1400 LT 14 July 1998 over an area of the Great Basin Desert. The Great Salt Lake and Lake Utah are outlined, and the inner box defines the location of the model inner grid (Fig. 9.3). Surface topography is displayed with the higher elevations shaded dark gray. (From Rife *et al.* 2002.)

small-scale structure to the wind field during both day and night. Even though the Great Salt Lake lake breeze seems to be apparent near the southern shore during the day, and strong downslope drainage flow can be seen in some places at night, it is generally difficult to attribute a wind vector to a particular type of forcing. Indeed, in general, different influences are likely superimposed at most places.

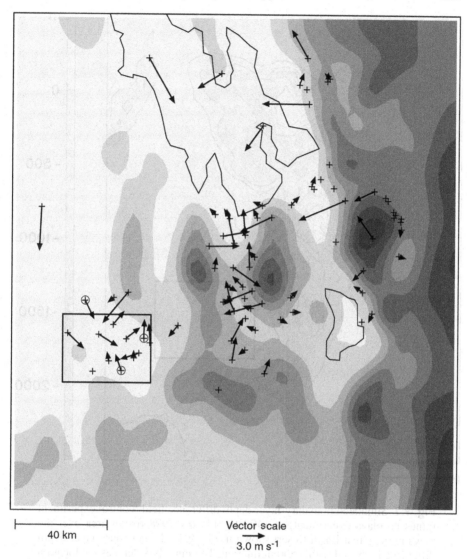

40 km

Vector scale
3.0 m s⁻¹

Fig. 9.6 Same as Fig. 9.5, but at 0500 LT 15 July 1998. (From Rife *et al.* 2002.)

To use the model to isolate the causes of the local winds, a control simulation was first performed with all of the effects represented: the lake breeze, the playa breeze, and the mountain-valley breeze. That is, the land-surface conditions used in the model represented the playa, the mountains, and the lakes, and thus the model simulation contained all of the prevailing processes. In order to isolate the playa's effect on the simulated atmosphere, it is only necessary to rerun the model with the playa surface conditions replaced by the conditions of the surrounding sandy, vegetated desert. In this case, the high albedo of the playa

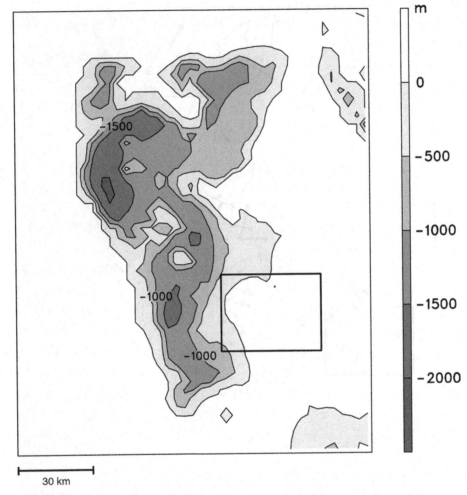

30 km

Fig. 9.7 Simulated boundary-layer depth difference field (control experiment minus no-playa experiment) at 1400 LT 14 July 1998 for an area over the playa in the Great Basin Desert shown in Fig. 9.4. The boundary-layer depth difference is shaded with a 500 m interval. The inner box denotes the location of the inner model grid (Fig. 9.3). (From Rife *et al.* 2002.)

was replaced with the lower albedo of the sparsely vegetated surrounding sand, and the properties of the moist, compacted playa substrate were replaced with the properties of the sandy substrate. The model solution from this "no playa" simulation is then subtracted from the control simulation, and the difference is the isolated playa breeze effect. Fig. 9.7 illustrates the boundary-layer depth difference at 1100 LT, with the playa causing a reduction in the depth of the well-mixed layer by greater than 1 km over a large area. Based on the same two simulations, Fig. 9.8 depicts the difference in the 2 m AGL temperature and

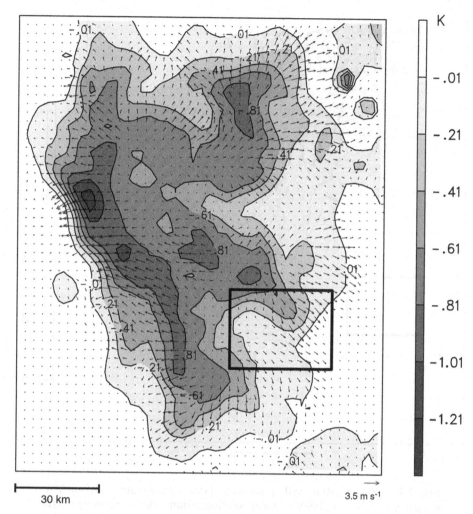

Fig. 9.8 Simulated 10 m AGL wind (see vector scale) and 2 m AGL temperature difference fields (control experiment minus no-playa experiment) at 1400 LT 14 July 1998 for an area over the playa in the Great Basin Desert shown in Fig. 9.4. The temperature difference field is analyzed and shaded with a 0.2 K interval, and wind vectors are plotted at every grid point. The inner box denotes the location of the inner model grid (Fig. 9.3). (From Rife *et al.* 2002.)

the 10 m AGL winds. Again, this represents the playa effect on the near-surface temperature and wind. As expected, the near-surface wind is directed away from the playa at this time, and the greater evaporation and higher albedo of the playa surface has caused a lower near-surface temperature.

In another experiment, the playa was placed back in the surface conditions, but both lakes were replaced with a sparsely vegetated sandy surface. Again, subtracting this model solution from that of the control simulation defines the

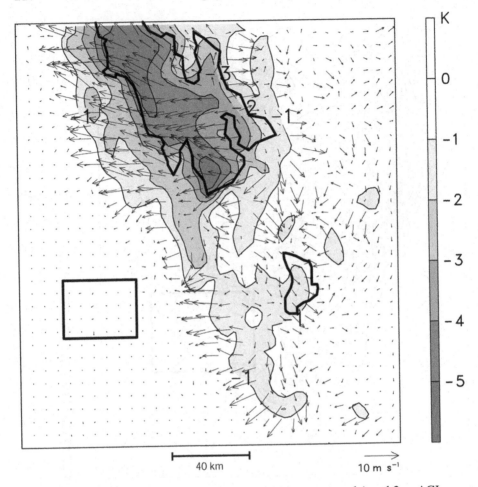

Fig. 9.9 Simulated 10 m AGL wind vectors (see vector scale) and 2 m AGL temperature difference fields (control experiment minus no-lake experiment) at 1900 LT 14 July 1998. The temperature difference field is analyzed and shaded with a 1.0 K interval, and wind vectors are plotted at every second grid point. The inner box denotes the location of the inner model grid (Fig. 9.3). Heavy lines outline the two lakes. (From Rife *et al.* 2002.)

lake-breeze circulation alone. Fig. 9.9 shows the 10 m AGL wind and the 2 m AGL temperature difference fields, representing the lake-breeze effects. In this figure are illustrated unexpected aspects of this lake breeze. For example, the use of a simple conceptual model might have led us to expect a symmetric pattern of winds flowing from the water to the land around the shores of the lakes. But, the mountains to the east of the Great Salt Lake have impeded the flow of the lake breeze in that direction, in contrast to the more extensive inland penetration over the desert to the west by the much stronger winds. Also, the flow to the west of Utah Lake is stronger than might have been expected from the relatively small size of the lake.

In summary, the use of the model has allowed us to develop an improved understanding of how the multiple causes of local winds in the desert have contributed to the complex flow patterns in this area on this particular day. It is worth noting that, to an observer on the ground at many locations within this study area, the landscape looks flat, relatively barren, and generally unremarkable, except for elevated terrain in the distance. Nevertheless, the model solution and observations illustrate the unanticipated nearby effects of the playa, and the far-reaching effects of distant mountains and lakes.

A study of boundary-layer circulations in the Arabian Desert

The objective of this case study was to better understand the boundary-layer structure and wind circulation in the Arabian Desert. This model simulation was six days in length, from 1200 UTC (1500 LT) 9 March to 1200 UTC 15 March 1991. Fig. 9.10 shows the computational grids, where the grid spacings were the same as used for the outer three grids of the previously discussed Great Basin Desert study. Thirty-five computation levels were employed. Some of the results from this study are presented in Chapters 11 and 12 in the context of intra-desert microclimates and their interaction.

This study will be used to make the point that model simulations must be compared with all available data, to confirm the accuracy of the model, before a simulation is used to learn about the atmosphere where there are no data. In this area there are few radiosonde soundings that can be used to estimate the vertical temperature structure of the atmosphere, from which the boundary-layer depth can be inferred. One location for which such vertical profiles were available twice per day was Hafar Al Batin in northern Saudi Arabia (Fig. 9.10). The boundary-layer depth estimated from the temperature soundings for this location for the six-day period were compared with the boundary-layer depth simulated by the model. Fig. 9.11 depicts a time–height section of the model-simulated and radiosonde-observed potential temperature and winds for the period. Where air is well mixed in the boundary layer, the potential temperature does not increase much with height. Thus, the top of the afternoon boundary layer can be estimated as the height at which the potential temperature begins to increase significantly with height. Below that height in the figure, the lines of constant potential temperature are approximately vertical; that is, the value is approximately constant with height, or does not increase rapidly with height. Above that level, the lines become more horizontal, and the potential temperature increases more rapidly with increasing height. From the figure (bottom panel) it can be seen that the top

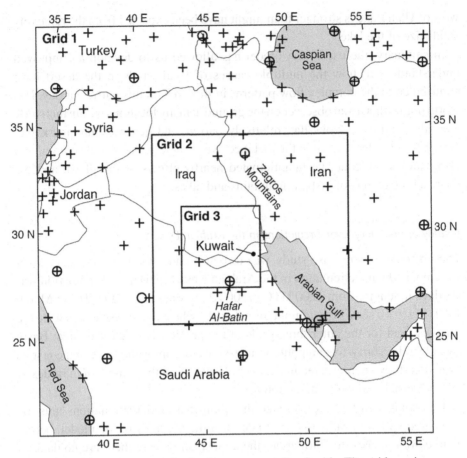

Fig. 9.10 Area coverage for the three computational grids. The grid spacings for grids 1 through 3 are 30 km, 10 km, and 3.3 km, respectively. The locations of observations used in the model are shown with circles (radiosonde observations) and plus signs (near-surface observations). (From Warner and Sheu 2000.)

of the model-simulated afternoon boundary layer (short, wide line) is slightly higher than the 800 mbar level for the first three days of the period. By the fourth day, and thereafter, the layer of relatively uniform potential temperature becomes deeper, with the top of the boundary layer being above 700 mbar.

For comparison, the potential temperature and afternoon boundary-layer tops estimated from the radiosonde-observed temperatures are plotted in the upper panel. Even though there was no discernible layer of uniform potential temperature on the afternoons of days two and three, from which to define a boundary layer, the model performed reasonably at estimating the boundary-layer depth on the other four days. The lower-tropospheric winds during the period were

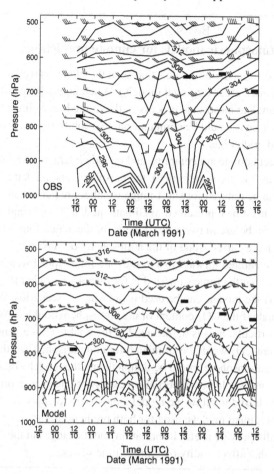

Fig. 9.11 Simulated (Model) and observed (OBS) vertical-time sections of potential temperature and wind for Hafar Al Batin on grid 3 in northern Saudi Arabia for the 6 day simulation period. Observed values were available every 12 h at 0000 UTC and 1200 UTC (0300 LT and 1500 LT), and the model-simulated values are plotted every 6 h. The isotherms are labelled in degrees Kelvin and are plotted at an interval of 2 K. Long barbs on the wind symbols correspond to 10 knots, while short barbs correspond to 5 knots (1 m s^{-1} = 1.94 knots). The estimated vertical extent of the 1500 LT boundary layer is defined by a short, wide line. Subjective estimates based on observed soundings are not plotted for two times because of the absence of a near-neutral layer. The abscissa is labeled as hour (Greenwich) over date (March 1991). (From Warner and Sheu 2000.)

Box 9.2

The use of atmospheric models for climate prediction

Climate predictions are performed with models that span the globe, as well as with models that cover only limited areas (regional models). When higher horizontal resolution is needed to simulate mesoscale processes, the limited-area climate models are employed because it would be computationally prohibitive to cover the globe with closely spaced grid points.

The processes that affect climate are complex, and include interactions among the atmosphere, the oceans, the cryosphere (ice), the biosphere, and the land surface. Processes that can be ignored for short-range weather prediction, such as circulations in the ocean, must be represented in climate models. Global and regional climate models can be use in two ways. One is for the actual forecasting of future climate based on the current state of the atmosphere, oceans, etc. That is, the model equations are integrated forward just as they are for a typical weather forecast. The difference is that the forecast period is for over a year rather than for a few days. The other way the models can be used is in "what if" scenarios. What would happen to the climate if the CO_2 content of the atmosphere doubles? What would be the result if a particular area is deforested? In the double CO_2 example, the climate with the current CO_2 would be simulated for comparison. Then the CO_2 content in the model atmosphere would be doubled, and the model would be run again. These simulations would be for periods in excess of one year in order to represent a full annual cycle and to allow the model atmosphere to adjust to the new forcing. Global and regional models, used for both predictions and what-if scenarios, could potentially provide useful information about changes in the rainfall and other factors that affect aridity on climate time scales.

also captured fairly well by the model, such as the shift from westerly to southerly flow below 700 mbar on 12 and 13 March.

Another test of model skill was a comparison of the observed and simulated 2 m AGL wind at Hafar Al Batin. Fig. 9.12 shows the observed and simulated wind speed and direction for the period. The model was able to simulate the overall day-to-day trends, as well as some of the diurnal oscillation.

Based on this reasonably good comparison of the observed and simulated conditions at Hafar Al Batin, and at other locations, it was concluded that the model could be trusted to reasonably simulate the conditions at other locations. Thus, the model was employed in an analysis of the boundary layer structure over the entire Arabian Peninsula, and some of the results are described in Chapters 11 and 12.

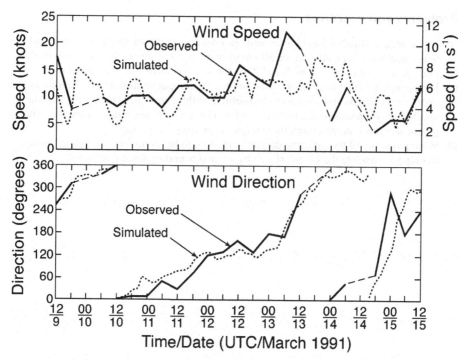

Fig. 9.12 Simulated and observed 2 m wind direction and speed at Hafar Al Batin in northern Saudi Arabia. Dashed lines indicate missing data. (From Warner and Sheu 2000.)

Suggested general references for further reading

Pielke, R. A., 1984: *Mesoscale Meteorological Modeling* – a classic text on numerical modeling of atmospheric processes on smaller scales. The basic concepts of numerical atmospheric modeling, the numerical formulations, and the treatment of physical processes are discussed.

Warner, T. T., 1989: *Mesoscale atmospheric modeling* – a very short tutorial on this subject.

Questions for review

(1) Explain the concept of model grid spacing.

(2) How could numerical-model experiments be used to isolate the effects of the Tibesti Mountains in the Sahara Desert on local weather? What results would you expect?

(3) Explain the difference between a model simulation and a model forecast.

(4) From where are model lateral boundary conditions obtained, for forecasts? For historical simulations?

(5) What physical processes in the atmosphere must be represented by numerical models? What surface processes need to be represented?

Problems and exercises

(1) If you need to simulate land-breeze and sea-breeze circulations near a coastline, and both circulations typically extend 50 km from the coast, estimate the minimum model grid spacing that could be employed. Explain how you arrived at your answer.

(2) It was explained that the model must complete its computations relatively rapidly in order for an operational forecast to be useful, and this limits the grid spacing (i.e. the number of grid points over a particular area). For research simulations, however, what practical controls do you think there might be on the grid spacing?

(3) What properties of the substrate do you think need to be represented in a model in order to properly simulate the land-surface physics and boundary-layer processes?

10

Desert boundary layers

. . . The desert: now awakened from its dream, and we have all left the Ark of this dream. As one man. But I am already waiting for the night to return. The same night, if possible. To hear the song of man and to reconcile myself, the shadow, with him who casts it. The night will return. I am a prisoner of all the rest. A prisoner captured by the desert, guarded by it inasmuch as we are all desert, all of us. In the very obscurity of our flesh. I feel myself invaded by its dry, white odour to the depths of my being. Desert of deserts. Dust of dust. Silence of silences. Maybe we have won and the world has lost. Perhaps the void has made its nest in you and you have become just anybody exposed to the four winds, with no substance or outer covering other than the void which can only become emptier and melt you in the blaze of day.

Mohammed Dib, Algerian novelist and poet
Le Désert Sans Détour (1992)

"Come and smell the sweetest scent of all", and we went into the main lodging, to the gaping window sockets of its eastern face, and there drank with open mouths of the effortless, empty, eddyless wind of the desert, throbbing past. That slow breath had been born somewhere beyond the distant Euphrates and had dragged its way across many days and nights of dead grass, to its first obstacle, the man-made walls of our broken palace. About them it seemed to fret and linger, murmuring in baby speech.

T. E. Lawrence, British writer and adventurer
Seven Pillars of Wisdom (1926)

Basic concepts of boundary-layer structure

Some of the basic concepts of boundary-layer structure were summarized in Chapter 2, and that section should be reviewed. The atmospheric boundary layer is defined as that region of the atmosphere immediately above Earth's surface wherein vertical turbulent transfers of heat, moisture, and momentum are large relative to the transfers in the troposphere above. It can be a few hundred meters to over a few kilometers in depth. This section will briefly review the

291

nature of turbulent exchange in the lower atmosphere, and the typical profiles of the meteorological variables in the desert boundary layer. Emphasis here will be on those aspects of boundary-layer structure and turbulence that are required for understanding discussions in later chapters. Texts on turbulence and boundary-layer structure, such as Stull (1988), should be consulted for a more thorough treatment.

Turbulent exchange of atmospheric properties

The airflow above the surface of Earth can be either laminar or turbulent. Given that the horizontal wind speed is zero at the surface, the speed immediately above must increase with height. In laminar (non-turbulent) flow, the layers of air slide over each other without much mixing between them. The only mixing that occurs is through the exchange of molecules between layers. Molecules from a slower-moving layer nearer the ground enter a faster-moving layer above, and the slower speed of the molecules represents a drag that slows down the upper layer. Analogously, faster-moving molecules move downward, with the drag effect causing the lower layer to speed up. Molecules of water vapor move between layers, causing a net transfer from moist layers to dry layers, and heat is exchanged by virtue of the different kinetic energies of the molecules exchanged between layers. In non-turbulent flow, this is how the layers of air "feel" each other. With turbulent flow, there are eddies that mix the air between the layers, with this type of mixing being much more efficient than the molecular mixing of laminar flow. One can imagine the vertical exchange of properties with turbulent mixing in the same way as with molecular mixing. For example, because water-vapor content near the surface generally decreases with height, upward-moving air in the turbulent eddies will contain more water vapor and downward-moving air will contain less water vapor.

The turbulence is generated through buoyancy as well as through the vertical shear of the horizontal wind. During the day, the heating of the lowest layer of the atmosphere, by virtue of its contact with Earth's surface, causes parcels of buoyant air to rise. These rising parcels, and ones involved in the compensating subsidence, mix the lower atmosphere. The vertical extent of this turbulent mixing defines the daytime (convective) boundary-layer depth. At night, the atmosphere near the surface cools, and the source of the buoyant energy is eliminated. Any new turbulent energy in this stable boundary layer near the ground must now be derived from the vertical shear in the horizontal wind. Unless the horizontal wind is exceptionally strong, this nocturnal boundary layer is much more shallow (see Fig. 2.3a). Fig. 10.1 illustrates distinctions between nocturnal turbulence that is generated from wind shear alone, and daytime turbulence that

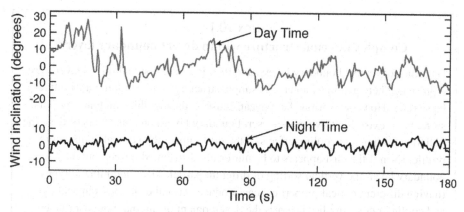

Fig. 10.1 The variation with time of the vertical inclination of the wind at 29 m AGL, based on bivane measurements, for daytime and nighttime conditions. The mean wind speed in the horizontal was 3–4 m s^{-1}. (From Priestley 1959.)

results from both buoyant motion as well as wind shear. Both curves show the variation with time of the vertical inclination of the wind at 29 m AGL, based on bivane[10.1] measurements. The lower curve, with relatively small-amplitude and high-frequency excursions from the horizontal direction, shows the effect of nocturnal turbulence that results only from the vertical shear of the horizontal wind. In contrast, the upper daytime curve shows similar high-frequency variability, but it is superimposed on a lower-frequency change with a period of perhaps 15–60 s. The longer-period changes during the day are associated with vertically moving eddies of buoyant air.

Boundary-layer structure

The turbulent mixing of heat, water vapor, and momentum within the boundary layer has a distinct effect on the vertical profiles of temperature, water-vapor content, and wind speed (see Fig. 2.3b). Wind speed and water-vapor content are relatively uniform with height in the well-mixed boundary layer. However, parcels of unsaturated air that are mixed upward or downward by turbulence, cool and warm, respectively, at the adiabatic lapse rate. Thus, the well-mixed vertical profile of temperature is the dry adiabatic lapse rate, which is known as a "neutral" temperature profile. In these conditions, the potential temperature is uniform with height. Exceptions to this simple boundary-layer structure are described in Boxes 10.1 and 10.2.

[10.1] A bivane is a wind vane with two axes of rotation, one horizontal and one vertical.

Box 10.1
Complex internal structure within desert boundary layers

Daytime convective boundary layers, and the nighttime residual mixed layers, are represented here and in Chapter 2 as simple structures with smooth temperature lapse rates. However, various factors can cause a considerable amount of internal structure to exist. First, when the desert boundary layer contains layers of dust, the radiative heating and cooling effects of the dust impact the vertical temperature profile. Even if the dust appears to be uniformly distributed throughout the boundary layer, the vertical sorting of different particle sizes and mineral types (having different optical properties) can produce vertical differences in heating and cooling rates. Another factor is the development of internal boundary layers that result from air flowing over surfaces with contrasts in properties such as the heat flux or roughness. For example, if the wind transports boundary-layer air from a smooth, hot desert surface to a cooler, rougher one, such as an oasis with vegetation, an internal boundary layer that is forced by the oasis surface develops within the original boundary layer. That is, the greater surface roughness and the different surface heat flux of the oasis would produce an internal boundary, within the upstream one, across which the wind and temperature profiles would differ. This boundary would intersect the surface at the upstream edge of the oasis, and rise with increasing distance downstream. There are always subtle to major contrasts in surface properties over deserts, so whenever the boundary-layer air moves horizontally, these internal boundary layers will complicate the structure.

Box 10.2
Boundary layers that are detached from the surface

All boundary layers originate because of the influence of a surface with which they are in contact. Even though boundary layers can develop against vertical surfaces such as canyon walls or the sides of buildings, by far the most common are those that form over the quasi-horizontal or moderately sloping surfaces of Earth. However, these boundary layers do not necessarily remain in contact with the surface over which they form, with important consequences for the weather. An example of one such situation is shown in Fig. 12.12, where a boundary layer first develops over the high desert plateau of northern Mexico. When southwesterly lower-tropospheric winds cause this heated layer of air to move toward the lower terrain elevations to the northeast, the boundary layer becomes detached from the surface. When this happens, a temperature inversion, i.e. a stable layer of air, forms at the base of the elevated mixed layer. This stable layer can inhibit the development of convective rainfall because parcels of air that are buoyant in the neutral lapse rates near the surface are no longer buoyant when they encounter this warm elevated mixed layer. This thermal "lid," represented by the hot boundary layer air, allows moisture and heat to build up in the layer between it and the ground. If the convective motions erode through the inversion, or leak out around it, the consequence is a sudden release of energy that can lead to severe convective storms. This process produces severe convective weather in the semi-arid Great Plains of the United States in the spring. There are many other areas in the world where mixed layers become elevated, with important potential consequences for downstream precipitation.

The turbulence and the well-mixed profiles of the different meteorological variables penetrate progressively upward during the daytime heating cycle. Fig. 11.7 shows measured profiles of the vertical structure of potential temperature at a few times during the daytime heating cycle in the Great Basin Desert. As the heating continued during the day, the depth of the constant potential-temperature layer progressively increased. At night, as the cooling surface in turn cools the lowest layer of the atmosphere, a temperature inversion forms.

Aerodynamic roughness and the vertical wind profile

The near-surface turbulent fluxes of heat, moisture, and momentum are influenced by the structure and spacing of surface-roughness elements such as rocks, plants, and soil grains. In general, rougher surfaces cause more intense turbulence. Expressions representing the effect of the turbulence on the vertical wind profile in the surface layer employ a parameter called the roughness length (z_0) to describe the roughness characteristics of the surface. In particular, it can be shown that, for conditions of neutral stability (strong convective mixing)

$$U = \frac{u_*}{k} \ln \frac{z}{z_0},$$
(10.1)

where U is the speed of the mean wind at height z, u_* is the friction velocity, k is the von Karman constant with a value that is thought to be 0.35–0.40, and z is the height above the ground. The friction velocity represents the drag of the atmosphere against Earth's surface, or the frictional stress. Recall from Chapter 2 that the surface layer, where this equation applies, is the lower 50–100 m of the mixed layer where the turbulent transport of heat, moisture, and momentum vary relatively little.

If z in Eqn. 10.1 is set to z_0, U is equal to zero. Thus, z_0 is the height above the ground at which the wind speed goes to zero in neutral conditions, and is proportional to the roughness of the surface. Because u_* is not a function of height in the surface layer and k is a constant, U increases logarithmically with increasing z. Fig. 10.2 shows a typical wind-speed profile with height for neutral conditions (i.e. a solution to Eqn. 10.1). If the equation is solved for u_*, it is clear

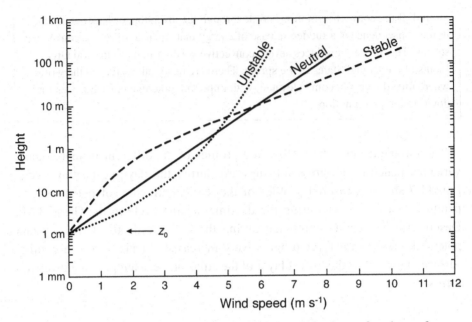

Fig. 10.2 Wind speed as a function of height within the surface layer, for neutral, stable, and unstable conditions. (From Stull 1988.)

that u_* is linearly related to the slope of the line $(U/\ln(z/z_0))$. The y-intercept of the line is z_0. Also shown are profiles of U with height for unstable and stable conditions; that is, with temperature lapse rates greater than and less than, respectively, the neutral (dry adiabatic) value. For stable conditions the profile is concave downward, and for unstable conditions it is concave upward.

The fact that the friction velocity is proportional to the drag of the moving atmosphere against the surface is why it is used in Chapter 16 as a measure of the ability of the atmosphere to drift and elevate sand and dust. That is, because it is a more direct measure of the drag of the atmosphere on surface particles, it is used instead of the wind itself.

Obviously it is important to be able to estimate the roughness length in order to apply this equation. Bagnold (1954) determined that the roughness length over a bare flat sand surface is approximately equal to the mean diameter of the sand grains, divided by 30 (roughly 10^{-5} m for typical sand). For small sand grains, the roughness length is only marginally larger than that estimated for a calm water surface (Oke 1987). Sullivan (1987) used wind-profile measurements to estimate that the roughness length over a bare, smooth, Mojave Desert clay playa is an order of magnitude larger, 10^{-4} m. When vegetation is present, the roughness lengths are larger and more difficult to obtain (Driese and Reiners 1997). Even though studies over agricultural fields with regularly spaced plantings are

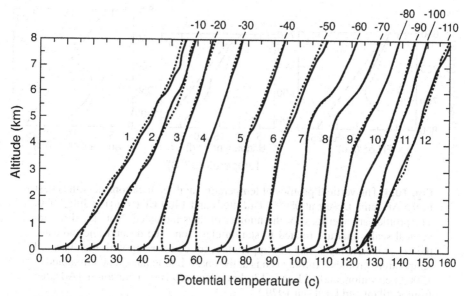

Fig. 10.3 Monthly means for 1989 of potential temperature for 0000 UTC (solid line, 0300 LT) and for 1200 UTC (dashed line, 1500 LT) at Riyadh on the Arabian Peninsula. The numbers on the curves indicate the month, and the numbers at the top should be subtracted from the potential temperature scale at the bottom. (Adapted from Gamo 1996.)

abundant, there are fewer studies of less homogeneous natural environments. Table 8.3 shows various roughness-length estimates for deserts.

Unique aspects of desert boundary layers

Desert boundary layers are unique from boundary layers in more-temperate climates, primarily in terms of their depth, dryness, and dustiness. Their depth is generally greater because the clear skies and low substrate water content allow higher surface temperatures, which can result in stronger sensible-heat fluxes to the atmosphere and deeper turbulent mixing. They are drier because of low substrate moisture and the resulting low evapotranspiration rates. And they are dustier because the lack of vegetation allows the wind speed very near the surface to be higher, and the desiccated substrate does not contain the water whose surface tension causes soil particles to adhere to each other, both factors that cause more dust to be elevated into the atmosphere (see Chapter 16 for further details).

The diurnal and seasonal variation in desert boundary-layer structure is shown in Fig. 10.3 for Riyadh on the Arabian Peninsula. During the warm season, from April through September, the depth of the afternoon boundary layer (dashed

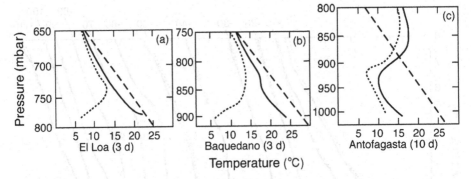

Fig. 10.4 The vertical profiles of temperature at three locations near the coast in the Atacama Desert at 0800 LT (dotted) and 1400 LT (solid) in July 1970. The profiles are averages for the number of days indicated. Note the different vertical scales that are related to station elevation. The dashed line indicates the dry adiabatic lapse rate of temperature, for comparison with the observed temperature profiles. Antofagasta is at the coast, Baquedano is 60 km inland at 1300 m elevation, and El Loa is 130 km inland at 2100 m elevation. (Adapted from Rutllant and Ulriksen 1979.)

lines) ranges from 4 km to 5 km. The stable temperature inversion at the top of the boundary layer is most evident in July and August. During the winter months, the afternoon depth is 1 km or less. The nighttime profiles (solid lines) are cooler below about 2 km AGL, with the neutral layer above continuing throughout the night during the summer. The most stable nocturnal layer is confined to the lowest 500 m AGL. As revealed by a comparison of the January and July profiles, there is a very great seasonal variability in the boundary-layer depth and the stability of the lower atmosphere.

As is the case with temperate climates, it is unwise to generalize too much about boundary-layer structure because there can be large variations with location and from one weather regime to the next. For example, Fig. 11.7 uses radiosonde measurements to show how the boundary-layer depth varies over the short distance between a playa surface and the surrounding sandy surface in the Great Basin Desert. For the same area, Fig. 9.7 illustrates a numerical simulation of the effect of the playa on the boundary-layer depth, and Figs. 11.12, 11.13, 12.7, and 12.10 illustrate model-simulated spatial variations in the boundary-layer depth over the Arabian Peninsula. An example of the variability in the boundary-layer structure near coastlines is shown in Fig. 10.4 in terms of the vertical temperature profiles at three locations near the coast in the Atacama Desert. Antofagasta, located on the coast, has a cool boundary layer that varies very little with time of day. A strong temperature inversion separates the cool marine boundary layer and the warmer air above. Baquedano is 60 km inland,

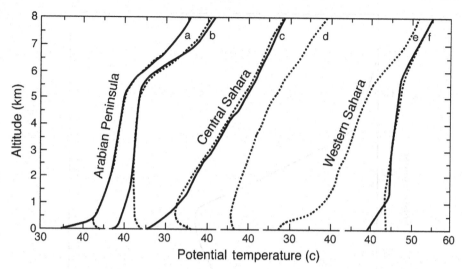

Fig. 10.5 Vertical profiles of potential temperature in July 1989 for three areas at similar latitudes (about 26–32° N) in north Africa and the Arabian Peninsula. For each area there is a coastal location and an inland location. On the Arabian Peninsula are Dhahran (a, coastal) and Medina (b, inland), in the central Sahara are Tripoli (c, coastal) and Sabhah (d, inland), and in the western Sahara are Nouadhibou (e, coastal) and In Salah (f, inland). Solid lines are 0000 UTC (local night) profiles and dashed lines are 1200 UTC (local day) profiles. Local times are zero to three hours after Greenwich times. (Adapted from Gamo 1996.)

and shows a strong diurnal signal associated with the surface heating and cooling. The daytime profile shows a slight hint of an inversion between 825 and 850 mbar. El Loa, even farther inland at 130 km from the coast, has a deep daytime boundary layer that shows no sign of an inversion. Thus, an upstream cold ocean current can have a dramatic effect on boundary-layer structure.

Variability in boundary-layer structure also occurs on regional scales as well as on the smaller local scales described above. This is shown in Fig. 10.5 for three desert areas at about the same latitude. The inland daytime profile for the Arabian Peninsula (b) shows a neutral layer to 5 km AGL. The more stable afternoon conditions at the coast (a) reflect the fact that the water temperatures are less than those of the land surface, and in particular the probable influence of the Gulf breeze transporting cooler maritime air across the coast at low levels. In the central Sahara, Tripoli (c) is on the Mediterranean coast while Sabhah (d) is well inland to the south. Exposure to the maritime airmass explains the stability and coolness of the Tripoli afternoon profile, where the boundary layer is less than 1 km deep. Inland at Sabhah, the trade winds (the harmattan) flow from the north and advect more stable air from the Mediterranean (Fig. 3.12), also

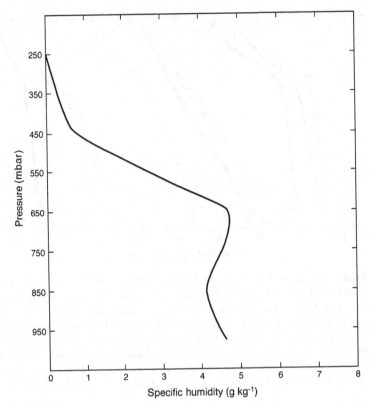

Fig. 10.6 A two-day-average daytime specific-humidity profile in the Empty Quarter on the Arabian Peninsula in May 1979. The potential-temperature profile indicated that the boundary-layer top was near 650 mbar. (From Blake *et al.* 1983.)

limiting the depth of the boundary layer there to less than 1 km. In the western Sahara, Nouadhibou (e) is on the Atlantic coast and In Salah (f) is far inland. The coastal station shows the expected stability of the maritime airmass. At In Salah, a near neutral to slightly stable profile exists up to about 6 km AGL. When the trades do penetrate that far south, the lower atmosphere has been destabilized by its long exposure to the hot surface.

An example of a summer moisture profile within a daytime desert boundary layer is shown in Fig. 10.6. This is an average profile for two days in May in the Empty Quarter in the Arabian Peninsula when a 3.5 km deep boundary layer extended to about 650 mbar. The specific humidity was relatively uniform up to the top of the boundary layer, above which it decreased rapidly within the dry air subsiding from the upper troposphere.

Suggested general references for further reading

Stull, R. B., 1988: *An Introduction to Boundary Layer Meteorology* – a general reference on turbulence and boundary-layer structure and processes.

Questions for review

(1) Explain in your own words how turbulent eddies mix the boundary layer such that high values are decreased and low values are increased. Provide a separate explanation for molecular transfer of properties in the boundary layer.
(2) What are typical depths of the daytime convective boundary layer for desert and non-desert (temperate-climate) locations?
(3) Explain the concept of an internal boundary layer.
(4) Define a residual mixed layer.

Problems and exercises

(1) Show, using plots of temperature as a function of height, how an elevated mixed layer causes parcels of rising buoyant air below to lose their buoyancy.
(2) For what arid areas of the world might we expect elevated mixed layers to be produced, and affect the precipitation downwind?
(3) In what situations would we expect to find the deepest boundary layers (time of day, time of year, latitude, proximity to mountains and coastlines, substrate types)?

11

Desert microclimates

It was as if there were no names here, as if there were no words. The desert washed everything away in the wind, obliterated everything. The men had the freedom of the wide open spaces in their eyes; their skin was like metal. The sunlight exploded everywhere. Ochre, yellow, grey and white, the light sand shifted, revealing the wind. It covered every footprint, every bone. It repelled light, drove water, life, far from a centre that nobody could recognize. The men knew that the desert didn't want them; so they walked without stopping, on paths other feet had already trodden, to find something else. As for water, it was in the aiun, eyes the colour of the sky, or in the damp beds of ancient mudstreams. But it was not water for pleasure or for rest. It was just a trace of sweat on the surface of the desert, the parsimonious gift of a dry god, the last spasm of life. Heavy water torn from the sand, dead water of the fissures, alkaline water that caused stomach pains and made people vomit. Keep on going, then, bent a little forward, in the direction the stars had given.

But it was perhaps the last and only free country, the country where men's laws had no importance. A country for the stones and for the wind, but also for scorpions and jerboas, creatures that know how to take refuge when the sun burns down and the night is frosty.

J. M. G. Le Clézio, French writer
Désert (1980)

Within deserts there are sometimes large horizontal variations in a variety of surface properties, such as the type of substrate, the depth of the water table, the type and density of the vegetation, and the topographic elevation, slope and aspect. Through effects on the surface energy, moisture, and momentum fluxes, these contrasts in surface properties lead to atmospheric variability, such as in the rainfall, temperature, and wind field. These near-surface and boundary-layer properties define the microclimate of the lower atmosphere.

For example, the impact of mountains and high plateaus on the wind-field climatology within deserts is very evident from the complex orientation of surrounding sand dunes. And within vast unvegetated areas of the Sahara are

islands of green grasslands where mountains focus the rainfall. These and other
surface-forced modulations in the large-scale atmospheric processes define the
characteristics of intra-desert microclimates. In addition, there are sometimes
strong contrasts at the margins of deserts, in the transition zone with oceans or
less arid ecosystems. Numerous examples of intra-desert microclimate variability were provided in the descriptions in Chapter 3 of the climates of the various
world deserts. This chapter will describe the general causes of desert microclimate variability and provide specific examples based on measurements and
numerical models. In the next chapter will be a discussion how this variability
leads to the development of thermally forced circulations in the wind field, which
are also part of the desert microclimate.

Causes of intra-desert microclimates

Climate can vary over a wide range of spatial scales. On the smallest, the shade
behind a plant may be defined as a microclimate. On a somewhat larger scale,
daytime temperatures within tens of meters above the surface can differ greatly
over a distance of 100 m between the opposite north- and south-facing slopes of a
hill, or between a moist playa surface and surrounding sand or rock. Small-scale,
shallow variations such as these will have significant impacts on the climate very
near the ground, which is important to desert-dwelling flora and fauna, but horizontal mixing will smooth out the atmospheric response before it reaches very
high into the boundary layer. Larger-scale surface variations would be needed
to influence the climate of the entire few-kilometer-deep warm-season daytime
boundary layer. Even larger-scale variations would be required to influence the
deeper circulations associated with cloud and rainfall formation. The following subsections summarize the different causes of microclimate variability in
deserts.

Vegetation

In temperate climates, the natural state is for vegetation to be ubiquitous. However, in the water-limited arid environment, many factors such as water-table
depth, substrate type, and terrain elevation, aspect and slope have important influences on the existence, type, density, and vitality of vegetation. Vegetation
can influence the microclimate in a variety of ways through the effects on the
surface-energy, water, and momentum budgets that are described in Chapter 6.
A good summary is provided in Table 6.3. Naturally, the area-average effect of
the vegetation will depend on the vegetation density. In general, vegetation decreases the daytime maximum temperature and increases the nocturnal minimum

temperature near the surface, it reduces the wind speed at the surface and increases the turbulence, and it increases the relative and specific humidities. These and other effects can significantly contribute to variations in the microclimate between vegetated and unvegetated surfaces. The vegetation–microclimate relationship is complicated by the fact that vegetation characteristics are greatly affected by the microclimate, in addition to the fact that they influence it.

Substrate

Desert substrates – sand, salt flats, gravel, solid rock, etc. – are sometimes quite variable from place to place. Even when the substrate appears to be uniform, just below the surface it is often not. For example, Ackerman and Inoue (1994) compared satellite-estimated albedo and long-wave fluxes over an apparently uniform area of the interior of the Sahara, and found no correlation between the quantities, implying to them that the substrate must have significant specific-heat variability.

Chapter 7 discusses how the various substrate types influence the energy and water budgets. For example, Figs. 7.5 and 7.6 compare the near-surface diurnal temperature variations for sand with those for desert pavement and a substrate of deep rock. The sand, with the lowest thermal diffusivity, had a much greater diurnal variation in surface temperature, and therefore a much different microclimate, than the other two nearby locations. Substrate hydraulic properties are important too. For example, the fact that rains can infiltrate to greater depths in sand than in clay sometimes means that the sand supports trees and shrubs while the clay grows primarily grasses. Substrate albedoes can also vary greatly and influence the microclimate. Playas with salt surfaces and nearby volcanic rock can have albedoes that differ by 50%. Lastly, the chemical properties of the substrate, and of the water that it contains, limit the nature and quantity of the vegetation, and this, in turn, influences the microclimate.

Terrain elevation, aspect, and slope

There are a number of ways in which terrain elevation influences the microclimate. A first-order effect is that temperatures near the surface typically decrease with increasing terrain elevation simply because the average tropospheric temperature decreases with height (Fig. 2.2). However, when a valley exists between two mountain ridges, air on the slopes that is cooled by the surface at night will drain downslope, and the lowest temperatures may be in valleys. The highest nocturnal surface temperatures may thus be between the highest elevations of the ridges and the lowest elevations of the valleys, i.e. along the mountain slopes.

Also, as shown in Chapters 3 and 13, precipitation is higher over elevated terrain, forming altitude oases, and Chapters 2 and 3 show that mountains cause rainfall deficits downwind, causing rain-shadow deserts. The effect of terrain elevation on microclimate, through rainfall and temperature, is so great that it is possible in some places to walk from a high-altitude Alpine forest to rocky cactus plains in a few hours.

Terrain slope, for example associated with alluvial fans, affects water infiltration and soil moisture (Fig. 7.3). And the combination of the slope angle and the aspect of the slope determines the angle at which the sun intercepts the surface, and therefore is a strong control on the solar energy flux and therefore the surface energy balance (Fig 4.3). The importance of this effect on the microclimate is illustrated by the fact that desert vegetation on north- and south-facing slopes is often greatly different in type and density. Logan (1968) discusses slope effects on microclimates in the Mojave Desert, documenting differences in daily maximum temperatures of over 13 K (23 °F) between north- and south-facing slopes separated by only a few hundred meters. Similarly, Fig. 5.11 shows large observed contrasts in the surface temperature on different dune faces, with the result being the existence of a number of dune and interdune microhabitats that support unique biota.

Temperature extremes are just as important a property of a desert microclimate as is the daily mean temperature. In fact, it is the extremes that are more likely to determine the prevailing vegetation than is the mean. Thus, topographic effects such as elevation, slope, and aspect, and cold air drainage into valleys, exert important controls on the vegetation that prevails in a desert microclimate.

Water-table depth

In spite of the existence of a desiccated condition at the surface, water tables in deserts are sometimes within a few meters of the surface. If the water is within reach of root systems, significant stands of vegetation can survive and influence the microclimate through various mechanisms. Where the water table is very near the surface, such as in salt flats or basin floors, enhanced evaporation can have a great effect on the microclimate.

Proximity to coastlines

We have seen that many deserts abut seas and lakes. In common with non-deserts, the land areas near the coastlines have microclimates that are quite different from those farther inland. One cause is the sea-breeze circulation that transports cooler and more-humid maritime air inland during the day, sometimes

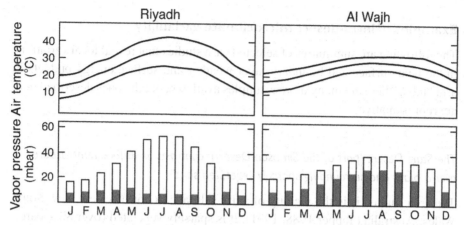

Fig. 11.1 Monthly average temperatures, temperature maxima, and temperature minima for Riyadh and Al Wajh, Saudi Arabia. Al Wajh is on the Red Sea coast, and Riyadh is near the center of the peninsula. Also shown are the monthly average vapor pressure (gray-shaded bars) and saturation vapor pressure (total bar length). (Adapted from Abd El Rahman 1986.)

for distances of over 100 km. In general, the maritime climates that prevail near coastlines are more humid, and the diurnal and annual variations in temperature are less. The lower temperature variability is partly a result of the fact that the water temperatures do not vary seasonally or diurnally to nearly the same degree as the land temperatures. Also, the greater water-vapor content of the coastal air absorbs more terrestrial infrared radiation, which maintains higher daily minimum temperatures. To illustrate this effect, Fig. 11.1 shows the monthly average vapor pressure and temperature for two locations in the Arabian Desert. Al Wajh is on the Red Sea coast and Riyadh is near the center of the peninsula, both at similar latitudes. The shaded part of each bar corresponds to the vapor pressure, and the total bar length indicates the saturation vapor pressure, which is a function of the temperature. The water-vapor content at the coastal city is greater, as is the relative humidity. The annual variation in the monthly mean temperature is almost twice as large at Riyadh as at Al Wajh. In every month there is a greater range of temperatures at Riyadh, as seen by the average monthly maxima and minima.

Also, when cold ocean currents exist near the coast, the adjacent coastal areas can be humid and foggy much of the year. The transition from the coastal microclimate to the less humid, hotter one inland was shown in Fig. 3.4 for the Namib Desert and in Fig. 2.28 for the Peruvian Desert. Figure 10.5 also compares coastal and inland boundary-layer temperature profiles for different arid areas.

Examples of intra-desert microclimate variability

The following are summaries of selected case studies that reveal local variations in the near-surface and boundary-layer weather and surface energy budgets in arid lands. There are many more examples available, but the ones described here are representative.

The Sturt Desert (part of the Simpson Desert): contrasts in surface temperature over a playa and surrounding sparsely vegetated sand

The area studied is in the vicinity of a dry salt lake in northwestern New South Wales, Australia (Tapper 1988, 1991). It is sparsely vegetated (over 80% bare), with substrate conditions that include sand, clay, silt and lake salt. The averaged sand and vegetation albedoes were 0.2–0.3 onshore of the salt lake, and the albedo of the salt surface was 0.6. The soil-moisture variation is primarily related to the fact that the salt lake comprises a shallow salt crust that overlays saturated clay. The salt lake itself has an area of approximately 70 km². During April 1985 (Austral autumn), surface temperatures were estimated by using portable infrared thermometers and satellite imagery. Both afternoon and nocturnal temperatures were observed along a 1.5 km transect across the margin of the salt flat. For a transect at 1500 LT, surface temperatures decreased monotonically from 40 °C (104 °F) in the sparsely vegetated sand about 1 km away from the salt flats, to 24 °C (75 °F) over the salt-lake surface, whereas the temperature of the vegetation varied only slightly from place to place (24–27 °C, 75–81 °F). A 0645 LT transect (approximately sunrise) recorded substrate surface temperatures of 0–3 °C (32–37 °F) inland of the salt lake, and 8 °C (46 °F) over the salt lake. The satellite imagery for 1451 LT showed a large amount of spatial variability in the daytime surface temperature over a 30 km² area that is centered on the lake, with numerous large "hot spots" over bare clay surfaces having **radiative temperatures** of almost 50 °C (122 °F). Fig. 11.2 shows maps of the satellite-derived temperatures for 1451 LT and 0213 LT. Tapper (1991) attributes the spatial temperature variation between the salt lake and its surroundings to be roughly equally attributable to substrate albedo and thermal-conductivity differences, a conclusion that was confirmed in numerical experiments by Physick and Tapper (1990). In terms of the soil thermal conductivity, estimates by Tapper were 2.32 W m^{-1} K^{-1} and 0.33 W m^{-1} K^{-1} for the salt-lake sediment and the sand, respectively. This large difference in conductivity, and the fact that Krusel (1987) measured evaporation rates to be very low for both this playa and its sandy surroundings, supports the conclusion that albedo and conductivity differences

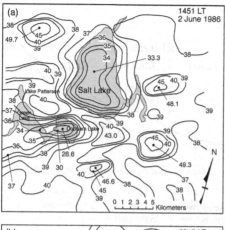

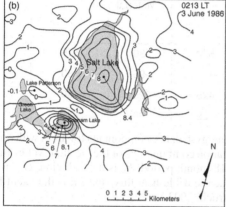

Fig. 11.2 Maps, based on satellite measurements, of the radiative surface temperatures in °C for 1451 LT and 0213 LT in the vicinity of a dry salt lake in northwestern New South Wales, Australia. The area is sparsely vegetated (over 80% bare), with substrates that include sand, clay, silt, and lake salt. The area of the salt lake is shaded. (From Tapper 1991.)

are the dominant factors responsible for the contrasting desert microclimates in this area. Clearly, the variation of over 16 K (30 °F) in daytime surface temperatures over a distance of 5–10 km (Fig. 11.2) represents a significant microclimate contrast.

The Great Basin Desert: contrasts in surface temperature and fluxes over a playa and the surrounding vegetated alluvial fan and mountains

Laymon and Quattrochi (2003) used satellite data to estimate mesoscale surface fluxes in a 16 km × 40 km area of the Great Basin Desert during an 18-month

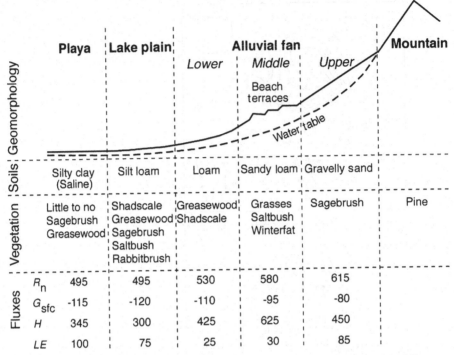

Fig. 11.3 Spatially averaged energy-balance components for areas with different geomorphology–soil–vegetation combinations of the Goshute Valley of the Great Basin Desert during an 18 month period. R_n is net radiation, G_{sfc} is the ground heat flux, H is the surface sensible-heat flux, and LE is the latent-heat flux. (From Laymon and Quattrochi 2003.)

period. After performing a survey of the heterogeneous desert soils and vegetation, a water- and energy-balance flux station was situated in each of the six major plant ecosystems. Relations between point and remotely sensed measurements were used to produce estimates of the spatial variability of the surface fluxes. Albedoes varied from 0.2 to 0.3 at the flux-station locations. Mid-morning (0939 LT 19 June 1994) surface temperatures at the flux stations that represented different ecosystems varied by 7–9 K (13–16 °F), depending on whether the estimates were based on infrared thermometers or satellite data. At this time, among the different soil/vegetation categories, sensible-heat fluxes varied by over a factor of two and latent-heat fluxes varied by a factor of four. Fig. 11.3 summarizes the spatially averaged energy-balance components for areas of the terrain profile having different geomorphology–soil–vegetation combinations. Large latent-heat fluxes and soil heat fluxes were located over the moist, thermally conductive playa. Lower latent-heat fluxes and higher sensible-heat fluxes were found over the alluvial fan. Malek *et al.* (1997) and Malek and Bingham (1997)

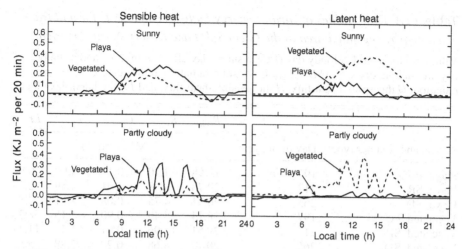

Fig. 11.4 Diurnal variation of the surface sensible- and latent-heat fluxes over a vegetated sandy surface and a nearby playa in the Great Basin Desert, on a sunny and a partly cloudy day. (From Malek *et al.* 1990.)

describe the energy-balance components at five of these surface stations, based on twenty months of continuous data. Among the stations for this period, the 20-month mean 2 m air temperatures varied by 1.4 K (2.5 °F) and the 20-month mean surface temperatures varied by 1.5 K (2.7 °F). Alluvial fans, such as the one in Fig. 11.3, occur commonly in deserts, and thus the impact on the microclimate of the varying soil, vegetation, and moisture characteristics should be considerable and widespread.

Another study (Malek *et al.* 1990) conduced in the Great Basin Desert also aimed at contrasting the surface energy budgets over a playa (150 km^2) and over a grass- and shrub-covered desert margin of the playa, with the separation distance being about 5 km. Under the thin salt crust of the playa, which reached a depth of 3 mm in a normal summer, the ground always appeared visibly moist. Fig. 11.4 shows the diurnal variation of the surface, sensible- and latent-heat fluxes for a sunny, summer day after a rain and for a partly cloudy day, at the playa and vegetated desert sites. Over the unvegetated playa, the latent-heat fluxes are much smaller and the sensible-heat fluxes are greater. The water evaporated over the playa is from upward diffusion through the crust, or rehydration, during the night, as evidenced by the fact that the latent-heat flux decreased during the sunny day as the surface dried. The latent-heat flux in the vegetated area does not show the morning maximum because most of the water loss is from transpiration rather than from surface evaporation. Table 11.1 shows the daily totals of the energy-budget components for the two sites, for three summer days: a sunny day after a rain (Fig. 11.4, sunny), a sunny day after a long dry spell, and a cloudy day (Fig. 11.4, partly cloudy). Regardless of the weather,

Table 11.1 *Daily values of various energy-budget components for a playa and a nearby vegetated area in the Great Salt Lake Desert, North America*

Values are for a partly cloudy day (*PC*), a sunny day after a rain (*SR*), and a sunny day after a long dry spell (*SD*). R_{solar} is the solar radiation at the surface, R_n the net radiation, G the ground heat flux, H the sensible-heat flux, and LE the latent-heat flux.

		R_{solar}	R_n	G	H	LE
Station and weather type	Day of the year			$(MJ\ m^{-2}d^{-1})$		
Playa, PC	197	17.98	4.40	0.17	4.00	0.57
Playa, SR	212	29.73	8.88	−1.91	4.59	2.38
Playa, SD	265	20.91	4.75	−0.08	3.81	0.85
Vegetated, PC	197	18.14	8.49	−0.90	−1.15	8.73
Vegetated, SR	212	29.51	13.96	−0.91	1.58	11.47
Vegetated, SD	265	20.73	8.68	−0.28	2.88	5.52

Source: Adapted from Malek *et al.* (1990).

Table 11.2 *Warm-season, total-energy-budget terms for vegetated and playa sites*

See the caption for Table 11.1 for definitions of the energy-budget terms.

Station	R_{solar} $(MJ\ m^{-2})$	R_n $(MJ\ m^{-2})$	G $(MJ\ m^{-2})$	H $(MJ\ m^{-2})$	LE $(MJ\ m^{-2})$	Rainfall (mm)
Playa	4000	1115	25	810	330	64
Vegetated	4000	1798	−88	510	1200	62

Source: Adapted from Malek *et al.* (1990).

the playa sensible-heat flux to the atmosphere was always greater than over the vegetated margin, whereas the latent-heat flux over the playa was always less. Even though there is no documentation of the playa water table on these days, as the warm season progressed the water level decreased from 3–5 cm above the surface to about 30 cm below it. After a dry spell, the latent-heat flux was significantly less than after a rain, at both the playa and vegetated sites. Table 11.2 shows the warm-season total energy budget terms for the vegetated and playa sites, and perhaps best illustrates the distinction in the microclimates. Even though the solar radiation and precipitation were virtually the same at the playa and vegetated locations, at the vegetated site the net radiation was over 60% greater, the sensible-heat flux was almost 40% less, and the latent-heat flux was almost four times as great. The higher sensible-heat flux over the

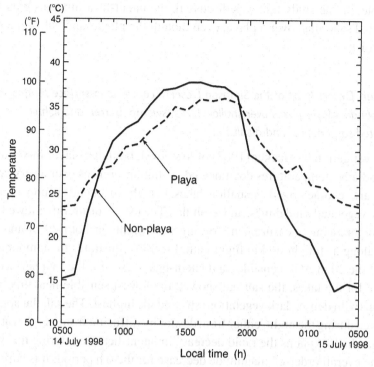

Fig. 11.5 Time series from 0500 LT 14 July to 0500 LT 15 July of the 2 m AGL temperature near the center of a playa and at a nearby sandy location outside of the playa. (From Rife *et al.* 2002.)

playa relative to the vegetated area contrasts with observations in other areas because the strongly transpiring vegetation in this case utilized a great deal of energy.

From another study (Rife *et al.* 2002), high-temporal-resolution plots of the 2 m AGL temperatures for a location within a playa and a nearby location in a surrounding sparsely vegetated sandy area in the Great Basin Desert are shown in Fig. 11.5 for a 24 h period in the summer. As discussed earlier, over the playa the nighttime temperatures are higher and the daytime temperatures are lower, relative to the surroundings. Thus, the diurnal temperature variation is much greater outside of the playa. Here, the diurnal temperature amplitude over the playa is 13 K (23 °F), while in the surrounding sandy area it is 23 K (41 °F).

These studies in the Great Basin Desert have demonstrated the potentially large and variable effects of playas on the microclimate. However, it is difficult to generalize about some aspects of the playa effect because the moisture availability at the surface varies greatly with season and geographic area. For example, in one study that was discussed, the latent-heat flux was greater over vegetated sand than over a playa, even though the playa sand below the surface was moist.

The Sturt Desert (part of the Simpson Desert): contrasts in surface temperature and boundary-layer fluxes over shallow lakes, swamp, barren and lightly vegetated sand dunes, and a salt flat

An investigation by Hacker (1988) of low-level, boundary-layer fluxes and surface radiative temperatures documented spatial differences over a very complex desert area characterized by shallow lakes, heavily vegetated swamp, barren and lightly vegetated sand dunes, and a salt flat. The use of an aircraft allowed almost contemporaneous measurements during four flights at both 20 m and 150 m AGL along a straight 60 km flight path. Fig 11.6 illustrates the temporal variation of the 20 m AGL sensible- and latent-heat fluxes for the four surface types. Except for the lakes, the salt flat showed the lowest sensible-heat flux, and the swamp (with dense, dark vegetation) showed the highest. The salt flat apparently hydrated during the night, either from dew or by diffusion from a high water table below, which explains the rapid decrease in latent-heat flux during the morning and the overall order-of-magnitude decrease for the 6 h period. It is curious that the morning latent-heat fluxes from the salt flat exceed those from the lake, in spite of the similar initial temperatures.

The Great Basin Desert: contrasts in boundary-layer vertical temperature profiles over a salt flat and a partly vegetated nearby surface

Observations were made in the Great Basin Desert in September 1997 on a relatively clear, calm day to determine how much the surface type affected the boundary-layer structure. Four pairs of simultaneous radiosonde ascents were made at two locations about 20 km apart. One location was in a dry salt lake and the other was at about the same elevation in a sparsely vegetated area (less than 20% desert shrubs) with sandy soil. Near-surface temperature and wind measurements were also made at the sounding locations. The winds in the lowest 500 m were relatively light, with speeds of 1–2 m s^{-1} in the early morning, increasing to 3–5 m s^{-1} by mid-afternoon. Therefore, the temperature advection by these weak winds should be relatively small, and thus the differences between the boundary-layer temperature profiles should be primarily related to the local surface characteristics. The average 2 m AGL temperature (observed hourly) for the two sites rose from about 7 °C (45 °F) at 0700 LT to about 26 °C (79 °F) by 1400 LT; however, the salt flat 2 m temperature was 2 K (3.6 °F) higher and

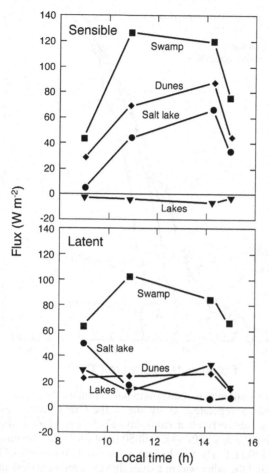

Fig. 11.6 Temporal variation of the 20 m AGL sensible- and latent-heat fluxes for four surface types as estimated from aircraft measurements over an Australian desert: shallow lakes, swamp, barren and lightly vegetated sand dunes, and a salt flat. The measurements were conducted on 20–21 May 1987, about one month prior to the Austral Winter Solstice. The four data points correspond to the times of the four flights. (From Hacker 1988.)

3.5 K (6. 3 °F) lower at 0700 LT and 1400 LT, respectively, as expected because of the differences in albedo and substrate thermal conductivity. Fig. 11.7 shows the potential temperature profiles from the surface to 2500 m AGL for the two sites, and for the four ascents at each site at 0830 LT (0850 for the salt-flat sounding), 1000 LT, 1200 LT, and 1400 LT. The temperature over the salt flat is 1–2 K (2–4 °F) lower in the noon and early afternoon profiles below about 500 m. The layer over which the temperature difference is observed increased as the boundary-layer depth increased up to about 1000 m.

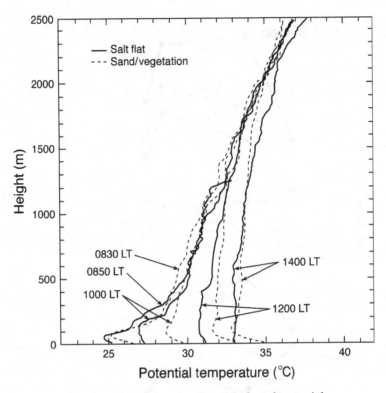

Fig. 11.7 Four pairs of simultaneous radiosonde-based potential-temperature profiles for a salt flat and a vegetated, sandy site in the Great Basin Desert in September 1997 on a relatively clear calm day. There were four pairs of radiosonde ascents during the day at 0830 LT (0850 for the salt-flat sounding), 1000 LT, 1200 LT, and 1400 LT. The two locations are about 20 km apart. One was over a dry salt lake and the other was in a sparsely vegetated area (less than 20% desert shrubs) with sandy soil. (Courtesy Elford Astling, United States Army Dugway Proving Ground.)

The Sahel: surface fluxes and boundary-layer structure over various partly vegetated surfaces

The Hydrological Atmospheric Pilot Experiment in the Sahel (HAPEX-Sahel) investigated land–atmosphere interaction in the semi-arid Sahel in southwest Niger through the wet season and the subsequent drying period in 1992 (Taylor *et al.* 1997; Goutorbe *et al.* 1994, 1997; Dolman *et al.* 1997; Gash *et al.* 1997; Said *et al.* 1997; Wallace and Holwill 1997). The study covered an area with dimensions of about 10 km × 10 km, where the vegetation consisted of areas of tiger bush, cultivated millet, and fallow savannah, interspersed with bare areas of sandy or crusted soil. Automated surface weather stations were located in

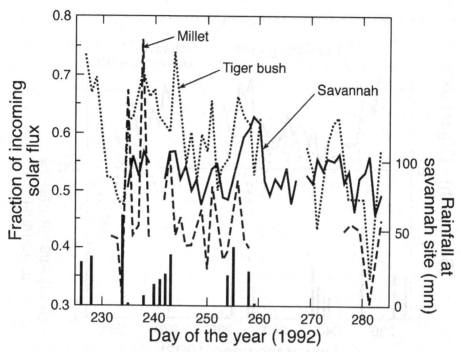

Fig. 11.8 Total of the daily-average surface sensible- and latent-heat fluxes
normalized by the incoming solar flux, for three differently vegetated arid sites
for a two-month period during the Hydrological Atmospheric Pilot Experiment
in the Sahel (HAPEX-Sahel). The abscissa represents the day of the year, 1992.
The study covered an area with dimensions of about 10 km × 10 km, where
the vegetation consisted of areas characterized by tiger bush, cultivated millet,
and fallow savannah, interspersed with bare areas having sandy or crusted soil.
Also shown is the rainfall for the savannah site, which had an accumulation of
about 450 mm for the period. (From Taylor *et al.* 1997.)

areas represented by each of the three types of vegetation. The surface data were
periodically augmented with radiosonde ascents, turbulence flux measurements,
and aircraft observations. Fig. 11.8 shows the sum of the daily-average surface
sensible- and latent-heat fluxes, normalized by the incoming solar flux, for the
three sites. Also shown is the rainfall for the savannah site, which had an accumulation of about 450 mm for the period, compared with about 350 mm for the millet
site and about 300 mm for the tiger bush site. Because of the proximity of the sites,
the flux differences in the figure should be largely due to surface-property differences rather than to large-scale meteorological effects. Even though it is difficult
to separate soil-moisture, vegetation, and substrate effects, there are major contrasts among the sites in terms of the sum of the latent- and sensible-heat fluxes.
Fig. 11.9 illustrates the local differences in the 9.5 m AGL specific humidity and

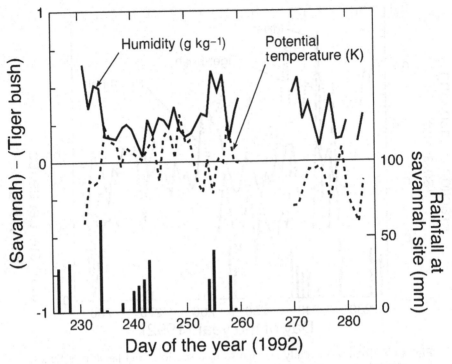

Fig. 11.9 Differences in the 9.5 m AGL specific humidity and potential temperature between the HAPEX-Sahel savannah and tiger bush sites. Also shown is the rainfall for the savannah site, which had an accumulation of about 450 mm for the period. (From Taylor *et al.* 1997.)

potential temperature between the savannah and tiger bush sites. Even though the temperatures are similar, the savannah site is typically about 0.3 g kg^{-1} more moist than the tiger bush site (which received less rainfall). There appears to be a tendency for the specific-humidity differences to be greater during the drier periods. The significant specific-humidity differences between the sites that still prevailed more than 20 d after the last rain event may have resulted from the transpiration of deep soil moisture by vegetation.

Aircraft were used to measure boundary-layer temperature and specific humidity in the region on three days. Fig. 11.10 shows the potential temperature and specific humidity deviations from the 80 km flight-leg average for flights that took place on 9 September 1992. Boundary-layer winds were very weak, and thus temperature and humidity variability along the path is presumed to be locally forced. Variations in total precipitation along the path during the previous 3 weeks were less than 20%. Flights were performed at three altitudes: about 50 m AGL; about 600 m AGL, near the boundary-layer top; and at about 300 m AGL, near the middle of the boundary layer. The temperature deviations and

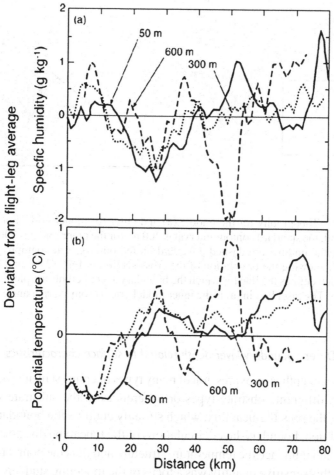

Fig. 11.10 Aircraft-measured boundary-layer (a) specific-humidity deviation and (b) potential temperature deviation from the flight means in the HAPEX-Sahel region. The flights took place between 1115 and 1215 LT on 9 September 1992. Boundary-layer winds were very weak, and thus variability along the path is presumed to be locally forced. Variations in total precipitation along the path during the previous three weeks were less than 20%. Flights were performed at the three altitudes shown. (Adapted from Taylor *et al.* 1997.)

specific-humidity deviations from the average tend to be out of phase at both 50 m and 300 m AGL. Deviations from the average are somewhat larger near the surface. Because the inversion capping the boundary layer was near the 600 m flight level (documented by radiosonde ascents), some of the large horizontal variability is a result of the aircraft flying through the large vertical gradients near the top of the boundary layer (for example, the 2 g kg^{-1} variation at about 50 km).

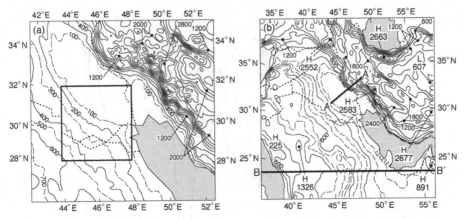

Fig. 11.11 Topography for the middle (a) and outer (b) model grids. For the middle grid, the standard contour interval is 200 m; for the outer grid it is 300 m. Intermediate contours are dashed. Overlaid on the outer grid topography are line A–A′, showing the orientation of the cross-section in Fig. 12.11, and line B–B′, which defines the line for which the boundary-layer heights are plotted in Fig. 11.13. The inset box in (a) is the inner model grid. (From Warner and Sheu 2000.)

The Arabian Desert: boundary-layer depth related to surface characteristics

Boundary-layer depth is an expression of many types of contrast in microclimate. For example, different substrate types or differences in the substrate wetness can influence the sensible-heat flux, which strongly controls the boundary-layer depth. In addition, boundary-layer circulations of the type to be discussed in the next chapter can influence the temperature structure and thus the boundary-layer depth. Thus, this sensitive quantity was a focus of the modeling study by Warner and Sheu (2000) who describe the influence on the Arabian Desert boundary layer of orography, coastal circulations, and variability in the characteristics of the desert surface. Six-day simulations were performed with a numerical model for the period 1200 UTC 9 March 1991 to 1200 UTC 15 March 1991, where the triply nested model grids employed grid increments of 30 km, 10 km, and 3.3 km. Fig. 11.11 depicts the terrain elevation on the outer two grids. Major terrain features include the Zagros Mountains in western Iran (upper right of Fig. 11.11a) and the high elevations over the western third of the Arabian Peninsula (lower left of Fig. 11.11b).

The variability in the boundary-layer structure will first be described using the model solution on the inner grid, which has the 3.3 km grid increment (inset box in Fig. 11.11a). The grid is located just to the northwest of the Arabian Gulf, in the south end of the Tigris–Euphrates Valley. Because the Arabian Gulf

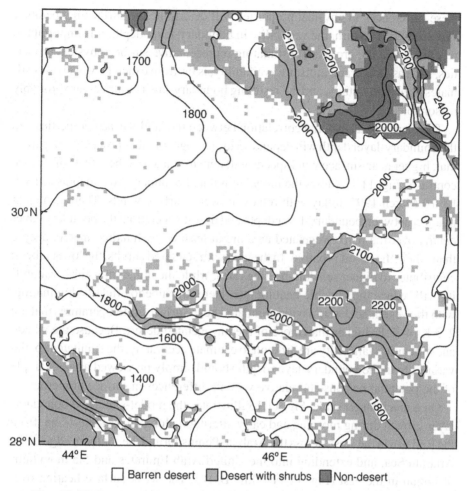

Fig. 11.12 Depth (solid lines) of the simulated boundary layer on the inner model grid (Fig. 11.11a, inset) over central Iraq at 1300 LT 10 March 1991. Depths are labeled in meters and are plotted at an interval of 100 m. Gray shading indicates land-cover characteristics, with white being barren desert, light gray being desert with scattered vegetation, and dark gray being non-desert. (From Warner and Sheu 2000.)

breeze does not extend inland very far in this area, the primary effect of surface forcing on the boundary-layer depth should be associated with differences in the substrate properties. Fig. 11.12 shows that most of the area is barren desert and desert with shrubs. In the model, the desert with shrubs had a 5% higher assigned moisture availability (resulting from transpiration) and a 10% lower albedo compared to the barren desert. There is also a small area in the northeast part of the grid that is agricultural land and range grassland. The net effect of these differences in physical properties will determine the surface-temperature variability and, in turn, the variability in boundary-layer depth. It is important to recognize that the process of estimating representative surface physical properties, such as albedo and moisture availability, is far from exact. However, the estimates used in this study, and the resulting boundary-layer contrasts, are probably reasonable.

The clarity of any spatial correlation between the land-surface properties and the boundary-layer depth will depend on the strength of other processes. In particular, higher near-surface wind speeds will tend to smear out the effects of surface contrasts. Fig. 11.12 shows the model-simulated boundary-layer depths at 1300 LT 10 March 1991, a day with relatively weak surface winds. There is a clear relation between boundary-layer depth and desert vegetation: the boundary-layer depths over the partly vegetated area are at least a few hundred meters deeper than where the desert is barren. In the area with scattered shrubs, the lower albedo contributes to higher sensible-heating rates, while the higher moisture availability will reduce the sensible-heating rate. Here, the albedo effect is dominating, with the vegetated desert having simulated near-surface air temperatures that are 2–3 K (4–5 °F) higher than those of the unvegetated desert. Because the albedo and moisture availability are difficult to estimate accurately, the simulated spatial variability in the boundary-layer depth should simply be viewed as an example of the potential response to relatively small differences in surface properties.

On a much larger scale, Fig. 11.13 illustrates the model-simulated boundary-layer growth during morning and early afternoon along an east–west line, from the Red Sea in the west, across the Arabian Peninsula and the southern edge of the Arabian Sea, and extending into the United Arab Emirates and the mountains of Oman in the east (Fig. 11.11b, line B–B'). The weak surface heating over the Red Sea and over the Arabian Gulf causes the boundary layer there to be very shallow. In contrast, the heating and resulting upward motion over the high terrain of the western Arabian Peninsula cause a deeper daytime boundary layer there than over the lower elevations to the east. These simulated regional differences in the boundary-layer depth reflect the general variability in the boundary-layer microclimate that can be expected from the existence of water bodies and mountains in desert areas.

The Sonoran Desert: influence of precipitation and vegetation on the temperature and sensible-heat fluxes

Humes *et al.* (1997) and Kustas and Humes (1997) describe the effects of rainfall on summer surface temperatures and sensible-heat fluxes, and therefore the

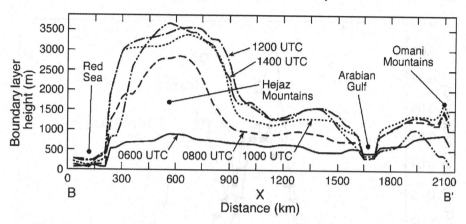

Fig. 11.13 The model-simulated boundary-layer depth at various times during the morning and early afternoon of 15 March 1991 along an east–west transect, from the Red Sea in the west, across the Arabian Peninsula and the southern edge of the Arabian Sea, and extending into the mountains of Oman in the east (Fig. 11.11b, line B–B′). (Adapted from Warner and Sheu 2000.)

microclimate, for the semi-arid, partly vegetated Walnut Gulch Experimental Watershed in the northern Sonoran Desert. As a summary of the effect of the precipitation on surface temperature, Fig. 11.14 shows the frequency of different temperatures across the watershed at about the same time of day on three relatively clear days, each with different antecedent precipitation distributions. The high surface temperatures on day-of-year (DOY) 213 reflect the fact that there had been no significant precipitation over the watershed for almost two weeks. The place-to-place variation in surface temperature in the watershed was 20 K (36 °F). During the subsequent three days, one large and several small precipitation events occurred over the watershed, which caused the most frequent surface temperature on DOY 216 to be approximately 7 K (13 °F) lower. Only moderate drying occurred between these first two dates because skies were mostly overcast. Between DOY 216 and DOY 221, only two extremely small precipitation events occurred, and thus the temperature frequencies should show a drying of the surface during the period. In fact, the frequencies imply a surprisingly slow drying process for the desert in the summer season. The sensible heat fluxes at eight surface-flux-measurement stations ranged from 195 to 241 W m^{-2} on DOY 213, from 78 to 129 W m^{-2} on DOY 216, and from 127 to 195 W m^{-2} on DOY 221. As a reference, the DOY 221 sensible-heat flux was 275 W m^{-2} at another site in the watershed that had received no precipitation during the entire period.

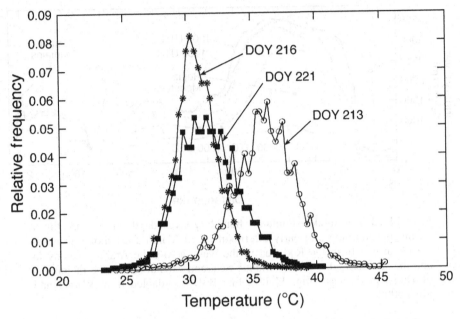

Fig. 11.14 Frequency of different aircraft- and ground-measured surface temperatures across the Walnut Gulch Experimental Watershed in Arizona at about the same time of day on three relatively clear summer days in 1990 each with different antecedent precipitation effects.(From Humes *et al.* 1997.)

Coastal microclimates of the Peruvian and Namib Deserts

The microclimate contrast in coastal deserts having cold, nearby ocean currents is generally greatest in the direction perpendicular to the coast. This contrast is not only related to distance from the cold water and associated fog, but also to the fact that the topographic elevation often increases rather abruptly with distance inland. For example, Fig. 2.28 illustrates temperature measurements at various locations between the coast and higher inland elevations in the Peruvian Desert. In particular, the temperature inversion in the atmospheric column that is transported inland is reflected in the fact that the temperature increases with elevation over the land near the coast. In addition, Fig. 3.4 illustrates the conditions near the coast of the Namib Desert. The foggy cool air that is advected inland across the coastline produces saturated, foggy, and cool conditions for most of the day near the coast. Only for a short period in the afternoon does the heating of the surface cause some of the fog to evaporate, and allow more efficient heating of the surface. With greater distance inland, however, the fog is dissipated for a longer period of the day. Clearly, microclimate contrasts near such desert coasts are exceptionally large.

Suggested general references for further reading

Geiger, R., 1966: *The Climate Near the Ground* – a classic reference on the physical causes of microclimates, including the effects of vegetation, substrates, and topography.

Hastings, J. R., and R. M. Turner, 1965: *The Changing Mile* – discusses the microclimates of an area in the northern Sonoran Desert, primarily as related to elevation differences.

Oke, T. R., 1987: *Boundary Layer Climates* – discusses physical processes at the surface and in the boundary layer. This is a good reference for students from a variety of backgrounds.

Rife, D. L., *et al.*, 2002: *Mechanisms for diurnal boundary-layer circulations in the Great Basin Desert* – discusses wind and temperature observations in an area with complex terrain, playas, and lakes in the Great Basin Desert.

Yoshino, M. M., 1975: *Climate in a Small Area: An Introduction to Local Meteorology* – describes the physical causes of microclimates for various surface types, for coastal areas, for cities, and in complex topography.

Questions for review

(1) Discuss the mechanisms by which vegetation influences the desert microclimate.

(2) Discuss the mechanisms by which the substrate type influences the desert microclimate.

(3) Discuss the mechanisms by which terrain-elevation, aspect and slope influence the desert microclimate.

(4) Discuss how the proximity to coastlines in deserts affects the microclimate. Distinguish between deserts that exist next to cold ocean currents, and deserts that abut warmer waters.

(5) By what mechanisms do playas influence the microclimate?

Problems and exercises

(1) Further speculate upon the causes of the relative magnitudes of the different sensible- and latent-heat fluxes shown in Fig. 11.6 for the different surface types.

(2) With reference to Fig. 3.4, explain why the more inland locations in the Namib Desert experience less fog.

(3) The differences in the boundary-layer structure shown in Fig. 11.7 were measured in relatively calm winds for nearby locations over a salt flat and a sandy, vegetated site. Why would we be less likely to see significant differences if the mean wind speed near the surface was greater?

(4) The spatial variations in temperature and specific humidity in Fig. 11.10 are out of phase at 50 m and 300 m AGL. Explain why this is reasonable.

(5) An example was shown of a situation where the latent-heat flux over a salt flat was greater than over the surrounding desert, whereas in another example the latent-heat flux was less over the salt flat. Discuss the factors that can determine the latent-heat flux contrast.

12

Dynamic interactions among desert microclimates

The desert landscape is always at its best in the half-light of dawn or dusk. The sense of distance lacks: a ridge nearby can be a far-off mountain range, each small detail can take on the importance of a major variant on the countryside's repetitious theme. The coming of day promises a change; it is only when the day has fully arrived that the watcher suspects it is the same day returning once again – the same day he has been living for a long time, over and over, still blindingly bright and untarnished by time.

Paul Bowles, American/Moroccan writer
The Sheltering Sky (1949)

The desert doesn't lie; everything is there for you to see. There's even something shameless about it, the naked earth. The sand covers it in places, but apart from that its skeleton can be seen wherever you look.

Theodore Monod, French naturalist
Deserts (1994)

The water reveals itself to the ground without reservation. And the dry ground waits, completely open with its bare rock and expectant passages like a lover who has no hesitation. The water tumbles wildly inside. The message is scrawled into the desert, a savage, but impeccable, signature.

Craig Childs, American naturalist and writer
The Secret Knowledge of Water (2000)

The previous chapter established the existence of microclimates in deserts, where near-surface and boundary-layer properties sometimes vary greatly over short distances. This chapter will discuss the processes that can result from interactions between these contrasting microclimates. When mesoscale wind circulations develop as a result of contrasts in heating between different microclimates, these winds also become a property of the microclimate. Most of this chapter will be devoted to discussions of how these local wind systems develop in deserts, and thus the material should be viewed as an extension of the previous chapter.

Thermally forced wind circulations in desert environments

There is abundant evidence from hundreds of observational and modeling studies that horizontal contrasts in surface energy budgets (i.e. microclimates) can cause the development of mesoscale circulations in the wind field that greatly influence the local climatology. These differences in the disposition of solar energy at Earth's surface, and the resulting circulations, can be associated with land–water boundaries (Abbs 1986), soil-moisture-availability gradients (Segal *et al.* 1989; Ookuchi *et al.* 1984; Yan and Anthes 1988), vegetation contrasts (Hong *et al.* 1995), urban–rural land-surface contrasts, and snow–bare-land boundaries (Segal *et al.* 1991).

Before discussing examples of these types of circulations in deserts, let us first show the mechanics of how they develop as a result of horizontal contrasts in the heating or cooling of the atmosphere by Earth's surface. Fig. 12.1 shows a cross-section through the lowest two kilometers of the atmosphere. In Fig 12.1a, there are no horizontal contrasts in temperature, the isobars are horizontal, and there are no horizontal pressure gradients. Therefore there are no winds. This equilibrium can be disturbed by any horizontal variations in the heating or cooling of the lower atmosphere by the surface. In Fig. 12.1b, a warm surface is heating the air on the right and a cool surface is cooling the air on the left. This causes the column of air on the right to expand, and the column on the left to contract, with the resulting slope of the isobars causing a pressure gradient and therefore a pressure force to the left above the surface. Because in this initial step of the process we have not allowed a wind to develop in response to the pressure force, the pressure at the surface has not changed. That is, simply expanding and contracting the columns does not change the amount of air in them, and thus the surface pressure has not yet changed. But in step two the pressure gradient above the surface causes a wind to develop, directed from the heated column to the cooled column. This moves air from the column on the right to the one on the left, which results in the development of a pressure contrast at the surface. This then causes a low-level wind to develop, directed toward the heated column. In order to satisfy continuity of mass, upward motion develops in the heated column and downward motion develops in the cooled column. In order for this type of circulation to develop, there simply needs to be a contrast in the heating or cooling of the atmosphere at the surface. For example, strong heating of the right column and weaker heating of the left column would produce a similar circulation. This is the mechanism by which sea breezes, lake breezes, and salt breezes develop. In all cases, there are two branches of the horizontal winds in the circulation: a low-level branch directed toward the more-heated or less-cooled column and an upper-level branch directed away from it. Mountain–valley

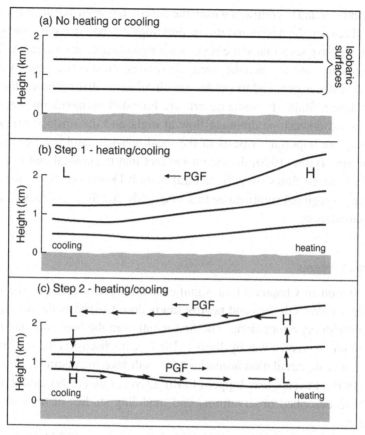

Fig. 12.1 The mechanism for the development of a thermally direct circulation as a result of different rates of surface atmospheric heating or cooling. (a) The atmosphere at rest, with no horizontal pressure gradients; (b) the development of an upper-level pressure gradient as a result of horizontally different heating of the air; (c) the final pressure distribution and wind circulation.

circulations develop through similar dynamics because the higher elevations undergo heating and cooling as part of the diurnal cycle, but the adjacent air that is elevated above the valley floor does not undergo similar thermal forcing because it is distant from the surface below. The result is upslope flow during the day and downslope flow at night. During daytime heating there is upward motion over the mountain tops (Fig. 2.6f) and downward motion over the valleys, with a reversal at night. Separation of the above argument into two steps (Fig. 12.1) was an instructional contrivance. In reality, the upper and lower horizontal pressure gradients develop simultaneously.

Thermally forced, boundary-layer circulations over deserts result from a number of different sources, and are related both to contrasts between the desert and its surroundings and to contrasts within the desert. The ones that arise because of contrasts between the desert and its surroundings can be especially intense. For example, we have seen that some deserts are dynamically associated with cold, ocean boundary currents, and the strong thermal contrast between the water and the desert surfaces can lead to sea-breeze circulations that penetrate far inland over the desert. Similarly, some deserts are bounded by mountain ranges, and the associated downslope drainage flow at night and the upslope breeze during the day are important aspects of the local mesoclimatology. For example, Lindesay and Tyson (1990) document the fact that the coastal and orographic thermally induced flows over the central Namib Desert on the west coast of Africa have a regional significance that frequently equals or exceeds that of the general circulation.

Salt or playa breezes

The discussion in Chapter 11 of significant differences between the surface energy fluxes over salt flats and nearby sandy desert suggests the development of thermally direct circulations. The first mention in the literature of such *salt breezes* or *playa breezes* was by Tapper (1988), who discusses the possibility of such a mesoscale circulation around a "dry" salt lake in Australia. Since then, Tapper (1991), Physick and Tapper (1990), Davis *et al.* (1999), and Rife *et al.* (2002) have described observational and modeling studies of this salt-breeze circulation.

The salt lake studied by Tapper (1991) has a surface area of 70 km^2, has a shallow salt crust overlying saturated clay, is located in virtually flat terrain, and is the local ground-water sink. Shrubs exist around the lake margin; the surrounding sand dunes are sparsely vegetated with native grasses. In the previous chapter is discussed the thermal contrast between this playa and its surroundings (Fig. 11.2). The near-surface wind climatology was mapped by using a five-month series of 12 000 hourly anemometer measurements from four sites around the salt lake during Austral autumn and winter. Fig. 12.2 shows 0900–1200 LT wind statistics for this period, after the days with strong synoptic forcing have been omitted and the regional mean wind vector has been subtracted for the time of each measurement. There clearly is a pattern in which substantial off-playa winds prevail during this period of the day. A similar plot for the period 0000–0300 LT (not shown) illustrates that on-playa winds of commensurate strength prevail at night. The divergence plotted in Fig. 12.3 was calculated from the near-surface wind data set, confirming that the nocturnal circulation is almost as intense as the one during the day. During the day the near-surface winds diverge over the area of the playa (are directed away from it), whereas at night the

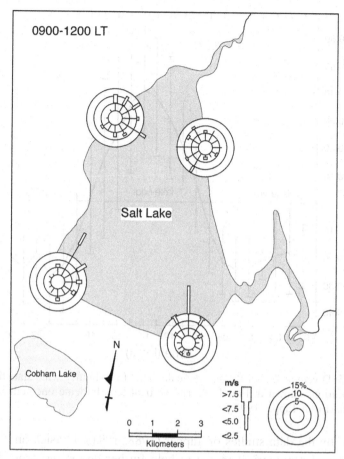

Fig. 12.2 Wind-rose statistics for 0900–1200 LT around the perimeter of a dry salt lake in Australia. The surface airflow climatology was mapped by using a five-month series of 12 000 hourly anemometer measurements from four sites around the salt lake during Austral autumn and winter. The days with strong synoptic forcing were omitted and the regional mean wind vector was subtracted for the time of each measurement. The salt lake has a shallow salt crust overlying saturated clay, and is located in virtually flat terrain. See the key for wind-direction and wind-speed plotting conventions. (From Tapper 1991.)

negative divergence indicated that the winds converge over the playa (are directed toward it). Pilot-balloon[12.1] measurements during one diurnal period showed the depth of the near-surface, daytime, off-playa breeze to be about 200–250 m and that of the nocturnal on-playa breeze to be about 100 m. Return circulations above were difficult to separate from the large-scale winds.

[12.1] Pilot balloons are tracked as they ascend in order to determine wind speed and direction. They do not have an electronic instrument package like radiosondes.

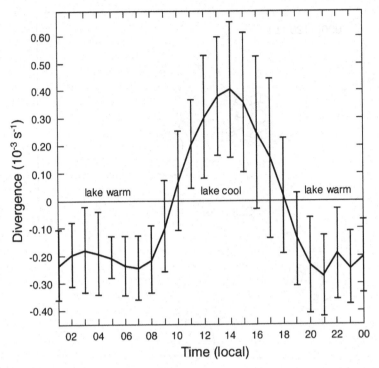

Fig. 12.3 Divergence over the salt flat calculated from hourly wind statistics at the locations shown in Fig. 12.2. The vertical bars indicate one standard deviation. (From Tapper 1991.)

Motivated by the data studies of Tapper (1988, 1991), Physick and Tapper (1990) employed a numerical model to help further define the properties of this salt breeze. For a westerly mean flow of about 5 m s^{-1} at 10 m AGL, a simulation for a salt lake with the dimensions of the one studied by Tapper (1988, 1991) showed the dominant effect of the lake at 1200 LT to be a large downwind subsidence region, bordered laterally by zones of upward motion. Vertical velocity in the ascent and descent regions exceeded 20 cm s^{-1}, as shown in Fig. 12.4. Qualitatively similar results were obtained for a larger dry lake, having dimensions of about a factor of ten greater.

Other modeling and observational studies document salt breezes in the Great Basin Desert (Davis *et al.* 1999; Rife *et al.* 2002). Fig. 12.5 shows the near-surface characteristics of an observed and model-simulated salt breeze. The playa is located in the upper left area of the figure; the rest of the area is dry desert with shrubs and rocky, elevated terrain. Fig. 12.5a shows the model-simulated 40 m AGL wind vectors and temperature. Fig. 12.5b shows the streamlines[12.2] based on the model winds in Fig. 12.5a, and observed 2 m AGL winds. The

---

[12.2] Streamlines are lines drawn parallel to wind vectors.

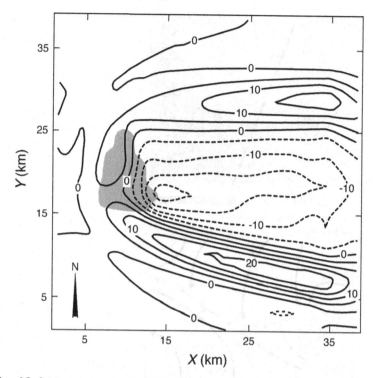

Fig. 12.4 Model-simulated, 1200 LT, vertical velocity showing ascent and descent regions downwind from a salt lake (shaded) having the dimensions of the one studied by Tapper (1988, 1991). There is a westerly mean flow of about 5 m s^{-1} at 10 m AGL. The dashed (solid) lines indicate downward (upward) motion. Vertical velocity is labeled in centimeters per second, and is plotted at an interval of 5 cm s^{-1}. (Adapted from Physick and Tapper 1990.)

wind-shift line, seen in both the observations and model winds, clearly defines the location of the mesoscale front associated with the leading edge of the cooler air from the playa. Fig. 12.6 depicts the model-simulated winds and potential temperature on a vertical cross section oriented along the line A–A′ in Fig. 12.5a. The entire simulated circulation is about 1.5 km deep, with the lower branch being limited to a few hundred meters above the ground, consistent with the results of Tapper (1988, 1991) for the Australian salt breeze. The strongest upward motion is located with the front, which is where the horizontal potential-temperature contrast weakens toward A′, near the 20 km point on the distance scale.

Desert circulations associated with oases, lakes, and seas

Desert environments ironically often occur in proximity with bodies of open water of various sizes. The contrasting atmospheric daytime heating and

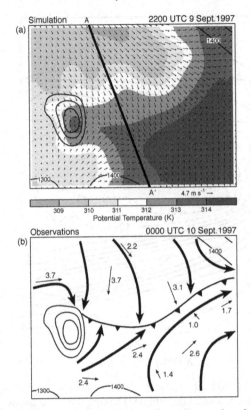

Fig. 12.5 The characteristics of an observed and simulated salt breeze in the Great Basin Desert. The playa is located in the upper left area of the figures; the rest of the domain is characterized by dry desert with shrubs and rocky, elevated terrain. The shading in Fig. 12.5a shows the near-surface temperature contrast, which drives the salt breeze, and the model-simulated 40 m AGL wind vectors. Figure 12.5b shows the streamlines (heavy lines) based on the model winds in (a), and 2 m AGL surface observations of the winds, plotted as vectors with the wind speed (m s^{-1}) noted beside. The wind-shift line clearly defines the mesoscale front associated with the leading edge of the cooler air from the playa. The wind-vector scales are different for the two plots. Both panels show the terrain elevation, with contours (light lines) plotted every 100 m. (From Davis *et al.* 1999.)

nocturnal cooling across the coastline, and the resulting pressure gradients, force mesoscale wind circulations that can penetrate considerable distances from the coast. When the wind penetrates inland from the coast during the day, it can influence the desert's near-surface and boundary-layer meteorology for hundreds of kilometers. For example, Steedman and Ashour (1976), based on observations, show that the coastal breeze on the eastern shore of the Red Sea penetrates over 200 km inland over the western Arabian Desert on some summer days. In addition, in a model- and observation-based study, Lieman and Alpert (1992)

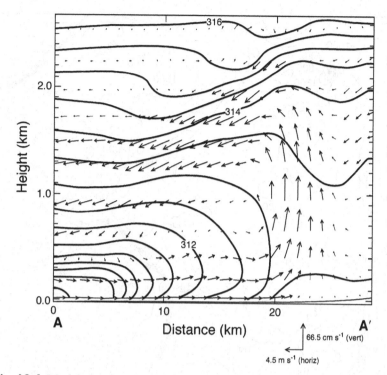

Fig. 12.6 Model-simulated winds and potential temperature on a cross-section oriented along the line A–A′ in Fig. 12.5a. See the scale for the horizontal and vertical components of the velocity vectors. The potential temperature is plotted in degrees Kelvin, at an interval of 0.5 K. (From Davis *et al.* 1999.)

show the effect of the Mediterranean Sea breeze and orography on the boundary-layer depth over Israel. The boundary-layer growth was suppressed after the passage of the coastal-breeze front as a result of the subsidence behind the front. For the same desert area, Bitan (1974, 1977) shows observations of the Lake Kinneret (Sea of Galilee) lake breeze.

Warner and Sheu (2000) use a model to document the penetration into the desert of the Arabian Gulf coastal breeze on the eastern side of the Arabian Peninsula. Fig. 12.7 shows the maximum inland penetration of the simulated Arabian Gulf breeze at 1700 LT 10 March 1991, in terms of its effect in suppressing boundary-layer growth. The boundary-layer depth suppression behind the front extends almost 100 km inland over the desert to the west of the Gulf. The modeled Gulf breeze was discernible on virtually all of the six days simulated with the model, but its strength, inland penetration, and effect on the boundary-layer depth depended on the large-scale weather conditions. There is some evidence of the boundary-layer depth suppression to the east of the Gulf,

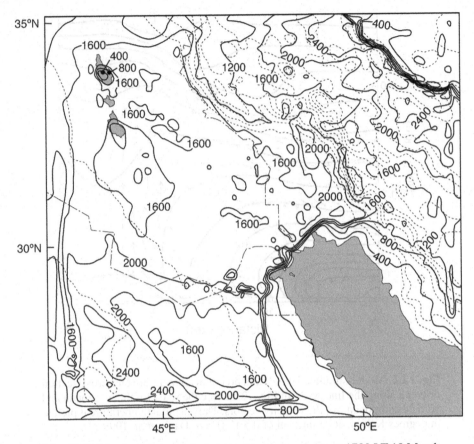

Fig. 12.7 Depth of the model-simulated boundary layer at 1700 LT 10 March 1991. Isopleths are labeled in meters and are plotted at an interval of 300 m. The effect of the Gulf breeze is seen to the west of the Arabian Gulf in terms of lower boundary-layer depths behind the Gulf breeze front. (From Warner and Sheu 2000.)

but it is not as pronounced because the proximity of the mountains retards the inland penetration of the Gulf breeze.

 The interaction of the sea breeze from the Mediterranean and the lake breeze from Lake Kinneret (dimensions of about 10 km × 20 km), which is about 40 km to its east, was studied by Bitan (1981). The annual rainfall in this semi-arid area is 400–500 mm. Topographic circulations are also important here because the lake is in the Jordan Rift Valley, and surrounded by mountains. The August near-surface wind-direction climatology for 0200–0300 LT is shown in Fig. 12.8a. This nocturnal near-surface wind, which is directed toward the lake in most places, is a result of both the land breeze and the drainage flow that descends from the nearby mountain slopes that abut most of the coast. By late

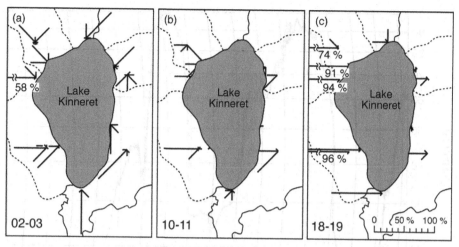

Fig. 12.8 Near-surface wind-direction frequency in the Lake Kinneret (Sea of Galilee) region at (a) 0200–0300 LT, (b) 1000–1100 LT, and (c) 1800–1900 LT during August 1973–76. See the frequency scale (c, lower right) to determine the percent of the time that the wind is from each direction. (Adapted from Bitan 1981.)

morning (Fig. 12.8b), most of the observations show the low-level wind to be from the lake toward the land. Again, there are probably contributions from the lake breeze as well as from the upslope flow that is forced by the heating of the surrounding mountains. However, after about noon, the local winds are overwhelmed by the Mediterranean Sea breeze entering from the west, as seen in Fig. 12.8c for 1800–1900 LT. During the winter, the Lake Kinneret circulation dominates the winds near the lake throughout the diurnal period because the weaker solar heating does not allow the Mediterranean Sea breeze to penetrate this far inland. Avissar and Pan (2000) describe model simulations of this complex and interesting combination of lake, sea, and mountain-valley wind systems.

Another interesting example of what is likely the diurnal interplay between a humid Mediterranean Sea breeze and a dry *khamsin* wind from the desert is shown in Fig. 12.9. Before the sea breeze reaches this location in the Negev Desert (80 km from the Mediterranean Sea) each day, dry air blows from the desert interior and the relative humidity is very low. When the moist breeze from the Mediterranean arrives in the early afternoon, the humidity rises from less than 10% to almost 100% during a 6 h period. By late evening, the sea-breeze forcing subsides, and the dry khamsin wind from the desert again dominates until about noon of the next day. More data would be needed for a definitive analysis of the processes that led to the diurnal pattern in the figure, but clearly there are diurnally varying local wind systems that expose Avdat to contrasting regional desert climates.

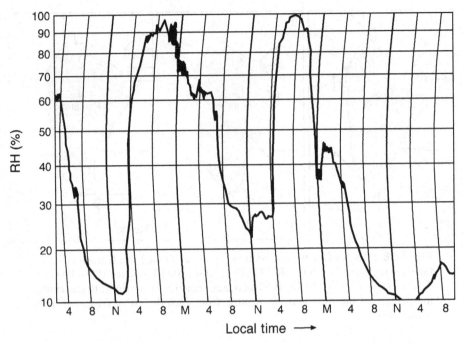

Fig. 12.9 Relative humidity (RH) at Avdat in the Negev Desert on three days: N, noon; M, midnight. (Adapted from Evenari 1985a.)

Even though near-surface winds associated with the sea breeze are generally thought of as being relatively weak, this is not always the case, especially when the land–sea temperature contrast is large. For example, along the coasts of the Peruvian and Atacama Deserts, the temperature difference is large between the cold near-shore water and the land, which is heated in the afternoon after the fog dissipates. This results in an onshore sea-breeze with speeds in excess of 10 m s^{-1}. Similarly strong sea breezes occur at the coast of the Namib Desert.

Topographically forced circulations

Many deserts contain, or are bounded by, mountainous areas. Indeed, we have seen that aridity can be a result of upwind mountains, and that many desert areas are hydrologically closed basins because they are surrounded by elevated terrain. Thermally direct circulations forced by the mountains can significantly modify the wind field and the boundary-layer depth over the adjacent desert. That is, upward motion over the heated, elevated terrain during the day (Fig. 2.6f) causes the vertical temperature profile to become less stable, while subsidence over adjacent lower elevations causes the vertical temperature profile to be more stable. This produces shallow boundary layers over valleys and deep boundary layers

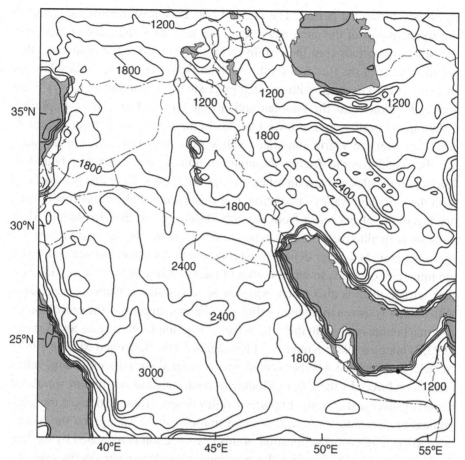

Fig. 12.10 Average daily maximum boundary-layer depth (m) for a six-day simulation period from 9 to 15 March 1991. Isopleths are labeled in meters and are plotted at an interval of 300 m. (From Warner and Sheu 2000.)

over higher terrain. Thus, the depth of the daytime boundary layer is correlated with the average daytime vertical velocities that are part of mountain–valley and coastal circulations. This is illustrated in the aforementioned study (Figs. 11.13 and 12.7) of the effect of surface forcing on the multi-scale spatial variability of the Arabian Desert boundary-layer. From the same study, Fig. 12.10 shows the average, for the six days of a model simulation, of the maximum daily boundary-layer depth. The largest gradients are along the coastlines of the Red Sea and the Arabian Gulf because the water temperature is nearly constant and remains cooler than that of the land during the day. There is less contrast along the coasts of the Caspian and Mediterranean Seas because the solar heating of the land is less at these higher latitudes, and the solar radiation in these areas was reduced by clouds during the period. The greatest boundary-layer depths are found in the southern part of the area, where the solar heating is the greatest, and over the higher elevations (see the terrain in Fig. 11.11). Over the mountains that are to the east of the Red Sea on the Arabian Peninsula, the boundary layer is in excess of 3 km deep, with the depths gradually becoming less toward the lower elevations near the Arabian Gulf to the east and the Tigris–Euphrates Valley to the northeast. The lowest depths in the desert, away from the coastlines, are over the Tigris–Euphrates Valley (less than 1500 m). Large values are also found over the Zagros Mountains in Iran. As noted, the deep boundary layers over the elevated terrain reflect the upward motion there that is associated with the mountain–valley circulation. Conversely, the shallow daytime boundary layers in valleys result from the subsidence part of the circulation. Most days show qualitatively similar general characteristics in the spatial variability in the boundary-layer depths, implying that the constant surface properties are important controls. In other studies of the boundary-layer structure in complex terrain, there is also much evidence presented that subsidence associated with thermally driven mountain–valley circulations suppresses the growth of the daytime boundary layer in the valleys (see, for example, Kuwagata and Kimura 1995; Whiteman 1982; Kimura and Kuwagata 1995; Bader and McKee 1983). In order to determine whether subsidence associated with the mountain–valley circulation forced by the Zagros Mountains contributed to the shallow simulated boundary layer in the Tigris–Euphrates Valley desert, cross-sections of modeled vertical motion were examined along a transect of the valley and the mountains to the northeast. The location of this cross-section is indicated by the line A–A′ in Fig. 11.11b. Because the maximum boundary-layer depths over the northern Arabian desert are generally similar to the height of the Zagros Mountains, subsidence that would have an effect on boundary-layer growth would be in the valley below the mountaintop level. Fig. 12.11 shows an example of the simulated vertical motion for 1500 LT along this cross-section, which spans the Tigris–Euphrates Valley and the edge of the mountains to the east. Downward vertical velocities (dashed lines) associated with the mountain–valley circulation clearly exist over this area of the desert valley at low levels, and are likely responsible for suppressed daytime boundary-layer depths there.

Circulations resulting from soil-moisture gradients

Soil-moisture gradients can be large in zones of transition between areas having different degrees of aridity, and within an arid area the distribution of recent precipitation and watertable depths can create soil-moisture contrasts. In either

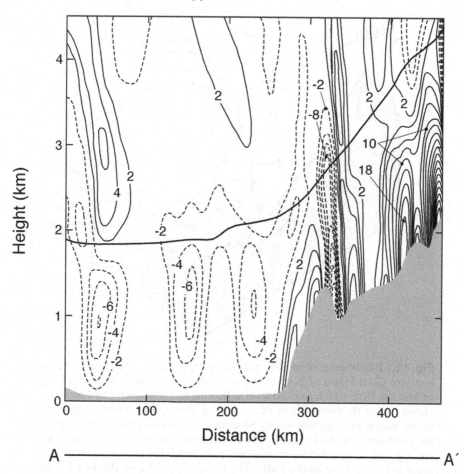

Fig. 12.11 Model-simulated vertical velocity at 1500 LT 10 March 1991 along the cross-section of the Tigris–Euphrates Valley defined by line A–A' in Fig. 11.11b. The isotachs are labeled in centimeters per second and are plotted at an interval of 2 cm s^{-1}. Dashed lines are downward motion and solid lines are upward motion. (From Warner and Sheu 2000.)

case, the horizontal variation in the amount of net radiation that is partitioned to latent heating can cause gradients in the sensible heat flux to the atmosphere. The wind circulations that result from such differential heating across soil-moisture gradients can be relatively benign or they can have important consequences. An example of an important one is seen in the semi-arid southern Great Plains of North America. Here, a boundary layer that forms over the elevated desert plateau of Mexico can be advected northeastward by southwesterly flow in the lower troposphere. As the boundary-layer air moves over the lower terrain of the Great Plains, it becomes elevated above the surface (Fig. 12.12), forming what is known

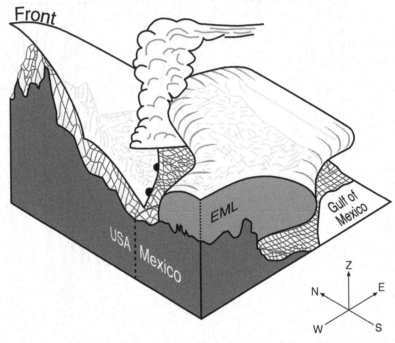

Fig. 12.12 Schematic of an EML-related severe-weather environment over the southern Great Plains of North America. In the foreground is the high plateau of Mexico, from which a mixed layer is flowing to the northeast and becoming elevated over the Great Plains of the United States. The northwest quadrant of the area is the southern Rocky Mountains. Convective clouds are seen on the northwest edge of the EML, associated with low-level southeasterly flow causing moist, unstable air to run out from under the convection-inhibiting inversion at the base of the EML. The meteorological case discussed in the text had a front positioned to the east of the Rocky Mountains, as illustrated. (Adapted from Lakhtakia and Warner 1987.)

as an elevated mixed layer (EML). Because this elevated airstream was strongly heated over the high Mexican desert, it is warmer than the air immediately below it, the result being a temperature inversion at its base. This stable layer represented by the temperature inversion traps the convectively unstable air below it, and prevents the development of convective rainfall there. However, if the convection in this moist unstable layer "leaks out" from around the edge of the inversion layer, or if it erodes through it, intense rainfall can result. The cumulus clouds depicted to the northwest of the EML in the schematic in Fig. 12.12 can result from a low-level southeasterly wind below the EML, which can be caused by the typical soil-moisture pattern in the area. This pattern is shown in Fig. 12.13a in terms of a soil moisture availability parameter that shows

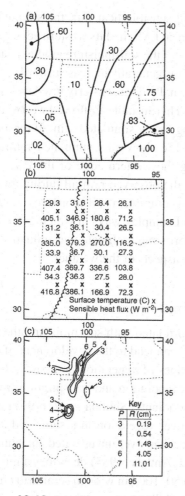

Fig. 12.13 (a) Climatological soil moisture availability distribution in the northern Chihuahuan Desert and the southern Great Plains of North America. (b) Model-simulated surface temperature and surface sensible heat flux at 1500 LT 9 May 1979 at selected model grid points, where the scalloped line is the western edge of the EML. (c) Model-simulated rainfall for the 3 h period from 1500 to 1800 LT 9 May 1979, where a logarithmic rainfall scale is used such that $P = \ln(R + 0.01) + 4.6$, where R is rainfall in centimeters. (Adapted from Lakhtakia and Warner 1987.)

a tongue of dryness extending from the northern Chihuahuan Desert into the southern Great Plains. During daytime heating, the sensible heat fluxes to the atmosphere will be greatest over the driest surface, and this will cause a low-level easterly flow over the Great Plains to the east of the dry tongue. Such a wind flow will cause the moist unstable air to run out from under the edge of the EML, potentially causing severe weather and heavy rainfall. Fig. 12.13b shows model-simulated surface temperature and sensible heat flux for selected grid points at 1500 LT 9 May 1979, a day when an EML had its western edge located by the scalloped line in the figure. Because of the soil-moisture gradient, surface temperatures and surface sensible-heat fluxes generally increase toward the west. As anticipated, the model simulated a thermally direct easterly low-level flow of convectively unstable air below the EML. This unstable air flowed westward, out from under the capping inversion, causing heavy simulated rainfall throughout the late afternoon and evening. Fig. 12.13c shows the simulated rainfall for the period 1500–1800 LT, aligned along the western edge of the EML. This simulated rainfall corresponds well with the observed rainfall during the same period. In a model simulation that employed uniform soil moisture, no rainfall resulted. There are many other simple and complex ways in which natural soil-moisture contrasts that are part of a desert's microclimate can influence local circulations and possibly other important aspects of the weather.

Circulations associated with vegetation contrasts

Even though desert vegetation is often correlated with soil moisture, the direct effects of vegetation contrasts themselves can lead to the development of circulations. In particular, differences in albedoes of vegetation and bare substrate, and the difference between bare-ground evaporation and transpiration from the vegetation, can create sensible-heating gradients. Shuhua *et al.* (1997) used a model to show that the boundary layer over vegetation in a semi-arid area of northwestern China was cooler and wetter than over unvegetated surroundings, and a boundary-layer circulation resulted; Anthes (1984) describes a modeling study that shows that bands of vegetation 50–100 km wide in semi-arid climates will generate circulations and enhance rainfall.

Oasis effects

In the context of the interaction among desert microclimates, the use here of the term oasis effect refers to the impacts on surface heat and moisture fluxes of hot, dry air advecting over modest-sized areas having high surface moisture. Natural areas of high moisture availability are desert oases with possible surrounding irrigation, lakes, or playas with standing water or moist surfaces. This anomalous effect on surface fluxes, described thoroughly earlier in Chapter 4, is important in the moist areas of deserts that are simply too small to generate mesoscale circulations like the salt breeze.

The surface energy balance over a small area of well-watered Sudan grass in the eastern Sonoran Desert is a good example. Here, warm, dry air from the

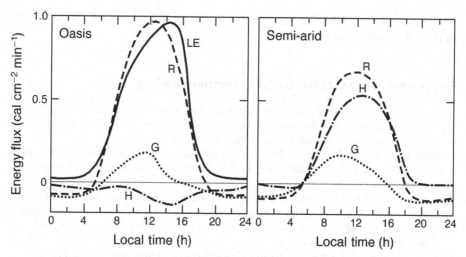

Fig. 12.14 The surface energy balance components for an oasis and a nearby semi-arid area of the Kyzylkum Desert. (From Budyko 1958.)

surrounding desert is continuously advected over the moist area, and the resulting large vertical vapor-pressure gradient near the ground causes rapid evaporation. Figure 4.18 shows the energy-balance components during a diurnal period. The rapid evaporative consumption of heat maintains a surface temperature that is lower than that of the overlying air that is advected from the surrounding desert, and this causes a downward flux of heat to the ground that supplements the net radiation. The result is a unique surface energy balance. Fig. 12.14 further demonstrates this oasis effect because it not only shows the energy balance for the oasis, but also for the upwind semi-arid area. It applies to an irrigated oasis in an arid area of Uzbekistan, to the southeast of the Aral Sea in the Kyzylkum Desert. The sensible heat flux (H) is positive in the arid area, but negative over the oasis, as with the Sonoran Desert budget in Fig. 4.18. In addition, the latent-heat flux is roughly equal to the net radiation over the oasis, but it is essentially zero over the arid area.

When oases or other irrigated arid areas are of sufficient size, their impact goes beyond the above local effect because their contrast with the surrounding dry microclimate can generate thermally forced wind systems. Yan and Anthes (1988) provide a general discussion of the development of boundary-layer circulations by areas of wet land surface. Chen *et al.* (2001) and Chase *et al.* (1999) use a model to show that irrigated areas in the semi-arid Great Plains of North America interact with distant areas through mesoscale circulations, in this case influencing the summer rainfall over the Rocky Mountains to the west. Lastly, Segal *et al.* (1989) employ model simulations to illustrate the development of

local circulations around the edge of irrigated agricultural land in the same part
of the semi-arid Great Plains.

Suggested general references for further reading

Davis, C., *et al.*, 1999: *Development and application of an operational, relocatable,
     mesogamma-scale weather analysis and forecasting system* – contains a discussion of
     a model simulation of a salt breeze in the Great Basin Desert.
Physick, W. L., and N. J. Tapper, 1990: *A numerical study of circulations induced by a dry
     salt lake* – a numerical simulation is used to illustrate the structure of a salt breeze in
     the area of a salt flat in Australia.
Rife, D. L., *et al.* 2002: *Mechanisms for diurnal boundary-layer circulations in the Great
     Basin Desert* – describes observations and simulations of wind and temperature that
     are associated with thermal circulations produced by complex terrain, playas, and
     lakes in the Great Basin Desert.
Tapper, N. J., 1991: *Evidence for a mesoscale thermal circulation over dry salt lakes* –
     discusses observations of the wind field associated with a salt breeze in the area of a
     salt flat in Australia.

Questions for review

(1) In your own words, explain how a wind circulation develops as a result of horizontal
     contrasts in the near-surface sensible heating or cooling of the atmosphere.
(2) How can soil-moisture contrasts be responsible for the development of thermally forced
     circulations?
(3) Define a salt-breeze front in a way that is analogous to how you would define a
     synoptic-scale front. Is the front in Fig. 12.5 a cool or a warm front?
(4) Summarize how vertical motions affect boundary-layer depth.

Problems and exercises

(1) How can the arguments in question (1) above be applied to explain a salt breeze?
(2) Apply the same argument to the development of a mountain–valley breeze that is
     caused by heating or cooling of a mountain–valley terrain profile.
(3) Speculate about what scale limitations there are regarding the development of thermally
     forced circulations. For example, why don't we observe the development of parking-lot
     breezes?
(4) Explain the cause of the model-simulated vertical motion field shown in Fig. 12.4,
     associated with the heating contrast around a salt flat.

13

Desert rainfall

In most deserts, mountains enhance rainfall and produce moist vegetated islands in the larger arid landscape. Here is an example from the northern Chihuahuan and Sonoran Deserts.

The location of forests in New Mexico and Arizona is largely a matter of the force and direction of the prevailing winds. These tend to draw along the chutes prepared for them by the cumbres of the Continental Divide. From the gulfs of California and Mexico, great wind rivers go over with enormous freightage of sunlit cloud. Surcharged, they pile and topple and carom against the raking ranges and give down the precious ballast of the rain. Or the wind leaves them in fleets, like great barges becalmed in mid-air, until they darkle and run together and reveal the true nature of clouds. On the miraculous floor of the air the rain stands upright between the mountains. In pure, shadowed grayness it stretches from cumbre to cumbre.

Mary Austin, American naturalist and writer
The Land of Journeys' Ending (1924)

I dabbed the surface of this tiny pool with a finger . . . because I remembered the Tohono O'odham people. . . ., and how it is their customary belief that water is not to be taken boastfully. . . To ask for too much water is to invite disaster. Only in a place like this would you bow your head and humbly request just the water you need and no more. Only here would you walk away from water when thirsty, but not thirsty enough.

Craig Childs, American naturalist and writer
The Secret Knowledge of Water (2000)

Depending on your definition of desert, rainfall events there are more or less very rare occurrences: in some places so rare that there are intervening rainless periods of years or decades. On some days a few cumulus congestus clouds, lightning, and virga may tease in the distance. Even though your waiting is generally not rewarded by rain, on the horizon there are a sparsely vegetated and relatively impervious few square kilometers of stony desert that are receiving torrential rain and being scoured by the runoff.

Table 13.1 *Example extremes of storm rainfall in arid areas, relative to the mean annual rainfall*

Location	Date	Mean annual precipitation (mm)	Single-storm precipitation
Chicama, Peru	1925	4	394 mm
Aozou, Chad	May 1934	30	370 mm/ 3 days
Swakopmund, Namibia	1934	15	50 mm
Lima, Peru	1925	46	1524 mm
Sharjah, United Arab Emirates	1957	107	74 mm/ 50 min
Tamanrasset, Algeria	Sept. 1950	27	44 mm/ 3 hours
Biskra, Algeria	Sept. 1969	148	210 mm/ 2 days
El Djem, Tunisia	Sept. 1969	275	319 mm/ 3 days

Source: Adapted from Goudie and Wilkinson (1977).

The characteristics of the rainfall can be foreign to the experience base of non-desert-dwellers in a few respects. The rainfall is often scattered and isolated in the sense that there generally are not large convective complexes or other storm systems that have widespread effects. When rainfall does reach the ground without first evaporating, it can appear to be very intense. And, of course, the rainfall is scarce, sometimes only occurring in a particular season, and sometimes being so scarce that there is no identifiable season.

The rainfall climatologies of arid locations often show relatively few events dominating the statistics for a given year or even decade. This is especially the case for extremely arid areas. Examples are easy to find.

- At Hurghada, a station on the Red Sea in the eastern Sahara Desert, 41 mm occurred in one day, whereas the annual mean is only 3 mm: 13 years of rain in a few hours.
- In an extreme El Niño year, the northern half of the coast of the Peruvian Desert had more rainfall in March than during the entire preceding 10-year period.
- Yuma, in the Sonoran Desert, experienced only 25 mm of rain in one year, but more than 280 mm a few years later, over a factor of ten difference.
- Tamanrasset, in the Sahara Desert, with 40 mm mean annual rainfall, received 44 mm in three hours, with three-fourths of that falling in about 40 minutes.
- In the Thar Desert at Doorbaji, 864 mm of rain fell in two days even though the annual average is only 127 mm.

Tables 13.1 and 13.2 show additional storm-total and 24 h total rainfall for desert stations, relative to the mean annual precipitation. Such situations where a large fraction of the annual rainfall is received in one event should not be considered anomalous. For example, Satchell (1978) points out that, for a location in the United Arab Emirates, one year in four has a single-day rain that amounts to more than half the annual average.

Table 13.2 *Maximum 24 h precipitation for African desert stations, relative to the annual mean, maximum and minimum*

A dash indicates data unavailability: tr refers to a trace of precipitation.

	No. of years of record	Mean annual precipitation (mm)	Max. annual precipitation (mm)	Min. annual precipitation (mm)	Max. precipitation in 24 h (mm)
Khartoum (Sudan)	30	164	382	76	80
Faya-Largeau (Chad)	30	17	48	tr	48
Nouakchott (Mauritania)	25	156	—	—	249
Bilma (Niger)	27	22	—	0	49
Dongola (Sudan)	30	23	60	0	36
Atar (Mauritania)	33	106	—	—	69
Etienne (Mauritania)	33	27	—	0	83
Wadi Halfa (Sudan)	24	3	33	0	19
Dakhla (Egypt)	25	0.5	11	tr	8
Quseir (Egypt)	25	4	34	tr	20
Cairo (Egypt)	25	24	63	3	44
Sollum (Egypt)	20	95	324	4	121
Galcaio (Somalia)	26	149	448	33	160
Berbera (Somalia)	40	49	178	2	132
Djibouti (Horn of Africa)	64	129	300	10	211
Luderitz, Namibia	20	18	59	1	31

Source: From Goudie and Wilkinson (1977).

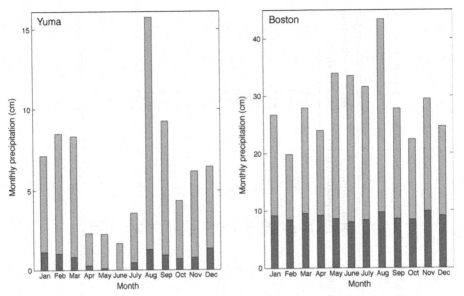

Fig. 13.1 Average, monthly precipitation (dark shading), and the greatest monthly precipitation observed during the period of record (light shading). Yuma is in the northern Sonoran Desert, and Boston is a non-arid location on the east coast of North America. Fifty years of data were used for Yuma, and 173 years for Boston.

Figure 13.1 illustrates the great degree to which precipitation amounts during a given year or month can depart from the average in arid areas. It shows the 50-year average precipitation amounts in each month for Yuma in the northern Sonoran Desert, and the maximum precipitation that has been recorded in that month during the same period. Shown for comparison is a similar plot for non-arid Boston on the east coast of North America, based on a 173-year record. The maximum observed monthly amounts for Yuma are sometimes over an order of magnitude greater than the average, a much greater anomaly than is characteristic of non-desert locations such as Boston.

An example of the ironies of such desert rainfall extremes is described by Meigs (1966). Piura, on the northern margin of the Peruvian Desert, has historically relied upon marginally adequate local rainfall (rather than irrigation) for some of its agriculture. During the 100 year period from 1791 to 1890, sixty-eight years were classified as "dry" relative to agricultural needs. Of the remaining years, ten were classified as "extraordinarily good". Unfortunately, during every one of these ten years, the rains that were "extraordinarily good" for agriculture produced extensive damage from flooding (see also Box 13.1).

Even though the surface-hydrologic consequences of desert rainfall are undeniably more extreme than in temperate areas, it is an interesting question whether

---

Box 13.1
The two-edged sword of desert rain

Childs (2000) points out that even desert dwellers have not reconciled themselves to the two-edged sword of the severity and scarcity of desert rain:

Don't pray for too much water in the desert, even if the crops demand it. It will come eventually, and it will bring its desideration with it. Catholic saints are often employed to call the rain for crops or drying wells. I've heard many stories of people running to hide the small ceramic or plastic figurines that they have placed, as lightning punctuates the ground around them, as outbuildings are lifted away in the wind, as the arroyos fill, then overflow with a raging dun-colored water that smells of all the villages and lives upstream that have been consumed. The displayed santos are quickly clutched up, hidden away as if pulling the plug on the rain, concealing the request. At that point it is too late.

---

desert rain, especially convective rain, is actually more intense than in other climatic regions. There are intriguing reports: for example, over 2.5 cm of rainfall in one minute in the western San Gabriel Mountains in semi-arid coastal California in the United States in 1926 (Jahns 1949).[13.1] Unfortunately, one sometimes gets the message from the literature that such occurrences are typical of the desert. This may result from the fact that our subjective impressions of desert rainfall intensity can be misleading for a number of reasons. First, desert rainfall is sometimes a rare event, so when it does come it has a greater psychological impact because of its uniqueness. In addition, contributing to the impression of the severity are (1) the associated flooding that is a consequence of the surface conditions and not necessarily the rainfall intensity, (2) the precursor dust storms that are sometimes generated by convective outflow boundaries, (3) the fact that desert visitors are sometimes forced to endure the storms without the benefit of shelter when caught on foot, and (4) our ability to often get an unobstructed view of the looming storm cloud. On the other hand, there are meteorological reasons why some desert rainfall events might actually be near the end of the severity spectrum. First, the tropopause is higher in the subtropics than in mid-latitudes, and hence convective storms will be deeper, possibly extending up to 18 km and containing more liquid water to rain out. Also, the fact that the westerlies above subtropical deserts are weak to non-existent means that the storms will move more slowly than their mid-latitude counterparts, and the rainfall will last longer. But the duration will still be less than in large mid-latitude convective complexes. In any case, the firsthand experiences of desert dwellers can certainly be convincing.

[13.1] Rainfall rates this large have been recorded in non-arid locations as well. For example, 3.2 cm was recorded in one minute in Unionville, Maryland, in the United States in 1956 (Costa 1987).

The image of a classic Tucson [northern Sonoran Desert] summer thunderstorm is one that goes straight up, has a nice mushroom (symmetric) anvil, and has such a rush of rainfall coming out of the bottom that it appears almost black. If you are caught in one of these driving, day or night, you have to pull off the road and stop because there is no way to drive safely. No visibility; windshield wipers can't keep up either. Lightning everywhere. Frequent power outages.

(Brant Foote, personal communication)

Later in this chapter some statistical evidence will be presented about the relative intensity of desert- and temperate-climate rainfall.

The relative contributions of small-scale convection and larger-scale storms to the total annual accumulation of desert rainfall is very geographically dependent. We have seen that Mediterranean-type desert climates receive most of their rainfall in winter, and much of this is from equatorward-intruding, large-scale, mid-latitude disturbances. In contrast, on the equatorward side of subtropical deserts, most of the precipitation is of summer convective origin. In other places, there are two seasons of rainfall. The seasonality, altitude, and latitude also control the relative contributions from rain and snow. In each geographic area, the prevailing physical causes of the precipitation, such as convective thunderstorms or synoptic-scale storms, will strongly control the statistical characteristics of the rainfall. Thus, one should be cautious about generalizing about "desert precipitation."

As a further step toward conceptualizing desert precipitation, it is useful to see examples of how many rainy days are experienced in different deserts, and how much rain is received on a typical rainy day. Table 13.3 provides this information for six different desert areas. The column listing mean annual rainfall shows the range over the various rain gauges for each area, indicating a distinct and expected lack of homogeneity in the rainfall climate. The next two columns list the range in the average number of rainy days per year at each station, and the average rainy-day rainfall for all stations in each area. There are many areas with a substantial number of days when at least some measurable rainfall is received, and the average rainy-day rainfalls are not inconsequential. Thus, even though we have seen in Chapter 3 that there are extreme desert areas where precipitation is infrequent to rare, this table shows that this is generally not the case.

Now that we are familiar with these *subjective* impressions of desert rainfall, the following section will review statistical measures that can be used to quantitatively contrast its characteristics with those of non-deserts. The next section will describe the meteorological processes that cause and modulate desert precipitation. Lastly, artificial rainfall enhancement in desert areas will be discussed, as well as dew and fog deposition in the desert.

Table 13.3 *Rainy days, and rainfall per rainy day, in arid areas*

Area	Number of stations	Range of mean annual rainfall (mm)	Range of number of rainy days receiving > 0.1 mm rainfall	Average rainfall per rainy day (mm)
Karakumy Desert	12	92–273	42–125	2.56
Gobi Desert	6	84–396	33–78	4.51
Patagonian Desert	11	51–542	6–155	5.41
Northern Sahara Desert	18	1–286	1–57	3.82
Northern Sahara Desert (west)	20	17–689	2–67	9.75
Kalahari Desert	10	147–592	19–68	9.55
Combined	77	1–689	1–155	6.19

Source: Adapted from Thomas (1997a).

Statistical characterization of desert rainfall

The nature of the rainfall is a major criterion by which we define a place as a desert. However, we frequently only consider the annual mean when comparing deserts with each other or with temperate climates. Other statistical measures will be discussed here that reveal additional information. These metrics represent ways of summarizing or encapsulating the characteristics of time series or frequency distributions of rainfall-related quantities. Frequency distributions show the frequency, or commonness of occurrence, of something. There are many different properties of rainfall that could be described, including duration of rainfall events, event-total rainfall amount, 24 h rainfall amount, annual rainfall amount, and time between rainfall events. A common distribution would be the annual rainfall for a particular location. The abscissa of the plot would be annual total rainfall, in increments of perhaps 1 cm, and the ordinate would reflect the fraction of the period of record that the total annual rainfall fell within each of the 1 cm intervals. The following is a brief glossary of some relevant statistical measures. The first list contains measures of the central tendency of the distribution: i.e. they are ways of defining the "middle."

- Mean – the arithmetic average.
- Median – the number such that half the values fall above it and half fall below it.
- Mode – the value, or range of values, that occurs most often.
- Interquartile range – the range of the middle 50% of the observations in the frequency distribution. The quartiles separate the distribution in terms of the number of observations. For example, the first (or lower) quartile is the point in the distribution

such that one-fourth of the observations occur below that point and three-fourths of the observations are above that point. The third quartile is the point such that three-fourths of the observation are below that point. The quartile range is defined as $Q_3 - Q_1$, where the Q values are the values of the property at the quartile points.

Additional measures represent the variability and the shape of the distribution.

- Standard deviation – the square root of the mean of the squared deviations from the mean.
- Coefficient of variation – the standard deviation divided by the mean, often expressed as a percentage.
- Interquartile range – as noted, this represents the range of the middle half of the observations. It provides information about the variability as well as the central tendency.
- Interquartile variability factor – defined as $100(Q_3 - Q_1)/Q_2$, where Q refers to the values of the property at the first, second and third quartile points in the frequency distribution. The Q_2 is obviously the median. Thus, the interquartile variability factor is the quartile range divided by the median, converted to a percentage.
- Skewness – a measure of the degree of asymmetry about the mean. A positively skewed distribution is one in which relatively small values are more likely than larger values, and visa versa. One simple formula for skewness is (mean − median)/mean.
- Range – the maximum minus the minimum value.
- Normalized range – the range divided by some factor such as the mean.

The following subsections describe desert rainfall in terms of some of these statistical properties.

Annual-total rainfall statistics

In general, the drier the climate the more positively skewed is the distribution of annual rainfall. That is, relatively small values are more likely than larger values. Another way of stating this is that, compared with a statistically normal distribution, there are some extreme values of large annual rainfall on the right tail of the distribution. A number of examples of this sort were noted in the introduction to this chapter. Fig. 13.2 illustrates the annual rainfall frequency distributions, and the means and medians, for three arid locations and for one location in a temperate climate. Arid Alice Springs, Yuma, and Tamanrasset show the positive skewness, whereas Philadelphia does not. Because of this skewness, the mean tends to be somewhat of a misleading indicator of the "center" of the distribution because the relatively few large annual totals cause the mean to move to the right in the distribution, away from the more prevalent small annual

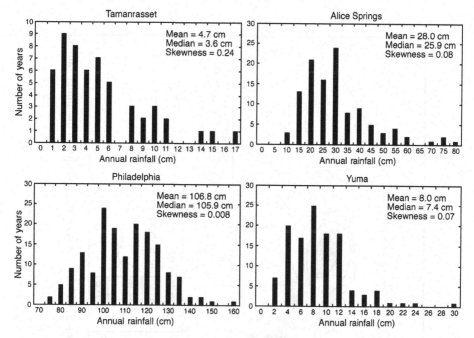

Fig. 13.2 Frequency distribution of annual rainfall for Alice Springs, Australia; Yuma, USA; Tamanrasset, Algeria; and Philadelphia, USA. The annual rainfall amounts are displayed such that the bar labeled 10 cm for Alice Springs represents amounts between 5 and 10 cm. Note that annual rainfall "bin sizes" vary for the different locations. The mean, median, and skewness (defined by the simple formula above: mean−median)/mean) are shown.

totals. In contrast, the median and the mode are little affected by the skewness in terms of representing the midpoint of the distribution. For example, the median represents the annual rainfall value that is exceeded one-half the time, regardless of how far out on the tail the large values are. And the mode, the most commonly occurring value, is not affected by extremes. For this reason, Katz and Glantz (1977), in the context of Sahelian rainfall, claim that the mode and median are preferable to the mean as indicators of "normal" rainfall. They also argue that the interquartile range gives a better idea of central tendency than can the single numbers of the aforementioned metrics.

It is interesting that this relation between the skewness of the annual rainfall distribution and the annual mean rainfall applies on very local scales as well as to variations from one large-scale rainfall climate to another (as in Fig. 13.2). To illustrate, Landsberg (1951) calculated the skewness of the rainfall frequencies for 22 locations on the island of Oahu, in the Hawaiian Islands. This island is similar to the island of Hawaii, for which the annual rainfall is illustrated

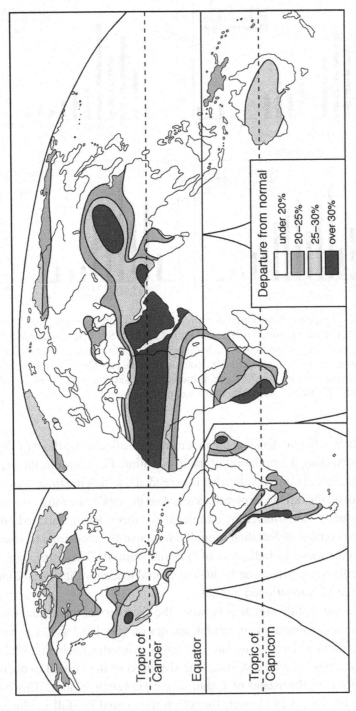

Fig. 13.3 The world distribution of the coefficient of variation of the rainfall, which is the standard deviation of the annual rainfall divided by the mean, multiplied by 100. (From Rumney 1968.)

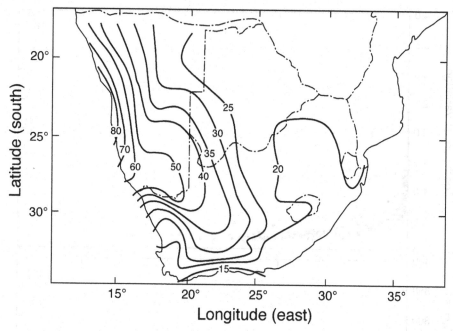

Fig. 13.4 The coefficient of annual rainfall variability (standard deviation divided by the mean) for southern Africa, showing the contrast between the Namib Desert, the Kalahari Desert, and non-desert areas. (Adapted from Schulze 1972.)

in Fig. 2.19, in that the orography and the regularity of the trade winds cause a large spatial variability in the annual total rainfall. The skewness, defined using the simple formula provided above, was four times larger for the locations with a semi-arid microclimate (annual means of 50 cm) in the rain shadow of the mountains than for upwind locations with seven times the average annual rainfall. Such contrasts occurred over distances of 20 km.

Regarding the variability of the distribution, the standard deviation and the coefficient of variation are good measures only when the observations are normally distributed, or at least symmetrically distributed, without extreme values. The extreme large values in arid-land distributions cause large squared deviations from the mean, and these contribute much more than do the squares of the many smaller deviations. In addition, the meaning of "standard deviation from the mean" becomes questionable when the mean itself does not reflect the "center" of the skewed distribution. Nevertheless, the coefficient of variation is commonly used as a measure of interannual rainfall variability in arid lands, with the value being directly proportional to the aridity. Fig. 13.3 illustrates the global distribution of the coefficient of variation, and the clear relation that it has to the degree of aridity reflected in the classification of Meigs (Fig. 3.1). Fig. 13.4

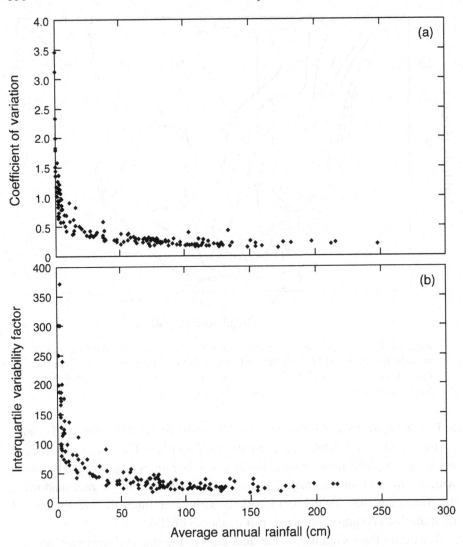

Fig. 13.5 (a) The coefficient of variation and (b) the interquartile variability factor for arid and non-arid locations in Africa.

shows this relation with higher spatial resolution. In this area of southern Africa, the west coastal Namib Desert is by far the driest area, with some locations receiving less than 25 mm per year. It is here that the standard deviation of the annual rainfall almost equals the mean. The much less arid Kalahari Desert to the east has a smaller coefficient, with the least normalized variability being in the regions with the highest rainfall along the south and southeast coasts. Fig. 13.5a shows the coefficient of variation plotted against annual rainfall for

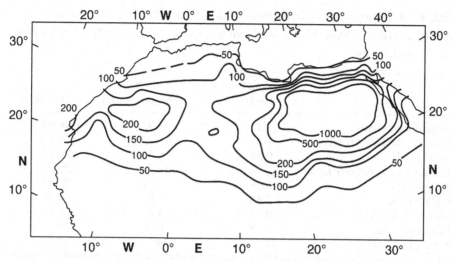

Fig. 13.6 Interquartile variability of annual rainfall for North Africa. (From Griffiths and Soliman 1972.)

many arid and non-arid stations in Africa. A clear increase in the coefficient is seen as the annual rainfall decreases.

The interquartile variability factor is an alternative to the standard deviation and the coefficient of variation, and is a better measure of the normalized variability for skewed, arid-land rainfall distributions. It is similar to the coefficient of variation in that, in both cases, the measure of the spread is normalized by a quantity that represents the central tendency, the mean for the coefficient of variation and the median for the interquartile variability factor. Fig. 13.6 shows the interquartile variability factor for the Sahara Desert, and Fig. 13.5b is a plot of it as a function of average annual rainfall for arid and non-arid locations in Africa. The largest values are in the eastern Sahara Desert where the annual rainfall amounts are smallest. These values (in excess of 1000) reflect the fact that the middle 50% of the distribution of annual rainfall spans a range of values that is more than ten times larger than the median value that splits the distribution. For example, the distribution of annual rainfall for Tamanrasset in Fig. 13.2 has a median of 3.6 cm, and the middle half of the distribution is roughly from 2 to 8 cm. Thus, the interquartile variability factor would be about 170, which is consistent with the fact that the location is near the northeast corner of the 150 line in Fig. 13.6, in the northwestern Sahara.

Table 13.4 provides a few specific examples of the annual rainfall statistics for arid and temperate locations. Clear distinctions exist. First, the standard deviation and range of annual rainfall totals are a much larger percent of the central tendency for arid locations. Yuma, in the northern Sonoran Desert, has

Table 13.4 *Properties of interannual rainfall variability at two desert locations (Yuma, Death Valley) and one temperate location (Philadelphia) in the United States*

	Yuma (Sonoran Desert)	Death Valley (Mojave Desert)	Philadelphia
Annual mean (cm)	3.12	2.32	40.93
Standard deviation (cm)	1.83	1.23	6.79
Standard deviation/mean	0.59	0.53	0.17
Interquartile variability factor	62	116	20
Range (cm)	8.98	4.21	27.11
Range/mean	2.88	1.81	0.66
Maximum (cm)	9.23	4.21	56.45
Minimum (cm)	0.25	0.00	29.34
Years of record	81	29	47

annual precipitation with a coefficient of variability that is almost four times larger than that for Philadelphia, and Death Valley in the Mojave Desert has an interquartile variability factor that is almost six times larger than Philadelphia's. Thus, annual rainfall in temperate areas is much more predictably close to the norm, in terms of percentage departure from it. Obviously many of the metrics are mathematically undefined for those extremely arid climates where annual totals are all zero during the period of record.

Variability of rainfall within seasonal and diurnal cycles

Many arid locations exhibit only a single season with rainfall. For example, Fig. 3.13 shows the monthly rainfall for various places in the Sahara and its surroundings. Locations in the Mediterranean-type climate on the northern margin of the desert experience mostly winter rainfall and very dry summers, whereas locations in the south have wet summers and very dry winters. Similarly, the semi-arid southwestern coast of the United States has wet winters and dry summers, but more tropical Mexico to the south and the semi-arid United States Great Plains receive most of their rain in the summer (Fig. 3.30). The dynamical causes of a large seasonal variability are very location-dependent, but include monsoon effects, the migration of subtropical high-pressure centers, and seasonal variability in the strength and track of mid-latitude storms.

As with seasonal variability in rainfall, the day–night partitioning of desert rainfall is very location-dependent. In some locations and seasons, there does not seem to be much preference. In others, rainfall during one part of the day

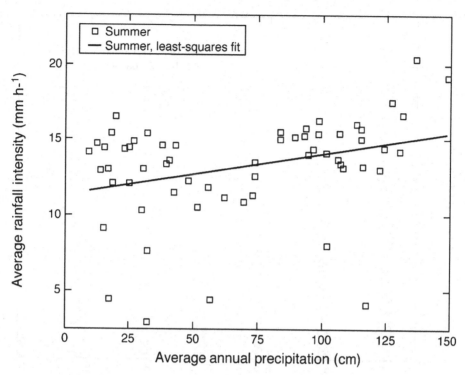

Fig. 13.7 The average summer (June, July, and August) rainfall intensity as a function of the average annual precipitation for locations in arid and humid areas of North America. The rainfall intensity is based on 15 min totals (see text).

dominates the statistics. For example, Otterman and Sharon (1979) document an afternoon rainfall maximum in the Negev Desert, with the greatest day–night difference prevailing for the heavier intensities. In contrast, Douglas and Li (1996) describe a nocturnal rainfall maximum in the northern Sonoran Desert.

Intensity of summer precipitation

An earlier discussion speculated upon the question of whether desert summer precipitation is more intense than summer precipitation in non-arid areas, or whether it is just that the hydrologic consequences of the rainfall are more severe. To provide some insight for one continent, precipitation statistics for many stations in North America were analyzed. The data were 15 min totals for stations with a long period of record (at least 30 years) and having a wide range of annual precipitation amounts. Fig. 13.7 shows the average hourly rainfall intensity as a function of the average annual rainfall for the summer months of June, July, and August. If there was any rainfall during a 15 min period, the value

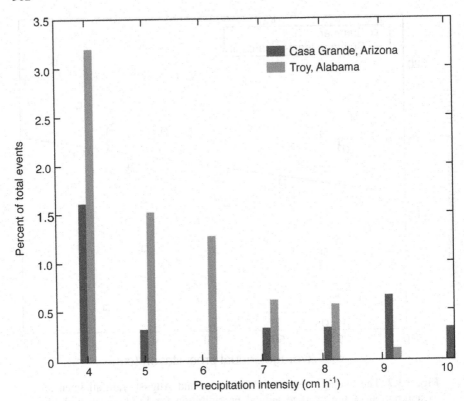

Fig. 13.8 Frequency of rainfall rates at two locations in North America. One lo-cation is in the northern Sonoran Desert (Casa Grande, Arizona) and the other is in the humid southeastern part of the continent (Troy, Alabama). Fifteen-minute precipitation data were used for both locations, with the amounts measured in increments of 2.54 cm (1 in). An "event" was defined as 2.54 cm or greater of precipitation in that period.

was included in the average. Trace amounts were not included. The figure shows that there is not a strong relation between average summer rainfall intensity and the degree of aridity, but there is a tendency for the average rainfall to be a little less intense in arid areas. Winter (December, January, February) precipitation intensities average 20–30% less than the summer average intensities for both arid and non-arid areas, where the average winter intensity is relatively independent of the annual average rainfall (not shown).

What this plot does not provide insight about is the probability distribution of rainfall rates at a given station. That is, an arid station may have an average rainfall rate that is similar to the average rate at a location with a humid climate, but the skewness of the rain-rate distribution might be such that the arid location has more extremely heavy events. Fig. 13.8 shows the frequency distribution of rain rates for two of the locations in Fig. 13.7: one is Troy, Alabama, in the humid southeastern United States, and the other is Casa Grande, Arizona, in the northern Sonoran Desert. The distribution plot is truncated on the low-precipitation-intensity side in order to better illustrate the frequency differences for the heavy intensities on the right tail. For rainfall rates of less than 1 cm h^{-1}, the desert location has a frequency of 80% and the humid location's frequency is 70%. For rates of 1–2 cm h^{-1}, the frequencies are roughly the same. Above that rate, the humid location has higher frequencies until the rates become very high, about 9 cm h^{-1}. For the aggregate of the two highest rainfall-rate categories, the frequency is about ten times greater for the desert location. Thus, for these two randomly selected locations, intense rainfalls make up a higher percent of the total number of events (15 min periods) at the desert location.

Event-total, or wet-period total, precipitation amounts

How a time series of precipitation data is subdivided into individual precipitation "events" is arbitrary. If the precipitation data are hourly, an event could be defined as spanning a series of contiguous non-zero hourly totals. If there is an intervening period of *n* hours or more with zero precipitation amounts, two events might be defined. In the analysis described here, an event represents a period containing non-zero hourly amounts, within which there is no interval with six or more hours without precipitation. Thus, it may be more appropriate here to refer to wet periods or periods of precipitation, rather than to precipitation events (which might imply some meteorological context). In this subsection, the average amount of precipitation associated with wet periods, thus defined, is compared for arid and non-arid areas of North America. This comparison is made for both the summer months of June, July, and August, and the winter months of December, January, and February. Fig. 13.9 illustrates that, for both winter and summer, the average wet-period precipitation totals are considerably less for arid than for non-arid areas. The figure does not provide information about how the intensity of the precipitation and the duration of the precipitation period individually contribute to the period total. It is interesting that the summer wet-period-total precipitation amounts are greater than the winter amounts for all rainfall climates. This may be related to the fact that synoptic-scale winter precipitation associated with cyclonic storms can be light and temporally sporadic, and thus precipitation from a single storm could be assigned to multiple wet periods in this calculation.

There is a large variability in the average wet-period-total precipitation, for similar average-annual-precipitation climatologies. For example, in arid areas

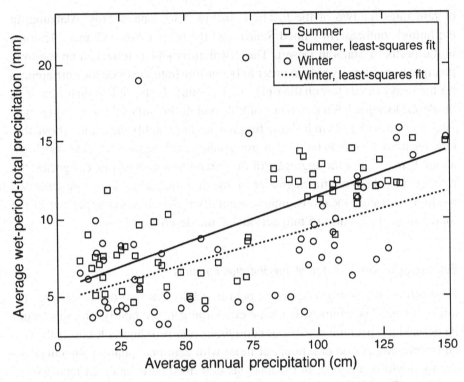

Fig. 13.9 The average summer (June, July, and August) and winter (December, January, February) wet-period-total precipitation as a function of the average annual precipitation for locations in arid and humid areas of North America.

the range in wet-period-total precipitation values exceeds a factor of four, even within a single season. This is obviously a consequence of the particular regional characteristics of the precipitation-producing weather events. A similar degree of location-dependent variability is also seen for non-arid areas.

Duration of precipitation periods

Using the same definition as above for precipitation periods or wet periods, Fig. 13.10 shows the relation between the average duration of the periods and the degree of aridity as defined by average-annual precipitation, for North American locations. For arid areas, the winter precipitation periods are approximately twice as long as the summer periods. In wetter climates, the percentage difference between the seasons is less. For both seasons, the average duration is less in arid areas than in wet climates. Here, also, the results are sensitive to how precipitation

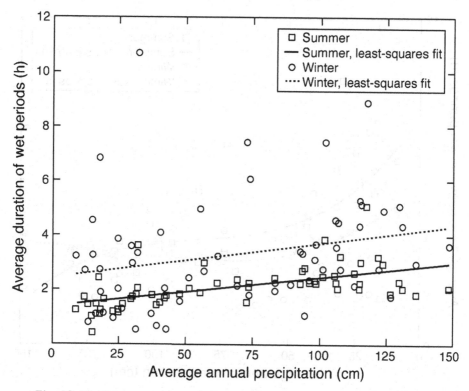

Fig. 13.10 The average summer (June, July, and August) and winter (December, January, February) duration of wet periods, as a function of the average annual precipitation for locations in arid and humid areas of North America.

periods are defined. For example, in the winter season the precipitation from large-scale storm systems is sometimes intermittent, and the data analysis will define a few wet periods associated with a single storm.

Time between precipitation periods

This statistic of the precipitation, as expected, is quite different for arid and non-arid locations. Fig. 13.11 shows that the time between precipitation periods ranges typically from two to eight days in both winter and summer, for North American locations with greater than 50 cm of annual rainfall. With increasing aridity, the summer and winter precipitation periods are less frequent at many locations, as expected. But at some locations, the precipitation periods are just as frequent as in wetter climates. In these places, the aridity must instead be related to the existence of less precipitation per wet period, resulting from the average wet-period duration and/or intensity being less. For arid locations, the

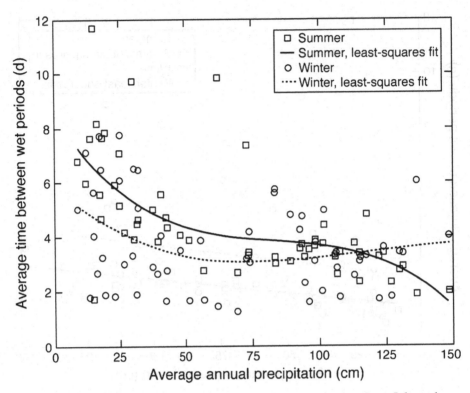

Fig. 13.11 The average time between wet periods for summer (June, July, and August) and winter (December, January, February), as a function of the average annual precipitation for locations in arid and humid areas of North America.

time between wet periods is generally greater in the summer than in the winter. The data-fitting curves shown are fourth-order polynomials; the summer curve is strongly influenced by the few arid locations having very long times between precipitation events. Local meteorological factors clearly play a role here also, in terms of producing a large scatter in the time between precipitation events for locations with similar average annual rainfall.

Spatial continuity, or "spottiness," of rainfall

To the extent that the rainfall in arid areas is convective, it is also very localized. In particular, arid-land convection is not organized into large complexes of thunderstorms, which means that the horizontal scale is of the order of individual cells: a few kilometers to tens of kilometers. An example of very fine-scale variability is described in Goudie and Wilkinson (1977), where 20 rain gauges sampled convective rainfall over 10 ha (equivalent to a square about

300 m × 300 m) in the Negev Desert. Over this very small area, the gauges recorded amounts that varied from 2.2 to 7.8 mm from the same storm. Sharon (1972) also discusses rainfall variability in the same desert. In the very arid southern Arava Valley in the Negev Desert, daily rainfall was measured for three locations that are within 15 km of each other, and that have similar average annual rainfalls. Over a three-year period, there was rainfall in excess of 3 mm at at least one of the stations on 21 days. During that period, there also were 29 days with rainfall of less than 3 mm. On four of the 21 days, the rainfall was similar at each of the three stations. However, on 11 days, there was a ratio of at least 20 : 1 between the greatest and the least rainfall amounts. On four of the days, the greatest amount exceeded 20 mm and the smallest amount was between zero and 3 mm. These larger amounts are not much less than the average annual total for this area. Thus, one station will receive virtually its annual average in a few hours, while a nearby station 10 km away will receive nothing. This provides a subjective impression of the spottiness of the rainfall.

Given that convective rainfall in arid areas is spotty, with spatial correlations dropping off rapidly, it is an interesting question whether correlations can again increase at certain distances. That is, are the convective cells organized in such a way that certain cell spacings are dynamically preferred? It is well known that convection in laboratory experiments in which there is an imposed uniform and sufficient heating of the lower boundary of a fluid (or cooling of the upper boundary), can form organized, rather than random, patterns. The regular polygonal convective cells, which are usually hexagonal, are called Bénard cells. There is also occasional evidence, primarily from satellite and radar imaging of the clear boundary layer, that large-scale equivalents of these regularly spaced cells exist in the atmosphere. For example, based on rainfall measurements in the Namib Desert, Sharon (1981) shows that correlations among rainfall measurements first decrease with distance, but then increase again, with maxima near 40 km and 80–100 km.

When the precipitation is organized by synoptic-scale disturbances (rather than convection), it covers larger areas. This is quantified in terms of the approximate length scales shown in Fig. 2.6a for convective causes of precipitation and in Figs. 2.6c and 2.6d for precipitation caused by large-scale, mid-latitude, winter-season storms. Fig. 13.12a shows an analysis of the rainfall from a cool-season, synoptic-scale disturbance near Tamanrasset in the Algerian Sahara. The rain is considerably more widespread than what would be produced by spotty convective events, such as the one that caused the rainfall in the northern Sonoran Desert shown in Fig 13.12b. Note the relative length scales of the two analyses.

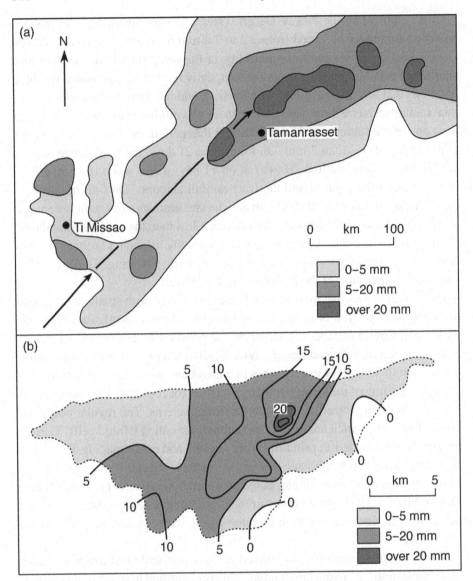

Fig. 13.12 Examples of rainfall from a single storm (a) near Tamanrasset, in the Sahara Desert, where the arrows indicate the estimated track of the center of the storm (from Goudie and Wilkinson 1977) and (b) in the Walnut Gulch watershed (dashed-line boundary) in the northern Sonoran Desert (adapted from Reid and Frostick 1997). Note the difference in the length scales between the two panels.

Figure 13.13 shows the dependence of the correlation coefficient of daily rainfall totals on the distance between observations, for arid and non-arid areas. The more rapidly the correlation decreases with distance, the more spotty is the rainfall. It is probably best to use these diagrams as an illustration of the general regional variability that can exist in rainfall correlation with distance, rather than as an indicator of desert–non-desert differences. The data were obtained in a number of different studies, so it is not straightforward to compare the curves. A common factor in Fig. 13.13a is that all curves apply to summer rainfall. There clearly is much case-to-case variability, and the arid locations show both high and low spatial correlations. Hershfield (1968) shows a similar diagram, with the single arid location having the most-rapid decrease in spatial correlation with distance. Fig. 13.13b illustrates the expected much stronger spatial correlation of frontal rainfall in winter.

It remains an open question whether desert rainfall is any more spotty than rainfall elsewhere. Russell (1936) agrees that desert rainfall is spotty, but claims that it is just as spotty in the humid tropics or in the humid mid-latitudes, commenting that perhaps the unobstructed visibility in arid lands makes rainfall events appear to be more solitary or isolated. Another factor is that the inarguable temporal rarity (temporal spottiness) of rainfall events in the desert may be what lay people really refer to when they loosely use the term spotty. It is clear, however, that some deserts do not experience broad-scale rainfall from cyclonic storms to the same degree as the mid-latitudes, nor do you find large areas that are wetted by the contiguous complexes of convective cells (mesoscale convective systems or complexes) that occur in many areas of the world. Even though rainfall accumulations over these large areas may themselves contain much irregularity, it is perhaps a matter of definition whether this should be equated to the isolated nature of some desert rainfall.

Reliability of rainfall

In rain-scarce areas, knowledge of the reliability of receiving some minimum amount of rain, which represents perhaps a threshold for agricultural or other activities, is much more useful than knowing a mean. The great importance of the reliability of rainfall in arid areas is not only a consequence of the fact that water demand and supply are often uncomfortably similar, but also results from the fact that the year-to-year variability is a much greater percent of the mean.

An example of a rainfall-reliability analysis is presented by Rijks (1971) for the semi-arid Senegal River Basin. Analyses are illustrated for two of the rainfall stations in the Basin: Kidira, Senegal, and Kayes, Mali. About 80 km apart, their

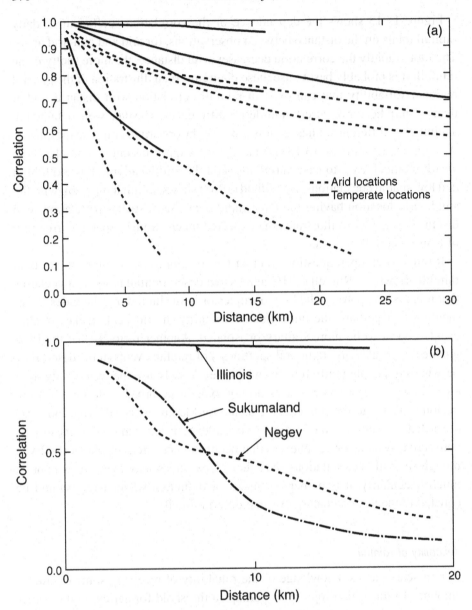

Fig. 13.13 Correlation of rainfall as a function of distance, for arid and temperate climates. (a) Summer rainfall only, where dashed lines are for arid areas and solid lines are for temperate locations. (Adapted from Sharon 1972.) (b) Winter, frontal rainfall in temperate Illinois, United States and cellular convective rainfall in semi-arid Sukumaland, in Tanzania, and in the Negev Desert of Israel. (Adapted from Sharon 1974.)

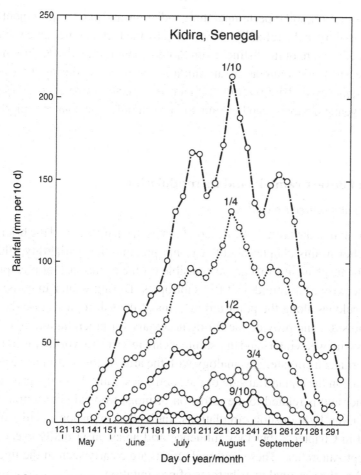

Fig. 13.14 Rainfall reliability analysis for Kidira, Senegal: the amount of rainfall that can be expected to be exceeded 9 years in 10, 3 in 4, 1 in 2, 1 in 4 and 1 in 10. (From Rijks 1971.)

annual rainfall totals are similar at 747 mm and 763 mm, respectively. Fig. 13.14 shows the reliability analysis for Kidira in terms of the amount of rainfall that can be expected to be exceeded nine years in ten, three in four, one in two, one in four, and one in ten. Rijks (1971) states that 15 – 20 years of daily rainfall totals are required for this kind of analysis. Given some decision about an acceptable level of risk, such an analysis can be useful in choosing sowing dates and crop types. For example, if 15 mm of rainfall are required every 10 d for germination and growth of a particular crop, and it is acceptable for a crop to be lost one in four years, seed should not be sown before the end of June. If a crop loss one year in two is acceptable, the crop could be sown near the beginning of June. It is best if such rainfall information is used in combination with data about potential evapotranspiration, soil type, temperature, radiation, and plant requirements. The spatial variability in the reliability is revealed by the fact that the same reliability curves for Kayes are quite distinct from those shown for nearby Kidira, in spite of the almost identical annual mean rainfall. For example, during the summer, the same quantity of 10 d rainfall that can be expected to be exceeded in nine out of ten years in Kayes, can be counted upon in only three out of four years in Kidira.

Causes of desert rainfall, and its modulation

Large-scale and convective causes

The discussion of the large-scale desert climates in Chapter 3, and the brief statements earlier in this chapter about seasonal precipitation variability, illustrate some of the mechanisms that are responsible for the occurrence of precipitation in the otherwise arid latitudes of the subtropics. During winter months, extratropical cyclones along the poleward margins of the subtropics cause the intrusion of moisture and provide the lifting necessary for precipitation production. Any areas with wet winters and dry summers are referred to worldwide as having a "Mediterranean climate," by analogy with the area on the mid-latitude edge of the Sahara. On the equatorial side of the subtropics, rainfall occurs primarily in the summer and is related to the intrusion of moist tropical convection, which is often driven by monsoon circulations. Individual monsoon circulations are described in Chapter 3, but they influence arid areas on virtually every continent except Antarctica. These seasonal effects are clearly seen in the figures in Chapter 3 of the seasonal distributions of precipitation.

Orographic influences

In Chapter 2, various physical mechanisms were described that contribute to the maintenance of aridity. Within these overall precipitation-suppressed regions, a local effect that can focus or enhance the sparse precipitation is the existence of terrain that is elevated relative to its surroundings. The elevated terrain can potentially cause precipitation, given the availability of sufficiently moist air, by (1) serving as a physical barrier that forces the large-scale flow of air to rise over it (Fig. 2.6b) and (2) producing a local pressure gradient around the heated elevated terrain during the day, which causes upslope flow (Fig 2.6f). In both situations, the upward motion created must have compensating subsidence elsewhere, which will reduce the likelihood of rain there. In any case, the local distribution of rainfall within deserts is such that there is often a correlation

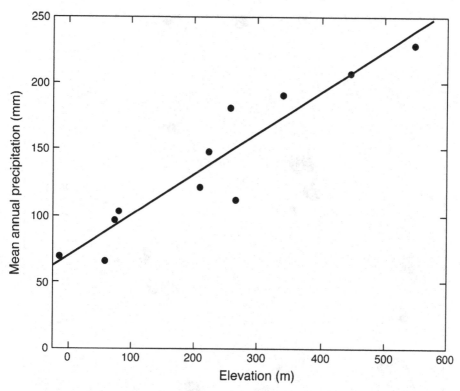

Fig. 13.15 Relation between station elevation and mean annual precipitation for eleven stations in the northern Sonoran Desert. (From MacMahon and Wagner 1985.)

of higher rainfall with higher terrain elevation. Fig. 13.15 clearly shows this in terms of the mean annual precipitation plotted against station elevation for eleven locations in the northern Sonoran Desert. In this geographic area, 88% of the spatial variation in annual average precipitation can be statistically explained by elevation (MacMahon and Wagner 1985). Similar plots of the annual number of rain days and the average annual rainfall as a function of elevation are shown by Wheater *et al.* (1991) for locations near the Red Sea on the Arabian Peninsula. Fig. 13.16 also shows an illustration of this orographic effect for an area with an annual rainfall of about 23 cm in the northern Chihuahuan Desert. During a three-month period in the summer (June, July, August) of 1998, a record was kept every 6 min of the location of radar echoes from precipitation over the arid Tularosa Valley and surrounding Sacramento and San Andres mountains. The figure shows that the greatest frequency of rainfall echoes is over the mountains or their slopes, with the lowest frequency in the valley between the mountains. As noted earlier, the upward motion associated with the rainfall over the mountain ranges causes enhanced subsidence over the lower elevations. In that respect,

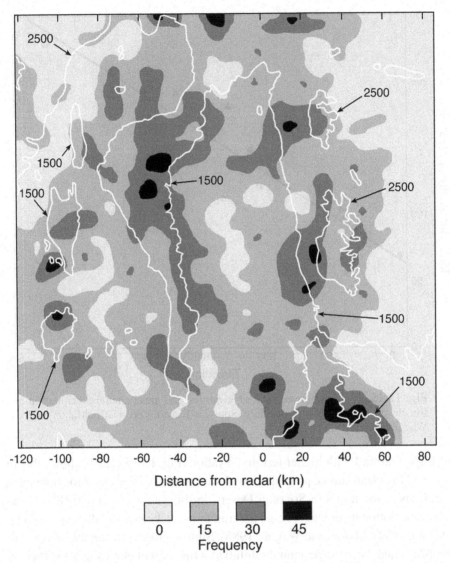

Fig. 13.16 The frequency of occurrence of vertically integrated liquid water of greater than 2.5 kg m^{-2}, based on 6 min radar reflectivity data for the months of June, July and August 1998 in the northern Chihuahuan Desert. The white lines outline terrain with elevations greater than 1500 m. (Courtesy Cindy Mueller, National Center for Atmospheric Research.)

the rainfall over the mountains can cause the aridity to be even greater over the Tularosa Valley and other lower elevations. In another respect, the thunderstorms that originate over the mountains are sometimes transported over the valley when the summer lower tropospheric winds are significant, and thus all areas benefit.

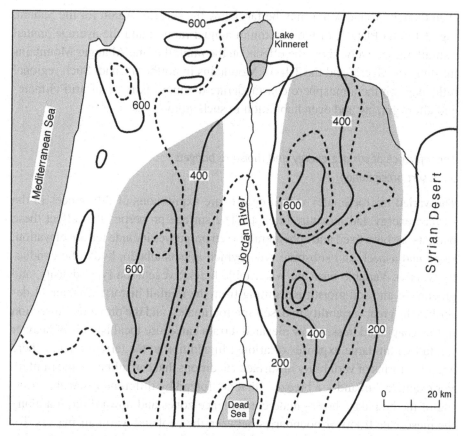

Fig. 13.17 Mean annual rainfall (millimeters; solid and dashed lines) for the area between the Syrian Desert and the Mediterranean Sea. The shaded areas are hills with elevations reaching 600–900 m. Jordan Valley elevations are generally 200–300 m below sea level. (From Sharon 1979.)

Another arid area for which orographic effects have been well documented by using rain-gauge data is the Jordan River Valley between the Mediterranean Sea and the Syrian Desert. Fig. 13.17 shows that the elevated terrain on both sides of the valley produces over a factor of three enhancement in the annual mean rainfall, compared with the southern valley near the Dead Sea.

Rainfall favoring, and caused by, higher elevations is observed even in extremely arid climates. For example, in the eastern Sahara, the Gilf Kebir Plateau is just high enough to allow rainfall there about once every eight years according to local observers. This allows the existence of some vegetation, unlike in the surrounding barren, flat, and rainless desert (Bagnold 1935). Other examples abound of isolated mountains or highlands in the desert causing a little water to be squeezed occasionally from the very dry air, and creating an oasis of sorts even though ground water may be far below the surface. Again for the Sahara, Fig. 3.15 and Plate 1, of the maximum annual rainfall and the average annual rainfall, respectively, show the effects on the rainfall of the Ahaggar Mountains in southern Algeria and the Tibesti Mountains in northern Chad. Such regional influences on the atmosphere can moderate the most extreme of arid climates and allow grazing and even habitation in such altitude oases.

The influence of surface energy and moisture budgets on desert precipitation

The varied microclimates within deserts are expressions of differences in the surface energy and moisture budgets. The surface properties that affect these budgets are both fixed and ephemeral. Fixed properties include terrain elevation, slope and aspect, and substrate type (which affects albedo, hydraulic conductivity, etc.). Variable ones are water table height; vegetation type, density, and greenness; and soil moisture resulting from the rainfall history. Chapter 11 describes the great variability in desert microclimates, and the previous subsection of this chapter addresses how elevated terrain can cause local islands of heavier rainfall within large expanses of aridity. In addition to the terrain effects, there are other forms of variability in surface conditions that can influence the rainfall. For example, "inland sea breezes" that are forced by differential sensible heating of the boundary layer create convergence zones and upward motion along the mesoscale fronts or boundaries between the cool and warm air masses (for example, see the discussion of the salt breeze in Chapter 12). This upward motion may help trigger moist convection when the large-scale conditions, such as relative humidity, are uncharacteristically favorable. Similarly, thermally forced circulations at coastlines of deserts (actual sea breezes) can initiate convection over the desert during the daytime inland penetration of the mesoscale front. However, there is little evidence in rainfall climatologies to suggest that such processes share the same importance as does elevated terrain in the modulation of rainfall.

One of the most important characteristics of the land that influences surface sensible- and latent-heat fluxes, and potentially moist convection, is the soil moisture. Unfortunately, it is difficult to generalize about its effect on rainfall because it can simultaneously influence the substrate albedo, the soil thermal conductivity, the greenness and density of the vegetation, and the partition of net radiation between sensible- and latent-heat fluxes. In some situations, high soil moisture can inhibit the development of convection because the available heat is consumed in evaporation rather than in heating and destabilizing the boundary layer. This effect is the basis of an old and simple forecasting rule that states

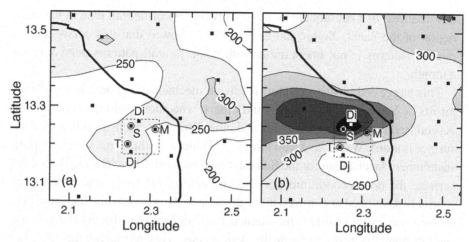

Fig. 13.18 Rainfall (mm) (a) for the first three months of the wet season (1 May – 30 July 1992) and (b) for the subsequent seven weeks in southwest Niger in the Sahel. (Adapted from Taylor *et al.* 1997.)

that airmass convection is less likely to occur where rain wetted the ground yesterday. On the other hand, where the convection is moisture-limited, such as in arid environments, the situation may be different. Taylor *et al.* (1997) provide some insight through the analysis of data obtained in southwest Niger, in the semi-arid Sahel, during the Hydrological Atmospheric Pilot Experiment in the Sahel (HAPEX-Sahel). Over the study area, there is little mesoscale variability in the long-term rainfall climatology. For the first three months of the wet season (1 May – 30 July 1992), the total rainfall was spatially fairly uniform (Fig. 13.18a), and consistent with the climatological values. However, from the 15 large-scale rainfall events of roughly similar intensity that passed through the area during the following seven weeks, there was consistently more rainfall over a narrow swath, as shown in the rainfall total for this period (Fig. 13.18b). The accumulation for the period was more than twice as much at location Di as at location Dj (see figure). For all but one of the fifteen rain events, the combined accumulation at the two northern gauges at Di and S was greater than the combined accumulation at the two locations T and Dj to the south. Without model-based studies it is impossible to confirm the mechanism involved. However, it is possible that atmosphere–surface feedback processes caused the soil-wetness path from a heavily raining cell within a large-scale rain event early in the period to produce enhanced convection over that area within the subsequent large-scale events. The process might have been initiated when an area of heavy rainfall produced an anomalously wet patch of surface, and this moistened the boundary layer, which was lifted by an ensuing large-scale disturbance. Thus, the rainfall anomalies might have been sustained through succeeding large-scale disturbances. In a subsequent analysis of two years of rainfall data in the same region of the Sahel, Taylor and Lebel (1998) showed that this persistence of rainfall patterns is not uncommon, with some rainfall patterns persisting for a month.

This suggested positive rainfall feedback mechanism is consistent with the results of Yan and Anthes (1988) and others, who show that enhanced rainfall is produced from sea-breeze fronts and other lifting mechanisms when the land surface is moist. Walker and Rowntree (1977) illustrate this same sort of rainfall-sustainment mechanism for the Sahara, based on model simulations. With a dry surface, the desert conditions persisted in spite of the large-scale storms that routinely developed and moved through the area. However, when the same land surface was made moist, it maintained itself in that state for weeks because rainfall regularly resulted from the depressions. This reinforces the idea that desert rainfall anomalies can persist, through effects on the surface fluxes, and that aridity also tends to persist.

Artificial rainfall enhancement in arid areas

Rainfall enhancement by various methods has been attempted in a variety of climates. It is, however, in the arid lands of the world that there is perhaps the greatest motivation. The approach that has been most often adopted is to enhance the efficiency of existing clouds by injecting into them (seeding) particular types and concentrations of particles on which cloud droplets and ice crystals can form. Even though the particles can be released from generators at the surface, with natural updrafts required to carry the material to cloud height, it is more common to release the particles from aircraft. In semi-arid and arid regions, cumulus (convective) clouds are often the primary precipitation producers. Randomized seeding experiments in semi-arid areas of South Africa and Mexico, using **hygroscopic** particles released from flares on aircraft, have been promising in the context of increasing the radar-estimated rain within the clouds. The seeded clouds rained harder and longer than did the unseeded clouds, with the results being statistically significant at the 95% confidence level (Mather *et al.* 1997).

Rainfall enhancement in arid areas faces special challenges compared with such efforts in other climates. Even though the cloud-microphysical processes may be successfully modified so that more rainfall is produced in the cloud, the practical benefit of the endeavor in arid climates depends on what happens to the rainfall after it falls from the cloud base. Here, the initial subcloud layer has low relative humidity, and therefore much of the incremental increase in rainfall may be lost through evaporation before the subcloud layer becomes saturated and the virga touches the ground. Moreover, an important further concern is how much of any additional rain that reaches the ground is consumed by the rapid evaporation before it reaches a reservoir or penetrates to recharge the groundwater. Thus, the cost–benefit analysis of rainfall enhancement in arid regions must involve surface-hydrologic considerations.

There have been other proposed approaches to rainfall enhancement in arid areas through modification of the surface. These have not undergone field testing. For example, Anthes (1984) demonstrates through numerical-model experiments that the planting of bands of low-water-use vegetation 50–100 km wide in arid areas could enhance convective rainfall through a decrease in albedo, an increase in evapotranspiration, and an increase in net radiation. An environmentally less palatable approach was proposed by Black (1963) and Black and Tarmy (1963), who suggested that the creation of large asphalt-covered areas could enhance convective rainfall through reduction in the albedo and the consequent increase in heating.

Bruintjes (1999) provides some advice that has special relevance to rainfall enhancement in arid regions. Often, such enhancement activities are not initiated until after a drought has begun and conditions are desperate. Also, the programs tend to be discontinued after the droughts have ended. It can be argued that this reactionary response to water resources management is doomed to failure for a couple of reasons. For example, during droughts there are often not enough clouds available to seed, regardless of the effectiveness of the seeding technology. A better approach would be to enhance the rainfall as a regular part of a water resources management strategy, and store the additional water in reservoirs or the water table in preparation for the inevitable dry periods.

Dew and fog deposition in the desert

Even though dew and fog are not precipitation in a conventional sense, they are considered in this chapter because these atmospheric sources of liquid water are qualitatively equivalent to precipitation in terms of their impact on the surface water budget and climate. Dew forms when the substrate or the foliage reaches the dew-point temperature of the air in contact with it. The nocturnal cooling of the surface of the substrate and the low-level air is greatest when the downwelling long-wave radiation is weak (last term in Eqn. 4.3); i.e. when the atmosphere is cloud-free with low water-vapor content. Near-surface winds should be in the range 1–3 m s^{-1}. If the winds are calm, there will not be any turbulence; which is required to produce a downward flux of boundary-layer water vapor to the surface. If the winds are much stronger, they will mix the cooling layer of air near the surface with the warmer air above, and the air temperature will not be maintained at the dew-point temperature.

Desert fog is generally coastal and advective in nature (Chapter 2). Air at low levels is advected across a cold, coastal ocean current, and fog forms when the air cools to its dew-point temperature. The fog is advected onshore, and eventually evaporates as the daytime heating over the land mixes the fog layer into the warm, deep boundary layer. Radiation fogs also form at night in deserts, especially after the surface has been moistened by rain.

The desert areas for which fog is an important contributor to the regional climate have been described in Chapter 3. They generally are coastal areas with a cold, adjacent ocean current, and include coastal areas of the Atacama, Peruvian, Namibian, Sonoran, Madagascar, Arabian, and Somali–Chalbi Deserts. As noted above, radiation fogs can also form. Evenari (1985a) and Tivey (1993) estimate that there is from 50 to 300 mm per year of natural fog deposition in foggy coastal deserts. This rate of collection of fog water depends on many factors, including the nature and density of vegetation foliage on which the droplets can be deposited. Fog-water recovery is often enhanced through the use of devices that employ wire screens or nets to collect droplets. Whether the collectors are natural (vegetation) or artificial, the water collection rate increases with distance above the ground, presumably because the wind speed increases with height and so therefore does the flux of droplets through the collector. In some cases, more water is gained from fog than from rainfall. For example, Table 13.5 shows the annual accumulation of water from rainfall and from fog for a 22-year period at Gobabeb, Namibia, about 70 km from the coast in the Namib Desert. On the average, about 25% more water was provided at the surface by fog than by rainfall. However, because the average daily amount of fog deposited at the surface is typically only 0.5 mm, unless the water is concentrated by running down plant stems or is absorbed directly by leaves, its availability to vegetation before evaporating is questionable. This is in contrast to the situation with rainfall, which occurs less frequently but generally provides greater amounts of water per occurrence.

There are few dew measurements for most of the world's deserts, but it is probably safe to say that many of world's hottest deserts are dewless, at least for some seasons. There are some exceptions, however, such as the Negev Desert, for which over 20 years of daily dew observations are available. At one measurement site in the Negev, the average number of dew nights per year was 195, with the average annual amount measured by a dew gauge being 33 mm (Evenari 1985a). Monteith (1963) computes the maximum possible dew deposition based on the surface energy budget, and arrives at values of 0.17–0.45 mm per night. Campbell

Table 13.5 *Sources of moisture at Gobabeb, Namibia*

Year	Rainfall (mm)	Fog (mm)
1963	12.8	na
1964	22.9	na
1965	18.4	na
1966	14.2	35.0
1967	26.6	30.4
1968	5.3	45.7
1969	29.8	na
1970	3.5	na
1971	20.6	28.4
1972	25.1	44.7
1973	7.9	29.0
1974	23.2	8.0
1975	7.0	26.7
1976	127.4	9.6
1977	14.3	25.8
1978	109.6	23.5
1979	31.0	33.5
1980	5.5	44.5
1981	4.6	48.6
1982	15.8	36.0
1983	11.6	41.8
1984	1.7	27.3
Mean	24.5	31.7

Source: From Thomas (1997a)

and Harris (1981) estimate that actual rates are perhaps one-quarter of this value. A complication to the general dew-deposition problem is that one-quarter to one-half of the water that condenses on foliage has evaporated from the soil (called distillation). But only the dew that forms from water that originates in the atmosphere represents a net gain in the surface water budget. This is not an issue in extremely arid areas with desiccated soils. Additional discussion of dew formation in arid areas can be found in Kidron *et al.* (2000), Kidron (2000), Wallin (1967), Oke (1978), and Slatyer and Mabbutt (1965).

Sometimes dew is so plentiful that it can serve as a source of drinking water. For example, when large, roughly horizontal, concave surfaces of non-porous rock exist in areas that are protected from direct sunlight and wind, dew can accumulate in pools. These small accumulations of water are sometimes called "kiss tanks" in North America, because of the typical way of drinking the water. They have been documented in the deserts of western North America (Ives 1962), but are common in many other deserts where dew is produced.

Suggested general references for further reading

Brown, B. G., *et al.*,1985: *Exploratory analysis of precipitation events with implications for stochastic modeling* – a good example of the analysis of the statistical properties of precipitation time series, including the concept of a precipitation "event."

Bruintjes, R. T., 1999: *A review of cloud seeding experiments to enhance precipitation and some new prospects* – a state-of-the-science review of rainfall enhancement procedures and prospects.

Dennis, A. S., 1990: *Water augmentation in arid lands through weather modification* – provides a short review of rainfall enhancement concepts, with sections on the seeding of winter orographic clouds and cumuliform clouds over flat country.

Katz, R. W., and M. H. Glantz, 1977: *Rainfall statistics, droughts, and desertification in the Sahel* – a good summary of appropriate statistics for representing rainfall properties in arid areas.

Wilks, D. S., 1995: *Statistical Methods in Atmospheric Sciences* – one of the best technical summaries of the use of statistics in the analysis of atmospheric data and for assessment of the skill of numerical models of the atmosphere.

Questions for review

(1) Describe measures of the central tendency of a statistical distribution. Do the same for measures of the variability.

(2) What can you say about the properties of the precipitation of semi-arid climates in which the precipitation is predominantly (a) in the winter and (b) in the summer?

(3) From Fig. 13.14, if a crop requires 25 mm of rainfall every 10 d, and the economics are such that one crop failure in four years is acceptable, approximately how early can the crop be planted? Such rainfall-reliability analyses are clearly intended to represent the "short-term" interannual variability in the rainfall. Would longer-term changes in the regional climate influence the validity of such methods?

(4) Identify from Plate 1 where elevated terrain enhances the average annual rainfall in the Sahara Desert. Use a geographic atlas to identify the terrain features.

(5) Use the monthly rainfall histograms displayed for the different deserts in Chapter 3 to identify semi-arid regions with dominant winter or summer rainfall.

Problems and exercises

(1) Are advective coastal fogs or radiation fogs more likely to make significant contributions to the surface water budget in the desert?

(2) There have been many proposals to enhance rainfall in arid lands through the construction of lakes with large surface areas. Discuss whether such proposals make sense, and would be likely to succeed.

(3) It has been informally claimed that the infiltration of dew over the millennia may be responsible for aquifer recharge in deserts. Comment on the likelihood of this process.

(4) The distinctions between the characteristics of precipitation in arid and non-arid areas of North America were described in this chapter. Propose how the analysis procedure might be designed differently, in terms of, for example, the way that wet periods are defined. How might such changes produce different results?

14

Anthropogenic effects on the desert atmosphere

The irrigation of the Ogallala region, which has occurred almost entirely since the Second World War, is, from a satellite's point of view, one of the most profound changes visited by man on North America; only urbanization, deforestation and the damming of rivers surpass it. In the space of twenty years, the high plains turned from brown to green as if a tropical rainbelt had installed itself between the Rockies and the hundredth meridian. From an airplane, much of semiarid west Texas appears as lush as Virginia.

Marc Reisner, American hydrologist and writer
Cadillac Desert (1986)

Regarding the knowledge needed to avoid the human destruction of arid environments:
. . . it is safe to say that we have today not all the knowledge we need, but a great deal more than we use.

Paul Sears, American environmentalist and writer
Deserts on the March (1935)

In terms of the "greening" by irrigation of the arid San Luis Valley of Colorado:
. . . even the most prosperous farmers there know that from nature's point of view this is cosmetology. The place is still a desert, harsher yet for their efforts than the one their ancestors settled. If the water table drops, if credit dries up, if prices fall, if soil salinity increases, if fuel and chemical costs rise, if exotic pests multiply, the green makeup will crack right off the hard dry face of it.

Sam Bingham, American environmentalist and writer
The Last Ranch: A Colorado Community and the Coming Desert (1996)

Human activity has often profoundly altered the characteristics of the desert land surface, and hence the surface moisture, energy, and momentum budgets. Even though the effects of such changes on the local weather and microclimate may be sometimes difficult to document, there is considerable positive evidence of an impact. Human modification of the surface of arid and semi-arid lands has taken many forms, including the following.

- Irrigation
- Livestock grazing, and overgrazing
- Lowering of the water table
- Introduction of non-native vegetation
- Dryland agriculture
- Urbanization
- Off-road vehicle use

Irrigation water imported to the desert surface through groundwater pumping and ditches has greened vast areas of arid land worldwide. It also has permanently destroyed vast areas through salinization. Agricultural exploitation in general – livestock overgrazing, dryland farming, and irrigation – in semi-arid areas or desert margins is collectively transforming 37 800 km^2 of land per year to desert (Campbell 1997). Water-table lowering has dried perennial and ephemeral desert playas and oases, and killed native vegetation whose roots can no longer reach water. Urbanization has had myriad effects such as grossly modifying surface conditions and placing huge demands on limited local water supplies. These processes, some of which are intended to tame arid lands for human exploitation, actually have the long-term effects of making them more arid. Sears (1935) points out that "the effect of human intervention with natural vegetation and soil nearly always takes the form of making rainfall less efficient and of introducing the weeds characteristic of drier climates . . .". It is ironic and perhaps unexpected that the most undisturbed desert areas, in North America at least, are often military reservations that have been off limits to grazing and recreation, while other areas have suffered from degradation from various sources. This chapter will discuss the short-term and long-term effects of these processes on the surface physics and the atmosphere.

There is some necessary overlap between the material in this chapter and that in Chapter 18 on desertification. In particular, desertification can result from both natural and anthropogenic causes, where normally a combination of the two is responsible. Thus, those anthropogenic effects that contribute to desertification are mentioned in both chapters, but hopefully the treatments are complementary rather than redundant. There are also some anthropogenic effects that decrease the desert-like condition of the surface. For example, irrigation of arid or semi-arid land produces a surface condition that is more like that of a temperate environment. It is only when the practice leads to salinization of the soil and an inability to support the natural pre-irrigation vegetation that it represents desertification. Also, in some cases, overgrazing and species invasion may simply cause one type of vegetation to be replaced by another. This could certainly affect the land–atmosphere interaction and microclimate, but it may not necessarily be called desertification.

Irrigation

For thousands of years desert dwellers have tapped surface and subsurface waters for agricultural use. This greening of the desert drastically changes the surface energy and moisture balances, which in turn influences the lower atmosphere. Natural surface waters are most easily exploited, but they are often ephemeral in deserts. Wells dug to water tables of accessible depth are alternatives. When necessary, massive projects have been undertaken to import irrigation water from other areas. Early predecessors of today's vast networks of irrigation canals, ditches, tunnels, and pipes existed almost 3000 years ago in the Middle East, where nearby mountains were favorite sources of surface and underground water. Tunnels to tap the water table under the mountains were dug near-horizontally through rock, by hand, for distances of tens of kilometers (Balland 1994). This water from beneath the mountains flowed down the very gentle tunnel slope to the base of the mountain, where it irrigated the surrounding desert. It is estimated that 30 000 of these tunnels (qanats or falaj) are, and have been, used to locally transform the desert and microclimate.

Numerous empirical and modeling studies have demonstrated that irrigation of agricultural crops can have a significant influence on atmospheric processes. The direct effects are at least three-fold: the agricultural crops that are reliant on the irrigation have different physical characteristics, such as albedo, compared with the semi-arid natural vegetation or dry-land agricultural crops; evaporation is enhanced because of the existence of the free-water surface during flood or sprinkler irrigation, and the transpiration from the agricultural crops growing in the moistened soil is likely greater than that from the natural vegetation growing in drier soil; and the heterogeneities in the surface energy budget between irrigated and non-irrigated land can generate mesoscale circulations. The specific mechanisms by which these irrigation effects influence regional weather, and therefore regional climate, are varied and speculative, but the consensus of modeling and empirical evidence seems to suggest a definite relationship.

For example, purely empirical studies indicate that irrigation of agricultural crops in a semi-arid area can significantly affect the precipitation climatology. Barnston and Schickedanz (1984) show that irrigation in the southern Great Plains of North America can increase precipitation under certain large-scale conditions. Rosenan (1963) documents increases in precipitation in Tel Aviv after irrigation in that area. In contrast, Fowler and Helvey (1974) could find no correlation between climate factors and irrigation in the Columbia Basin in the northwest United States; however, they point out that large-scale subsidence during the months of their study likely mitigated any effects of moisture increases. Strahler and Strahler (1973) and Landsberg (1970) cite studies that estimate that irrigation has caused increases of 10–50% in summertime precipitation in some areas of the southern Great Plains of North America. Beebe (1974), in a study relating irrigation and severe-storm enhancement, correlates increased irrigation in the Great Plains with changes in the regional climatology, such as increased tornado frequency, a doubling in the number of hail days, and an increase in the surface dew-point temperature of over 3 K (5 °F).

Modeling studies also seem to confirm the correlation between irrigation and atmospheric effects, including enhanced precipitation. Segal *et al.* (1989) employed a mesoscale model and observations to study the atmospheric effects of irrigation in eastern Colorado in the United States, and demonstrated a significant impact. Yeh *et al.* (1984) used a simple global-scale model to show that large-scale irrigation has an effect on regional climate, and especially precipitation. Chang and Wetzel (1991) employed a mesoscale model to show that spatial variations in soil moisture and vegetation affect the evolution of the prestorm convective environment in the eastern Great Plains of North America. Beljaars *et al.* (1996) document that model precipitation forecasts of extreme rainfall events in July 1993 in the Midwest United States were strongly related to soil-moisture anomalies about one day upstream. Paegle *et al.* (1996) relate model-simulated rainfall for the same period to local evaporation, where the link was through effects on the low-level jet. Chen and Avissar (1994) demonstrated that landscape discontinuities (such as those associated with boundaries between irrigated and non-irrigated land) enhance shallow convective precipitation. Chase *et al.* (1999) and Chen *et al.* (2001) show how conversion of semi-arid grasslands to dry farm land and irrigated farm land in northeastern Colorado in the United States affected the local atmospheric conditions as well as the rainfall in the mountains to the west. There is also a large body of literature that correlates soil-moisture gradients with the low-level jet in the Great Plains of North America, and it is well known that the low-level jet in that area is related to moist convection.

A very well documented case of the effects of desert irrigation on the regional climate is that of the Aral Sea area, which is bordered by the Karakumy and Kyzylkum Deserts and the Ustyurt Plateau (Fig. 3.50). In the 1960s, this sea was Earth's fourth largest inland water body, smaller only than the Caspian Sea, Lake Superior, and Lake Victoria. Its water supply was replenished by the Syrdarya and Amudarya (*darya* means river, in Turkic). However, massive irrigation projects siphoned off river water into huge canals for irrigation of cotton and rice in arid Uzbekistan, Turkmenistan, and Tajikistan. As a consequence, from 1960 to 1995 the surface area of the lake decreased by over 50%, shorelines receded over 70 km in some places, the water level dropped 19 m, and the salinity tripled. Over 50 lakes in the Amudarya delta dried up, and wetlands shrank to less than 5% of their original area. This has changed the regional climate in many ways. The large lake historically moderated the seasonal variations in temperature in the area; warmer and drier summers and colder winters now result from the shrunken and segmented water surface. Dust storms from the desiccated surface of the dry lake and the surroundings annually transport 75 million metric tons of dust, salt, and heavy metals for distances of up to 1000 km. The toxic effects of these "salt storms" are devastating to ecosystems, and have led to a variety of disorders in the human population. Naturally, much of the irrigated land is now so salinized that it appears snow-covered. The aggregate impacts on the atmosphere of these multiple effects of irrigation are immense, complex, and not well documented. A summary of the situation is provided in Glantz *et al.* (1993), with thousands of other papers having been written about various aspects of the problem.

Lastly, the oasis effect is discussed in terms of the surface energy budget in Chapter 4 and microclimate interactions in Chapter 12. In summary, the rapid evaporation from the wet surface to the dry air that is advected from the desiccated surrounding area causes a rather unusual surface energy budget. The heat loss from evaporation produces a low surface temperature, which leads to a downward sensible heat flux. Thus, the evaporative heat loss is balanced by the net radiation and the sensible heat flux, ignoring the flux into the substrate. In terms of the effect of this process on the local and downwind weather, the air will naturally be cooler, the specific and relative humidities will be higher, and the more stable temperature profile near the surface may lead to reduced wind speeds.

Water-table deepening

There are at least a few ways in which human activity can lower the water table. First, groundwater pumping for industrial, residential, and agricultural use has lowered the water table in some regions by over 200 m. Understandably, natural vegetation, both riparian and non-riparian, has suffered greatly. In addition, the desert surface often subsides as the water is removed, and the ground compacts and settles. The second mechanism is related to the fact that the water table in arid areas is often not far below the surface. Streams and rivers, whether they are seasonally ephemeral, only run after a rain, or are continuously flowing, often have a bed that is not far above the water table. Removal of the surface vegetation by overgrazing, dryland agriculture, fire, etc., can cause rainfall runoff to increase and infiltration to decrease. This increases the discharge in the waterways after rain events, and causes the water to scour deeper into the substrate, which may be gravely and easily erodible. If the bed of the waterway is cut deep enough that it is below the water-table level, the water table may empty into the eroded waterway, and its depth below the surface will be increased. The deeper the waterway cuts, the more subsurface water is drained off. A positive feedback can develop, in principle, in that lowering the water table sufficiently can cause native vegetation to die, which increases the severity of the runoff and the rate of streambed erosion. Thirdly, non-native phreatophytic vegetation introduced into a desert region can consume additional groundwater, and lower the water table.

The lowering of the water table by any of these mechanisms can eventually cause native vegetation to die. In the desert southwest of North America, if the water table drops by only 0.3 m, one-quarter of the undergrowth can die off, and a 1 m drop may kill all but the large trees (Childs 2000). It is not hard to imagine the impact of the water-table drops of over 200 m that have resulted in some places from human activity. In Arizona, in the United States, many of the once perennial desert streams and their riparian vegetation no longer exist because of human water extraction.

The loss of this vegetation has a substantial impact on the surface moisture and energy budgets, through the decrease in transpired water and the decreased heat consumed by the evaporation. Even though conditions are arid, such transpiration of water supplied by the water table can be substantial. For example, sometimes riparian and other general vegetation draws sufficient water during the daylight when transpiration is taking place that running surface water disappears, to return at night when the water table rises after transpiration shuts down for the day (Childs 2000). Similarly, Slatyer and Mabbutt (1965) report regular diurnal variations in the difference between discharge measurements at two locations on the same stream. The discharge differences were a result of both transpiration by the riparian vegetation and evaporation from the water surface between the measurement locations. The loss of vegetation also changes the albedo, as well as the wind speed near the surface.

Grazing and overgrazing

Native grasses, shrubs, and trees reduce the wind speed at ground level, and inhibit wind erosion of the soil. In addition, the vegetated surface better retains surface moisture, and the soil grains are less likely to be elevated by wind. Thus, removal of native vegetation through agricultural practices, either the use of the plow or grazing, can enhance wind erosion. Removal of vegetation also encourages erosion of the topsoil by water, and has numerous other effects that have already been discussed. Near-surface temperatures are also affected because exposing more bare soil through grazing changes the aggregate albedo of the surface. Recall the discussion in Chapter 6 of the temperature and evaporation contrasts across the Mexico – United States border caused by differences in the degree of overgrazing (Bryant *et al.* 1990), where the objective there was to illustrate vegetation effects on the surface energy budget. The more grazed Mexican side had a 5% higher albedo, and during the drying process after a rain there were 1–4 K (2–7 °F) near-surface air-temperature differences across the border and significant differences in the latent-heat flux (see discussion of Fig. 6.10). As additional evidence of the albedo impacts of overgrazing, Table 18.6 shows albedoes for overgrazed desert and nearby protected areas based on satellite data for various parts of the world.

Consider North America as an example of the evolution of the desert surface that results from overgrazing. Before the introduction of cattle and domestic sheep, there were significant parts of the Chihuahuan and Sonoran Deserts with abundant grasses (Campbell 1997). However, within 20 years in the latter half of the nineteenth century, cattle driven into the area denuded millions of hectares of grassland. The native grasses such as grama, that once grew thick and up to a horse's belly, and that were harvested by ranchers, were replaced by cactus and shrubs such as creosote bush. An additional result of the overgrazing was severe erosion, creating arroyos where once stable stream beds existed. This deepening of the water courses caused a lowering of the water table, which had the aforementioned effect of killing much vegetation whose roots needed to reach the level of the water. Cattle have also caused mesquite trees (*Prosopis*) to spread far beyond their normal range in the desert southwest of North America, with dramatic changes in the ecosystem. The mesquite once had a range that was limited to floodplains, where dense localized stands spanned relatively small areas. However, free-ranging grazing cattle and horses, and cattle being driven along the various cattle trails, ate the bean-like pods, and excreted the, now even more fertile, seeds intact. This spread billions of mesquite seeds into new range that, also because of the cattle, had been denuded of grass that could have competed with the mesquite. Now these large shrubs dominate over 340 000 km^2 of the desert southwest of North America. By 1932, in Texas alone, 120 000 km^2 that were once covered with grass had reverted to only mesquite and thorny weeds (Reisner 1986), primarily because of cattle grazing. Sears (1935) reminds us that

weeds, like wild-eyed anarchists, are the symptoms, not the real cause, of a disturbed order . . . the native vegetation had already been destroyed by the plow and the thronging herds . . .

There are many other feedback mechanisms, within the complex environmental systems of arid lands, by which overgrazing triggers other processes that accelerate the modification of the surface, and, in turn, the atmosphere. Sears (1935) points out one.

We have heard much of the plague of grasshoppers, crossing the country like a wave of devastation and consuming every green thing in their path. Yet a fence of three barbed strands of wire has been known to stop them. In the Wichita National Forest is such a fence. On one side the herbage is heavily populated with various types of destructive grasshoppers. On the other side the species present are somewhat different, and the numbers very much less. Actually, of course, the fence served to prevent overgrazing. But the truly surprising thing is that the hungry pests did not occur to serious degree on the side with the large amount of potential food. Like scavengers and troublemakers who have no place in an ordered existence, they found their opportunity only after the natural balance had been practically destroyed.

Introduction of non-native vegetation

An example of a non-native plant species that has altered a desert biome is the tamarisk, or salt cedar, of the North American Desert. It is unclear whether it was first introduced from the Mediterranean area by the early Spanish explorers in the hay for their horses, or by more contemporary sources. In any case, it is known that between 1899 and 1915, the United States Department of Agriculture further introduced eight varieties of tamarisk into the United States for ornamental use, erosion control, and windbreaks (Tweit 1992). Unfortunately, it has long roots, grows and spreads rapidly, and consumes vast quantities of water. Where large thickets of tamarisk have colonized the desert along perennial or intermittent waterways, large quantities of groundwater are transpired and the water tables have been lowered. Native plant species have been crowded out by this aggressive invader.

Dryland agriculture

Conversion of semi-arid lands with native vegetation to dryland agriculture has major and potentially disastrous consequences. This is discussed at greater length in the chapter on desertification, because that is one of the potential results. Dryland agriculture generally involves plowing under the native vegetation, and relying on marginal rainfall to sustain a crop. This greatly alters the energy and moisture budgets of the surface, allows the invasion of non-native species of vegetation, and enhances wind and water erosion. It is generally agreed that the great "dust bowl" dust storms and associated loss of millions of metric tons of topsoil in the semi-arid Great Plains of North America in the 1930s were a consequence of the plowing of the prairie and the planting of wheat. Even without erosion by wind and water, other effects of replacing natural vegetation with crops would be albedo, roughness, and possibly transpiration changes.

Urbanization

Importation of water to arid lands, either through groundwater pumping or by aqueduct construction, can allow dramatic population growth and urbanization because the climates are often very favorable for habitation (except for the lack of rainfall). This regional metamorphosis is sometimes so dramatic and complete that it is eventually forgotten by many that the area, by most standards, is arid. For example, consider California, the most populous state in the United States, which is more intensely irrigated than any other area of the world. San Francisco receives only slightly more rain than Chihuahua, Los Angeles is drier than Beirut, Sacramento is as arid as the Sahel (Reisner 1986), but we find these facts surprising because the introduction of water during the twentieth century allowed an immense artificial greening of the environment, and adoption of lifestyles that are often oblivious to water conservation.

It is hard to overstate the land–atmosphere interaction differences between natural deserts and urban areas. Evaporation is probably less in cities because water from any precipitation runs into storm sewers; transpiration may be more or less depending on the relative amounts of natural vegetation versus watered, decorative urban vegetation; the concrete substrate has thermal properties that are very different from those of most natural desert substrates; the roughness of the surface (buildings) greatly alters the wind field; the building geometry changes the orientation of the surfaces that absorb solar and emit infrared radiation, and this affects the energy budget; and aerosols are injected into the atmosphere.

One known consequence of urbanization in non-desert areas is the "urban heat-island" effect. Here, cities have higher temperatures, especially in the night, than do their rural surroundings. There is also evidence of enhanced rainfall around cities. However, these effects have been less well documented for urban areas embedded within desert environments. Lastly, the development and associated land disturbances within and on the fringe of desert cities leads to more frequent and more severe dust storms. Phoenix, in the Sonoran Desert, has been one of the best studied North American cities. For example, Balling and Cerveny (1987) show that, as the city has grown during the previous 38 years, there has been a statistically significant effect on the mean monthly wind speeds. There has also been an increase in the temperature contrast between the city and the surroundings. Satellite estimates of the Phoenix heat island are described in Balling and Brazel (1989).

Off-road vehicle use

Off-road vehicle use in deserts can be related to recreational, commercial, or military activities. In each case there is a need or desire to not use the available established roads. The resulting first-order impacts on the environment are as follows.

- Shrubs and grasses are dislodged.
- Crusts held together by microflora are destroyed.
- Crusts of desert pavement are destroyed.
- Soils are compacted.
- Desert varnish is disturbed.
- Tires or tank tracks produce gullies.

Derivative effects that are a consequence of the above are the following.

- Water erosion is accelerated; hill slopes are destabilized.
- Wind erosion is accelerated.
- Water infiltration is affected.
- The aggregate surface albedo is changed.

There is, unfortunately, plentiful evidence worldwide of all of the above types of degradation of the desert land-surface. One of the easiest impacts on the atmosphere to document is the increase in dust storms after extensive traffic over a particular desert area. For example, increased frequency and severity of dust storms have been noted in the Sahara Desert after the North African campaign in World War II (Oliver 1945), in the northern Sonoran Desert after a desert training camp was established there during World War II (Clements *et al.* 1963), and in the Middle East after the 1991 Gulf War (Thomas 1997a). In southern Tunisia, the tracks of World War II tanks and wheeled vehicles can still be seen in the sand, and vegetation has never returned to the area. Wind erosion in a tank-maneuver area in the northern Chihuahuan Desert is discussed by Marston (1986).

Deforestation

Because deserts generally do not contain forests, it may seem strange that deforestation can lead to changes in the desert surface and atmosphere. However, remote effects need to be considered in which removal of forests in a temperate area may influence the aridity in a nearby or remote desert area. For example, Zheng and Eltahir (1997) used model experiments to show that deforestation of the coastal areas of the Gulf of Guinea affects the strength of the monsoon flow that provides most of the rainfall for the Sahel and the southern Sahara. In particular, the replacement of all the coastal forest by savanna caused a collapse of the monsoon circulation, with much less rainfall for the arid interior of the continent. The annual rates of deforestation in the Ivory Coast, Nigeria, and

Ghana in the 1990s were, in fact, the largest of any region on Earth, reaching 15% of the forested area per year.

Suggested general references for further reading

Changnon, S. A., Jr., and R. G. Semonin, 1979: *Impact of man upon local and regional weather* – a general and brief review of the topic of human impacts on weather.

Reisner, M., 1986: *Cadillac Desert: The American West and its Disappearing Water* – describes irrigation's impacts on arid areas.

Webb, R. H., and H. G. Wilshire, (Eds.), 1983: *Environmental Effects of Off-Road Vehicles* – describes the various types of environmental damage that result from the use of off-road vehicles in natural areas.

Questions for review

(1) Summarize the effects of off-road vehicle use on the desert surface, and the atmosphere.
(2) What are the potential mechanisms by which the existence of cities in desert environments can affect the local atmosphere?
(3) How does irrigation affect the temperature, humidity, and wind speed, locally and downwind?
(4) How can water-table deepening affect the desert atmosphere?

Problems and exercises

(1) Discuss various mechanisms by which irrigation can influence rainfall amount and distribution.
(2) Speculate on the physical mechanisms by which urban areas in deserts might affect precipitation.
(3) Is the urban heat island effect likely to be more or less intense for desert cities, compared with cities in temperate climates?
(4) Discuss the possible physical mechanisms by which the replacement of coastal forest by savanna could affect the Guinea monsoon of Africa.

15

Changes in desert climate

In deserts, that which seems eternal may change overnight, and that which is least expected is always possible.

Susan Arritt, American natural-science writer
Deserts (1993)

He went on to talk about the colors of the desert, how they had been formed through the cooling of the earth; . . . he spoke about the . . . earth moving into strange new zodiacal realms and flopping over on its axis; about the great climatic changes, sudden, catastrophic changes burying whole epochs alive, making deserts into tropical seas and pushing up mountains where once there was sea, and so on. He spoke fascinatingly, lingeringly, as if he had witnessed it all himself from some high place in some ageless cloak of flesh.

Henry Miller, American author
The Air-conditioned Nightmare (1945)

May sickened into June. The short-grass curled.
Of evenings thunder mumbled 'round the sky;
But clouds were phantoms and the dawns were dry,
And it were better nothing had been born.

John Neihardt, American poet
The Twilight of the Sioux (1925)

Deserts have developed and disappeared on the grand geological time scales of continental drift, spanning hundreds of millions of years. However, the episodic expansions and contractions of the deserts during the present Quaternary Period will be the focus of this chapter. The Quaternary Period encompasses the Pleistocene Epoch, the previous 1 000 000 years of glacial and interglacial periods, and the most recent 10 000 years, the Holocene Epoch.

The limitation of this discussion to the *natural* variability of aridity narrows the focus, excluding the anthropogenic changes that are discussed in Chapters 14 and 18 and Box 15.1. On interannual time scales we have seen that natural precipitation variability is extremely high in arid areas. And we also know that

Box 15.1
Olive trees, and recent changes in northern Sahara climate

There has been great debate about whether the demise of agriculture in the
northern margin of the Sahara, once known as the granary of Rome, has been due
to climate change or to human factors. Lowdermilk (1953) reports the ruins of
great olive presses at Timgad in northern Algeria, where he observed not a single
olive tree from horizon to horizon. Some scholars say the climate changed so
much that people were forced to give up their agricultural livelihood. But others
have challenged that idea, claiming that the natural environment has not changed
appreciably since Roman times. As an experiment, olive trees were planted and
maintained at Timgad in the manner prescribed by Roman literature, and the trees
are thriving. Similarly, early travelers found olive presses at Sfax on the coast of
Tunisia, but again no olive trees. As an experiment, 40 years ago olive trees were
planted there. Today there are over 60 000 ha in Sfax that are planted with olive
trees. Even though it may be argued that an intervening drought devastated
agriculture, the farming never returned. Lowdermilk reports on olive trees that
were over 1500 years old growing in Sousse, also on the coast of Tunisia. This
circumstantial evidence supports the view that the major change on the northern
margin of the Sahara has been a result of human abuse of the land, and not the
climate.

periodic swings in global processes such as the El Niño – Southern Oscillation
(ENSO) have sometimes profound impacts in desert regions. This chapter will
expand upon this subject and describe what is known more generally about
longer-term climate variation in arid areas.

Time scales of climate and aridity change

As noted, we have seen examples of the high degree of variability of annual total
precipitation from one year to the next in arid regions. However, we would likely
agree that this type of variability should not be defined as climate change. For
precipitation deficits of a few months to a few years, we might call it a short-term
drought. One could imagine such a precipitation anomaly resulting from purely
statistical factors. For example, if most of the annual rainfall for a location is
associated with a few heavy convective rain events, it is reasonable that, in some
years, a particular area will be missed by these randomly located storms. On the
other hand, a precipitation deficit of the same duration could result from a weak
monsoon or an extreme in the ENSO cycle, in which case the cause of the aridity
variability would be related to some regular oscillation in the physical system.
In either case, we would refer to the precipitation deficit as a drought, but not

likely a change in the climate. The terminology here is clearly murky, and can become confusing and misleading. Various terms such as climate noise, climate fluctuations, short-term weather changes, climate variability, and climate change are commonly used without much definition. Because of its human impacts in semi-arid areas, drought, or a precipitation deficit, is often the most important kind of weather or climate anomaly. This term drought is loosely applied for deficits of a few months to thousands of years.

Climate, by definition, is the average weather in some sense (with no firm stipulation about the averaging period), where the climate can be quantified in terms of dozens of possible parameters. For arid areas, some of the more important metrics would be (1) the number of days per year with measurable rainfall, (2) the average annual rainfall, (3) the average seasonal distribution of the rainfall, (4) monthly average daily maximum and daily minimum temperatures, (5) average annual temperature extremes, (6) average near-surface wind speed, and (7) the number of dust-storm days. Changes in any of these quantities could affect the degree of surface desiccation and the vegetation.

In the interest of clearly defining our terminology in this book, we will limit the use of the term climate *change* to changes on time scales significantly longer than some of the typical oscillations associated with the ENSO cycle and the sun-spot cycle: i.e. for changes that occur on time scales that are much longer than a decade. For less enduring changes, we will call it climate variability.

A summary of recent climate periods

The most recent series of wet (pluvial) periods began with the Pleistocene Epoch one million years ago. At this time, climates throughout the world changed markedly. A rhythm of glacial periods (ice ages) and interglacial periods affected higher latitudes, with the Pleistocene encompassing over five episodes in which the ice sheets grew and advanced Equatorward, and then retreated (Fig. 15.1a). The glacial parts of the cycle lasted from 20 000 to 40 000 years in duration, with glaciation in the two hemispheres being in phase. The last glacial period extended from 80 000 to 12 000 years BP (before present), with the maximum ice extent at about 18 000 years BP.

The Pleistocene Epoch was considered to end about 10 000–12 000 years ago, when the Holocene or Recent Epoch began. During the Holocene (Fig. 15.1b), there have been about five periods of cooling superimposed on a gradual global warming. The first cooling period of the Holocene was the Younger Dryas event, from about 11 000–10 000 years BP. (Fig. 15.1b). This event not only affected mid-latitudes but also caused a period of enhanced subtropical aridity, as reflected in east African lake-level records (Roberts *et al.* 1993). A lengthy warm period,

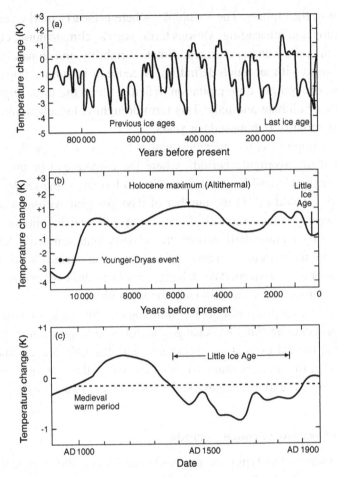

Fig. 15.1 Schematic diagrams of the global temperature variation for the Pleistocene and Holocene Epochs, on three time scales. (Adapted from Folland *et al.* 1990.)

called the Altithermal, prevailed from about 8000 to 4000 years ago. The most recent cool period was the Little Ice Age, which began about AD 1300 and extended into the 1800s (Fig. 15.1c). The overall period of Holocene warming came to an end in the twentieth century (Perry 1984), and was followed by an abrupt cooling. A warming trend has prevailed since the 1970s.

Methods of estimating climate change in arid regions

There are a number of techniques for estimating the history of climate in general, and the degree of aridity in particular. Application to arid regions of some of the approaches summarized below, based on Smith (1968), can be more challenging than normal. Techniques not discussed below include analysis of marine-sediment cores and ice cores.

Historical records

Actual meteorological measurements are less spatially dense in arid areas, and the records are generally shorter than is typical in temperate climates. There are other quantitative or qualitative historical records that can sometimes serve as surrogates for actual meteorological measurements, however. These include recordings of lake levels, drought periods, water levels in wells, and caravan routes.

Archeological data

The finding of abandoned buildings; agricultural infrastructures such as terraces; hunting, fishing, and farming implements; and paintings and engravings on rocks, can provide circumstantial evidence of previous wetter climates in areas that are now arid. Of course, desertification as a result of human activities, such as deforestation, needs to be ruled out in order to interpret the changes in terms of climate variability. The particular characteristics of archeological evidence may be of value in narrowing the estimate of the prevailing climate. For example, the contemporary fauna of the type depicted in the rock engravings in what are now lifeless parts of the Sahara are known to have certain habitat ranges. Included in this category of archeological evidence would also be the analysis of more contemporary paintings for evidence of past flora and fauna, glacier size, etc.

Tree-ring studies (dendrochronology)

A sufficiently large sampling of the annual variations in the thicknesses of tree rings and their chemical composition in an area can provide some qualitative insight into climate variations that have prevailed during the lifetime of the vegetation. Many complications exist in such an analysis, one of which is finding vegetation of sufficient age. Well-preserved dead vegetation can also be employed.

Existence of isolated communities of relict flora and fauna

When climate changes, sometimes relict flora and fauna of the old climate survive in ecological niches. Examples include tropical fish in some Saharan oases, blind fish that have been found to abound in subterranean water exposed by wells dug in the Sahara (de Villiers 2000), crocodiles that exist in waterholes in the Saharan highlands, and elephants that survived in the Atlas Mountains until they were exterminated (Smith 1968; Gautier 1935).

Fossils of flora and fauna

Fossilized remains of plants and animals, and the strata in which they are embedded, provide evidence of past climates and their chronology. Dramatic examples are the fossilized remains of contemporary-looking hippopotamuses, rhinoceroses, and elephants in the central Sahara. Pollen grains that have been deposited in sedimentary rock strata can also be used to estimate the meteorological conditions that prevailed at the time they were deposited.

The term fossils in this context does not have to imply mineralized material. Because of the slow decay rate in deserts, the remains of dead vegetation can persist for thousands of years as evidence of earlier wetter times. For example, poplar forests existed in some areas of the Taklamakan Desert until about 1500 years ago, when the water table dropped (Breed *et al.* 1979). Contemporary exploration of that desert reveals ancient skeletal remains of stunted trees in interdune lowlands. Similarly, rodent middens have survived for over 40 000 years, with the contents of seeds and other debris providing insight about the prevalent vegetation and the associated climate at the time. For example, Betancourt *et al.* (2000) describe now fossil rodent middens have been used to document monsoonal precipitation in the Atacama Desert, over the past 22 000 years.

Relict shorelines, beaches, and salt deposits

There are many examples of lakes being much smaller now than they once were, not because of human impacts, as is the case with the Aral Sea, but because the climate has changed. In such situations, the earlier size of the lake is evident from the existence of shorelines high on the mountainsides. For example, the Great Salt Lake in North America was once ten times larger than its current size; the original lake was called Lake Bonneville. Salt deposits can also reveal where water once existed, and then receded. Another type of circumstantial evidence of lake-level changes is described by Bowman (1935). As Goose Lake receded in Oregon–California in North America, wagon tracks were exposed on the bed, presumably made about 1850 by gold miners heading into California when the lake was smaller. A summary of global lake-level fluctuations is found in Street and Grove (1979).

Desiccated watercourses

Erosion channels, now permanently dry, where permanent or ephemeral water once flowed, indicate that water, at least upstream, was once more abundant. There are many examples of such features in virtually all deserts. Interpreting such features requires that some assessment be made about the rarity of the rainfall events that created them.

Stabilization or reactivation of sand dunes

Where sand dunes have been stabilized by the growth of vegetation, the climate must have been more arid in the past. That is, when the dunes originally formed, the sand was mobile. Subsequently, the environment changed to allow the vegetation to develop and arrest the dune movement. Reactivation of stable dunes points to a recent increase in aridity.

Soil profiles

The existence of buried layers of soil where soil-forming processes are now weak or non-existent indicates a trend toward greater aridity. For example, there are fossil soils in Egypt that are thought to be inconsistent with the present amount of moisture available.

Floods

The flooding of river channels is partly dependent upon the amount of *heavy* precipitation or rapid snowmelt (and not overall wetness), as well as the ability of the soil surface to absorb precipitation rather than create runoff. Evidence of flooding during the Quaternary has correlated positively with high lake levels, and thus should be considered as one indicator of a humid climate. The dating of floods can be performed by radiocarbon dating of organic matter in sediment that is deposited in backwater zones where the flow speed of the river was small. An example of the use of this technique can be found in Ely *et al.* (1993).

Examples of changes in desert climate during the Late Pleistocene and Holocene

The complexities of climate variations in desert regions are illustrated in Fig. 15.2, which shows the estimated transitions among humid–subhumid, arid–semi-arid, and hyperarid–arid climate regimes for five locations in Africa to the

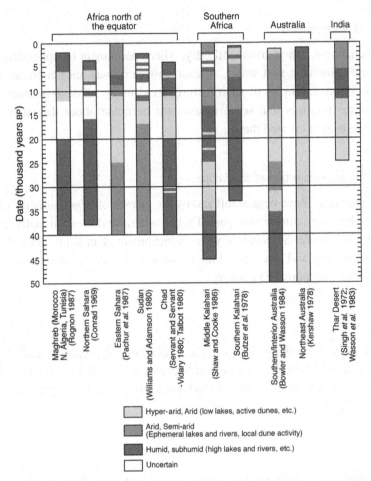

Fig. 15.2 Simplified chronologies of climate variations during the Holocene and Late Pleistocene in Africa, Australia, and India. Transitions were less abrupt than indicated, and changes with durations of less than 1000 years are not shown. (From Thomas 1997d.)

north of the Equator, two locations in southern Africa, two locations in Australia, and one location in India. The period spans the late Pleistocene and the Holocene Epochs. There is a general tendency for climates in some of these areas to have been humid earlier than 15 000–20 000 years BP, but there are clear exceptions to this. In general, there is only modest consistency in climate from place to place during a given time period, illustrating the complexity of the processes. Contributing to the regional contrasts are latitude differences, as well as degrees of continentality. Another illustration of the episodic nature of the precipitation climate in arid areas since the last glacial maximum is shown in Fig. 15.3, for five locations in Africa and Asia that are currently arid. This evidence

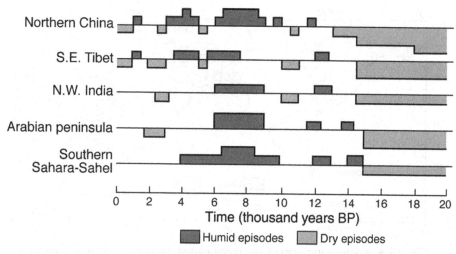

Fig. 15.3 Climate changes for locations in Africa and Asia since the last glacial maximum. (Adapted from Yan and Petit-Maire 1994.)

suggests that all locations were drier during the last glacial maximum, and for a few thousand years after. A humid period prevailed from about 6000 to 8000 years BP.

The Altithermal period was mentioned before as an extended warming during the Holocene Epoch. An illustration of the possible precipitation anomalies during this period is shown in Fig. 15.4. Parts of the Sahara, Middle Eastern, southern North American, and Australian Deserts were wetter during this period, while the mid-latitude desert areas of North America were drier. Figure 15.5 shows an example from one of the many studies that are consistent with the map in Fig. 15.4. It shows the water levels of Lake Lahontan in the Great Basin Desert of North America during the Late Pleistocene and Holocene Epochs, which reveal the distinctly arid conditions during the Altithermal period.

Manifestations and causes of climate change in deserts

Because the details of how climate changes in response to external forcings on the system (such as differences in solar output) are a result of complex interactions among the ocean, atmosphere, biosphere, cryosphere (ice), and land surface, it is generally very difficult to sort out the many system interactions and feedbacks. For example, even though it is useful to show how aridity in different areas responded to the glacial cycle, the more fundamental question is what caused this Pleistocene cycle to begin with. Similarly, there is some satisfaction in showing

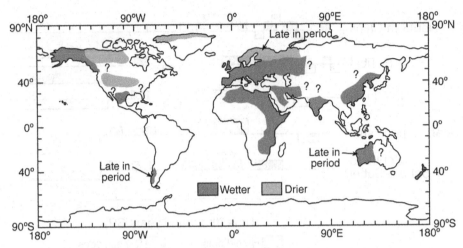

Fig. 15.4 A schematic map of the precipitation anomaly, relative to current amounts, during the Altithermal period from about 8000 to 4000 years BP. Blank areas may indicate either a lack of data on which to judge precipitation amounts, or regions in which the precipitation change was small. (From Kellogg 1978.)

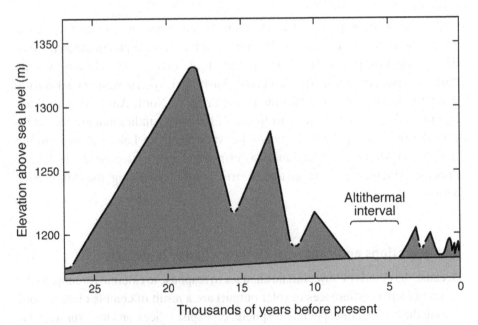

Fig. 15.5 Estimated fluctuations in the level of Lake Lahontan in the Great Basin Desert during the Pleistocene and Holocene. (From Morrison 1964.)

how El Niño and La Niña episodes greatly modify the aridity, but this simply leads to the more fundamental question of what controls the ENSO cycle. On even shorter time scales, droughts have been associated with temporary anomalies in the atmospheric circulation pattern. Such analyses, for example for Sahel wet and dry periods (Tyson 1981, 1984), for drought in the semi-arid prairies of Canada (Dey 1982), and for floods in the desert southwestern United States (Ely *et al.* 1993) are useful, but for a complete understanding of the whole system it is necessary to know what determined the circulation pattern. An example is when drought in a particular area is ascribed to a "blocking" ridge in the upper-atmospheric circulation, which steers moisture-producing storms to the north or south for an extended period. The real question is what caused that particular circulation pattern to persist for so long, and produce a drought (or flooding in other latitudes), in contrast to the typical situation where the upper-atmospheric wind and temperature patterns are changing regularly.

This said, many very useful studies have documented how particular manifestations of climate change, for example glacial cycles or El Niño episode frequency, have affected aridity, without addressing the larger question of what caused the manifestations. Links between some of these global-scale climate responses and the external forcing of the climate, such as solar output or the Earth–sun geometry, have been suggested. But some climate change may simply be a natural very long-term internal oscillation in the climate system, or a periodic instability that is occasionally realized. It is thus difficult to say sometimes whether the causes of change are internal or external. The following subsections contain discussions of both manifestations of climate change, such as glacial cycles, and possible external causes of change, with the understanding that there is sometimes a connection that we do not yet well understand. The effects on climate of volcanic eruptions are not discussed.

Glacial cycles

During the colder, glacial periods, the rainfall in lower-latitude continental-interior deserts tended to increase. This may have been because the region of strongest horizontal temperature gradient moved farther south, and so did the tracks of the mid-latitude cyclones that derive their energy from the temperature gradients. As a consequence, there were increases in precipitation in the arid southwestern United States, the Mediterranean region, and North Africa.

There were also complex successions of pluvial and interpluvial periods, related to the glacial cycle, in more-equatorial latitudes. At least part of this low-latitude response must be attributed to the migration of the ITCZ, and its associated rainfall, in harmony with the glacial cycle. For example, during glacial

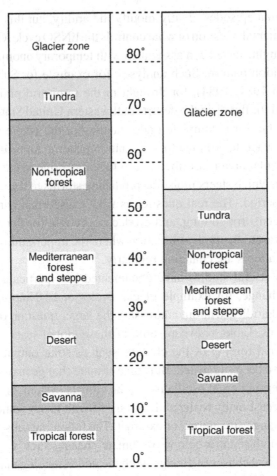

Fig. 15.6 One estimate of climate zones during the Würm glaciation near the end of the Pleistocene (right) and during the following Holocene (left). Latitudes are indicated in the center. (From Büdel 1957.)

periods the Sahara extended farther south, as evidenced by the existence of low lake levels and broad belts of activated dunes (see, for example, Thomas 1997d).

In Fig. 15.6 is one estimate of the latitudinal shift in climate zones associated with a glacial phase, the Würm glaciation near the end of the Pleistocene (right), and the following Holocene (left) after the glacier receded. During glaciation, in higher latitudes the climate zones shifted southward and in the tropics they shifted northward, with the result being a reduction in the size of the belt of desert latitudes. A schematic of a somewhat different possible change in the aridity associated with the equatorward shift of mid-latitude disturbances

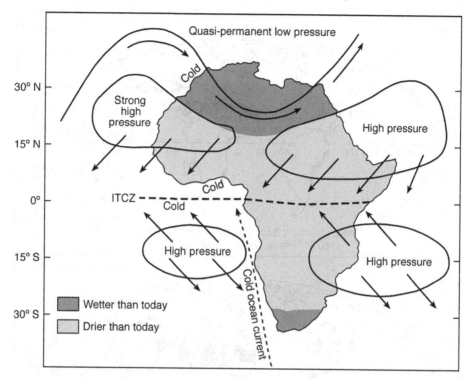

Fig. 15.7 Conceptualization of the change in the subtropical and tropical precipitation climate of Africa associated with glaciation. (From Nicholson and Flohn 1980.)

during a glaciation is shown in Fig. 15.7 for Africa. In both hemispheres, the climate becomes wetter on the mid-latitude side of the subtropical highs, and the intensification of the subsidence of the contracted Hadley circulation causes an increase in aridity in the subtropics and tropics. There is much evidence of this increased subtropical and tropical aridity. For example, Sarnthein (1978) shows that sand dunes were much more widespread 18 000 years ago at the glacial maximum than they are today. Between 30° N and 30° S, presently about 10% of the area is covered by sand desert, but 50% coverage is the estimate for the time of the glacial maximum. Fig. 15.8 illustrates the distributions of active sand dunes for the present and for the last glacial maximum, 18 000 years BP.

The aridity variation in Africa and Asia since the last glaciation can be interpreted in terms of the northward penetration of the monsoon. The northward range of the monsoon at the present, at the last glacial maximum, and at the Holocene maximum (Fig. 15.1) is shown in Fig. 15.9. Williams (1975) provides a further summary of Late Pleistocene tropical aridity.

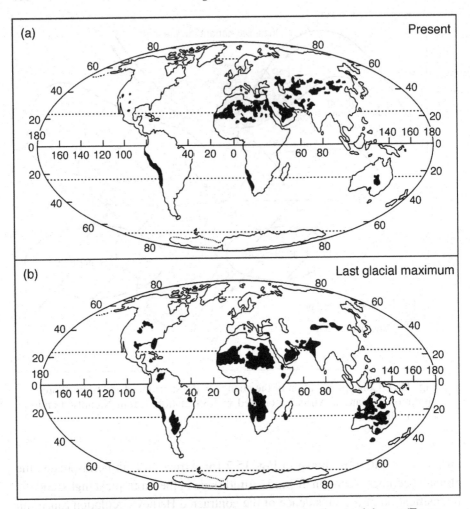

Fig. 15.8 (a) The present-day global distribution of active sand dunes. (From Goudie 1983a.) (b) The global distribution of active sand dunes during the last glacial maximum, 18 000 years BP. (From Sarnthein 1978.)

There would also be multiple effects of glaciation on wind speed, which affects the aridity of the climate through evaporation rate. One response would be that the stronger mid-latitude thermal gradient during glacial phases would cause wind speeds to be higher in that region on the north side of the subtropical highs (Petit *et al.* 1981). This could have compensated somewhat for the effect of the lower glacial-period temperatures on the evaporation rate. Also, because the trade winds are responsible for the equatorial tongue of cold water, potentially stronger trades associated with a stronger Hadley cell would have caused the tropical sea-surface temperatures to be up to 8 K lower.

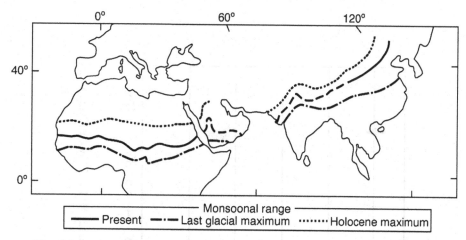

Fig. 15.9 The northward range of the African and Asian monsoons for the current climate, and for the last glacial maximum and the Holocene maximum. (Adapted from Yan and Petit-Maire 1994.)

The ENSO cycle

The reader should already be familiar with the general characteristics of the ocean–atmosphere variability and interaction during this cycle. El Niño typically occurs annually, generally in December, when the east-Pacific equatorial temperatures are anomalously warm for a short period of time. However, at irregular intervals of three to seven years, the warm anomaly is more extreme, persists for much longer than normal, and affects weather worldwide. In terms of the impacts on deserts, it can produce anomalously wet years, as has been noted in Chapter 13 in the context of the northern Peruvian Desert. It can also produce dry anomalies. A particularly dramatic example of rainfall increases during El Niño years is depicted in Fig. 15.10, which shows the annual rainfall for a 19 year period for Kiritimati Island in the central Pacific (about 2° N, 157.3° W), with El Niño years indicated by the arrows. This is a location whose normal aridity is hinted at and poorly resolved by the coarse-resolution satellite analysis in Fig. 3.60, which shows a narrow tongue of aridity extending westward along the equator from South America. Similarly dramatic impacts are found in northern Peru, where areas of the Peruvian Desert that normally receive little rainfall experience 1000–4000 mm in less than six months during El Niño events (Douglas *et al.* 2001).

As an illustration that ENSO effects on rainfall are not limited to the Pacific area, Fig. 15.11 shows the seasons and locations where El Niño causes conditions that are consistently drier and wetter than normal. Arid areas of central Australia, northern South America, the Indian subcontinent, and southern Africa become

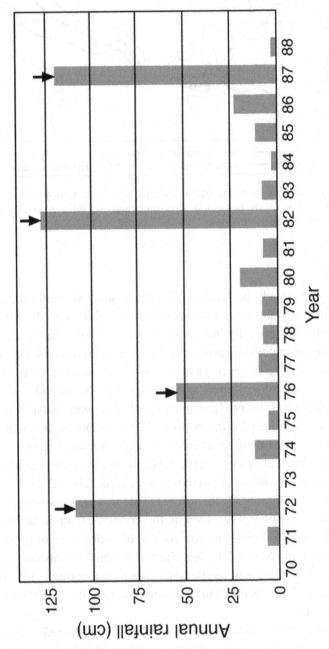

Fig. 15.10 Annual rainfall at Kiritimati Island in the equatorial central Pacific Ocean (2° N, 157.3° W). El Niño years are indicated with an arrow. (Adapted from UCAR 1994.)

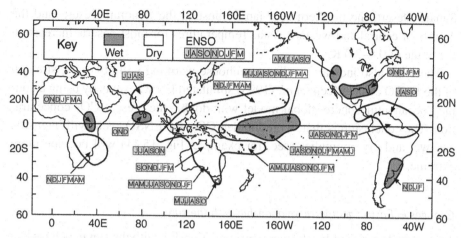

Fig. 15.11 Schematic diagram of the areas and months of the year with a consistent warm-phase (El Niño) ENSO precipitation signal. (Adapted from Ropelewski and Halpert 1987.)

drier, while some arid areas of North America and southeastern South America become wetter.

Greenhouse warming of the Earth climate system

Kellogg and Schneider (1977) point out that when a hemisphere warms it is likely that the poles will warm more than the tropics. The resulting decreased latitudinal temperature contrast weakens the westerlies and shifts them poleward. And it is likely that the Hadley circulation broadens, covering a wider latitudinal belt. Conversely, when the hemisphere is colder, the Hadley cell is more concentrated, and the downward motion in the subtropics is more narrow and vigorous. This general conclusion is borne out by present-day observed differences between the summer and winter hemispheres and by the evidence of climate shifts during warm and cold phases of the glacial cycle shown in Fig. 15.1. These observations may have some general applicability to the changes that might be expected by greenhouse-gas warming.

Global-model simulations of temperature and precipitation change from greenhouse-gas increases are plentiful, but should be viewed with some caution given that the models only approximately reproduce the current climate. Given that caution, projections (see, for example, Greco *et al.* 1994) indicate that some arid regions will experience an increase in precipitation of a few centimeters per year, but for most the aridity will remain about the same or increase. Associated with greenhouse warming, decreased rainfall is predicted for large areas of the

Sahara Desert, the northern Arabian Peninsula, the Sonoran Desert, and the deserts of central and western Asia. Temperatures for most deserts increase in the range of 0.5–2.0 K.

There are also indications that greenhouse warming will affect the strength of the ENSO cycle. Meehl *et al.* (1993) simulated ENSO events by using a coupled ocean–atmosphere global model with double the present CO_2 concentration. The conclusion is that, with doubled CO_2, the ENSO cycle will intensify, and wet and dry anomalies (see, for example, Fig. 15.11) will become more extreme.

Changes in Earth's geometric relationship with the sun

Two characteristics of Earth's geometric relationship with the sun may have had significant effects on the past climates of arid lands. First, Earth's orbit around the sun is elliptic, and changes in the ellipticity affect the seasonal distribution of the intensity of the sun's radiation. This degree of non-circularity of the orbit is its eccentricity. Secondly, the angular relation between Earth's equatorial plane and the plane on which Earth revolves around the sun (the ecliptic plane) affects the seasonal distribution of radiation. Presently, the angle between Earth's equatorial plane and the ecliptic plane (the obliquity of the ecliptic) is about 23.5°. Said another way, the axis of rotation of Earth is tilted by this amount out of the perpendicular to the ecliptic plane (see Fig. 4.5). This relation changes through precession, wherein Earth's axis of rotation executes a conical motion around the perpendicular to the ecliptic plane, much as the rotation axis of a spinning top evolves. Another periodic change results from the fact that the obliquity of the ecliptic follows a cycle: this process is called nutation. The periods of the precession- and nutation-related cycles vary somewhat with time, but typical values would be about 25 000 years and 40 000 years, respectively (Berger 1984). The eccentricity of Earth's orbit about the sun varies with a period of about 100 000 years.

An example of the effect of the cycles in the Earth–sun geometry on the climate of the arid areas of Africa and Asia is described in Prell and Kutzbach (1992). The average, summer, Northern Hemisphere solar radiation varies as a result of the superposition of the above cycles, and this affects the thermal contrast between the land and sea. The result is a variation in the strength of the summer monsoon that brings rain to the subtropical deserts of the continents. Fig. 15.12 shows the variation during the past 150 000 years of the June, July and August average Northern Hemisphere solar radiation intensity. Intensity values have changed from 12.5% above the current value (126 000 years BP), to 5.8%

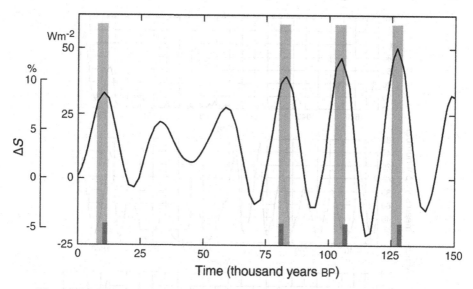

Fig. 15.12 The variation (ΔS) during the past 150 000 years of the June, July and August average Northern Hemisphere solar radiation intensity in terms of the percent and intensity change from the current value. The long shaded bars align with the four strongest positive excursions in summer radiation intensity, and the short solid bars at the bottom identify when marine-sediment cores indicate the existence of strong monsoons. (Adapted from Prell and Kutzbach 1992, with the marine-sediment data based on Rossignol-Strict 1983.)

below it (115 000 years BP), with the period of the variation being dominated by that of the precession cycle: 25 000 years. During that period, marine sediment cores have identified four distinct maxima in the strength of the monsoon in this area during interglacial periods (short, solid bars at bottom of Fig. 15.12). These maxima are strongly correlated in space across Africa and Asia, and coincide closely with the four largest precession-caused peaks in the summer radiation intensity. Global-model simulations were performed, and quantified the response of the monsoon and the associated hydrologic system to the variations in solar input. In particular, the system responded non-linearly in the sense that a certain percentage increase in the solar-radiation input produced a percentage increase several times larger in the rainfall. This large response is partly attributed to the non-linear relation between temperature and saturation vapor pressure. Prell (1984) shows how this variation in the intensity of the southwest monsoon flow over the Indian Ocean produced a documented historical response in the coastal upwelling strength and the resulting sea-surface temperatures off the southeast coast of the Arabian Peninsula and in the eastern Gulf of Aden.

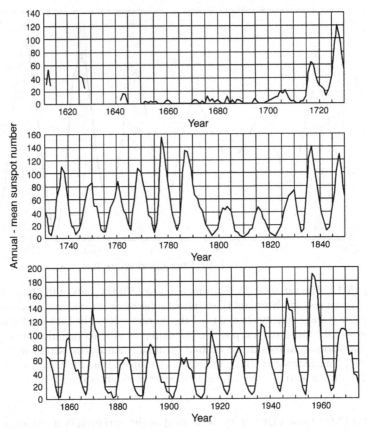

Fig. 15.13 The annual mean sunspot number from AD 1645 to 1980. The minimum from 1645 to 1715 is known as the Maunder Minimum. (From Eddy 1977, based on Waldmeier 1961 and Eddy 1976.)

Changes in solar output

The cyclic behavior of solar activity is most often discussed in the context of the number of dark spots on the sun; this number has a dominant periodicity of about 11 years. Other longer-period cycles are also apparent in the sunspot record. It is speculated that this sunspot number is a visible symptom of the solar-energy output. This number, along with other types of evidence of longer-term variation in solar output, provides us with qualitative information about the energy input to Earth's climate system. A recent history of the sunspot number is shown in Fig. 15.13. The virtual absence of sunspots from 1645 to 1715 is known as the Maunder Minimum. The period of the Maunder Minimum corresponds with the coldest extreme of the Little Ice Age, and the rise in the level of solar activity during the first half of the nineteenth century parallels the observed rise in global-mean temperature during that period.

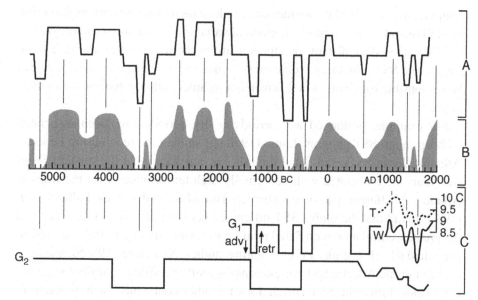

Fig. 15.14 Correlation of carbon-14, an indicator of solar activity, with glacial cycles and weather in Europe. Curve A shows the carbon-14 variation, based on tree-ring analysis, during the past 7500 years, with downward excursions of the curve corresponding to higher carbon-14 concentrations and lower solar activity. Curve B represents a schematic of the smoothed possible solar-activity level that corresponds to curve A. The curves in C represent four estimates of past climate. Curve G_1 shows the periods of advance and retreat of Alpine glaciers (based on Le Roy Ladurie 1967), and G_2 shows the same for worldwide glaciers (based on Denton and Karlén 1973). Curve T is an estimate of mean annual temperature in England (scale on right, based on Lamb 1972), and curve W is a winter-severity index (downward is colder) for the Paris–London area. (From Eddy 1977.)

Longer-term evidence of solar activity is provided by carbon-14 concentrations in tree rings. Carbon-14 anomalies, which have an inverse relation to solar activity, are shown in Fig. 15.14 (curve A) for roughly the past 7500 years, together with some estimates of past climate. The advance of glaciers, and of cold weather in western Europe, both show good positive correlation with solar-activity minima. Given the previous discussion of the effects of glacial cycles on the climates of arid areas, solar activity should have important controls.

Sea-surface temperature variation

It should not be surprising that the temperature of Earth's surface, whether it be land or sea, can have profound effects on atmospheric processes through the associated heat-flux patterns. For example, we have seen how high surface temperatures over land in summer can produce monsoon circulations that bring rainfall to arid areas. Similarly, horizontal variations in sea-surface temperatures (SSTs) can produce distant signatures in atmospheric processes. Because some large-scale SST structures evolve on seasonal and longer time scales, they can be responsible for changes in downstream weather patterns that last for months or years.

For example, prolonged wet periods or dry periods have been associated with upstream SST patterns. Lamb (1978a,b) studied the relationship between Atlantic Ocean SSTs and Sahel rainfall. He concluded that the western Sahel wet years are associated with abnormally high temperatures over the tropical Atlantic. A different pattern was characteristic of Sahel dry years. Folland *et al.* (1986) found that the global SST difference between the two hemispheres had the greatest correlation with Sahel rainfall; Fontaine and Bigot (1993) should be consulted for a list of many corroborating studies. Nicholson (1989b) discusses the rainfall and Atlantic SST correlations for southern Africa. Many investigators have studied drought–SST correlations for other continents, such as Ratcliffe (1981) for Europe, and Namias (1989, 1991), Palmer and Brancovic (1989), and Hong and Kalnay (2000) for North America.

Suggested general references for further reading

Berger, A. (Ed.), 1981: *Climate Variations and Variability: Facts and Theories* – describes the mathematical and physical basis of climate, mathematical techniques used in climate reconstruction, reconstruction of past climates, theories of climate variations and their modeling, humans' impact on climate, and climate impacts on people.

Berger, A., *et al.* (Eds.), 1984: *Milankovitch and Climate: Understanding the Response to Astronomical Forcing*, volumes 1 and 2 – contains papers dealing with orbital and insolation variations, geologic evidence for long-term climatic variations at astronomical frequencies, modeling long-term climatic variations in response to astronomical forcing, and a summary of climatic variations at astronomical frequencies.

Gribbin, J. (Ed.), 1978: *Climate Change* – discusses past climates, the global heat budget, astronomical influences on climate, climate modeling, and human effects on climate.

Lamb, H. H., 1972: *Climate: Present, Past, and Future.* Volume 1: *Fundamentals and Climate Now* – reviews radiation, atmospheric circulations, seasonal variability, the oceans, and the water cycle.

Lamb, H. H., 1977: *Climate: Present, Past, and Future.* Volume 2: *Climatic History and the Future* – describes human awareness of climate changes, evidence of past weather and climate, climate and the long history of the Earth, the Quaternary ice ages and interglacial periods, postglacial times, climate in historical times, and climate since instrument records began.

Pittock, A. B., 1983: *Solar variability, weather and climate: an update* – summarizes progress in the study of the effects of solar variability on climate, and contains a good list of references.

Thomas, D. S. G., 1997d: *Reconstructing ancient arid environments* – provides discussions of various types of evidence of arid-zone expansions and contractions, dating arid-zone fluctuations, and the characteristics and causes of Late Quaternary arid-zone extensions.

Trenberth, K. E. (Ed.), 1992: *Climate System Modeling* – a general introduction to climate modeling, including sea-ice and ocean modeling.

Wilhite, D. A., 2000a,b: *Drought – A Global Assessment*, Volumes 1 and 2 – discusses droughts from the context of background and concepts, causes and predictability, monitoring and early warning techniques, impacts and assessment methodologies, adjustment and adaptation strategies, and drought management.

Williams, M. *et al.*, 1998: *Quaternary Environments* – contains a chapter on Quaternary climate change in arid areas.

Questions for review

(1) Explain the dynamics by which solar output affects the strength of monsoon circulations and their effect on arid areas. Are both the summer and winter components of the monsoon affected?

(2) Summarize the different forensic methods by which past climates can be estimated.

(3) Describe the dynamical reasons by which the penetration of ice sheets into mid-latitudes causes a change in the horizontal and vertical wind circulations.

(4) Based on information from other sources, explain the atmospheric and oceanic changes that are associated with El Niño and La Niña phases of the ENSO cycle.

(5) Describe orbital precession and nutation, and illustrate your discussion.

Problems and exercises

(1) Speculate on how continental drift, and the resulting spatial arrangement of the continents, could have possibly affected the distribution of deserts.

(2) By what mechanisms has mountain-building, and erosion of mountains, affected the distribution of deserts on geological time scales?

(3) Transitive and intransitive climate systems were discussed in an earlier chapter. Review the concept in terms of climate change.

(4) Draw a schematic latitude-vertical cross-section of the wind circulation from the pole to the equator, for current conditions and what it may look like in a glacial maximum.

(5) How might the development of deep glaciers influence the development of downwind aridity?

16

Severe weather in the desert

A paleontologist and an adventurer describe their experiences with the brutality of Gobi and Sahara Desert sand storms, respectively, and reveal some of the psychological effects of being in one.

I could hardly breathe. Seemingly a raging devil stood beside my head with buckets of sand, ready to dash them into my face the moment I came up for air out of the sleeping bag. There was something distinctly personal and living about the storm. All of us felt it. It was not just a violent disturbance of the unthinking elements. It acted like a calculating evil beast. After each raging attack, it would draw off for a few moments' rest. The air, hanging motionless, allowed the suspended sand to sift gently down into our smarting eyes. Then with a sudden spring the storm devil was on us again, clawing, striking, ripping, seeming to roar in fury that any of the tents still stood.

Roy Chapman Andrews, American paleontologist
The New Conquest of Central Asia (1932)

. . . when the camels, craning their long necks, sniff high in the air and utter a peculiar cry, the garfla (caravan) *men know well the ominous signs; far off on the horizon, creeping higher and higher, the sky of blue retreats before a sky of brass . . . high in the air great flames of sand reach out, then the lurid sand cloud, completely covering the sky, comes down upon the garfla. The torment at times is indescribable, and some poor fellow, like the camels, will run maddened into the hurricane. The sand storm lasts from a few hours to six or seven days, and during it the men lie thus, occasionally digging themselves to the surface, as they become partially covered with sand.*

Charles Wellington Furlong, American adventurer and writer
The Gateway to the Sahara (1914)

If one defines severe weather as simply conditions that are threatening to life, even the daily desert sky, barren of clouds, would be included because of its causal association with the temperature extremes, the desiccating humidity, and the lack of liquid water. Acknowledging that even these undisturbed conditions would be severe for the unprepared resident or traveler of the desert, we will limit the discussion here to more disturbed weather that is especially challenging for survival. This disturbed weather falls into two categories: high winds with associated dust and sand storms, and rains that produce flash flooding.

Dust storms and sand storms

Dust suspended in the atmosphere, and dust storms, are common in arid and semi-arid areas because most of the surface is covered by mobile, loose sediments. Even though most of us think we know a dust or sand storm when we see it, there is a formal international meteorological definition. When the visibility is less than 1000 m, and when the dust is being entrained into the atmosphere within sight of the observer, then a dust storm is reported. However, the *popular* usage of the term includes a dense dust cloud that has been *previously* produced somewhere upstream. In spite of this international standard, many national weather services and investigators have their own particular definitions. For example, Tao *et al.* (2002) report on the criteria used in Inner Mongolia:

- *Dust storm* – at least three stations reporting horizontal visibility of less than 1000 m and an average wind speed of 10.8 to 20.7 m s^{-1};
- *Strong dust storm* – at least three stations with horizontal visibility of less than 500 m and an average wind speed of 17.2 to 24.4 m s^{-1};
- *Very strong dust storm* – at least one station with horizontal visibility of less than 50 m and an average wind speed of 20.8 m s^{-1} or greater.

When the horizontal visibility is *greater than* 1000 m and moderate winds are causing dust to be elevated into the atmosphere at a location, the condition is defined as *rising dust* rather than as a dust storm (Abdulaziz 1994). When dust has been entrained into the atmosphere at some upstream location, it is locally defined as **suspended dust**.

Dust elevated into the atmosphere by dust storms has numerous environmental consequences. These include contributing to climate change; modifying local weather conditions; producing chemical and biological changes in the oceans; and affecting soil formation, surface water and groundwater quality, and crop growth and survival (Goudie and Middleton 1992). Societal impacts include disruptions to air, land and rail traffic; interruption of radio services; the myriad effects of static-electricity generation; property damage; and health effects on humans and animals.

The extremely arid desert is almost exclusively an **aeolian** environment, where particles are transported primarily by the wind rather than by a combination of wind and water. The particulate material originates from the chemical and mechanical weathering of rock. Most of the aeolian particles in deserts are silt and not sand, silt being composed of particles of much smaller size and mass (see Chapter 7). The finer clay and silt particles, with diameters of up to 0.05 mm, are elevated into the atmosphere as "dust" and suspended for long or short distances, depending on the amount of wind energy. Larger particles (e.g. sand) are entrained into the airstream only at higher speeds, do not remain airborne for long, and sometimes accumulate as dunes. The particles that are too heavy to be entrained remain behind and form an armor on the surface. Thus, in spite of the popular symbolic connection between sand storms and deserts, the laden winds of a "sand storm" are in fact generally heavy with the smaller silt particles.

Dust storms and sand storms require both a dry, granular surface substrate, as well as winds to elevate the grains. Chapter 3 describes the meteorology and names of the specific local wind systems of the different deserts that are associated with dust storms. In spite of the varied dynamics, it is possible to divide the wind-generating processes into convective-scale and large-scale. On the large scale, the pressure gradients associated with synoptic-scale weather systems force the winds, and the high winds and dust or sand storm can last for days. On the convective scale, cold-air downdrafts are produced by the frictional drag of falling rainfall and the cooling associated with precipitation evaporation in the unsaturated sub-cloud layer. When these downdrafts near the ground, they spread horizontally outward from the thunderstorms, as shallow fast-moving density currents with strong winds. The durations of such sand or dust storms would be typically an hour or less.

The dangers associated with sand storms and dust storms are numerous. First, the amount of solid material in the air is sufficient to cause suffocation if inhaled and accumulated in the respiratory tract. Secondly, the high wind exacerbates the normal desiccating effects of the high temperature and low relative humidity, with the clothed human body losing up to a liter of water per hour in such conditions (Chapter 19). Bodies recovered soon after such storms have been completely mummified. A third risk is that of live burial if the sand is allowed to drift over the body.

The physical manifestation of dust storms and sand storms depends upon whether dust or sand particles are involved, and whether the winds are of large scale or convective origin. If sand only is involved, and not dust, the particles will generally not be elevated through as deep a layer. In addition, the air will clear more rapidly after the high winds cease. As wind speeds increase early in the event, a few grains of sand begin to move within the first centimeter above the surface. With higher winds, individual narrow streams of sand merge into an unbroken thin carpet that moves at great speed and obscures the ground from horizon to horizon. The depth of the current of moving sand depends upon the wind speed and the size of the sand grains. George (1977) states that depths of about 2 m are typical of his experience in the Sahara, with the top of the layer being quite distinct. His account follows, of one long-lived sand storm event in the Sahara.

The spectacle I beheld was unique. I stood under a blazing sky and looked down at the top of a seemingly infinite sea of sand flowing along at high speed. The brow of the hill on which I stood formed an island in the midst of this surging golden yellow sea. At some distance I could make out the brows of other hills, other "islands." But the most curious part of the experience was the sight of the heads and humps of several camels that appeared to be floating on the surface of the sand, like ducks drifting in the current of a river. The camels' heads and humps, all that showed above the upper limit of the sandstorm, were moving in the same direction as the drifting sand, but a great deal more slowly.

With very high winds, the depth of the layer of elevated sand can far exceed the 2 m in this example. The existence of clear sky above the sand in the situation just noted means that no dust was elevated throughout the boundary layer, before the winds reached the speed at which the sand was entrained into the air. Also, because the sky was cloud-free and the sand storm was long-lasting, these winds were likely forced by a large-scale pressure gradient, and not by a convective event. In contrast, a convectively forced sand/dust storm in the Gobi Desert of Mongolia is described in Man (1999).

Now the storm was coming for us with a sort of focused rage that had not been there when I first noticed it. To our left, that hard blade was the leading edge of a great grey-brown dome of dust that formed a semicircle, perhaps 15 km across and nearly a kilometer high. It was the mouth of the beast, and it was ringed above by a mantle of charcoal clouds, and inside it thunder boomed and lightning flashed. Round us, the desert stirred with the storm's first menacing breath. It was coming at us fast, at a gallop. As the edge of the disc ate up the desert to either side of us, the sky darkened, the distant mountains vanished, and the clouds began to close around us.

Andrews (1932) describes another brief but extreme dust storm in the Gobi:

. . . the air shook with a roar louder than the first, and the gale struck like the burst of a high-explosive shell. Even with my head covered, I heard the crash and rip of falling tents . . . For 15 minutes we could only lie and take it. The sleeping bag had been torn from under me and the coat of my pajamas stripped off. The sand and gravel lashed my back until it bled . . . Suddenly the gale ceased, leaving a flat calm.

Because of the shorter life time of convective events, the most severe part of the latter storms was over within a half hour.

The immense amount of sand and dust transported in these storms, over short and long distances, is a testimony to their severity and energy content. A dramatic example is related to the great wealth of unprecedentedly well preserved dinosaur bones that have been found in the Gobi Desert. It is speculated that suffocation in sand storms is an explanation for the fact that many of the deaths seem to have occurred in some cataclysm, even though with apparently none of the trauma associated with floods and other severe events. In the Museum of Natural History in Ulaanbaatar, Mongolia, are the fossilized bones of two dinosaurs that are still locked in combat, the way they died from some other act. The theory is that, in the Late Cretaceous, 80 million years ago, the area was a richly populated mixture of desert and savanna, with lush marshlands, mudflats, and lakes. However, periodically, the placid environment of the mammals and reptiles was abruptly interrupted by a tidal surge of sand from the sky, carried aloft by some severe Cretaceous storm system. The weight of the winds and the sand deluge caused dunes to collapse and sand to accumulate rapidly, repeatedly suffocating and preserving a wealth of fauna. There are sand storms today of this apparent severity.

The physics and characteristics of sand and dust transport

Sediment movement is a function of both the power of the wind and the characteristics of the grains that hold them in place. The effect of the wind is twofold. First, the air flowing over the grains causes a decreased pressure on top. Because the pressure under the grain is not affected, there is an upward force. This causes the grains to be "sucked" into the airflow. There is also a frictional drag effect between the wind and the grain that creates a force in the direction of the wind. In opposition, three effects tend to keep the particles in place: weight, packing, and cohesion. Packing is simply the physical forces of particles on each other when they are compacted into a shared volume. Cohesion represents any process that causes particles to bind to one another. One example would be surface tension effects among particles in damp soil. There are also organic and chemical compounds that can cause grains in desert surface crusts to adhere to each other. The weight simply represents the effect of gravity, where the force is proportional to the mass of a particle. As the effect of the wind overcomes the cohesive forces, particles begin to shake, and then lift into the airstream. Chepil (1945) shows that particles with greater diameter and with greater density have a higher wind threshold required for entrainment into the airstream. However, particles with diameters smaller than about 0.06 mm (about the size boundary between silt and sand) also have a high threshold because there are greater electrostatic and molecular forces of cohesion, the smaller particles retain moisture more efficiently, and the smaller particles are shielded from the wind in the spaces among the larger particles. Thus, the particles that are most effectively entrained into the airstream have diameters between about 0.04 and 0.40 mm (Wiggs 1997). There are many factors that complicate this analysis, however, such as substrate moisture, the mixtures of particle sizes, surface slope, vegetation, and surface crusting.

Substrate moisture is an especially poorly understood factor with respect to its effect on the particle entrainment threshold. The general concept is that the surface tension associated with pore moisture binds the grains together, and the entrainment threshold is higher for a wet than for a dry substrate. However, there is considerable disagreement among studies in terms of the particular sensitivity of the entrainment threshold to the moisture content. This is at least partly due to the fact that the moisture effect is sensitive to the grain-size distribution. Wiggs (1997) has a more detailed discussion of this effect.

There are two general mechanisms by which particles can be transported by the wind. For small particles with diameters of less than about 0.06 mm, the settling velocity is low and turbulence can mix the particles upward throughout the boundary layer, where they can remain suspended for many days. Such suspensions of dust can be sustained by convective boundary-layer turbulence, even in the absence of significant large-scale winds for elevating more dust. Because this dust can limit visibility to a few meters, and in a gentle breeze it moves like a mist, it has been called a "dry fog" (George 1977). Of course, the term fog is inappropriate in a literal sense because there is no condensate of water, so the sometimes-used terms dust haze, dry haze, or harmattan haze might be preferred. If a traveler is reliant on the sun or landmarks for navigation, the threat can be very great indeed.

Larger particles move through three different modes of **contact**, or **bedload**, **transport**: **saltation**, **creep**, and **reptation**. Saltation is the process by which grains are elevated from the grain bed, gain horizontal momentum through exposure to the higher-velocity airstream above the ground, descend back to the grain bed and impact other grains, and bounce upward again into the airstream (Anderson 1987). Some of the momentum from the original grain is transferred to the other grains by the impact, and they may be ejected from the grain bed into the airstream. When the newly ejected grains from this "splash" effect have sufficient momentum to enter the airstream, they may begin saltation themselves. When these splashed grains take only a single hop, the process of grain movement is called reptation. Fig. 16.1 shows a schematic of the reptation process. Lastly, surface creep describes the rolling of larger particles as a result of wind drag and the impact of saltating or reptating grains. Thus, most of the larger-particle transport takes place near the surface, with the particle concentration decreasing exponentially with height. For hard surfaces, Pye

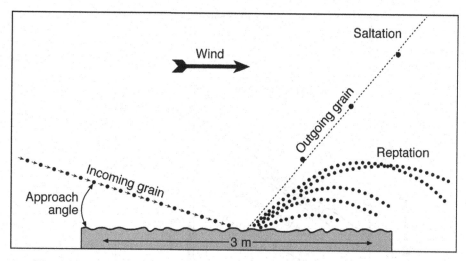

Fig. 16.1 The process of reptation, where a high-speed saltating grain ejects other grains from the surface. The vertical scale in this schematic is exaggerated for graphical clarity.

and Tsoar (1990) suggest a maximum saltation height of 3 m, with a mean height of 0.2 m. Heathcote (1983) estimates that 90% of sand is transported in the lowest 50 cm of the atmosphere, Wiggs (1992) measured up to 35% of sand transport taking place in the lowest 2.5 cm, and Butterfield (1991) states that 80% of all transport occurs within 2 cm of the surface. Clearly the wind field and the particle size strongly control the depth over which large particles are transported.

Figure 16.2 summarizes the type of particle transport in terms of the grain diameter and the friction velocity (sometimes called shear velocity) of the wind (see Chapter 10). The ratio of the fall velocity of the particles to the friction velocity of the wind is a criterion that can be used to separate the typical degree of particle suspension in the airstream (Chepil and Woodruff 1957; Gillette *et al.* 1974). Suspension of particles clearly is favored for smaller particle sizes and higher friction velocities. The short-term suspension category implies time periods of a few hours, whereas long-term suspension is days to weeks. In spite of the implication from this figure that sand particles only remain in suspension for very short periods, even for large friction velocities, there have been numerous observed instances of the long-distance transport of large particles. For example, Glaccum and Prospero (1980) found 0.09 mm diameter quartz grains over Sal Island in the Cape Verde Group, more than 500 km off west Africa; and Betzer *et al.* (1988) collected large numbers of quartz grains with diameters greater than 0.075 mm over the Pacific Ocean, more than 10 000 km from where they were

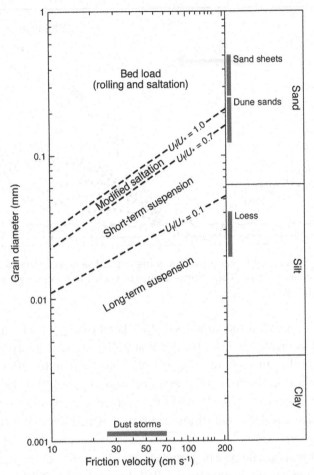

Fig. 16.2 The dependence of the type of sediment transport on the grain diameter and friction velocity. U_f is the settling velocity and U_* is the friction velocity. The modified saltation region represents a boundary zone between suspension and saltation processes. (Adapted from Tsoar and Pye 1987.)

entrained in an Asian dust storm. Sarre (1987) summarizes the threshold friction velocities required for entrainment of larger sand particles, based on a number of literature sources. There is much yet to be understood about the entrainment and transport of large particles. The correlation between the measured rate of sand transport and the friction velocity is shown in Fig. 16.3. There is a lot of scatter in the data points, because of the factors discussed above. Further discussion of the relations between threshold friction velocity and surface characteristics such as soil particle size and surface roughness can be found in Marticorena *et al.* (1997).

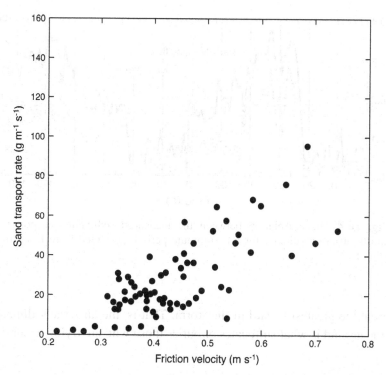

Fig. 16.3 Measured relation between friction velocity and sand transport rate. (Adapted from Wiggs 1997.)

Convective-scale and turbulence-scale variations in the wind speed have a strong effect on sediment transport, as can be anticipated from Fig. 16.3. Convective-scale **gust fronts** are caused by downdrafts emanating from the base of precipitating convective clouds, and these wind-speed spikes are likely responsible for much of the elevated sand and dust associated with haboobs. Also, turbulence intensity is strong near the surface when wind speeds are high, whether from synoptic-scale or mesoscale forcing, and the turbulent eddies produce much irregularity in the wind speed and direction. So, even though the time-averaged wind speed, as might be reported by an anemometer, is below the threshold for particle entrainment into the airstream, there might still be considerable transport as a result of the wind-speed peaks. This is illustrated in Fig. 16.4 in terms of the correlation between the measured friction velocity and sediment transport rate during a two-minute period. During this short time interval, the sediment transport rate varied by a factor of ten.

It should be noted that the near-surface transport of large grains, even though perhaps not as visually dramatic as the kilometers-deep clouds of dust, may

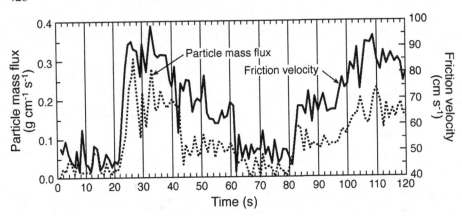

Fig. 16.4 The correlation between the measured sediment transport rate and the friction velocity for a two-minute period. (Adapted from Butterfield 1993.)

represent the greatest hazard of the storm. That is, the airborne sediment load near the ground is what suffocates or buries those who are trying to survive.

The effects of sand and dust on storm characteristics

A relatively subtle effect of the grains ejected from the surface into the atmosphere is an additional drag on the atmosphere. Recall that saltating grains gain momentum from the wind when they enter the airstream, but this means that there is a compensating decrease in the momentum of the air. This modifies the velocity profile, effectively adding an additional "roughness" effect beyond that associated with the surface itself (McKenna-Neumann and Nickling 1994). See Wiggs (1997) for a summary of this effect.

In Chapter 2 are described the radiative effects of desert dust. However, the radiative time scales involved are generally too long to significantly influence the local environment of a dust storm. However, another effect is thought to be more important. In Chapter 5 it is shown how much higher the daytime substrate temperatures are than the near-surface air temperatures. When dust and sand are elevated from this extremely hot surface, some of their heat is transferred by conduction to the atmosphere, raising the temperature. This effect is discussed by El-Fandy (1949). Also, Chen *et al.* (1995) discuss the effect of dust radiative heating on low-level frontogenesis within the same dust-elevating storm, and show that the dust can have a great impact on the dynamics of the storm that generated it.

Electrical effects within dust storms and sand storms

It is well known that lightning discharges occur in the dust clouds over volcanic eruptions, and there is evidence of lightning within large dust storms on other planets (Uman 1987). Within dust storms and sand storms on Earth, there have been numerous scientific observations of strong atmospheric electrical charges (see, for example, Kamra 1972), as well as much anecdotal data on the response of humans and equipment to electrical effects in deserts. Stowe (1969) reports potential gradients of 20 000 to 200 000 volts per meter in the lowest meter above the ground in a Saharan dust storm. Within sand storms, point discharges of up to 1 m in length have been observed (Uman 1987), and Kamra (1969) reports a "feeble lightning discharge" within a dust storm in India. Note that these electrical phenomena can occur within dust storms that are not associated with cumulus clouds, and thus any discharges observed are not related to the traditional mechanisms for the production of lightning in thunderstorms.

There are numerous personal accounts of the electrical effects of dust storms and sand storms. Clements *et al.* (1963) report that ignition systems of automobiles have failed to work unless the frame was grounded, telephone systems have failed, and workmen have been knocked to the ground by electrostatic voltages that have developed. George (1977) reports an experience wherein walking in a sand storm produced an extreme headache, that was largely relieved when a metal rod was used as a walking stick (and an electrical ground).

Dust devils

Dust devils are rotating updrafts of buoyant air that are otherwise unrelated to tornadoes or waterspouts. Indeed, they commonly form in fair-weather daytime conditions, over strongly heated surfaces. Occurrences tend to be most common during the hottest time of the day, between noon and mid-afternoon; when skies are relatively cloud-free; on the hottest days; and when the lower-tropospheric lapse rates are most unstable (Sinclair 1966). Horizontal velocities in dust devils are typically about 10 m s^{-1}, relative to 50 m s^{-1} for waterspouts and 100 m s^{-1} for tornadoes. Depths of dust devils are characteristically 100 m, and diameters are tens of meters (Stull 1988). They are visible because the horizontal wind speeds are high enough to entrain surface dust, and the mean upward motion in the outside of the vortex, and the turbulence, cause the dust to rise. A weak downdraft, consisting of less-dusty air, is generally seen in the center of the vortex. It is likely that invisible forms of these vortices exist just as frequently over non-dusty ground. The dust devil has a translational speed that is determined by the mean environmental wind speed averaged over its depth, with the vortex tilting in the direction of motion because of the increase in the speed of the mean wind with height. There is apparently no preferred direction of dust-devil rotation, presumably because they result from enhancement of random occurrences of rotation in the environmental flow. Path lengths and lifetimes of dust devils tend to be short. Small dust devils (diameter < 3 m) generally last less than one minute, while large ones (diameter of 15–30 m) have typical lifetimes of two to four minutes (Sinclair 1966). But, there are exceptions. For example, one dust devil reported over the broad alkali flats of the Great Basin Desert was over 600 m tall, lasted 7 h, and traveled a distance of over 65 km (McNamee 1995).

Dust-storm hazards

Dust storms have detrimental physiological effects (see Chapter 19), and can represent hazards to road, rail, and air transportation. A few of the many possible examples, in the latter category, follow. Thirty-two multiple-vehicle traffic accidents resulted from haboob-related dust storms on Interstate Route 10 in Arizona, in North America, between 1968 and 1975 (Brazel and Hsu 1981). The safety and efficiency of the aviation system is also seriously impaired by dust storms. Severe pre-frontal dust storms in 1988 in South Australia caused many airport closures (Crooks and Cowan 1993). In 1973, a Royal Jordanian Airlines aircraft crashed in Nigeria, with 176 fatalities, as a result of thick harmattan-related dust (Pye 1987). A very severe dust storm, called a "black storm" in China, occurred in northwestern China in 1993. The direct economic costs resulting from structural damage to buildings, broken power lines, traffic accidents, etc., were equivalent to US$66 000 000. In addition, there were 85 fatalities, 31 missing, and 264 injured. Over long periods, blowing sand can cause serious abrasive damage to wood structures such as telephone and electric poles. The greatest abrasion is often found at a distance of about 0.20–0.25 m AGL, that is the height of the optimal combination of wind speed and particle concentration (Oke 1987).

Surface and meteorological conditions, and geographic areas favorable for the occurrence of dust storms and sand storms

Dust production is most favorable for surfaces that are free of vegetation; do not have crusts; are composed of sand and silt, but little clay; have been disturbed by human or animal activity; are relatively dry; and are free of chemical (e.g. salts) or organic particle cements (Middleton 1997). Such conditions are often found where recent climate change, water action, or human activity have mobilized or concentrated dust or sand. Significant natural dust sources are floodplains,

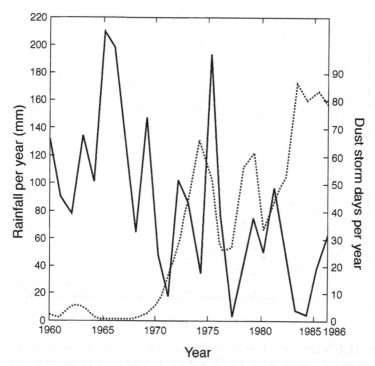

Fig. 16.5 The 25-year annual total rainfall trend (solid line) plotted against the number of dust-storm days per year (dotted line) for Nouakchott, Mauritania. (From Middleton *et al.* 1986.)

alluvial fans, wadis, glacial outwash plains, salt pans, other desert depressions, former lake beds, active dunes, devegetated fossil dunes, and loess (Middleton 1997). Many of these dust-conducive conditions are caused by the action of water.

The relationship between recent or long-term precipitation amounts and entrainment of dust has been extensively studied (see, for example, Brazel *et al.* 1986). Goudie (1983a,b), for example, found that dust-storm frequencies are highest when mean-annual precipitation is in the 100–200 mm range. Greater precipitation would generally cause (1) more vegetation, which would bind the soil and reduce the wind speed near the surface, and (2) a more cohesive moist soil. Where precipitation is below 100 mm, human disruption of the soil might not be as great as it is in the 100–200 mm range, where marginal agriculture is attempted at desert edges (Pye 1987). There are also clear correlations between short- and long-term droughts and dust-storm frequency. For example, Fig. 16.5 shows the 25-year rainfall trend plotted against the number of dust-storm days per year for Nouakchott, Mauritania. Increased anthropogenic disturbance of the surface during the period might have contributed to the rising

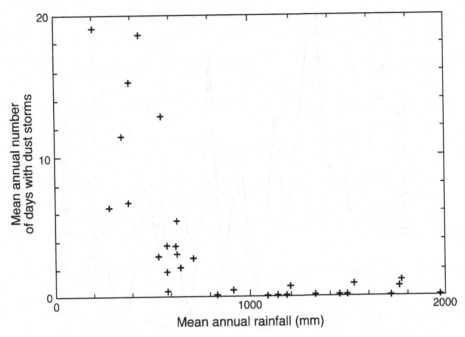

Fig. 16.6 Mean annual number of days with dust storms, as a function of mean-annual precipitation, for 30 stations in China. (Adapted from Goudie 1978.)

number of dust storms, but the rainfall deficit was likely the major factor. Similar plots for other locations are found in Goudie and Middleton (1992). Fig. 16.6 shows the appearance of a rough threshold effect in the relation between dust-storm frequency and mean annual precipitation for different locations in China. Goudie (1978) and Littmann (1991) show similar plots for other locations, with all illustrating a negative correlation between mean annual precipitation and dust-storm occurrence. Some have a similar threshold in the relationship. Holcombe *et al.* (1997) also show the effect of prior precipitation on threshold wind velocities.

In contemporary times, many geographic areas have been under sufficient climate stress and anthropogenic stress (e.g. agriculture) to have experienced at least mild to moderately severe dust storms. Some areas experience such storms with sufficient regularity that special names are associated with them. Middleton (1986) lists 70 such local names, and Middleton (1997) notes the geographic areas where the storms are the most frequent. Fig. 16.7 shows the major global dust trajectories that result from various atmospheric processes. In the desert southwest of North America, haboobs are a major source, whereas farther to

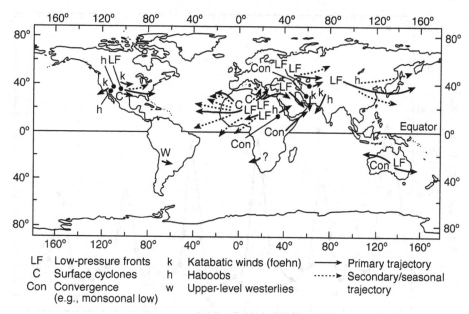

Fig. 16.7 Major global dust trajectories and the various atmospheric processes that cause the dust storms. (From Middleton *et al.* 1986.)

the north high-speed downslope winds (katabatic winds) are important. On the larger scale, strong pressure gradients around fronts and surface cyclones entrain dust in the semi-arid Great Plains. In South America, strong upper-level westerly winds may mix down to the surface, for example through the dynamic effects of the Andes Mountains, to elevate dust. Asia and Africa are major dust-storm regions, with fronts and surface cyclones being responsible for many of the storms. Monsoonal winds, katabatic winds, and haboobs are locally important. The dust-storm-day frequencies for some of the source regions indicated in the figure range from 20 to 80 per year. Further discussion of favored dust-storm regions may be found in Middleton (1997), Goudie (1978), and Littmann (1991).

Dust storms often have a strong seasonal variation in their occurrence (Littmann 1991) that is related to the annual distribution of precipitation (Yu *et al.* 1993), vegetation cover, freezing of the soil, meteorological systems with high winds, and agricultural activities that disturb the surface. Fig. 16.8 shows the number of official meteorological reports of dust at 750 stations for a six-year period in Asia. For this area, the maximum monthly dust frequency in spring is about five times greater than in other seasons of the year. The minimum in winter is because of frozen ground and snow cover. Duce (1995) shows the monthly distribution of dust storms for selected locations in Australia, illustrating about

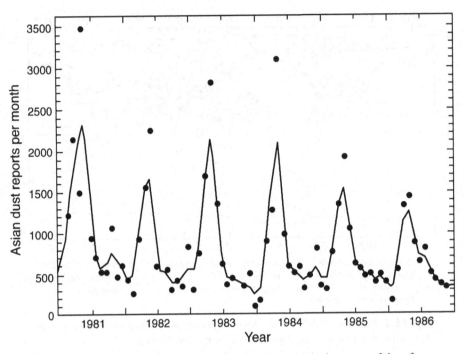

Fig. 16.8 The number of monthly official meteorological reports of dust for a six-year period in Asia. (From Prospero *et al.* 1989.)

a five-fold greater frequency in summer. A factor of five to ten greater area of global ultraviolet-absorbing aerosol has been measured for summer relative to winter (Herman *et al.* 1997). There is also a distinct diurnal maximum in dust-storm frequency in many places, when strong afternoon convection mixes higher-speed air to the surface or when thunderstorms cause haboobs (Hinds and Hoidale 1975). Membery (1985) shows the diurnal occurrence of dust storms in the Tigris–Euphrates flood plain in the summer (Fig. 16.9). Kuwait, near this source, commonly experiences low visibilities between about 1200 and 1900 LT. Bahrain experiences low visibilities from Iraq's dust in the early morning, after it has been transported 500 km to the southeast by the northwesterly shamal (Houseman 1961; Khalaf and Al-Hashash 1983).

Description of a haboob, a convection-generated dust storm

In a mature thunderstorm, the latent heat generated by condensation causes buoyancy-driven upward motion in only part of the cloud. In another part of the cloud, the frictional drag of the downrush of rain and hail causes air within some areas of the cloud to subside. This subsidence is enhanced because the rain falling

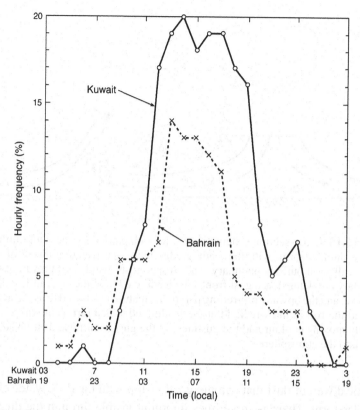

Fig. 16.9 The frequency distribution of hourly visibilities of 1500 m or less at Kuwait (near the dust source in the Tigris–Euphrates Valley), and of visibilities of 3000 m or less at Bahrain, about 500 km to the southeast, for June, July, and August 1982. These data are representative of the conditions in other years. Note that the Bahrain time axis is shifted from the Kuwait axis to account for the typical advective times scales associated with the dust transport from Iraq to Bahrain in the northwest wind. (Adapted from Membery 1985.)

through unsaturated air below the cloud cools and becomes more dense. When the cool, dense, rapidly subsiding air (called a **downburst**) reaches the ground, it moves outward away from the storm that created it, flowing rapidly along the ground as a **density current**. Fig. 16.10 shows a cross-section schematic of the cool outflow, with the leading edge called an **outflow boundary** that is propagating ahead of a mature thunderstorm. The strong, gusty winds that prevail at the boundary are defined as a **gust front**. The high wind speeds within this cold outflow region entrain dust into the atmosphere, and the turbulence associated with the large vertical shear of the wind mixes the dust upward, sometimes through a layer thousands of meters deep. The "dust front" propagates at up to 15 m s^{-1}, with an associated advancing, seemingly opaque,

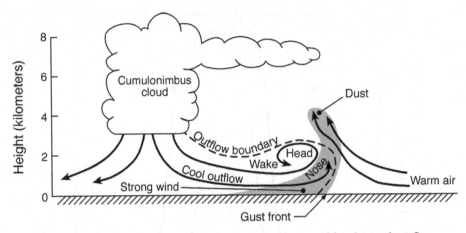

Fig. 16.10 Cross-section schematic of a haboob caused by the cool outflow from a thunderstorm, with the leading edge that is propagating ahead of the storm called an outflow boundary. The strong, gusty winds that prevail at the boundary are defined as a gust front. The leading edge of the cool air is called the nose, and the upward-protruding part of the feature is referred to as the head. Behind the roll in the windfield at the leading edge is a turbulent wake. The rapidly moving cool air and the gustiness at the gust front raise dust (shaded) high into the atmosphere.

near-vertical wall of dust that can merge at its top with the dark water-cloud of the thunderstorm. There is sometimes lightning visible through the dust cloud and above it, probably related to the thunderstorm but possibly also to the electrification of the dust cloud itself. When the dust storm arrives, visibility can be reduced to zero, but it commonly is 500–1000 m. Additional information about the outflow boundaries that are responsible for haboobs can be found in Wilson and Wakimoto (2001), Wakimoto (1982), and Droegemeier and Wilhelmson (1987).

Typical outflow boundaries last less than an hour, and this time scale thus represents the lifetime of most haboobs. This was confirmed in a study by Freeman (1952) of 82 haboobs in the Sudan. There are rare haboobs, however, that last much longer. A description of two long-lasting ones on the eastern Arabian Peninsula is provided by Membery (1985). Both occurred within a two-day period, reduced visibility to less than 500 m for more than 4 h, had propagation speeds of about 15–20 m s^{-1}, and produced dust storms that lasted for at least 9 h as they traversed hundreds of kilometers. That the motion of these dust storms resulted from the propagation of a density current is supported by the observed vertical profile of wind speed (Fig. 16.11). The large-scale wind speed at the 850 mbar level above the cold density current was only 2 m s^{-1}, whereas within the density current the air was moving at 15 m s^{-1}.

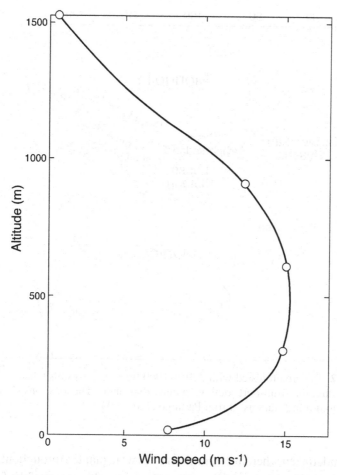

Fig. 16.11 Vertical profile of the low-level wind speed within a density current in the eastern Arabian Peninsula. (From Membery 1985.)

Mitigation of dust storms

In Chapters 14 and 18 are discussed how human impacts have made some places more like deserts, with the effect of increasing dust-storm frequency and severity. Thus, reversing the various factors that have led to desertification would contribute to a reduction in dust storms. There are also situations where dust storms from naturally occurring deserts affect populated areas, and measures have been taken to reduce their impact. For example, when cold fronts move southward from Siberia, the strong frontal winds entrain dust from the Gobi and Taklamakan Deserts, and the dust-laden air causes air-quality problems in downwind urban areas such as Beijing. With the intent of reducing the entrainment of dust at its source, and dust transport in the lower boundary layer, a massive forestation

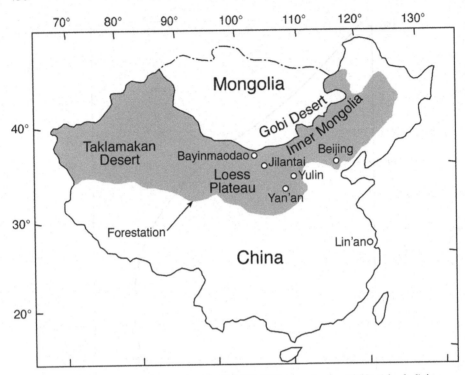

Fig. 16.12 The area forested with 300 000 000 trees in the 1950s (shaded) in order to reduce dust transport southward into urban areas. The locations shown are dust-monitoring stations. (From Parungo *et al.* 1994.)

project was undertaken wherein 300 million trees were planted throughout northern China in the 1950s. Fig. 16.12 illustrates the geographic extent of the forestation project, known as the "Great Green Wall." The map is indicative of only the general area of tree planting, since clearly trees will not grow in the central Taklamakan Desert. In any case, the intent of planting the new vegetation was to reduce the wind speed near the ground and increase soil moisture, and thus reduce the entrainment of dust in the source regions. Also, the trees would reduce the low-level wind speed downwind of the dust source, and thus some of the dust might settle out of the atmosphere more quickly. After the forestation, the dust storm frequency and duration in Beijing decreased dramatically. The frequency decreased from 10–15 events per month to almost none, and the duration decreased from over 50 h to less than 10 (Fig. 16.13). During this period, the overall rainfall and wind-speed climate for the upwind area did not change appreciably. At Bayinmaodao, near the upwind, northern edge of the forested area, there was very little change during this period in the dust storms' frequency

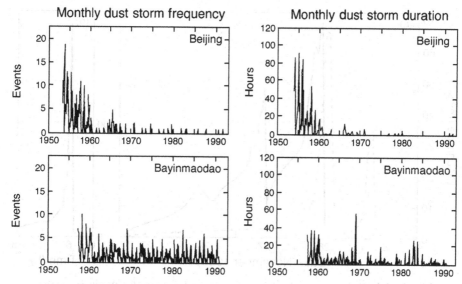

Fig. 16.13 Monthly dust-storm frequency and dust-storm duration for Beijing and Bayinmaodao, China. See Fig. 16.12 for locations. (From Parungo *et al.* 1994.)

and duration. There have been many other more modest attempts to modify the land surface so as to reduce the severity of dust storms.

Rainstorms, floods, and debris flows

Though desert weather is considered inherently severe, primarily because of a lack of rain, weather events that mitigate the dryness are ironically also often severe, both meteorologically and hydrologically (see Box 16.1). When rain does reach the ground, the surface is relatively impervious to water and devoid of vegetation that is sufficient to slow the runoff. The rapid concentration of this runoff from thunderstorms causes flash floods that account for considerable loss of life. It was noted earlier that more people currently perish in the desert from an excess of surface water, i.e. drowning in floods, than from a lack of it, i.e. thirst (Nir 1974). In general, compared with floods in humid climates, the river **hydrographs** rise more quickly, they have a sharper peak, and they recede more quickly. As evidence of the "flashiness"[16.1] of arid watersheds, in the United States the 12 largest flash floods all occurred in arid or semi-arid areas (Costa 1987), and more than half of railroad track washouts occur in the driest states.

[16.1] This refers to how prone the watershed is to the occurrence of flash floods. One definition of a flash flood is an event in which river discharge rises from normal to flood level within 6 h.

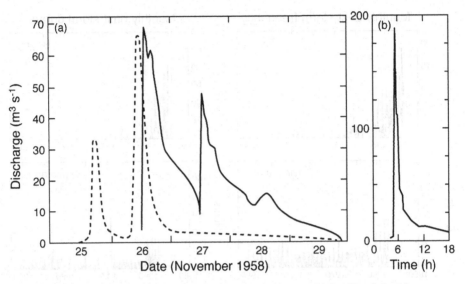

Fig. 16.14 Discharge as a function of time (hydrograph) for floods in (a) two central Saharan wadis associated with the same storm (adapted from Goudie and Wilkinson 1977), and (b) Tanque Verde Creek, Arizona, in the northern Sonoran Desert (adapted from Hjalmarson 1984.)

Box 16.1
Eyewitness to a Sahara flash flood

In seldom-visited areas of the arid world, the isolated rains and their hydrologic consequences often go unnoticed. For example, Cooke and Warren (1973) report on a flash flood in the Sahara witnessed by Medinger (1961). Storm-total rainfall that fell on the flat pavement surface of the Tadmaït Plateau was estimated by a nearby rain gauge to be about 16 mm (the annual average there is about 18 mm). It was calculated that 28% of the rain falling on the catchment contributed to the flood. The maximum flood discharge was estimated to be 1600 $m^3 s^{-1}$. For comparison, this discharge in a relatively small wadi was only about a factor of two smaller than low flows at the delta of the Mississippi River, and was similar to the winter base flow of the Nile River at Aswan before construction of the Aswan High Dam.

Fig. 16.14 illustrates hydrographs that are typical of desert flash floods, even though the rate of rise and fall of the curve and the width of the peak depend on many factors such as the size of the contributing watershed. Other examples of flood hydrographs in arid lands are found in Reid and Frostick (1997) and Knighton and Nanson (1997). It should not be assumed that most floods and flash

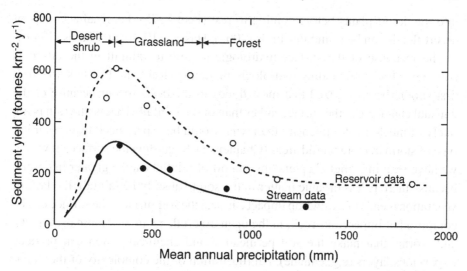

Fig. 16.15 Sediment yield in runoff, plotted as a function of mean annual precipitation. The empty circles are data based on rates of sediment accumulation in reservoirs, and the solid circles are based on sediment content of stream water. (From Langbein and Schumm 1958.)

floods in desert areas result from thunderstorm rainfall. For example, Kahana *et al.* (2002) show that 80% of the most severe floods in the Negev Desert during a 30-year period were associated with synoptic-scale precipitation events.

In desert floods, there is a spectrum of material involved. Some floods can consist of almost exclusively water, with little solid material. Others can be composed of 90% mud, boulders, and vegetation. Fifty-tonne boulders have been carried for many kilometers (Childs 2002). Toward the latter end of the spectrum, the floods are called mud flows or debris flows. A single flood can have both types of flow at different stages: mostly water, and mostly debris. In general, for a variety of reasons, the sediment load in the flood water in ephemeral desert rivers is often many times higher than in rivers in humid climates. This is shown in Fig. 16.15, where the sediment content in runoff is plotted against mean annual precipitation. The curves are extrapolated to show low sediment yield for extremely arid climates simply because the sediment must go to zero as the rainfall does.

When the flood material consists mostly of debris, it is much more viscous than water and moves much less rapidly, almost with the consistency of wet cement. In large debris flows, the leading wall is almost vertical and can be 10–15 m high. After the flow has exited its wadi or canyon, and spent its energy as it spread across a flood plain, it can dry with its near-vertical wall intact, sometimes five meters in height. Some floods have been known to run 400–500 km into the desert plains before coming to rest (Peel 1975). Personal accounts of desert floods can be found in Childs (2002), Hassan (1990), and Jahns (1949).

The response of the surface-hydrologic system to rainfall in the desert is complex, almost precluding hydrologic modeling (Reid and Frostick 1997). It first should be recognized that most floods in the desert occur because of the unusual character of the surface rather than of the rainfall. That is, the rain is not likely of much greater intensity than what would be experienced from a similar type of storm in more humid areas (Chapter 13). Regarding the surface, however, we have seen in earlier chapters how runoff of rainfall can be greater in deserts because there is less organic matter in the soil to absorb the rainfall; the lack of vegetation means that raindrop impact can seal the soil surface; the surface can be composed of impervious rock; in the open desert there are few animals, insects, and worms that make the soil permeable; and chemical action can produce impervious layers (e.g. caliche). An illustration of the complexity of the flood-generation process is how runoff is influenced by slope in the desert. Many studies have shown that slopes of 1–2° generate twice the runoff per unit area than do slopes of 20°. One contributing factor is that steeper slopes have had much of the fine material washed away, whereas this accumulated fine material on more horizontal surfaces permits the development of low-permeability crusts. Another counter-intuitive observation is that rain intensities in excess of 15 mm h^{-1} can produce less proportional runoff than does a rainfall of 3–10 mm h^{-1}. Again, the factor is surface-sealing-related in that heavy rainfalls can break up certain crusts (Peel 1975). Even though these and other factors make it perhaps difficult to be quantitative about desert runoff and flood production, the fact remains that flash floods are much more common in the desert, in general because more water runs off.

Other factors help determine the severity of floods that result from desert rain events. If the storm movement and the channel routing conspire so that the water arrives nearly simultaneously in the main channel from many directions, a major flood will result. In contrast, if the tributaries feed the water into the main channel over a longer period of time, the flood will not be as great. This factor is in addition to the obvious one that storms that remain stationary for significant periods, and pour water onto the same watershed, are more dangerous than storms that move rapidly and spread the water over a larger area.

Flood frequency and severity in the desert vary from year to year as much as does the rainfall that causes the floods. The extreme interannual variability in precipitation in deserts was documented in Chapter 13, and the longer-term climatic variation was discussed in the previous chapter. One of the cyclic patterns of the atmosphere that strongly controls rainfall and floods in some desert areas is the ENSO cycle. For example, 11 of the 14 major floods of the past century in the desert southwest of North America occurred in El Niño years (Childs 2002). On longer time scales, Ely *et al.* (1993) show that, over the past 5000 years, by far the greatest number of floods in Arizona (northern Sonoran Desert) and southern Utah (Great Basin Desert) occurred in the past 200 years when there have been more frequent strong El Niño events.

Suggested general references for further reading

Clements, T., *et al.*, 1963: *A study of windborne sand and dust in desert areas* – reviews the causes of dust storms and their properties, primarily for the northern Sonoran and Mojave Deserts.

Costa, J. E., 1987: *Hydraulics and basin morphometry of the largest flash floods in the conterminous United States* – describes hydraulic characteristics of large flash floods, and meteorological and physiographic controls on flood runoff.

Goudie, A. S., 1978: *Dust storms and their geomorphological implications* – primarily for the Middle East and Asia, discusses the frequency of dust storms, the dust content of the air, the distribution of dust deposition, the rates of sediment removal and deposition in dust storms, and the chemistry of dust.

Idso, S. B., 1976: *Dust storms* – a nontechnical description about dust storm properties and occurrences.

Knighton, D., and G. Nanson, 1997: *Distinctiveness, diversity and uniqueness in arid zone river systems* – discusses arid- and humid-region contrasts in river systems, river-system diversity within the arid zone, and the uniqueness of arid-zone river systems using Australia as an example.

Middleton, N., 1997: *Desert dust* – discusses the nature of desert dust, threshold velocities for dust entrainment, global geography of dust-storm frequency, meteorological systems, changes in dust-storm frequency, long-range transport, and environmental hazards of airborne dust.

Nickling, W. G., 1986: *Aeolian Geomorphology* – a collection of papers that includes discussions of particle transport in desert environments, dune formation, and the frequency and source areas of dust storms.

Peel, R. F., 1975: *Water action in desert landscapes* – provides a summary of some of the unique characteristics of desert landscapes that lead to runoff and flash floods.

Reid, I., and L. E. Frostick, 1997: *Channel form, flows and sediments in deserts* – describes rainfall and river discharge in terms of storm characteristics, the flash-flood hydrograph, and drainage basin size and water discharge; ephemeral-river channel geometry; and sediment transport.

Tsoar, H., and K. Pye, 1987: *Dust transport and the question of desert loess formation*

Wiggs, G. F. S., 1997: *Sediment mobilisation by the wind* – summarizes the nature of wind flow in the desert, sediment in the air, modes of sediment transport, sand transport modification of the wind profile, prediction and modeling of bulk sediment transport, and the role of turbulence. There is a substantial list of references.

Questions for review

(1) Explain the difference between suspension, saltation, and reptation.
(2) Describe the different meteorological phenomena that can produce dust storms.
(3) What causes diurnal and seasonal variability in dust-storm frequency?

(4) Explain the differences between typical dust storms and sand storms.
(5) What are the characteristic surface conditions that are conducive to dust-storm formation?
(6) How does the sediment load in rivers relate to the degree of aridity?
(7) Describe the meteorological characteristics of a haboob.
(8) Why did Bayinmaodao, China, show no effect of the Great Green Wall on dust-storm frequency and duration, whereas a significant impact was observed in Beijing?

Problems and exercises

(1) Speculate on how the "Great Green Wall" of China may have contributed to the reduced transport of dust downwind.
(2) What measures other than vegetation planting might be employed to reduce dust-storm severity in particular cases?
(3) Speculate on why hydrographs rise and fall more rapidly for desert rivers than for rivers in humid climates.

17

Effects of deserts on the global environment and other regional environments

Whether the flesh is forgotten or the mind becomes doubly sensitive to its prickings, whether the spirit grows drowsy or gains in lucidity, the desert is first and foremost a mirror in which one can see the world and maybe also glimpse the face of God. The only certainty is that sooner or later you will see yourself.

Mahin Tajadod, Iranian poet
A place of trials (1994)

These hours of the desert night, unsoftened by mild moonlight, have always seemed to me somehow terribly lonely, cold, and cruel . . . That night I did not feel as though I were looking up into the universe, but as though a new dimension had become visible and lay beneath me. I had the dizzying sense of plunging downward into infinite space. The merciless sensation of the coldness of space, of the vast void out there, was intensified by the absolute stillness of the desert night until it became inhuman, unbearable.

Uwe George, German naturalist and desert explorer
In the Deserts of This Earth (1977)

Heaven knows that flood can be frightening enough. But . . . The dust is different. . . It does not bring the sharp, quick, desperate terror of flood, but instead a slow, chilling, and pervasive horror, perhaps out of keeping with any immediate damage, but right enough in the long run.

Paul Sears, American author and environmentalist
Deserts on the March (1935)

We tend to think of deserts as isolated and remote from other parts of the planet. But, desert environments are actually intimately related in perhaps unexpected ways to distant places. This chapter describes some far-reaching processes that connect the world's deserts to the rest of the planet. In Chapter 15 is described the effect of large-scale, global processes, such as El Niño, on particular desert climates: the signal is from the global scale to the regional scale. In contrast, here will be described the impacts of deserts on non-desert areas in the same region as well as on global climate.

Global and regional transport of desert dust: background

One of the most widely studied physical connections between arid and non-arid areas is the airborne transport of desert dust to non-arid places that are oceans and continents away. Not only does elevated dust have extremely negative effects on air quality near the source, but there are a variety of major global impacts. Distant effects, both positive and negative, include the following.

- Visibility is greatly reduced, impacting surface and air transportation.
- Modifications to the vertical temperature structure of the atmosphere take place through radiative effects on the energy budget, influencing convective precipitation.
- The biogeochemistry of terrestrial and maritime ecosystems is influenced.
- Precipitation chemistry is affected.
- Cloud microphysical processes are impacted.
- The Earth–atmosphere albedo is modified through effects of suspended dust as well as dust deposited on snow and other surfaces.
- Human health is negatively affected through respiration of the finer dust particles.
- Soil formation is influenced.

For further reading, Perry *et al.* (1997) and Guerzoni and Chester (1996) provide a thorough list of references to these effects.

Satellite remote sensing and chemical analysis of samples obtained from surface-based networks (many are air-quality-oriented) allow reasonably accurate documentation of dust source–receptor relations. Dust sources can be related to deserts, as well as to anthropogenic effects such as biomass burning and wind erosion from tilled agricultural land, dried-up lakes, and construction sites. Most deserts are sources of some dust, but the degree to which they contribute depends on the local meteorology and the nature of the surface. Fig. 17.1 shows a satellite estimate of the occurrence of significant aerosols during a three-month period of the Northern Hemisphere summers of 1987 and 1988. In this figure, aerosols of desert origin extend across a wide latitude belt from the Caribbean to Mongolia, where the sources are the Sahara, the Sahel, the Middle Eastern deserts, and the Asian deserts east to the Gobi. The aerosols observed in South America and southern Africa during this period are primarily smoke from biomass burning. Over large areas of the Atlantic, dust was observed for more than half of the days in this period. In other satellite-derived aerosol maps it is common to see dust from other sources, such as in Australia and associated

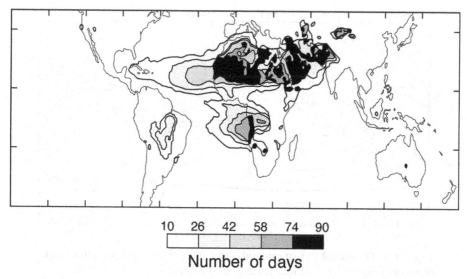

Number of days

Fig. 17.1 Distribution of ultraviolet-absorbing aerosol during July–September 1987 and 1988. The key indicates the number of days during which significant aerosol was observed by using the Nimbus 7 Total Ozone Mapping Spectrometer. (Adapted from Herman *et al.* 1997.)

with large dry lake beds such as near Lake Chad in the Sahel and the Aral Sea in Asia.

There are many examples of the long-range transport of desert dust.

- Australian dust is transported to both the east and the west, with a documented case in which over two million short tons were transported over 3000 km from western Queensland to New Zealand, where it blanketed the snowfields of the glaciers on South Island (Kiefert *et al.* 1996).
- Saharan Desert and Sahel dust from North Africa are transported westward to North and South America, and northward to Europe. This example of the global transport of desert dust is discussed in detail later in this section.
- Gobi Desert dust is transported in large amounts to the east and south. Such transport over geological time scales is why China owes its richest agricultural land to the Gobi winds and dust. Gobi dust even reaches the west coast of the United States.
- Dust from the Great Plains of the United States was transported as far to the east as Europe during the dust-bowl years of the 1930s.
- From the Thar Desert in eastern Pakistan and western India, dust is transported to the coast of east Asia.

In addition, there are hundreds of examples of regional rather than global-scale dust transport, such as dust from the dry bed of Owens Lake, California, being transported to Los Angeles, and there affecting the air quality. The global pattern of dust deposition is seen in Fig. 17.2, which illustrates the calculated fluxes of

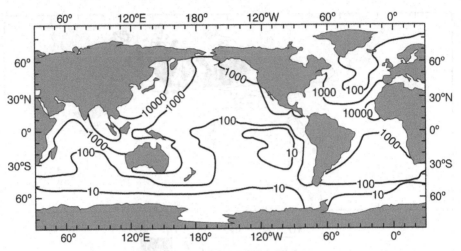

Fig. 17.2 The global flux of mineral dust from the atmosphere to the oceans, in units of mg m^{-2} y^{-1}. (From Duce *et al.* 1991.)

mineral dust to the world's oceans. The highest fluxes prevail over the Northern Hemisphere, where the most expansive deserts are located. In particular, maxima are to the east of Asia where the westerlies transport the dust, and to the west of North Africa where the trade winds are responsible for the transport.

Because the Sahara and Sahel are such prolific sources of dust, we will use North African deserts as a focus of discussion here. An early account of the westward movement of Saharan dust in the trade winds was given by Charles Darwin (1845), who describes his own observations from HMS *Beagle* in the Atlantic as well as those from others traveling the seas to the west of Africa in the early part of the nineteenth century. According to this early account, large amounts of dust fall in the Atlantic, to the point of greatly discoloring the water in a latitudinal belt of over 2500 km. The dust was observed by vessels as far as 1600 km out in the Atlantic, one-half the distance to South America. More recent evidence shows that, each year, large amounts of dust are transported from North Africa to Europe (Stevenson 1969; Guerzoni and Chester 1996; Schwikowski *et al.* 1995), North America and the Caribbean (Carlson and Prospero 1972; Prospero and Carlson 1972; Karyampudi and Carlson 1988; Westphal *et al.* 1987, 1988; Prospero 1996a,b, 1999), South America (Swap *et al.* 1992, 1996; Prospero *et al.* 1981), and the Middle East (Schütz 1980). The total amount of dust elevated into the atmosphere over northern Africa has been estimated to be as much as one billion short tons per year (Moulin *et al.* 1997). Over the Sahara, elevated dust is mixed throughout the boundary layer, which can be up to 6000 m deep in the summer season. The generally westward transport of the

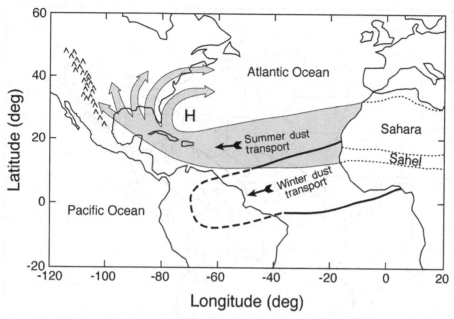

Fig. 17.3 Dust transport over the Atlantic Ocean from North Africa in the Northern Hemisphere summer and winter. (Adapted from Perry *et al.* 1997.)

dust across the Atlantic within the trade winds is periodically interrupted by easterly waves, which create a pulsation in the direction of the transport that diverts some of the flow and dust to Europe. Long-term analyses of particulates in Miami, Florida, and on Barbados show that Saharan dust is the dominant aerosol species in the months of June, July, and August. This summer maximum in Miami is related to the strength of the Northern Hemisphere southeasterly trade winds that prevail during this period. In the Northern Hemisphere winter, the dust is located farther south in the North Atlantic, and the source seems to be the semi-arid grasslands to the south of the Sahara, in the Sahel. Fig. 17.3 shows the seasonal evolution of the dust source–receptor relation across the Atlantic Ocean. A Bermuda High that is shifted westward from its normal location will cause dust intrusion into the central rather than the eastern United States. Fig. 17.4 shows the measured concentration of particulates with diameters of less than 2.5 μm in the eastern United States on 7 July 1993. Chemical analysis of the particulates, airmass trajectories, and the fact that the United States had been experiencing a moist period with no local sources of dust, confirm that the fine particles were from North Africa (Perry *et al.* 1997). This distribution is typical of large North African dust events, which can impact approximately 30% of the land mass of the United States. For the last half of the same month, Fig. 17.5

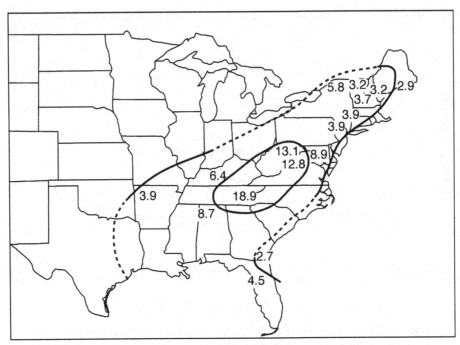

Fig. 17.4 Soil-dust concentrations for particles with diameters of less than 2.5 μm in the eastern United States on 7 July 1993. The observations are plotted in μg m^{-3}, and the isopleths are shown for 3 and 10 μg m^{-3}. (Adapted from Perry *et al.* 1997.)

illustrates the concentration of the 2.5 μm and smaller particles measured at Great Smokey Mountain National Park (where the 18.9 μg m^{-3} observation is located in Fig. 17.4). The four days of elevated concentrations during the last week correspond to an influx of African dust. Almost an order of magnitude increase above the background concentration is observed.

During especially dusty episodes in the southeastern United States and the Caribbean region, the dense mineral-dust haze has a yellow-brown color that is distinct from the bluish-gray of hazes from local pollution (Prospero 1999). Rainfall during these periods is turbid, and produces a filtrate of red-brown mud. There is some interannual variability in the concentration of the Saharan dust in Miami, which seems to be anticorrelated with annual rainfall in the Sahel (Prospero and Nees 1986). Nevertheless, it is sufficiently regular that, in an analysis of eight years of data for Miami, for each year the annual mean North Africa mineral dust loading was only exceeded by that associated with sea salt.

The impacts in South America of the dust plume are less well documented, except that from December through May the African dust is an important source

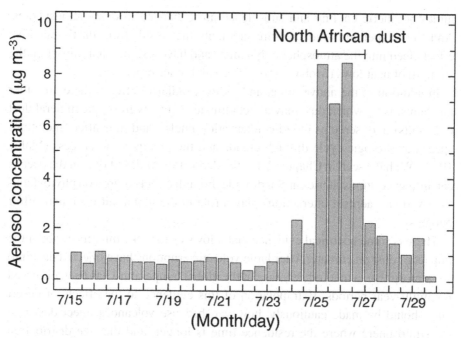

Fig. 17.5 Concentration of the 2.5 μm and smaller particles measured at Great Smokey Mountain National Park (where the 18.9 μg m^{-3} observation is located in Fig. 17.4). The four days of elevated concentrations during the last week correspond to an influx of African dust. Almost an order of magnitude increase above the background concentration is observed. (Adapted from Perry *et al.* 1997.)

of aerosol over the Amazon Basin, with 13 million short tons per year deposited there (Husar *et al.* 1997; Swap *et al.* 1992, 1996; Prospero *et al.* 1981; Talbot *et al.* 1986). In this area, the dust is considered to be an important source of soil nutrients.

Global climate effects of desert dust

Atmospheric mineral matter has a potentially important, but poorly understood, role in climate forcing. Dust radiative effects include the scattering and absorption of solar radiation, and the absorption of terrestrial infrared radiation. The effects on solar radiation are thought to be of greater importance. Radiative effects can be manifested in a number of ways. For example, the mean tropospheric temperature may be affected, and the vertical distribution of the heating may change the temperature lapse rate and therefore the stability of the atmosphere. This can have consequences in terms of precipitation. Recall from Chapter 2 that Bryson and Baerreis (1967) and Bryson *et al.* (1964) claim that midtropospheric cooling and subsidence are enhanced in dusty atmospheres. And the temperature and pressure gradients that result from dust's radiative effects feed into the atmospheric dynamics and have a significant impact on the strength of heat lows (Mohalfi *et al.* 1998) and other processes.

In addition to the above direct and indirect radiative effects, there are other mechanisms by which dust may affect climate. Some hygroscopic mineral dust (e.g. salts) may serve as cloud condensation nuclei, and thus affect the atmospheric hydrologic cycle through clouds and the precipitation process (Maley 1982). We have seen in Chapter 2 that the deposition of desert dust in the oceans can influence atmospheric carbon dioxide through effects of iron on phytoplankton. And gas–aerosol interactions play a role in the global nitrogen and sulfur budgets.

There is some historical evidence that allows us to make inferences about the impact of dust on climate. After large volcanic eruptions have injected dust into the atmosphere, a marked decrease in the global temperature has been observed for multi-year periods. Extrapolation of this evidence to the effects of desert dust should be made cautiously, however, because volcanoes inject dust into the stratosphere where the residence time is longer, and the size distribution and mineral composition of volcanic and desert dust are different. Dust effects have also been invoked to explain many changes in Pleistocene climate (see, for example, Overpeck *et al.* 1996).

Numerical modeling studies have been used in attempts to better understand the influence of dust on Earth's current climate. Coakley and Cess (1985) use a global model, run with and without dust radiative effects included. Surface temperatures were lower in some areas by about 1 K, with the dust effects included. A reduced solar heating at the surface decreased cloud and precipitation slightly between 30° and 60°. Joseph (1984) shows a stronger impact of mineral aerosols on lowering surface temperatures in experiments with a similar model. More recent modeling studies show larger impacts of dust radiative effects (Tegen *et al.* 1996; Miller and Tegen 1998). Alpert *et al.* (1998) compared estimated dust concentrations over the eastern tropical North Atlantic Ocean with significant errors in a global-model simulation that did not have dust effects included, and found a very strong spatial correlation. Sokolik *et al.* (2001), in a review paper on the subject, state that the sign and magnitude of dust radiative forcing of regional and global climate remain as major unresolved problems.

Regional and global human-health effects of desert dust

The human-health implications of the transport of dust from distant and local sources have been well documented. In North America, much of the dust is smaller than the 2.5 μm diameter threshold set by the United States Environmental Protection Agency. Such small aerosol particles can efficiently penetrate into the lungs. With high concentrations from North African and Asian desert sources, some locations violate clean-air health standards (Prospero 1999). The Great Green Wall of China, the vast area of trees planted to the south of the Gobi Desert, was created partly because of the health problems caused by the dust storms in downwind cities such as Beijing (Parungo *et al.* 1994). Pathogenic bacteria can also be transported by desert dust. It is thought that the attenuation of ultraviolet radiation by dust at higher altitudes shields the bacteria at lower altitudes during the approximately one week journey from North Africa. The bacteria may also be shielded from ultraviolet radiation when lodged within crevasses of inorganic dust particles (Griffin *et al.* 2001). See Griffin *et al.* (2001) for discussion of various mineral-dust-related diseases and allergies.

Distant ecological effects of desert dust

It was mentioned earlier that, in the context of the global carbon dioxide budget, iron in desert dust enhances ocean phytoplankton productivity. This iron can also cause blooms of toxic algae, called red tides, that have been known to kill large numbers of fish, shellfish, marine mammals, and birds. Sahara dust has been shown to cause such algal blooms in the Gulf of Mexico (O'Carroll and Carlyon 2001). Coral reef mortality in the Caribbean has also been linked to the effects of Saharan dust (Shinn *et al.* 2000; Griffin *et al.* 2001). Soils are also often formed primarily of dust transported from distant sources. For example, islands in the Caribbean have soils with parent material that is primarily of Saharan origin, and soils in east Asia are composed of dust from desert sources far to the west. Hawaiian rainforests are sustained by nutrient-rich dust from the Gobi and Taklamakan Deserts (Chadwick *et al.* 1999).

Dynamic effects of deserts on meteorological processes in other regions

It should be clear to the reader by now that there are many regional and global dynamic mechanisms that can link meteorological processes in one area with those in another. A few examples will be noted here of how deserts can influence meteorological processes in nearby less arid regions.

Monsoon circulations strongly influence the precipitation climatology of many regions of Earth, with the dynamic forcing being the latitudinal surface-heating gradient between the land and the sea. Given that the distinct characteristics of the desert surface have a great effect on the surface heat fluxes, there is no controversy that monsoons worldwide would be different without the dynamic effects of the deserts. Yang *et al.* (1992) have shown that *longitudinal* heating gradients may also be important. In particular, the largest longitudinal heating gradient in the tropics is found between the Sahara Desert and the Asian convective region during the South Asian monsoon, and there is a correlation between the strength of the rain-producing monsoon during individual years and the strength of the heating gradient. Thus, the conditions in the Sahara, one of the driest areas of Earth, affect the amount of rainfall in one of the rainiest places on Earth. Rodwell and Hoskins (1996) describe additional dynamic links between the Sahara and the South Asia monsoon.

A very well documented dynamic effect of deserts on the downwind meteorological processes is related to the plume of hot, dusty air from the Sahara that moves westward in the trade winds. Called the Saharan air layer (SAL), its evolution and influence on the meteorology over the Atlantic have been described by Carlson and Prospero (1972), Karyampudi and Carlson (1988), and Karyampudi *et al.* (1999). As the boundary-layer air moves westward over the African continent and the Atlantic Ocean, it becomes elevated above the surface. Observational and modeling studies illustrate the following effects. The SAL enhances (1) the strength of a midtropospheric easterly jet, or wind-speed maximum; (2) the strength of wave disturbances in the easterly trade-wind flow (called easterly waves), and (3) the intensity of cumulus convection in the equatorial zone. Numerical experiments were performed for a period in which an easterly wave developed into a hurricane. A sensitivity experiment showed that, when surface heat fluxes over the Sahara were shut off so that the SAL was not allowed to develop realistically, the wave did not develop into a hurricane (Karyampudi and Carlson 1988). Karyampudi *et al.* (1997) further discuss the possible influence of the SAL on tropical cyclogenesis.

A previously described potential effect of hot desert boundary layers that become elevated above the surface as they are transported downwind is the suppression of precipitation. Because the downwind area may be generally less arid, this represents a mechanism by which arid areas can influence the weather in different neighboring climates. The best-studied example of this process is the effect of the boundary layer from arid Mexico on the precipitation in the less arid Great Plains to the northeast.

Suggested general references for further reading

Charlson, R. J., and J. Heintzenberg, 1995: *Aerosol Forcing of Climate* – a collection of papers on sources, distributions, and fluxes of mineral aerosols; and the mechanisms and modeling of aerosol effects on climate.

Duce, R. A., 1995: *Sources, distributions, and fluxes of mineral aerosols and their relationship to climate* – describes dust sources and source strengths; the atmospheric distribution, transport, and deposition of dust; and the effects of dust on climate.

Gerber, H. E., and A Deepak (Eds.), 1984: *Aerosols and Their Climatic Effects* – a selection of papers on the climate effects of aerosols.

Hobbs, P. V., and M. P. McCormick (Eds.), 1988: *Aerosols and Climate* – a collection of papers on aerosol sources, distribution, properties, radiative effects, and impacts on climate.

Péwé, T. L., 1981: *Desert Dust: Origin, Characteristics, and Effect on Man* – describes the origin and transport of desert dust, the characteristics of the dust, and the effects of dust on climate change and hazards to people.

Westphal, D. L., *et al.*, 1987: *A two-dimensional numerical investigation of the dynamics and microphysics of Saharan dust storms* – an example of the use of a numerical model to simulate long-distance dust transport.

Questions for review

(1) Provide examples of the long-distance transport of desert dust by the trade winds and the westerlies.
(2) List the various meteorological, environmental, and health impacts of the transport of dust beyond desert boundaries.
(3) Explain the structure of the Saharan air layer over the Atlantic Ocean, and its meteorological consequences.
(4) Describe how elevated mixed layers can influence precipitation downwind of deserts.
(5) Describe how desert dust deposited in the ocean can affect the carbon dioxide content of the atmosphere.
(6) By what mechanisms does desert dust influence Earth's atmospheric and surface radiation budgets?

Problems and exercises

(1) Refer to Karyampudi and Carlson (1988) and describe the detailed structure of the Saharan air layer.
(2) Search for additional environmental effects of the transport and deposition of desert mineral dust.

18

Desertification

The desert lies in wait for arable land and never lets go.

Fernand Braudel
*The Mediterranean and the Mediterranean World in
the Age of Philip II* (1972)

*I think, if I may say so, we have been considering the Sahara rather from the
wrong point of view. All the stress has been laid on "the encroachment" of the
Sahara, but I would rather like to put it that the Sahara has seized the
opportunity of man's stupidity.*

Sir Arthur Hill, Director of the Royal Gardens
The encroaching Sahara: The threat to the West African colonies
(Edward Percey Stebbing 1935)

*Our land, compared by what it was, is like the skeleton of a body wasted by
disease. The plump soft parts have vanished and all that remains is the bare
carcass.*

Plato, Greek philosopher
Critias (4th century B.C.)

*We know that the white man does not understand our ways. . . He treats his
mother, the earth and his brother, the sky as things to be bought, plundered,
sold like sheep or bright beads. His appetite will devour the earth and leave
behind only a desert.*

Chief Seattle, Native American, in a letter to
the Great White Chief in Washington (1854)

*. . . everybody knows that the using up still goes on, perhaps not so fast nor so
recklessly as it once did, but unmistakably nevertheless. And there is nowhere
that it goes on more nakedly, more persistently or with a fuller realization of
what is happening than in the desert regions where the margin to be used up
is narrower.*

Joseph Wood Krutch, American author and conservationist
The Voice of the Desert (1956)

457

If the inability of the land and/or the climate to sustain abundant life is the primary criterion for naming a place desert, there can be no argument that many areas are now more desert-like than they were in past times. Of the effects of natural processes like changes in climate, it is clear from archaeological evidence that animals that are now found only in humid Africa, and humans that hunted them, once thrived in areas that are now part of the Sahara. Of anthropogenic causes, one need only recall images of the midwestern United States in the 1930s where plows exposed the underbelly of the natural High Plains grassland to winds, and the land ceased to produce anything for years except meters-deep piles of dust. In the same vein, Bingham (1996) refers to "virtual deserts" of human origin, such as urban areas, acid-rain-dead lakes, charred rain forests, and rivers and oceans depleted of life by poison and exploitation. He points out some of the ramifications of human-caused desertification over the past few millennia.

Even Robert Malthus, the great pessimist who predicted that population would outrun the productivity of land, did not factor in the active and evidently permanent destruction of its underlying fertility, but that destruction has occurred so uniformly in so many otherwise unrelated situations that one might suspect that desertification is the dark companion of all human progress.

Much of the rich history of the Mediterranean was bought at the price of desert, from the vanished cedars of Lebanon to the long-barren fields of Carthage and the naked Illyrian coast, whence came, once upon a time, the masts of Venetian ships. The legendary Timbuktu, the vanished myrrh-rich forests of Arabia and Yemen, the bitter shriveled Aral Sea, the lost gardens of Babylon, the scorched plains of Sind, the scoured fringes of the Gobi, the once-teeming veldt of southern Africa, and of course the vanished cattle and sheep empires of the American West all testify to the amazing diligence of humankind.

From earlier chapters, it should be clear that desert lands interact with the atmosphere very differently than do those that are more moist and more vegetated. Thus, surfaces that have been temporarily or permanently rendered more dry and barren of vegetation by either natural or anthropogenic forces, i.e. desertified, will have different associated microclimates and boundary-layer properties. The question of the feedbacks between surface conditions, the regional climate, and the human response to the physical system, is thus important to the topic of desert meteorology. Literature abounds on this subject of desertification, but regardless of whether it is caused by climate, or humans, or both in concert, the existence of a local effect on the atmosphere is without dispute.

It should be noted that more has been written about desertification than about any other environmental degradation process, probably because the actual and potential human consequences are so terribly immense. This chapter will simply summarize some of the major issues.

What is desertification?

The working definition of desertification to be used here is intuitive and simple; it is the development of any property of the climate or land surface that is more characteristic of a desert, whether the change is natural or anthropogenic. More formal and elaborate definitions are plentiful, and include a

change in the character of land to a more desertic condition [involving] the impoverishment of ecosystems as evidenced in reduced biological productivity and accelerated deterioration of soils and in an associated impoverishment of dependent human livelihood systems

(Mabbutt 1978a,b).

Glantz and Orlovsky (1983), Odingo (1990), and Rozanov (1990) provide a thorough review of various definitions, where the former source lists over 100. A popular and encompassing one is offered by Dregne (1977):

Desertification is the impoverishment of arid, semi-arid and some subhumid ecosystems by the combined impact of man's activities and drought. It is the process of change in these ecosystems that can be measured by reduced productivity of desirable plants, alterations in the biomass and the diversity of the micro and macro fauna and flora, accelerated soil deterioration, and increased hazards for human occupancy.

In this definition is included deterioration in the *soil or the biomass*, by *natural and/or human processes*, in *arid as well as subhumid areas*. He later expands the definition even further, referring to the "impoverishment of terrestrial ecosystems" in general, thus allowing the term desertification to be used even for humid climates (Dregne 1985). Dregne (1977) points out that desertification in the more subhumid zones is of greater importance than in truly arid areas because the impact is greater: a greater number of people are affected, there is more soil to be lost, and there is more of a local economy to be impacted. Such liberal extension of the term "desertification" to humid areas is in keeping with the first use of the term by Aubreville (1949) who referred to the creation of desert conditions by deforestation in humid parts of west Africa.

A caveat that is sometimes implicit in the definition of desertification is that the process leads to long-lasting, and possibly irreversible, desert-like conditions (Hellden 1988). Thus, the reduction in vegetation along a desert margin, associated with normal interannual or interdecadal rainfall variability, would not be referred to as desertification unless, perhaps, resulting wind or water erosion of the soil was so great that the process was not reversible on similar time scales: that is, vegetation did not return with the rains. The term "irreversible" should be used carefully because soils can regenerate on geological time scales, and it would be a rare situation indeed where life did not return in some form if the climate eventually became more favorable or the negative human impact ceased.

Desertification is popularly thought of as a problem of only third-world countries, such as those of the Sahel. Indeed, if the above mentioned degree of impoverishment of the human condition is a primary criterion, it *is* mainly a third-world problem. Without the social-support infrastructure that is common in many prosperous countries, starvation and malnutrition of millions of people result. Nevertheless, there is significant desertification in first-world countries, and it has many immediate implications. It portends very sobering long-range problems in areas such as food and energy supply, the environment, the economy, and the quality of life. Even though Sears (1935) and others make the distinction between *desertification*, that is caused by humans, and *desertization*, that is natural, we will adopt the former term for both processes.

A list of basic indicators of the desertification process is provided by Grainger (1990): ground conditions, climatic indicators, data on agricultural production, and socioeconomic indicators. A more specific list of major symptoms of desertification includes the following (Sheridan 1981):

- declining water tables
- salinization of topsoil and water
- reduction of surface waters
- reduction of substrate moisture content
- unnaturally high soil erosion (by wind and water)
- desolation of native vegetation

Additional signs are loss of soil organic matter, a decline in soil nutrient concentrations, and soil acidification (Bullock and Le Houérou 1996). There are many geographic areas that exhibit all six symptoms, but any one symptom is sufficient to qualify the area as undergoing desertification. The overall process often first becomes apparent in one aspect, but frequently progresses to more. For example, salinization or drying of the substrate may lead to a reduction of the vegetation cover, which permits greater erosion by wind and water. The wind-driven sand can then be responsible for stripping remaining vegetation, and for further scouring the surface.

Some erroneously imagine desertification as an advance of the desert into surrounding less desert-like areas. Rather, it seems to be a situation in which marginal areas, near a desert but not necessarily on its edge, degrade as a result of the combined effects of drought and human pressures of various sorts. Such areas will possibly grow and merge with each other, to the point that the patchy zones of destruction will become part of a nearby desert. Except for the special case of drifting desert sand encroaching on surrounding non-desert, this is closer to the real process of desertification (Dregne and Tucker 1988). The concept is further complicated by the fact that deserts exist in degrees, so that the further impoverishment of the vegetation in an existing vegetation-impoverished desert would also be rightly called desertification. This has been discussed earlier in the context of anthropogenic change, where, for example, the Sonoran and Chihuahuan Deserts have become markedly more barren of vegetation and wildlife in the past 100 years.

As noted earlier, there have been numerous attempts at quantifying and mapping the status, or degree of, desertification. For example, Dregne (1977) presents a classification system with the following subjective criteria that are generally related to vegetation, erosion, and salinity.

Slight desertification
- Little or no deterioration of the plant cover and soil has occurred.

Moderate desertification
- Plant cover has deteriorated to fair range condition; or
- hummocks, small dunes, or small gullies indicate that accelerated wind or water erosion has occurred; or
- soil salinity has reduced crop yields by 10–50%.

Severe desertification
- Undesirable forbs and shrubs have replaced desirable grasses or have spread to such an extent that they dominate the flora; or
- wind and water erosion have largely denuded the land of vegetation, or gullies are present; or
- salinity that is controllable by drainage and leaching has reduced crop yields by more than 50%.

Very severe desertification
- Large, shifting, barren sand dunes have formed; or
- large, deep, and numerous gullies have developed; or
- salt crusts have developed on nearly impermeable irrigated soils.

It is important to recognize that these criteria refer to a change in the state of the system, are evidence of a process, and do not reflect the actual condition. Thus, there must be some period of time implied, during which these changes have occurred relative to some base or reference state. For the maps of Dregne (1977) that show the worldwide status of desertification, this original *pristine* state is that which prevailed before human impact, a condition that is more difficult to estimate for the Old World, through which so many civilizations have passed. In addition, if the desertification definition is limited to human impacts, the assumption is that the climate has been stable for that period. In Dregne's classification, areas of extreme aridity were automatically placed in the "slight desertification" category because the reference state was already extremely arid and only slight additional degradation of the vegetation and soil was possible. Indicators of desertification have been discussed by many authors, including Reining (1978) and Warren and Maizels (1977).

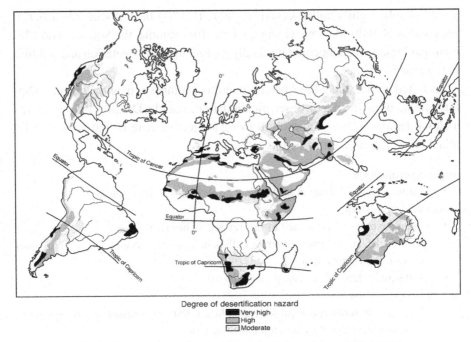

Degree of desertification hazard
■ Very high
▨ High
□ Moderate

Fig. 18.1 The global desertification hazard. (Adapted from UN 1977.)

Many of the definitions of desertification agree that, most often, the process occurs when human and natural pressures on the land coincide, for example during periods of drought. Thus, the natural drought cycle can be one of the contributors. On the other hand, significant interannual variability in the rainfall apparently can reduce the potential for desertification if it discourages agricultural use of an area. For example, Dregne (1986) points out that desertification has been limited to a moderate level in semi-arid northeastern Brazil by the large interannual variability in the rainfall there. That is, if droughts had been less prevalent, there would have been more agricultural pressure and land degradation.

The extent of desertification

Some desertification maps depict the areas that are susceptible to desertification, whereas others show the areas that are actually undergoing the process. Fig. 18.1 displays an estimate of desertification hazard that is based on climate conditions, the inherent vulnerability of the land, and the human pressures. Based on this map, at least 35% of Earth's land surface is now threatened by desertification, an area that represents places inhabited by 20% of the world population. There are also a number of estimates of the degree of actual worldwide desertification

Table 18.1 *Extent of desertification with different degrees of severity, for different continents*

The area is given in 1000 km², and the percent area is the fraction of all drylands in that region.

Region	Light area	%	Moderate area	%	Strong area	%	Severe area	%
Africa	1180	9	1272	10	707	5.0	35	0.2
Asia	1567	9	1701	10	430	3.0	5	0.1
Australia	836	13	24	4	11	0.2	4	0.1
N America	134	2	588	8	73	0.1	0	0.0
S America	418	8	311	6	62	1.2	0	0.0
Total	4273	8	4703	9	1301	2.5	75	0.1

Source: After Bullock and Le Houérou (1996); based on Oldeman *et al.* (1990), Le Houérou (1992), Le Houérou *et al.* (1993), and UNEP (1992).

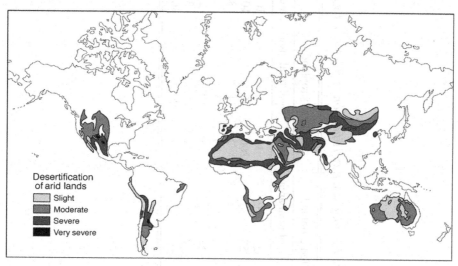

Fig. 18.2 Desertification severity. (From Dregne 1983.)

(not desertification hazard). Examples of such desertification-severity maps can be found in UNEP (1992), Dregne (1977), and Dregne (1983), with the latter depicted in Fig. 18.2. Table 18.1 lists the areas and percents of lands in different regions with light, moderate, strong, and severe desertification. Africa and Asia are comparable in terms of the percent of the land in each severity category that has been desertified. Of the remaining areas, North America has the greatest percent of its arid land with moderate or worse desertification, and this is followed by South America. If all severity categories are included, the order of decreasing

Table 18.2 *Areas suffering at least moderate desertification (1000 km²), by region*

	Rain-fed cropland		Irrigated cropland		Rangeland		Total	
	area	% of total area	area	% of total area	area	% of total area	area	% of total area
Africa	396	23	14	5	10 268	34	10 678	33
Asia	912	53	206	76	10 889	36	12 007	37
Australia	15	1	2	1	3 070	10	3 087	10
Europe (Spain)	42	2	9	3	155	1	206	1
N. America & Mexico	247	14	28	10	2 910	10	3 185	10
S. America	119	7	12	5	3 194	10	3 325	10
Total	1731	100	271	100	30 486	100	32 488	100

Source: From Dregne (1983).

Table 18.3 *Areas and numbers of people affected by at least moderate desertification, by region*

	Affected area (1000 km^2)	% total area affected	Affected population (millions)	% total population affected
Africa	7 409	37	108	38
Asia	7 480	37	123	44
Australia	1 123	6	0.23	0
Med. Europe	296	1	16.50	6
N America	2 080	10	4.50	2
S America and Mexico	1 620	8	29	10
Total	20 008	100	281.23	100

Source: From Mabbutt (1984).

Table 18.4 *Soil degradation in the susceptible drylands by process and continent, excluding degradation in the light category (million hectares)*

	Africa	Asia	Aust.	Europe	N. Am.	S. Am.	Total
Water	90.6	107.9	2.1	41.7	28.1	21.9	292.3
Wind	81.8	72.7	0.1	37.3	35.2	8.1	235.2
Chemical	16.3	28.0	0.6	2.6	1.9	6.9	66.3
Physical	12.7	5.2	1.0	4.4	0.8	0.4	23.9
Total	201.4	213.8	3.8	86.0	66.0	37.3	617.7
Area of susceptible dryland	1286.0	1671.8	663.3	299.7	732.4	516.0	5169.2
% degraded	15.6	12.8	0.6	28.6	9.0	7.2	11.9

Source: From Thomas and Middleton (1994).

percent of the area affected is Africa, Asia, Australia, South America, and North America. Of the arid, semi-arid, and dry-subhumid climate zones that were considered susceptible to desertification, 70% of the area, representing 3592 million hectares, has been actually affected (UNEP 1992). A similar table can be found in Dregne (1983). Table 18.2 shows that rangeland is most affected by moderate or worse desertification, in comparison with rain-fed and irrigated cropland. Seventy percent of the desertified area is in Africa and Asia. An estimate of the number of people affected by at least moderate desertification is provided in Table 18.3, with roughly four-fifths being in Africa and Asia. The physical causes of soil degradation, which is an eventual symptom of desertification, are estimated in Table 18.4. Even though this partitioning sheds no light on the complexity of factors that have led to the wind and water erosion, for example, the numbers are interesting. Both Africa and Asia have roughly equal contributions from wind and water erosion. Chemical degradation is primarily salinization, and physical effects include soil compaction. Note that these data do not represent areas with "light" degradation. The total areas reportedly affected by desertification on each continent depend considerably on the sources of data and the methods used, as indicated in these tables compiled by different investigators.

Anthropogenic contributions to desertification

Based on climatic data, more than a third of the Earth's surface is desert or semi-desert . . . If we go by data on the nature of soil and vegetation, the total area is some 43 percent of the Earth's land surface. The difference is accounted for by the estimated extent of the man-made deserts (9.1 million square kilometers), an area larger than Brazil

(UN 1978).

Most human contributions to desertification are related to agricultural exploitation of lands that are semi-arid. Movement of agriculture into such areas, whose natural condition is already agriculturally marginal, may be caused by a number of economic or social factors. These include the need to provide food for a larger population, the decreasing productivity of existing agricultural land, and greed. We will see that human attempts to use arid lands in ways that disregard their fragility, generally result in the land becoming even more unproductive, perhaps even permanently damaged. Even land in non-arid climates that is sufficiently abused can take on the characteristics of a desert. For example, tillage of steep slopes can cause all the topsoil to be lost by erosion during heavy rain events. And pollution can poison the soil, leaving it bare for generations.

There is much literature that emphasizes the importance of the human contribution to the problem, for a wide range of geographic areas. The following are examples (Bullock and Le Houérou 1996).

- China: Chao (1984a,b)
- Australia: Perry (1977), Mabbutt (1978b)
- South America: Soriano (1983)
- North America: Dregne (1983), Schlesinger *et al.* (1990)
- Europe: Lopez-Bermudez *et al.* (1984), Rubio (1987), Katsoulis and Tsangaris (1994), Puigdefabregas and Aguilera (1996), Quine *et al.* (1994)
- North and West Africa, and the Sahel: Pabot (1962), Le Houérou (1968, 1976, 1979), Depierre and Gillet (1971), Boudet (1972), Lamprey (1988), Nickling and Wolfe (1994), Westing (1994)
- East Africa: Lusigi (1981), Muturi (1994)
- South Africa: Acocks (1952), Hoffman and Cowling (1990), Bond *et al.* (1994), Dean and McDonald (1994)

Table 18.5 *Human causes of desertification, in percent of desertified land*

Regions or countries	Overcultivation	Overstocking	Fuel and wood collection	Salinization	Urbanization	Other
Northwest China	45	16	18	2	3	16
North Africa and Near East	50	26	21	2	1	—
Sahel and East Africa	25	65	10	—	—	—
Middle Asia	10	62	—	9	10	9
United States	22	73	—	5	—	—
Australia	20	75	—	2	1	2

Source: After Le Houérou (1992).

Table 18.5 lists estimates of the various causes of desertification in different areas of the world.

In Chapter 14 were discussed a number of anthropogenic impacts on deserts, where all the changes described do not represent a trend toward desertification. The following paragraphs elaborate on those human activities that can contribute to the desertification process.

Exposing soils to erosion through plowing and overgrazing

Any process that enhances the wind or water erosion of topsoils can have an aridifying effect, especially in an area that is already semi-arid. As such, plowing and overgrazing of semi-arid lands worldwide have been a common cause of human-made "natural" disasters that have temporarily or permanently rendered the land relatively uninhabitable and desert-like. An example of one of the most rapidly occurring, if not the worst, such disaster is the evolution of the Great Plains of North America from a lush (by semi-arid standards) grassland to a "dust bowl." During the teens and twenties of the twentieth century, rainfall was extraordinarily high in the area, and millions of acres of already overgrazed shortgrass prairie were plowed up and planted in wheat. This was done in spite of the knowledge that the 50 cm rainfall isohyet has historically wandered capriciously over a large distance between the Rocky Mountains and the Mississippi River. In fact, the semi-arid grasslands of the Great Plains of west-central North America were known as the Great American Desert in the nineteenth century. Exacerbating the problem were the pronouncements of the land companies and railroads that "rain follows the plow:" that by some miraculous physical mechanism, the rainfall would follow the farmer into the semidesert. In contrast to these optimistic, self-serving statements, a federally appointed Great Plains Committee evaluated the condition of the Plains before the dust-bowl debacle and made the following assessment:

Current methods of cultivation were so injuring the land that large areas were decreas-
ingly productive even in good years, while in bad years they tended more and more to
lapse into desert. . .

(Great Plains Committee 1936).

However, predictably, the wet years waned in the early 1930s, and the consequences were disastrous. The wheat stubble had been turned under, and the loose,
dry soil lay exposed to the first of a long series of wind storms that began on 11
November 1933. Some farms had lost virtually all of their topsoil by nightfall
of the first day. During the second day, the dust-laden skies were black. As more
storms stripped away the land's productivity, houses and machinery were buried
under meters of silt, some of which reached high in the atmosphere to the jet
stream, to be transported as far away as Europe. The drought and the damage
continued through the decade. There are few that claim that the drought alone
would have caused a similar calamity in an undisturbed prairie, without the un-
witting collaboration of the farmer. Since this experience of the 1930s, droughts
in the Great Plains in the 1950s and 1970s caused soil loss of equal or greater
magnitude in some areas.

There are myriad other examples of such desertification by agricultural abuse,
on both grand and small scales, throughout the world. A small-scale one, to
contrast with the large-scale experience of the Great Plains, is the semi-arid Rio
Puerco Basin of New Mexico in North America. This area was so agriculturally
productive 100 years ago that it was referred to as the "bread basket of New
Mexico." But by then, the land had already begun to erode and desiccate as a
result of long-term overgrazing. By 1950, all of the agricultural settlements had
been completely abandoned, and the process of extraordinarily severe soil loss
and arroyo deepening and widening continues today. Sheridan (1981) reports that
the Rio Puerco supplies less than 10% of the Rio Grandes water, but contributes
over 50% of its load of sediment. As with the attribution of the causes of the
phenomenally rapid erosion of the Great Plains in the 1930s, there are those who
claim that the climate of the Rio Puerco changed a bit in terms of the seasonal
distribution of the rainfall, and that this could help explain the severe erosion
during the past 100 years. Nevertheless, most analysts suggest that overgrazing
was the primary if not the only cause. Schlesinger *et al.* (1990) discuss a study in
the same area in which desertification of grassland is attributed to the effects of
overgrazing on an increase in the heterogeneity of soil resources, including water.
This leads to an invasion of the native grasslands by shrubs. Another example of
regional desertification is the semi-arid San Joaquin Valley of California, one of
the most productive agricultural areas in the world. Here, 80% of the privately
owned rangeland is overgrazed (Sheridan 1981).

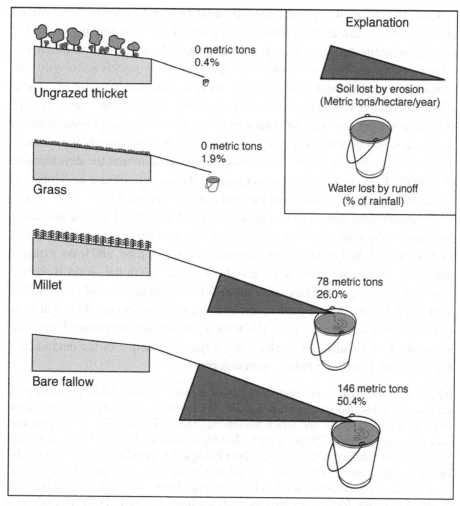

Fig. 18.3 Water and soil loss from runoff in four areas of about the same size and slope in semi-arid Tanzania. (Adapted from Goudie and Wilkinson 1977.)

Figure 18.3 illustrates the extent to which disturbing the natural vegetation and soil structure contributes to water and sediment runoff. In a semi-arid area of Tanzania, the percent of rainwater lost to runoff and the amount of soil loss were compared for four areas of about the same size and slope. The ungrazed grassland and shrubland had very little loss of soil and water, whereas the area planted with millet and the area that was bare, for example from overgrazing, had high rates of soil and water loss.

It is ironic that the "rain follows the plow" proclamation is not only untrue, it describes the opposite of the actual probable response of the Earth–atmosphere system. That is, plowing under natural vegetation and exposing bare ground produces effects that could diminish the rainfall. First, stripping the ground of vegetation diminishes its water-retention capacity, allowing more rainfall to run off or evaporate rapidly. Also, bared ground freezes more quickly and deeply than ground covered by vegetation, and thus water is more likely to run off. These effects of natural vegetation removal will reduce soil moisture, and therefore evaporation from the soil. In addition, when the surface is plowed, there is no transpiration of water vapor by vegetation. This lower water-vapor flux to the atmosphere by evaporation and transpiration could inhibit the development of rainfall. Thus, modification of semi-arid land by agricultural activities can potentially affect the regional climate and make it *less* suitable for agriculture.

The topsoil loss that results from erosion of plowed land by wind and water can naturally limit the ability of vegetation to survive, even if the rainfall amounts are not affected. In the extreme, the topsoil loss may be total, and leave exposed only unproductive subsoils, or even bedrock. Nowhere in the world is there a greater abundance of extreme examples of desertification by soil erosion than in the Middle East, where civilization after civilization, and conflicts that have laid waste to civilizations, have taken their toll on the environment over the millennia. Lowdermilk (1953) describes a typical example of the degradation that has taken place long before contemporary times.

We crossed the Jordan again [into Syria] into a region famous in Biblical times for its oaks, wheat fields, and well-nourished herds. We found the ruins of Jerash, one of the 10 cities of the Decapolis, and Jerash the second. Archaeologists tell us that Jerash was once the center of some 250,000 people. But today only a village of 3,000 marks this great center of culture, and the country about it is sparsely populated with seminomads. The ruins of this once-powerful city of Greek and Roman culture are buried to a depth of 13 feet with erosional debris washed from eroding slopes . . .When we examined the slopes surrounding Jerash we found the soils washed off to bedrock in spite of rock-walled terraces. The soils washed off their slopes had lodged in the valleys. . . Still farther north in Syria, we came upon a region where erosion had done its worst in an area of more than a million acres . . . French archaeologists, Father Mattern, and others found in this man-made desert more than 100 dead cities. . . Here, erosion had done its worst. If the soils had remained . . . the area might be repeopled again and the cities rebuilt. But now that the soils are gone, all is gone.

Salinization of soils through irrigation, and excessive well pumping in coastal areas

Rainwater that provides for the needs of native vegetation and agricultural crops does not increase the salinity of the soil. The rainwater is relatively free of minerals because its source is atmospheric water vapor. However, irrigation water, whether it is obtained from groundwater or surface water, has dissolved minerals.

Even when this groundwater or river water does not have an especially high original mineral content, the high evaporation rate in semi-arid and arid regions causes the minerals to become more concentrated (1) as the water evaporates from the surface after it is applied; (2) while the water is in transit through long irrigation canals; or (3) while waiting for use, in impoundments. We know from earlier discussions of the "oasis effect" that water surfaces or moist soils exposed to the ambient extremely dry, hot desert air evaporate very rapidly. The ultimate reservoir for the salts that remain is the soil from which the water evaporates. When these salt concentrations become excessive, agriculture can become impossible, and, unfortunately, even the original native vegetation can often no longer survive. An additional problem is that applying water to many arid soils causes salts already in the soil to rise toward the surface and concentrate. Salinization is especially a problem where downward drainage of mineralized irrigation water is blocked by an impermeable layer near the surface. After the land is sufficiently salinized that it is unable to support any vegetation, it thus becomes more susceptible to erosion by wind and water, and unable to provide any water vapor to the atmosphere through transpiration.

Large areas of semi-arid land have reverted to a more barren state because of this effect. For example, over 25% of the land of Iraq has been rendered unsuitable for producing native, or any other kind of, vegetation (Reisner 1986). Eckholm (1976) points out the following:

The first recorded civilization, that of the Sumerians, was thriving in the southern Tigris-Euphrates valley by the fourth millennium B.C. Over the course of two thousand years, Sumerian irrigation practices ruined the soil so completely that it has not yet recovered ... Vast areas of southern Iraq today glisten like fields of freshly fallen snow.

There are many other examples throughout history of a decline in agriculture as a result of salinization. Thomas and Middleton (1994) list the settlements of the Khorezm oasis in Uzbekistan, many oases in the Taklamakan Desert, and numerous other sites in China, as having been abandoned for this reason. In the Nile Valley, virtually the only agriculturally productive area of Egypt, yield reductions in excess of 25% are attributed to salinization (Dregne 1986). It is interesting to note that the natural annual flooding of the Nile has historically flushed salts from the soil. This has allowed it to remain one of the world's most agriculturally productive and densely populated areas for thousands of years, in contrast to the situation in Iraq where the Tigris–Euphrates system does not flood the soil in a similar way. However, the Aswan Dam now prevents the natural soil-cleansing Nile flooding. In addition, some areas of the immensely productive San Joaquin Valley in North America are beginning to show lost productivity through salinization.

Salinization of soils also results from excessive pumping of groundwater near coastlines with the sea. When the (fresh) water table drops, hydraulic pressure allows salt water to encroach inland, and this results in soil salinization. Oman and Yemen, among other areas, are experiencing this problem (Grainger 1990).

Groundwater mining

Groundwater mining is the use of non-renewable, or only very slowly renewable, reservoirs of underground water for agricultural and other human needs in arid lands. This exploitation can lead to desertification through a few different mechanisms. Perhaps the most damage is done when the water is used for irrigation. When the water is gone, not only will the land lose its greenness and revert back to the desert that the rainfall statistics say that it is, but the condition of the desert will be the worse for its brief impersonation of farmland. The soil will be more saline; the monoculture style of farming will have made it easier for invader species to replace native ones; the lower water table will make it impossible for some deep-rooted native plants to survive as they once did; streams and marshes and wet salt flats that relied on a high water table will not return; and the loss of soil through wind and water erosion of the bare surface will mean that it is less hospitable for the original vegetation. If the desert surface is not physically disturbed to a great degree, as is the case if the wells are used only to supply water for human consumption in urban areas, the permanent damage may be less severe. However, there are still the impacts of the lowered water table on streams and vegetation. In addition, the ground compacts and settles as water is removed from its interstitial spaces, and this means that the aquifer may never again naturally recharge, even if left alone for many generations. Because the settling is often differential, the surface can develop deep fissures and the slope can change. This severely alters the surface hydrology, and, in turn, the surface's interaction with the atmosphere. Subsidence of over 7 m has occurred in some places (de Villiers 2000).

There are many examples, worldwide, of how groundwater mining has changed the complexion of the desert, and how major human, ecological, and meteorological adjustments will be inevitable when the water can no longer be economically retrieved. For example, in the second half of the twentieth century, the Ogallala Aquifer beneath the Great Plains in North America (Fig. 18.4) supported the irrigation of over 60 000 km^2 of agricultural land that was once semi-arid. But the prairie-grass sod that once prevented serious erosion no longer exists. And the dryland farming that may possibly ensue after the Ogallala Aquifer is no longer usable, in the first half of the twenty-first century, will allow wind erosion that could cause the transformation to a desert. Reisner (1986) and others suggest that this could easily lead to another dust bowl. This is not an

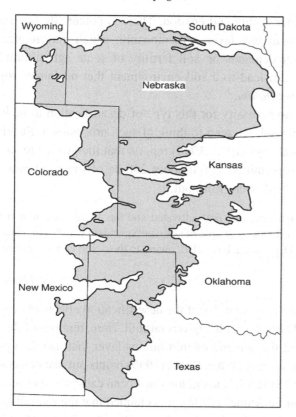

Fig. 18.4 Geographical extent of the Ogallala Aquifer in North America. (Adapted from Gutentag and Weeks 1980.)

isolated "doomsday" example. The remaining groundwater in many areas of the world will not be economically recoverable for irrigation in the near future, and it is an open question in many cases what the subsequent "natural" conditions will look like.

Deforestation

Removal of trees over large or small areas can cause desertification. The de-vegetation process is generally motivated by a need for more agricultural land, or by a need for wood for fuel and construction. A classic example is that hardly a tree survives within a radius of 100 km of Khartoum, Sudan. A variety of mechanisms may contribute to the resulting desertification. Erosion of soils by water may increase because there is less vegetation canopy to intercept rainfall. There is less leaf litter and other decaying biomass to replenish the organic material in soils, which means that rainfall absorption is less and runoff is greater. Removal of vegetation exposes soils to higher wind speeds and more sunlight, causing greater drying and erosion. There is potentially less transpiration to supply the boundary layer with moisture, which may reduce rainfall. Lastly, the greater demands on soil fertility of some agricultural crops that replace the trees can lead to a soil environment that no longer supports crops *or the original vegetation.*

There seems to be no necessity for this type of desertification to be limited to large scales, even though we tend to think of such processes in Sahel-sized areas. For example, Stebbing (1935, 1938) reports that indigenous residents in northern Nigeria were permitted to fell about 240 ha of forest for new farms because their existing lands were worn out.

Much of the disafforested land was not cultivated and has since become a complete waste of drifting sandy soil now incapable of cultivation, and on which a first attempt to reafforest by the Forest Department has failed, owing to the desert conditions becoming so rapidly established

(Stebbing 1935).

There are even numerous examples of the desertification effects of deforestation in humid tropical climates with copious rainfall. Here, removal of the virgin tropical forest exposed the sometimes thin humus layer that had always been protected by the dense canopy above. Beard (1949) points out that deforestation of most of the Lesser Antilles Islands in the Caribbean caused numerous symptoms of desertification. In Anthes (1984), it is claimed that the rainfall over the deforested islands appears to be significantly less than that over the few islands that were mostly spared the process of agricultural exploitation.

Where deforestation and other processes have led to desertification, it has been suggested that conserving, planting, or replanting trees might be a way to inhibit the further spread of the desert. For example, Stebbing (1935), referring to northern Nigeria, states

Plans have been drawn up to create belts of plantations (of trees) across the countryside, selected where possible from existing scrub forest or assisting the latter. . . also to sow seed of the Dom palm along the International Frontier to create a thick belt of the palm. This is a wise step and taken none too soon in the interests of the Province. . . the northern belt is required to counter one of the most silent menaces of the world, if not the most silent one, the imperceptible invasion of sand.

This shelterbelt, of proposed 25 km width and 2200 km length, was intended to inhibit sand encroachment from the desertified areas. This grand plan was never implemented, but more recently Anthes (1984) proposes that bands of vegetation in semi-arid regions would be useful for a different reason, claiming that convective rainfall might be increased. In contrast, Box 18.1 describes a reclamation technique that is actually being implemented.

> ## Box 18.1
> ## Reversing desertification in China
>
> There are areas in China, as elsewhere, where it is thought that land abuse has created entire deserts from grasslands. Overuse by pastoralists for thousands of years has transformed fragile steppe environments into deserts. One example is the Mu Us Shadi area, near the Huang He River south of the Gobi Desert. Here, the ruins of over ten large, ancient cities are covered by moving sands. In order to reclaim the area for agriculture, local residents have developed a method for stabilizing the dunes. The first step is to plant shrubs on the lower one-third of the windward side of the dunes. This lowers the speed of the wind on the windward face of the dune, preventing sand from moving up the slope. At the top, where the wind speeds are not diminished, the sand blows off and the dune is leveled. Trees are then planted on the flattened surface.

An interesting description is provided by Lowdermilk (1953) of how deforestation created moving sand dunes over a vast area, and how reforestation was able to reverse the process.

It is recorded that the Vandals in A.D. 407 swept through France and destroyed the settlements of the people who in times past had tapped pine trees of the Les Landes region and supplied resin to Rome. Vandal hordes razed the villages, dispersed the population, and set fire to the forests, destroying the cover of a vast sandy area. Prevailing winds from the west began the movement of sand. In time, moving sand dunes covered an area of more than 400,000 acres that in turn created 2¼ million acres of marshland [through blockage of rivers]. . .

Space will not permit my telling the fascinating details of this remarkable story – of how the dunes were conquered by establishing a littoral dune and reforesting the sand behind. . . Now this entire region is one vast forest supporting thriving timber and resin industries and numerous health resorts.

Fortunately for comparison, one dune on private land was for some reason left uncontrolled. This dune is 2 miles long, ½ mile wide, and 300 feet high. It is now moving landward, covering the forest at the rate of about 65 feet a year. As I stood on this dune and saw in all directions an undulating evergreen forest to the horizon, I began to appreciate the magnitude of the achievement of converting the giant sand dune and marshland into profitable forests and health resorts.

Off-road vehicle damage to soil structure and vegetation

This subject was discussed earlier in Chapter 14 about anthropogenic effects on the desert surface and atmosphere. In summary, operating vehicles on desert terrain that has fragile vegetation and substrates can cause long-term damage. Not only is vegetation removed, but the soil surface is destabilized through disturbance of desert pavement or other crusts. Both processes lead to increased wind and water erosion. Naturally, military as well as civilian-recreational off-road vehicles contribute to the problem.

Urbanization and industrial exploitation

Much arid and semi-arid land has undergone substantial anthropogenic change through urbanization and exploitation of mineral resources. Arid lands provide 82% of the world's oil, 86% of the iron ore, 79% of the copper, and 67% of the diamonds (Lines 1979; Heathcote 1983). In terms of urbanization, it was noted in the introduction that population increases have often been far greater in semi-arid areas than in others with harsher winters. One example is that urbanization consumed almost 200 000 ha of the immensely agriculturally productive San Joaquin Valley in California during the last quarter of the twentieth century.

Natural contributions to desertification

Some definitions of desertification include only anthropogenic factors; however, most allow for the fact that the negative impact of agriculture on the biophysical system will be greatest during periods of low rainfall. Thus, drought (defined in Chapter 15) is the most commonly discussed natural contribution to desertification. In general, any drought that degrades the vegetation cover or causes any of the aforementioned desertification symptoms would be sufficient.

The encroachment of drifting sands into surrounding vegetated areas, and the suffocation of the vegetation there, could also be considered as a natural desertification process. This occurs on local scales, such as associated with mobile sand dunes, as well as over long distances through accumulation of aeolian dust.

Additional selected case studies and examples of desertification

In the above discussion of the causes of desertification, some examples were provided for each category. This section will briefly review some additional dramatic instances of desertification in different areas. An immense amount has been written describing various incidences of desertification, and much debate has resulted regarding the relative contribution of drought versus human factors. What is not in doubt is that there are very few areas of Earth that have not experienced some degree of land degradation, so the following discussions cover only a few of the problem areas.

The Thar Desert of India and Pakistan

The deforestation and agricultural exploitation of this $1\,300\,000$ km^2 area of western India and eastern Pakistan began millennia ago, and the possible atmospheric consequences are described further in the next section about desertification feedback mechanisms. This Indus Valley area was once the cradle of the Indus civilization, a highly developed, agriculturally based society. About 3000 years ago, the monsoon rainfall on which it depended began to fail, and the ensuing dry period lasted for almost a millennium. Subsequently, the rains increased somewhat, but they are today only about one-third of what they were estimated to be 3000 years ago (de Vreede 1977). This may not be entirely natural variability. It is proposed in Bryson and Baerreis (1967) that the troposphere-deep layers of dense dust caused by the soil desiccation have augmented the natural subsidence in these latitudes, which has enhanced the desertification process. Hora (1952) summarizes the consensus of the New Delhi Symposium on the Rajputana Desert:

One thing, which was pointedly brought out in the Symposium, was that the Rajputana Desert is largely a man-made desert . . . by the work of man in cutting down and burning forests . . . [and by] the deterioration of the soils.

Punjab, India: north of the Thar Desert

Stebbing (1938) provides the following description of the large-scale conversion of a forest to the bare rock of a desert-like surface.

Perhaps one of the best-known often-quoted examples of this type of damage is the case of the Hoshiarpur Chos in the Punjab, India. This part of the Siwalik range of hills consists of friable rock. The hills were formerly covered with forest. In the latter half of the last century, cattle owners settled in the area, and under the grazing and browsing of buffaloes, cows, sheep, and goats, all vegetative growth disappeared and the trampling of the animals on the slopes loosened the already loose soil. Heat and the annual monsoon rains helped to carry on the process of erosion commenced by the animals. Gradually, ravines and torrents were formed which have cut the hill range into a series of vertical hollows and ridges of the most bizarre shapes; the material thus removed and carried down to the lower level forming fan-shaped accumulations of sand extending for miles out into the plains country, covering up extensive areas of valuable agricultural land. The loss has been enormous . . .

The Sahel, North Africa

The Sahel is a region on the southern margin of the Sahara that is subject to recurrent drought. Agnew (1990) reviews the various definitions of its extent in terms of different annual-isohyet belts between 200 and 750 mm. Its annual

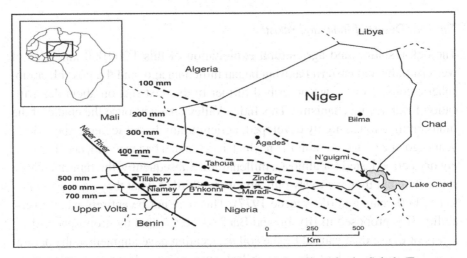

Fig. 18.5 Average annual rainfall (millimeters) for Niger, in the Sahel. (From Agnew 1990.)

precipitation is associated with the summer monsoon (see Fig. 3.14). When the monsoon is weak, the result is a dry year. It is during these drought years that human contributions to desertification can become especially important because of the combined stress of the natural, periodic rainfall deficit and the human pressures. Fig. 18.5 shows a close-up view of the large climatological gradient in annual rainfall for one part of the Sahel in Niger. Clearly, modest variation in the location of the large gradient would have significant consequences in terms of annual rainfall.

The works survive of Arab historians who describe the condition of the present Sahara and Sahel a number of centuries ago (Stebbing 1935). For example, an author recounts a pilgrimage from Mali, through central Niger, and eastward to Mecca, in 1496–97, where horses and donkeys were used for riding. It is suggested that this would have been possible only if the area had considerably more water and pasture than is the case today. In this area are ruins of permanent villages that had been occupied into the eighteenth century, until they were overwhelmed by sand. More recently, in a journey up the Nile in 1820 and 1821, Linant de Bellefonds reported on the "woodedness" of northern Sudan, which today is complete desert with hardly any vegetation away from the river (Cloudsley-Thompson 1993).

More recently, there have been substantial variations in the rainfall, with the most well-documented being in the last third of the twentieth century. During the wet years of the 1960s, human population and livestock numbers increased. But when rainfall amounts decreased after that, famine and large-scale loss of livestock resulted, in spite of massive international aid. An essential question is the degree to which human factors contributed to the decrease in green vegetation, through overgrazing for example. Fig. 6.2 illustrates the decrease of Sahel green vegetation during the drought. This figure also shows that the amount of green vegetation rebounded during subsequent wetter years, which were still drier than the twentieth-century normal. Clearly the desertification, defined in terms of green-vegetation change, was not irreversible in this case. The Sahara "advanced" during dry years, but it "retreated" when rain was more plentiful. This natural ebb and flow of the Sahara's vague edge has been taking place for centuries, without irreversible effects. In this regard, Nicholson (1989a,b) discusses the long term variability of Sahelian rainfall, noting that there were numerous other centuries with severe drought. Walsh *et al.* (1988) state that, in the context of the long-term Sahel climate, the real twentieth-century anomaly was the mid-century wet period, not the late-century dry one.

There are numerous areas in Africa in addition to the Sahel that have experienced desertification in the twentieth century. In west Africa, the annual rate of forest clearing was 4% during some periods. In terms of grasslands, the rich savanna, a vast area that extended from western Africa to the Horn in the east, was only 35% of its original extent in the early 1980s (Xue and Shukla 1993).

The arid western United States

A comprehensive survey was conducted of the land and water resources in this area by the Council on Environmental Quality of the United States (Sheridan 1981). The following is an excerpt from the conclusions of this report.

Desertification in the arid United States is flagrant. Groundwater supplies beneath vast stretches of land are dropping precipitously. Whole river systems have dried up; others are choked with sediment washed from denuded land. Hundreds of thousands of acres of previously irrigated cropland have been abandoned to wind or weeds. Salts are building up steadily in some of the nation's most productive irrigated soils. Several million acres of natural grassland are, as a result of cultivation or overgrazing, eroding at unnaturally high rates. Soils from the Great Plains are ending up in the Atlantic Ocean. All total, about 225 million acres of land in the United States are undergoing severe desertification – an area roughly the size of the 13 original states.

. . . The long-term prospects for increased production from U.S. arid land look unpromising, however. The rich San Joaquin Valley is already losing about 14,000 acres of prime farmland per year to urbanization and could eventually lose 2 million acres to salinization. Increased salinity of the Colorado River could limit crop output in such highly productive areas as the Imperial Valley. Economic projections in Arizona indicate a major shrinkage in cropland acreage over the next 30 years. On the High Plains of Texas, crop production is expected to decline between 1985 and 2000 because of the depletion of the Ogallala Aquifer. And certainly the end is in sight for irrigation-dependent increased grain yields from western Kansas and Nebraska as their water tables continue to drop.

Other assessments of desertification in North America are equally bleak. For example, Dregne (1986) states that about 90% of the arid lands of North America are moderately or severely desertified.

Hastings and Turner (1965) conducted an interesting study in which they made numerous photographs of the landscape at various places in the northern Sonoran Desert, where they tried to duplicate the viewing location and angle of photographs that were taken earlier in the twentieth century, perhaps 75 years before. Comparisons were revealing in terms of how the vegetation had changed during the period. Sheridan (1981) summarizes their evidence that all the plant species had undergone a change during the period that could reflect a warming and drying. The lower-elevation desert shrub and cactus communities are now thinner, the higher-level desert grasslands have receded upslope, areas that were originally oak woodlands are now primarily mesquite, and the timberline has moved upward. The pattern is one in which species have moved upslope toward more favorable environmental conditions. Hastings and Turner (1965) state the following:

Taken as a whole, the changes constitute a shift in the regional vegetation of an order so striking that it might better be associated with the oscillations of Pleistocene time than with the "stable" present.

Even though the impact of long-term overgrazing quite likely contributed to some aspects of the documented change, there is evidence that regional climate change acted in concert. Sheridan (1981) reports that, since near the beginning of the twentieth century, rainfall in Arizona and New Mexico has decreased by about 2.5 cm every 30 years, with most of the reduction in the winter season. Also, during the twentieth century, the mean annual temperature appears to have risen by 1.7–1.9 K (3.0–3.5 °F). Mechanisms by which the surface environmental change associated with the overgrazing may have contributed to these meteorological changes are discussed in the next section.

Groundwater mining is also creating major changes in southwestern United States' ecosystems. For example, in the San Pedro River Basin in southern Arizona, grasses, cottonwoods, willows and sycamores are being replaced by tamarisk and mesquite, and it is reasonable to assume that this is a result of a lowering water table. The taproots of mesquite can extend to depths of greater than 30 m, so it can thrive with deep water tables where other shallower-rooted species such as cottonwood cannot. But even the deep-rooted mesquite has suffered from the precipitous overdrafts that prevail in this region. South of

Tucson, Arizona, in the Santa Cruz Valley, about 800 ha of mesquite have died. Here, overdrafts are greater than in virtually any other area of the North American deserts (Sheridan 1981).

In the chapter on anthropogenic changes to deserts, the mining of the Ogallala Aquifer in the semi-arid High Plains of North America is mentioned as a prime example of how groundwater mining can temporarily green the face of the desert. This area was originally grassland, but the conversion to irrigated crops on an immense scale has desertification consequences. Figure 18.4 shows the great extent of this aquifer, and, by implication, the associated irrigation area. The exposed soils are often susceptible to wind and water erosion. Sheridan (1981) quotes a US Soil Conservation Service representative in Texas:

We are creating a new Great American Desert out there, and eventually the basic resource, soil, will be exhausted.

A specific example: in 1977 high winds in eastern Colorado and New Mexico scoured plowed wheat fields to depths of greater than 1 m, with the silt and sand being visible over the mid-Atlantic Ocean in satellite imagery. This event followed a period of prolonged drought. Another consequence of the conversion to plowed fields that the temporary use of the Ogallala has permitted is that, when the plow abandons the land for lack of water, there will likely not be anyone willing to pay to convert it back to grass, and it will lie open to invasion by non-native plant species and to erosion.

The middle Asian plains

In the 1950s and 1960s, food shortages in the Soviet Union caused the government to open up grasslands in northern Kazakhstan and western Siberia to settlement and farming. Forty million hectares of virgin land were brought under cultivation by hundreds of thousands of settlers. Before farming practices became more conservation-conscious, 17 million hectares were damaged by wind erosion, and 4 million hectares were entirely lost to production (Eckholm 1976). Zonn *et al.* (1994) further describe the environmental consequences of this program.

The loess soil areas of eastern China

Loess is a loamy type of soil consisting of particles that were deposited by the wind, and is very susceptible to water erosion. Because it is one of the most productive types of soil in the world, there is a great deal of economic and social pressure to exploit this land regardless of the environmental consequences.

Among the large areas of Asia where desertification borders on being classified as very severe, there are parts of China that have large deposits of loess 100 m deep that are cultivated. In the Huang He (Yellow) River watershed, over 300 000 km^2 experience major soil loss into the river. The sediment load in the river is sometimes as high as 46% (by mass). In some areas, soil erosion has eaten out gullies that are almost 200 m deep. Of the 600 000 km^2 of loess, it is estimated that 26 000 km^2 are gullies (Kuo 1976). However, the land areas around temples, where the natural vegetation is untouched and abundant, have not significantly eroded. This leads Lowdermilk (1953) to conclude that human misuse of the land caused the damage, rather than climate variability.

Caribbean islands

This is an illustration of the possible desertifying effects of deforestation that can occur even in the humid tropics (Anthes 1984). Beginning in the fifteenth century, many of the Windward and Leeward Islands of the Lesser Antilles were colonized. A consequence of the development of the agricultural industry was a deforestation of most of the islands. This caused a marked desiccation of the soil, drying up of many ponds, and increased erosion and runoff (Beard 1949; Bridenbaugh and Bridenbaugh 1972). However, Dominica retains much of its virgin forest, and Anthes (1984) points out that its average annual rainfall is greater by over a factor of three compared with other islands in the Lesser Antilles that were extensively deforested. It is, however, difficult to separate the deforestation effects from differences in the size and elevation of the islands.

The Tigris–Euphrates River Valley, Iraq

This area was discussed earlier in the context of soil salinization. As a result of millennia of irrigation, 20–30% of the country's irrigable land is now unsuitable for agriculture, or even native vegetation. Eckholm (1976) summarizes the situation well.

That Mesopotamia is possibly the world's oldest irrigated area in not an encouraging observation. The end result of six millennia of human management is no garden spot. The region's fertility was once legendary throughout the Old World, and the American conservationist Walter Lowdermilk has estimated that at its zenith Mesopotamia supported between seventeen and twenty-five million inhabitants (Lowdermilk 1953). One early visitor, Herodotus, wrote that "to those who have never been in the Babylonian country, what has been said regarding its production will be incredible." Today Iraq has a population of ten million, and on the portions of these same lands that have not been abandoned peasants eke out some of the world's lowest crop yields.

The Mediterranean region

It is hard to estimate the pristine state of this region, which has been disturbed by occupation and exploitation for so many millennia. But there is plenty of evidence that it at least looks more desert-like now than before, even though there is little indication of any significant climate change. Deforestation is one of the greater changes. Over 4500 years ago, Egyptians depended upon forests in Phoenicia (currently Syria, Lebanon, and Israel) to supply them with heavy timber for ships and temples, and for the resins that were used in mummification. Continued demand for eastern Mediterranean cedars, pines, and firs was so great that, by 500 years ago, timber was a scarce material (Mikesell 1969). The same degradation has occurred throughout the lands that neighbor the Mediterranean Sea, as reported by Tomaselli (1977), Mikesell (1960), Brandt and Thornes (1996), and Beals (1965).

Physical-process feedbacks that may affect desertification

It is worth being reminded that the term "feedback" implies that a series of cause and effect relationships exist, such that an initial perturbation sets in motion a series of responses wherein the original perturbation is either enhanced or reduced. In the context of desertification, for example, an initial natural or anthropogenic nudge of the system toward greater aridity would cause a series of responses that might contribute toward greater aridity (positive feedback) or less aridity (negative feedback). No enhancement or mitigation of the original perturbation of the system would be neutral feedback. Chapter 2 reviews atmosphere–surface feedbacks that might contribute to the maintenance and formation of deserts, and this section will review some of them in the context of desertification.

Societal feedbacks may be intertwined with physical feedbacks, to the same initial perturbation. Fig. 18.6 depicts this situation for the vegetation–albedo/transpiration feedback discussed below. On the biophysical side, the vegetation reduction contributes to reduced rainfall, which leads to a further reduction in vegetation, and so on. On the societal side, the initial perturbation toward desertification, the drought, results in overgrazing that contributes to the vegetation reduction. Other feedback diagrams treat different aspects of system interactions. For example, Schlesinger *et al.* (1990) present a diagram that links ecosystem properties with global biogeochemistry during desertification. Hulme and Kelly's (1993) feedback diagram includes global warming, and the ocean as part of the mechanism. The Walsh *et al.* (1988) diagram focusses on the hydrologic response.

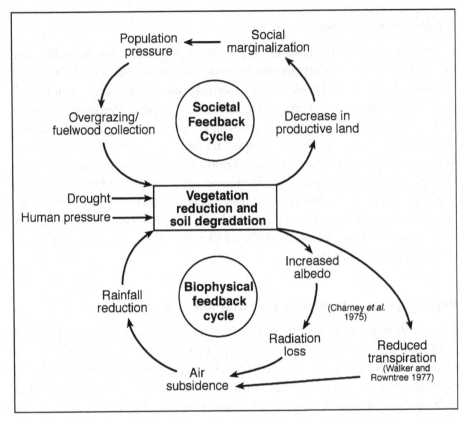

Fig. 18.6 An example of combined biophysical and societal feedback mechanisms. The biophysical component is the vegetation–albedo feedback. (Based on Scoging 1991, with modification by Thomas and Middleton 1994.)

Vegetation – albedo/transpiration feedback

One of the most discussed desertification feedback mechanisms is the albedo feedback proposed by Otterman (1974), Charney (1975), and Charney *et al.* (1977). The concept is that removal of vegetation by overgrazing in arid lands with light-colored sandy soils will cause an increased albedo because the removed vegetation is less reflective than the bare soil that is exposed. This will result in lower net radiation, and lower surface and boundary-layer temperatures. The more stable vertical profile of temperature will inhibit convection, and less rainfall will result, thus increasing the loss of vegetation. This further baring of the soil will cause the process to continue, which makes it a *positive* feedback loop. The work by Otterman (1977a,b, 1981), discussed in Chapter 6, and later in this chapter in the context of remote sensing of desertification, provides convincing documentation of the effects on albedo of grazing and other

Table 18.6 *Ratio of albedoes for anthropogenically impacted desert and nearby protected areas based on satellite data for various parts of the world*

Wavelength (μm)	0.5–0.6 (vis.)	0.6–0.7 (vis.)	0.7–0.8 (IR)	0.8–1.1 (IR)
Sinai/Negev	1.55	1.88	1.87	1.73
Afghanistan/Russia	1.16	1.21	1.21	1.19
Sahel, overgrazed/Ranch	1.20	1.37	1.41	1.39

Source: Adapted from Otterman (1981).

disturbances of arid lands. It is shown that disruption or removal by humans of the dead vegetation debris between live plants in deserts can have as large an effect on average albedo as the removal of the live vegetation. Table 18.6 shows the ratios of the albedoes of overgrazed areas on one side of a boundary to the albedoes of less damaged areas only a few hundred meters away on the other side of the boundary. The ratios are for bands in the visible and solar infrared. Comparisons are for the Sinai–Negev Desert boundary, where the Sinai is under heavier human pressure; for the Afghanistan–Russia border, where the Afghanistan side is more overgrazed; and for a border between the overgrazed Sahel and a protected ranch. The ratios, which range from about 1.2 to almost 1.9, clearly illustrate the albedo response to human pressure, and the potential effect on the surface energy budget and atmospheric processes.

Vegetation removal in the form of deforestation can have similar possible feedback effects, as implied in the above reference to Anthes (1984) regarding the relation of forests to precipitation. Additional evidence is found in Eltahir (1996) who has shown in a modeling study that deforestation reduces the surface net radiation, which leads to tropospheric subsidence that reduces the rainfall. Also, in Chapter 14 were discussed model experiments by Zheng and Eltahir (1997) who show that deforestation of the coastal areas of the Gulf of Guinea reduces the strength of the monsoon flow that provides most of the rainfall for the Sahel and the southern Sahara.

Dust–radiation feedback

A possible positive feedback that could lead to desertification is related to the radiative effects of desert dust. Bryson and Baerreis (1967) and Bryson *et al.* (1964) were among the first to discuss how anthropogenic factors such as deforestation and agricultural exploitation can lead to desiccation of the soil and the elevation of dust throughout the troposphere, and how this can lead to a positive feedback. In a study near the Thar Desert of India and Pakistan, they showed that the dust has the effect of increasing the midtropospheric subsidence rate by about 50%. The enhanced subsidence decreases the depth of the monsoon layer and reduces the monsoon's penetration into the desert. By this mechanism, a natural drought period or human factors could cause an increase in tropospheric dust, and this could sustain or enhance the initial perturbation toward soil dessication and desertification. Bryson and Baerreis (1967) describe archeological evidence of earlier thriving civilizations and agriculture in this region, suggesting that the aridity has increased dramatically over the last few thousand years. They speculate that the widespread documented deforestation and overgrazing that has occurred in this area may have been the initial perturbation that started the feedback that led to a self-sustaining desert.

Water-table deepening – vegetation – runoff feedback

A positive feedback that does not directly involve atmospheric processes was described earlier in the discussion of water-table deepening as an anthropogenic effect on the desert surface, and ultimately the atmosphere. Here, partial devegetation of the surface, often a result of agricultural exploitation such as cattle grazing, increases runoff of rainfall because the rain is not intercepted by the foliage and there is less biological litter on the surface to absorb it. This, in turn, causes water courses to be scoured out and deepened. Where water tables are high, as is sometimes the case in deserts, this results in a lowering of the water table, and a reduction in vegetation whose roots can no longer reach the water table. This reduction in deep-rooted vegetation causes increased runoff. Thus, we have a positive feedback in that a reduction in vegetation can lead to a further reduction in vegetation through the increased runoff and water-table deepening.

Satellite-based methods for detecting and mapping desertification

Dregne (1987) emphasizes the importance of satellite data in shedding new light on some major questions related to desertification:

Claims that the (desert) is expanding at some horrendous rate are still made despite the absence of evidence to support them. It may have been permissible to say such things ten or twenty years ago when remote sensing was in its infancy and errors could easily be made in extrapolating limited observations. It is unacceptable today.

Prior to the availability of satellite data, identification of desertification trends relied upon the knowledge of local inhabitants, written historical references, and ground reconnaissance by scientists. Indeed, to this day there is often no substitute for close first-hand examination, on the ground, of changes in the vegetation or soil characteristics. Nevertheless, the availability of remotely sensed imagery from satellites allows for subjective and objective large-scale analysis of trends in surface conditions. In the simplest approach, spatial and temporal contrasts in the surface brightness (albedo) in visible imagery can be revealing because differences in the amount of vegetation can be easily seen. That is, a vegetation canopy over a sandy substrate surface has a lower albedo than does the sand. Thus, a more vegetated surface appears darker in the visible wavelengths.

Satellite radiances can also detect other disturbances of desert surfaces, in addition to the direct effects on live vegetation. For example, the plant debris and dead plants that often cover the areas between the live vegetation can be removed through livestock grazing and trampling, cultivation, or fuel-wood collection. Otterman (1977a,b, 1981) explains that satellite imagery in the reflective (solar) infrared and visible wavelengths dramatically shows these human effects on the dead vegetation, and how they are important to the surface energy budget. Fig. 6.5 shows reflectivity for different wavelengths in the visible and reflective infrared based on Landsat satellite imagery for surfaces in the adjacent Sinai and Negev Deserts. The Sinai has been much more heavily impacted by overgrazing than has the Negev, and both live vegetation, and dead vegetation in the large interstitial spaces, are less prevalent in the Sinai. An exception is an area in the Sinai that was enclosed three years before the measurements, and natural vegetation had partly regrown. Also shown are the spectral reflectivities, based on a hand-held field radiometer, for live plants, dead plant material, and bare, disturbed soil. The crumbled soil referred to in the figure is what remains after the plant debris has been removed by the above-mentioned human processes. Having multispectral imagery, rather than just visible imagery, leads to the conclusion that the low reflectivities in the Negev Desert result from the fact that the dead plant material had not been disturbed to the extent that it had been in the Sinai, rather than because the Negev had more live vegetation. It is interesting that none of the satellite imagery shows the low reflectance in the visible red part of the spectrum (0.65 μm) that would be associated with the strong absorption by chlorophyll (see "plant" reflectance in Fig. 6.5). This is presumably because vegetation is sparse in both deserts. Thus, visible imagery alone would only have left uncertainty about differences in human impact on the two deserts. However, the satellite-observed reflectivities in the solar infrared are revealing. In this band, live vegetation is very reflective (see figure), as is bare soil, but the dead plant material has very low reflectivity. Thus, one of the major differences between the two deserts is the degree to which human activity has denuded the large areas between the live vegetation.

In an effort to use satellite radiances in the visible and solar infrared for detection of changes in desert vegetation, and associated desertification, approaches have combined information from different wavelength channels to construct indices such as the "Normalized Difference Vegetation Index" (NDVI) and the "green vegetation fraction." Both indices are proportional to the amount of green vegetation. For example, Dregne and Tucker (1988) and Tucker *et al.* (1991) describe the use of the NDVI to trace annual changes in the boundary between the Sahara and the Sahel: that is, the expansion and contraction of the Sahara. It is first demonstrated that mean precipitation and NDVI are approximately linearly related for the area, and the Sahara–Sahel boundary is defined as coinciding with the 200 mm y^{-1} isohyet. Thus, the NDVI field can be used to define the boundary location, from which insight can be gained about desertification. To provide a Sahel-average view of vegetation changes, Fig. 6.2 shows the average annual NDVI from 1980 to 1990 for the area of the Sahel having a long-term mean rainfall of 200–400 mm y^{-1}. Also shown is the corresponding NDVI for a zone in the central Sahara. These significant variations in the NDVI correspond well with the Sahel rainfall variability for the period. The year 1984 was one of the driest this century, and the vegetation clearly reflects this. In order to map the advances and retreats of the Sahara during this period, the NDVI was used to estimate the location of the 200 mm y^{-1} isohyet, defined as the Sahara's southern boundary. The 1984 longitudinal mean position of the "boundary" was mapped to be over 200 km farther south than in 1980. By 1988, it had receded northward by most of this distance.

Satellites can also be used to estimate soil moisture in arid areas. There are various possible approaches; Milford (1987) and Bryant *et al.* (1990) summarize some applications and limitations. Additional discussions of the use of satellite data to monitor land-surface conditions in arid areas can be found in Chen *et al.* (1998), Robinove *et al.* (1981), Justice and Hiernaux (1986), and Nicholson and Farrar (1994).

Suggested general references for further reading

Brandt, C. J., and J. B. Thornes, 1996: *Mediterranean Desertification and Land Use* – for the Mediterranean, contains a history of desertification and land use, a treatment of climate and climate change, a discussion of various field programs that have studied desertification processes, and a description of the modeling of different components of the desertification process.

Eckholm, E. P., 1976: *Losing Ground: Environmental Stress and World Food Prospects* – a discussion of deforestation and various other causes of desertification, with a treatment of land degradation in the humid tropics.

Glantz, M. H. (Ed.), 1994: *Drought Follows the Plow: Cultivating Marginal Areas* – discusses drought, desertification, and food production, and offers a number of case studies of desertification in the West African Sahel, Somalia, northeast Brazil, the dry region of Kenya, Australia, Ethiopia, northwest Africa, the former Soviet Union, and South Africa.

Grainger, A., 1990: *The Threatening Desert* – discusses the causes and definition of desertification, its scale, and its control.

Hellden, U., 1991: *Assessing desertification* – reviews the processes and status of desertification.

Lowdermilk, W. C., 1953: *Conquest of the Land Through 7000 Years* – a short summary, based on field observations, of the effects of agriculture, as practiced over the past seven millennia, on soil erosion and desertification.

Mainguet, M., 1991: *Desertification: Natural Background and Human Mismanagement* – contains summaries of the various meanings of desertification. An extremely comprehensive list of references is provided.

Thomas, D. S. G., 1997c: *Science and the desertification debate* – discusses the social and science dimensions of the desertification problem, and why scientific findings may have been misinterpreted by policy makers. It also identifies areas where science still has a role to play.

Thomas, D. S. G., and N. J. Middleton, 1994: *Desertification: Exploding the myth* – a general reference on desertification, with a comprehensive description of the institutional and historical aspects of the problem. It does justice to its title. An extremely comprehensive list of references is provided.

Warren, A., *et al.*, 1996: *The future of deserts* – describes the many pressures on arid lands, from urban and rural population increases, oil development, and water development, and the impacts on soil, vegetation, and climate.

Questions for review

(1) Summarize the various human causes of desertification.
(2) Review the types of physical feedback process that can initiate or enhance the desertification process.
(3) How can physical and societal feedbacks interact to enhance desertification?
(4) Why are extremely arid areas not considered at much risk of desertification?

Problems and exercises

(1) In the discussion of Fig. 18.6, which depicts societal and biophysical desertification feedbacks, references to other feedback diagrams are provided. Refer to these other diagrams and summarize them.
(2) Read the paper by D. S. G. Thomas (1997c) entitled *Science and the desertification debate*, and summarize what you consider to be the continuing role of science in better understanding desertification and how it can be controlled.

19

Biometeorology of humans in desert environments

Just as the Irish are said to have 40 words for the color green, desert dwellers have many ways of expressing nuances of thirst. The following are Arabic expressions.

al-'atash   thirst
al-Zama'   thirst
al-Sada   thirst
al-Ghulla   burning thirst
al-Luhba   burning thirst
al-Huyam   vehement thirst (or passionate love!)
al Uwam   burning thirst, giddiness
al-Juwad   excessive thirst (this is the thirst which kills)

E. S. Hills, arid-land researcher
Arid Lands (1966)

The psychological effects of desert heat and wind are described.

In the case of the Santa Ana winds, high pressure over Utah and Nevada causes air to spill off the Mojave Desert, rushing over the Pacific coastal range and onto the coastal lowlands. The coastal air is robbed of humidity by this thirsty invader and fills with static electricity. As it envelopes desert and littoral alike, the Santa Ana creates a weird atmosphere of impending doom. During its season, as Raymond Chandler wrote in his famous short story Red Wind, "Meek little wives feel the edge of the carving knife and study their husbands' necks. Anything can happen."

Gregory McNamee, American author
The Sierra Club Desert Reader (1995)

Biometeorology is the study of the response of living organisms to weather and climate. In particular, this chapter will address the effects of the desert environment on humans. First will be described the various mechanisms by which heat can be gained and lost by the body. This will be followed by a discussion of the ways in which the body attempts to maintain the thermal balance that is required to sustain life.

491

The heat balance of the human body in the desert

The different components of the body's heat balance are the same for deserts as for non-deserts. That is, the ways in which it exchanges heat with its surroundings are identical. We will see, however, that the high environmental temperatures cause most of the heat-transfer paths to be ones in which the organism gains heat rather than loses it. The mechanisms by which the human body gains and loses heat are explained in the following subsections. They include the emission and absorption of long-wave radiation by the body; conductive and convective exchanges of heat between the body and the atmosphere; the metabolic or internal generation of heat by the body; the absorption by the body of diffuse, direct, and reflected solar radiation; and the cooling of the body by evaporation of perspiration.

Long-wave radiation

In Chapter 4 was described how all matter emits infrared radiation with an intensity that is proportional to the fourth power of its absolute (or Kelvin) temperature. The normal skin-surface temperature of the human body is about 306.3 K (33.3 °C, 92 °F), and it thus emits infrared accordingly. Where the body is loosely clothed, some of the skin's radiation is intercepted and absorbed, heating the clothing. The clothing radiates to both the environment and the skin. Where the skin is exposed to the environment, its emitted radiant energy is absorbed by the surrounding substrate, vegetation, or atmosphere, or it is transmitted through the atmosphere. The body also receives infrared energy that is emitted by its surroundings; this received energy has an intensity that is proportional to the fourth power of the temperatures of the emitters. If the skin is clothed, the energy is absorbed by the covering, and if it is not, the skin absorbs the energy. Whether there is a net heat loss or gain from emission and absorption of long-wave radiation by the body depends, primarily, on the temperatures of the emitters in the environment relative to the temperature of the skin. In the simplest case for skin that is exposed (not clothed), if the dominant emitter in the environment has a temperature that is greater than the skin temperature, the intensity of the absorbed long-wave radiation will be greater than the intensity of that emitted by the skin, and there will be a net heat gain. The situation is complicated by the fact that many different emitters in the environment, including the air, bathe the skin in infrared, and the total rate of radiative heat absorption is a function of the temperatures of all of them. The body's proximity to the source of the emitted infrared also determines the intensity, as anyone can attest who has walked near an equatorward-facing rock wall or brick building in the afternoon or evening. Lee (1964) estimates that, for individuals walking during the day in the Sonoran Desert in summer, almost 50% more energy was gained by long-wave absorption from the terrain than was lost by long-wave emission from the skin.

Conductive and convective exchange with the atmosphere

In Chapter 2 is described how heat is transferred between Earth's surface and the atmosphere through conduction, within a laminar sublayer that is a few molecules to a few millimeters in depth. Outside that layer, turbulence is responsible for the transport of heat away from the source. Analogously, above the skin of the human body is a similar thin layer of air through which conductive heat exchange takes place with the air. The rate of heat exchange depends on a flux–gradient relation of the sort represented by Eqn. 5.9. Thus, the heat will flow from high to low temperature, which means that the body will gain heat by this mechanism if the air temperature is greater than the skin temperature, normally 33.3 °C (92 °F) in the shade and 35–38 °C (95–100 °F) in the sun. When the skin is not exposed to wind, the thin laminar sublayer of air is warmed or cooled to near the skin temperature, reducing the temperature contrast at the surface and the resulting conductive heat flux. When the skin is exposed to wind, the laminar sublayer is stripped away and the skin is exposed to air that is near the ambient temperature of the atmosphere. To non-desert dwellers, this concept is of greatest importance in the context of the "wind chill factor" that accounts for the cooling effect on the body of cold air ventilating the exposed skin. From experience and from wind-chill charts we know that the cooling effect is directly proportional to wind speed. The situation is analogous in the desert, where the heating effect of air that is warmer than the skin temperature is proportional to the wind speed. Here, clothing brings benefits by preventing the thin layer of air near the skin from being removed by the wind. Extensive discussions of the boundary layer around the human body can be found in Lewis *et al.* (1969) and Gates (1972). Conduction of heat also takes place between the body and solid objects such as the ground on which we may be standing or reclining.

Figure 19.1 shows the results of experiments where human test subjects were exposed to different air temperatures. The objective was to determine how the air temperature affected the rate at which heat was gained from the atmosphere by conduction. Other heat-budget factors such as exposure to the sun, the wind speed, clothing, and physical activity (metabolic heat production) were the same for all air temperatures. Because the air temperature (and the associated conduction) was the only factor that varied, it was assumed that differences in the cooling rate by perspiration were equal to differences in the heat gain by conduction.

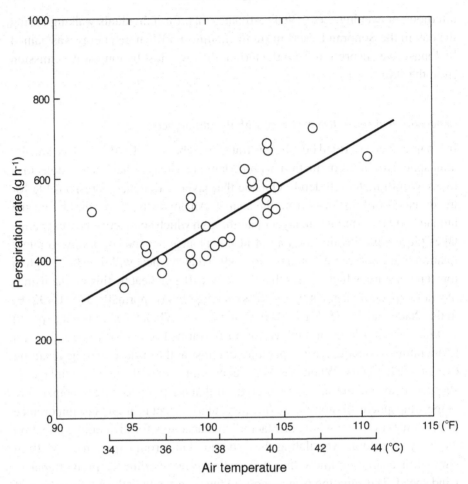

Fig. 19.1 Influence of air temperature on perspiration rate. The line is the least-squares fit to the data. (From Gosselin 1969.)

Perspiration/cooling rates, and therefore the rate of heat gain by conduction, varied by over a factor of two over a 10 K (18 °F) range of temperatures. Note that perspiration rates are generally estimated by weighing test subjects hourly, and accounting for the mass of the other gains and losses.

Metabolic source of heat

The body at rest generates about 80 kilocalories (3.34×10^5 J) of heat per hour (see Appendix C for conversion factor). This amount of heat is enough to bring a liter of ice water to the boiling point. The greater the level of exercise, the more

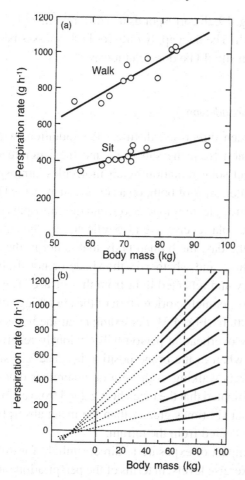

Fig. 19.2 (a) Relation between perspiration rate and body mass, for walking and sitting humans, based on field program data, and (b) schematic family of lines that relate perspiration rate to body mass for different work rates and degrees of radiative heat exposure. That is, a given line represents a particular rate of work and thermal exposure. The lines in Fig. 19.2a are least-squares fits to the data. (Adapted from Gosselin 1969.)

heat is generated. For example, walking at a normal to fast pace generates about 280 kcal per hour. At high work rates over a typical work day, the 24 h total metabolic energy generation would be about three times the value for a person at rest (Lee 1964).

Figure 19.2, based on field experiments, shows the strong dependence of perspiration rate, and therefore metabolic heating rate, on the level of physical activity. The greater the activity level, the more metabolic heat is generated and the greater is the rate of perspiration cooling required for maintaining a heat balance. Each of the lines in Fig. 19.2b corresponds to a different rate of work or degree of heat exposure, or both. The data apply only for body masses between 50 and 100 kg, so extrapolation should be done cautiously.

Direct, diffuse, and reflected solar radiation

Figure 4.4 distinguishes between direct and diffuse solar radiation, with the former being the direct emission from the sun itself and the latter resulting from the scattering of the direct solar radiation by air molecules, atmospheric particulates, and clouds. The intensities of both, on a surface such as the human body, depend on the sun's zenith angle, the cloud cover, and the atmospheric dust content. The direct and diffuse solar energy are also reflected by the substrate and vegetation. Thus, during the day, the human body that is not in the shade absorbs the direct solar radiation, and whether in the shade or not it absorbs scattered radiation from the sky and reflected light from the surface. The total-body absorption depends on the geometric orientation of the absorbing surfaces of the body relative to the emitters (Fig. 4.3). For example, at the hottest time of the day with the sun nearly overhead, a person will obviously receive less radiation when standing than when in the prone position because less surface area is exposed. For test subjects walking in the sun on a summer day in the Sonoran Desert, Lee (1964) estimated that the mean heat gained from absorption of direct solar energy was about six times greater than the mean absorption of solar radiation that had been reflected from the terrain.

Figure 19.3 illustrates the impact of exposure to direct sunlight for different air temperatures and for two exercise levels, in terms of the perspiration rate that is required for maintaining the heat balance. At high air temperatures, exposure to direct sunlight requires about a 30–40% greater cooling rate than if the same activity is undertaken in the shade or at night.

It is worth noting that the body's radiant heat gain directly from the sun in the desert is no greater than on a clear day in any other area of similar altitude and latitude. But reflection from a high-albedo sand surface could cause the gain from this source to be greater in the desert.

Evaporation

The effects of solar radiation and metabolic heat generation on the body's heat budget always represent an energy gain to the body, and long-wave radiative exchange and conductive heat exchange between the body and its environment can be either a gain or loss. In a *cool environment*, evaporative heat loss from the

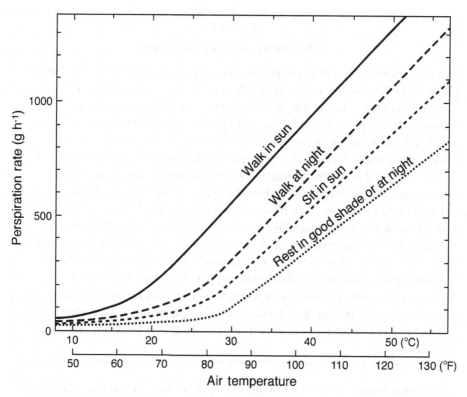

Fig. 19.3 Perspiration rate for different types of activity at different air temperatures. The perspiration rate of clothed test subjects has been normalized to a body mass of 70 kg. The data were obtained from field programs in the Sonoran Desert, supplemented by laboratory studies. (Adapted from Brown 1969a.)

respiratory system is equivalent to about 25% of the metabolic heat gain in a resting individual. There is no heat loss from perspiration. The rest of the metabolic heat is lost through conduction and radiation to the cooler environment. In the desert during the daytime, however, the above four processes generally all contribute to a heat gain by the body. The only mechanisms for cooling the system are the evaporation of perspiration from the skin (see Box 19.1), evaporation from the moist respiratory passages, and a slight subcutaneous evaporation that releases vapor through the skin, where the body supplies the latent heat of vaporization in each case. The latter two mechanisms can evaporate only about 60 g h^{-1}, and therefore produce cooling at a rate of about 35 kcal h^{-1}. Each gram of water that is vaporized consumes 0.58 kcal of heat. So, for each 100 kcal of heat gained, evaporating 173 g (0.173 l) of body water would balance it.

The rate of evaporation of perspiration on the skin depends on the same factors that were discussed earlier for evaporation from any water surface: high

---

Box 19.1
Thermoregulation in a sauna

Saunas employ dry air that is heated to temperatures of as high as 110 °C (230 °F).
How can human bodies survive in such heat for short periods of time? After all, a
pan of water taken into a sauna of this temperature will boil, and a steak will cook
thoroughly within the period of time that a person remains in the heat. One factor
certainly is that the human body develops its own shallow boundary layer of air
whose temperature remains close to that of the skin. Gates (1972) tested the
importance of this insulating effect by blowing on the back of his hand while
in a sauna. This disturbance to the boundary layer exposed the skin to the direct
effects of the hot air, and a red blister immediately formed. The tips of his nostrils
also became raw because breathing disturbed the boundary layer there. Rapid
evaporation of the perspiration on the skin and the moisture in the respiratory canal
are also obviously important in preventing those surfaces from receiving burns.
In spite of these protections, clearly the body could not provide sufficient
thermoregulation to prevent its core temperature from rising if one remained in the
sauna room for very long. But, the fact that such temperatures can be tolerated for
30–45 min in a sauna attests to the efficiency of the body's insulation and cooling
mechanisms.

---

evaporation rates result from low humidity, high wind speed near the surface,
and high temperature. In deserts these factors are generally sufficiently favorable
for rapid evaporation so that perspiration is often not visible on the skin. That is,
it evaporates as fast as it is formed, in contrast to the situation in humid climates
where profuse perspiration is obvious. Table 19.1 shows how much perspiration
evaporates under different conditions, based on measurements in the northern
Sonoran Desert. In most situations, a very large percentage of the perspiration
evaporates. An exception is that large rates of perspiration at high temperatures
(e.g. 45 °C) result in a significant fraction of the perspired water not benefiting
the body's heat budget through evaporative cooling (i.e. the perspiration drips
off the skin). In more humid conditions, such as in foggy deserts and in the hot
deserts downwind of water bodies, even more body water is wasted.

Another factor that affects the evaporation rate from the skin is the existence
of dissolved salt in perspiration. The salt accumulates on the skin as the water
evaporates, and can reach sufficient concentrations that it saturates further perspired water. With the saline solution at saturation, the saturation vapor pressure
is less than that over a surface of pure water. That is, a vapor pressure that corresponds to a relative humidity of 75% with respect to pure water is equal to the
saturation vapor pressure over water that is saturated with salt. This reduces the
rate of evaporation.

Table 19.1 *Perspiration rates of individuals in the northern Sonoran Desert in summer*

Fans created air movement of about 5 m s^{-1} to provide a controlled, simulated wind speed. The walking speed was 1.5 m s^{-1}, and the vapor pressure varied from 7 to 20 mbar.

Conditions	Rate of perspiration production (g h^{-1})			Rate of perspiration evaporation (g h^{-1})		
	35 °C	40 °C	45 °C	35 °C	40 °C	45 °C
Nude, marching, sun	965	1210	1450	935	1165	1375
Clothed, marching, sun	—	—	—	640	910	1120
Clothed, marching, shade	—	—	—	490	730	960
Nude, sitting, sun	455	685	915	385	615	800
Clothed, sitting, sun	280	500	730	280	460	610
Nude, sitting, shade	220	380	540	220	360	475
Clothed, sitting, shade	245	435	620	245	305	375
Marching, shade						
Shorts, fans	690	930	1120	675	910	1100
Shorts, no fans	640	850	1015	—	—	—
Clothed, fans	570	805	900	500	735	900
Clothed, no fans	575	775	940	575	740	860

Source: Adapted from Lee (1968).

The process of maintaining the heat and water balance

The body's heat balance may be represented symbolically as

$$Q_M + Q_{LR} + Q_{SR} + Q_H + Q_E + Q_G = \Delta Q_S, \tag{19.1}$$

where Q_M is the metabolic heat gain (always positive), Q_{LR} is the long-wave net radiative gain or loss of heat, Q_{SR} is the solar radiative gain of heat (always zero or positive), Q_H is the sensible heat exchange with the atmosphere by conduction and convection (positive or negative), Q_E is the evaporative loss of heat (always negative), Q_G is the heat gain or loss by conduction with the ground, and ΔQ_S is the net heat gain (storage) or heat loss as a result of an imbalance among the terms.

The ΔQ_S term must remain relatively small, or be large and of the same sign for only a short period, in humans because of the small range of tolerable body temperatures. In particular, the core body temperature must remain close to 37 °C (98.6 °F) for the long-term health of the organism. However, just the metabolic heat generated in one hour by a person at rest is sufficient to raise the body temperature by about 1 K (1.8 °F) if no heat is lost, and in the desert in the daytime there are the other aforementioned sources. Thus, the body has a couple of mechanisms by which it regulates its temperature. One of course is the rate at which perspiration is produced. Another is dilation of the arteries that control the capillary blood flow near the surface. The capillary blood that flows near the skin surface is cooled if its temperature is greater than the air temperature, and this allows an exchange of heat with the interior. The blood supply to the skin is about $0.16\ \mathrm{l\ m^{-2}\ min^{-1}}$ for a nude person resting with an air temperature of 28 °C (82.4 °F), but increases to about $2.6\ \mathrm{l\ m^{-2}\ min^{-1}}$ in a very hot environment (Robinson 1963). Also, heat conduction through the body tissue allows the surface cooling to lower the internal temperature. However, all but the metabolically generated heat is gained at the surface of the skin, so the loss of heat by evaporation at the surface does not require heat transport from the deep tissue in order to maintain the necessary temperature.

Given sufficient water consumption, is there any limit to the ability of the body to cool itself and maintain a heat balance? It appears that there is, because strenuous physical activity in the desert has been shown to cause rectal temperatures to increase by 2.5 K (4.5 °F) in one-half hour (Adolph 1969), even with unlimited water. Apparently the transport of metabolic heat from the interior of the body to the skin where the heat is lost is not of the necessary efficiency to allow a steady state to be attained. In addition, adequate water consumption does not necessarily lead to perspiration rates that provide sufficient cooling. This is supported by the fact that perspiration rates in the desert have not been observed in excess of $1700\ \mathrm{g\ h^{-1}}$ (sustained for 1 h), whereas in humid conditions at 34 °C (94 °F) perspiration rates as rapid as $3500\ \mathrm{g\ h^{-1}}$ have been sustained for several hours (Adolph 1969). The reason for the lower, and sometimes insufficient, perspiration rate in the desert is apparently that the skin temperature does not rise enough (because of efficient evaporation of perspiration) to produce a maximum response from the sweat glands.

One way of illustrating the effect of environmental temperature and the rate of metabolic heat production on the perspiration rate and the pulse rate is through a diagram of the type shown in Fig. 19.4. From the slopes of the lines that separate the temperature regimes it is clear that the perspiration rate increases more rapidly with increasing work rate (A, B, C) for higher temperatures (1–5). At a given work rate, the pulse rate also increases gradually with increasing temperature.

Figure 19.5 is a schematic that relates the rate of heat production or loss in humans to the environmental temperature. The quantity T_{b} is the deep-body temperature, which remains relatively constant within the environmental temperature range BE that represents the zone in which the system can adequately regulate its heat balance. The heat-production curve represents the rate at which metabolic heat (Q_{M}) is generated in order to maintain body temperature. Point B

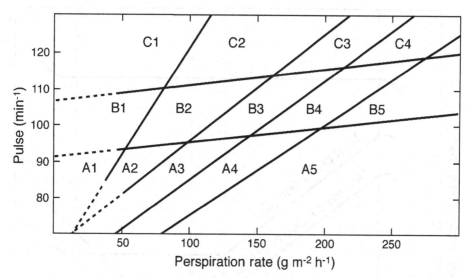

Fig. 19.4 Perspiration rate and pulse rate as a function of environmental temperature (T) and work (metabolic) rate (WR). A, WR < 100 Cal h^{-1}; B, WR $= 100$–200 Cal h^{-1}; C, WR > 200 Cal h^{-1}; 1, $T > 22$ °C; 2, $T = 22$–27 °C; 3, $T = 27$–32 °C; 4, $T = 32$–37 °C; and 5, $T > 37$ °C. (From Lee 1968.)

is the maximum rate of metabolic heat production. As the environmental temperature increases from B to C, less metabolic heat production is needed to maintain body temperature, and non-evaporative heat losses (Q_{LR}, Q_H, Q_G) decrease. In the temperature range BC, the heat loss associated with evaporation from the respiratory tract and vapor transport through the skin are largely independent of environmental temperature. At temperatures between C and D, the heat loss from non-evaporative mechanisms and from evaporation without perspiration approximately balance the minimum metabolic heat generation (and whatever is gained from Q_{SR}). This is called the zone of least thermoregulatory effort. As the environmental temperature approaches the skin temperature, above D, the non-evaporative loss plummets and the evaporative loss must increase to compensate. That is, perspiration begins just to the right of D. As non-evaporative heat-transfer mechanisms begin to represent very large gains in the range E–F, and the body is unable to export sufficient heat from the core, T_b begins to rise significantly and hyperthermia results.

To illustrate the response of T_b in the critical D–E–F area of Fig. 19.5, Fig. 19.6 depicts data on the body's core temperature for three different work rates at different environmental temperatures. Over a wide range of temperatures between D and E, the body is able to regulate the heat. But when a critical temperature is exceeded, the body is unable to effectively cool its core, and deep-tissue

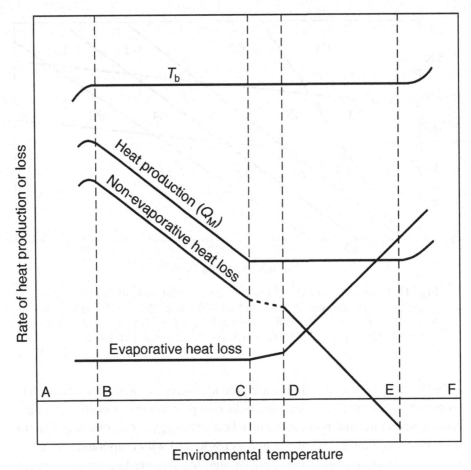

AB - zone of hypothermia
BE - zone of thermoregulation
CD - zone of least thermoregulatory effort
EF - zone of hyperthermia
B - temperature of summit metabolism and incipient hypothermia
D - temperature of marked increase in evaporative loss
E - temperature of incipient hyperthermia

Fig. 19.5 Schematic that relates the rate of heat production or loss to the environmental temperature. The quantity T_b is the core body temperature. See the text for details. (From Mount 1974.)

temperatures rise markedly. The critical environmental temperature is slightly higher for lighter levels of work (lower metabolic rates, Q_M).

The previous discussion and figures refer to different hourly perspiration rates, and hence water requirements, for different situations in the desert. However, it is also useful to consider the fluid requirement during a complete 24 h period,

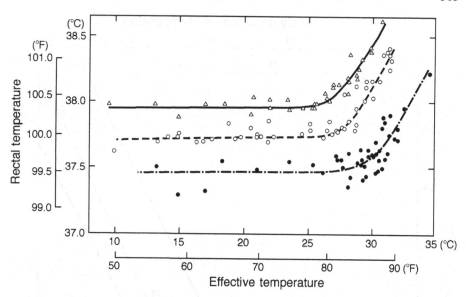

Fig. 19.6 The rectal temperature at three work rates for different environmental temperatures. The upper curve is for the highest work rate, and the lower curve is for the lowest work rate. (From Leithead and Lind 1964.)

allowing for typical diurnal variations in environmental temperature, sun angle, and activity level. Fig. 19.7 shows this daily fluid requirement for different environmental, daily mean temperatures. The daily maximum temperature is assumed to be 11 K (20 °F) higher than the daily mean temperature. Fluid requirement curves are shown for hard work during 8 h of the day, the same work during 8 h of the night, and resting in the shade for the entire period. The fluid saving from performing work at night is also shown. As an example of how computed daily fluid requirements vary within a geographic area, Fig. 19.8 is provided for North Africa in July. The values are based on climatological estimates of the meteorological conditions, and the assumption of an acclimatized person at rest. Analogous maps of water requirements for the other desert areas of the world are found in Adolph (1969).

Acclimatization to the desert environment

Even though it has been shown that humans cannot develop an ability to survive with less water, there is the issue of whether one can become "acclimatized" to the desert heat in some other sense. In fact it has been shown that repeated or continuous exposure to desert conditions leads to a reduced pulse rate under stress, reduced internal body temperature, an ability to work faster with comfort,

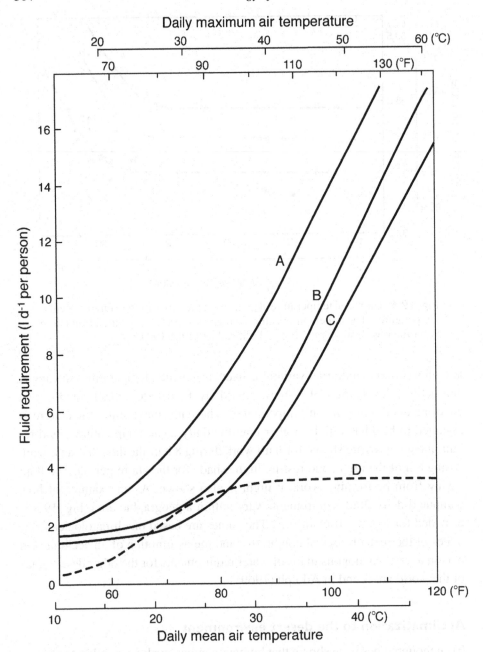

Fig. 19.7 Dependence of daily human fluid requirement on daily mean air temperature and activity level: A, hard work during 8 h of the day; B, same work as A during 8 h of the night; C, rest in the shade; D, water saved by night work (A − B). (From Brown 1969a.)

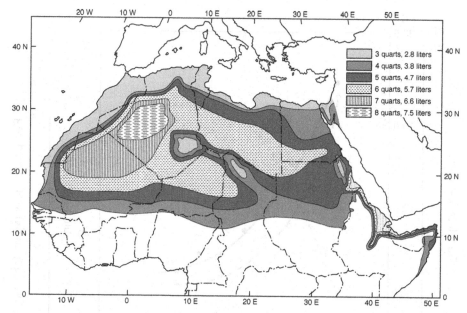

Fig. 19.8 Estimated daily water requirement based on average July climate, for an acclimatized person at rest. (From Brown 1969a.)

and an increased feeling of wellbeing (Adolph 1969). For example, Bass (1963) describes the changes that accompanied acclimatization by individuals who were asked to walk at 5.6 km h^{-1} for one hour at 49 °C (120 °F). Early in the acclimatization period, the subjects experienced nausea, dizziness, a very rapid pulse, and an inadequate secretion of perspiration. After a few days, the symptoms of discomfort diminished, and perspiration became more copious.

Figure 19.9 shows this effect quantitatively in terms of the changes in the core body temperature, the rate of sweat loss, and the pulse rate during a nine-day period in which test subjects became acclimatized to desert conditions. The data for each day apply for a 100 min period of work at an energy expenditure of 350 W, in an environment with a temperature of about 49 °C (120 °F). During the nine-day period, core temperature dropped by over 1 K (1.8 °F), pulse rates decreased by over 40 beats per minute, and the rate of sweat loss (cooling rate) increased by almost 20%. Kerslake (1972) cites a study that shows almost a factor of two increase in peak perspiration rate as a result of acclimatization. The volume of circulating blood may also increase by 15% (Lee 1968), reducing some of the side effects of heat regulation described later, and the salt concentration in the perspiration decreases.

Ingram and Mount (1975) mention another effect of acclimatization to desert heat. In the unacclimatized individual under heat stress, thirst is not great and

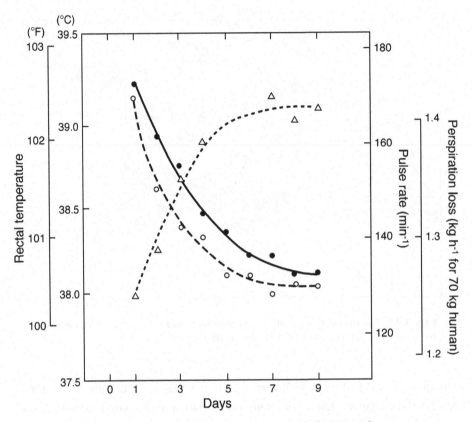

Fig. 19.9 Typical changes in the pulse rate (open circles), rectal temperature (solid circles), and perspiration loss (triangles) during a nine-day period in which individuals were worked for 100 min at an energy expenditure of 350 W. Day-zero activities were in a cool climate, but on days 1–9 they were repeated in a desert climate with a temperature of 48.9 °C (120 °F). (From Leithead and Lind 1964.)

water intake is not adequate to replace that lost. The acclimatized person, in contrast, is much more thirsty for a given water deficit, and better maintains the water balance by voluntary drinking. Even then, only 50–70% of the water deficit is consumed when water is freely available (Tromp 1980). Another adaptation to the heat is that the body conserves salt by reducing the concentration in the perspiration and the urine.

It is also an interesting issue whether some individuals have innate, genetic, or acquired (beyond acclimatization) characteristics that make it easier or more difficult for them to work comfortably in the desert. One factor, shown in Fig. 19.2, is that individuals with high body mass require more water, and they are less efficient at lowering their internal body temperature. There is some evidence that persons indigenous to hot climates have perspiration rates that are less than those of acclimatized Europeans in the same environment (Edholm 1972). Another study of different ethnic groups living in a particular hot climate showed very little difference in heat tolerance among the groups (Strydom and Wyndham 1963). The benefits of heat acclimatization are lost rather quickly when individuals are removed from the hot environment, or the work rate in the desert lessens. Estimates are that reversion to the unacclimatized state takes place in a few weeks.

Another aspect of acclimatization to desert environments is not physiological. Rather, it relates to the development of an economy of movement while doing required work. By becoming more efficient in undertaking physical activity, less metabolic heat is generated and more work can be accomplished with less water required.

Economizing water and reducing the heat load

Three proven ways of economizing on our need for water in the desert are to (1) cease activity, (2) remain clothed, and (3) avoid direct exposure to the sun. Beyond these measures, there are many questions about conserving water that relate to how it is consumed. For example, should the available water be consumed at a steady rate during the period when it is needed, should it be saved (conserved) in the canteen and only consumed at long intervals or when thirst becomes extreme, or should it be consumed as needed based on thirst? Old superstitions abound of the sort that state that water should not be consumed during the day or during hard work in the desert, because "drinking water only makes you more thirsty." There is also the idea that self-deprivation can actually "train" the body to require less water in the long term; that is, that the acclimatization process includes adjusting toward a decreased need for water. Numerous studies have addressed these issues. The consensus is that there is no long-term saving of water by deferring its consumption. Water loss through perspiration occurs at a rate dictated by the heat budget, not by how much water is stored in the body (see the approximately straight lines in Fig. 19.10), and there is no significant additional loss through urination if water is consumed as needed. In fact, it is beneficial to consume water when thirst signals a need for it, regardless of whether it is during the heat of the day or during a period of hard work. By doing so allows one to operate more efficiently and comfortably, and avoids the possible severe consequences of too much deprivation of water. That is, it is better to carry the water in your body than in your canteen.

Figure 19.11 shows daily perspiration output as a function of daily water intake, with the difference being urine output. Water was consumed as needed.

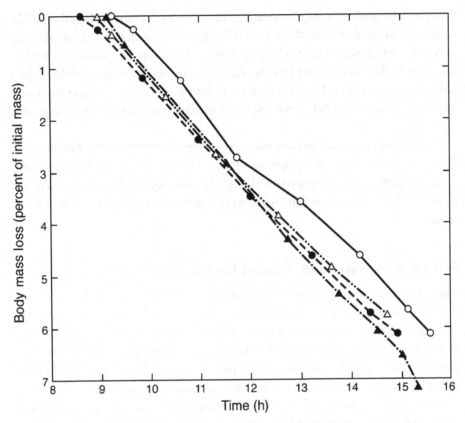

Fig. 19.10 Loss of body mass by each of four test subjects dehydrating at a controlled air temperature of 50 °C (122 °F). (From Gosselin 1969.)

Even for high water intake rates, nearly all the water consumed was used for production of perspiration. Urine volume tended to be about $0.7\,l\,d^{-1}$ for a wide range of intake volumes.

When expending energy at a high rate and sweating profusely in the desert, there can be a false sense that one is economizing on water because thirst in those conditions is ironically experienced less than when at rest in the same environment. The water deficit that develops during this period needs to be made up for eventually, however.

It goes without saying, then, that one should consume as much water as possible before entering a desert for a short period, if for no other reason than to avoid carrying extra canteens. Even if some of the water from this pre-hydration is lost through urination, before the body demands it for perspiration, there is still some net gain. If conditions are already hot and the perspiration rate is high, most of the additional water will not be wasted.

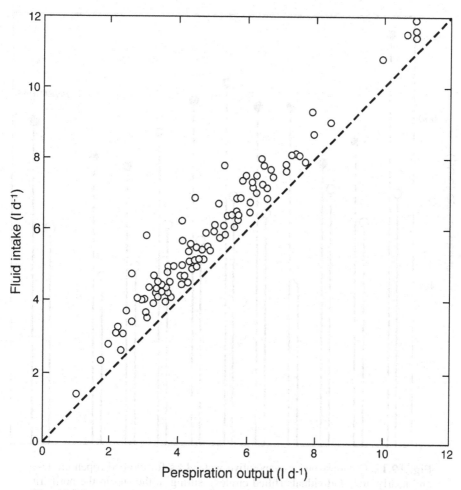

Fig. 19.11 Relation between fluid ingested and the output of perspiration. The diagonal represents a one-to-one correspondence. The offset of the observations from the diagonal represents fluid loss by urination. (From Gosselin 1969.)

Of the three ways of conserving water that were mentioned at the beginning of this section, perhaps the easiest one to adopt concerns remaining clothed. That is, if one is in the desert to enjoy it, it is hardly possible to cease activity and to remain in the shade all of the time. Fig. 19.12 shows the effect of clothes on the perspiration rate for test subjects that were fully clothed, and for the same subjects when they were nearly nude. In many cases, removal of the clothes led to a 50–100% increase in the cooling required from perspiration.

If a certain distance must be traversed on foot, there is great advantage to traveling at night to avoid the heat gained by direct exposure to the sun. An estimate is that, for a given amount of water available, an average person will

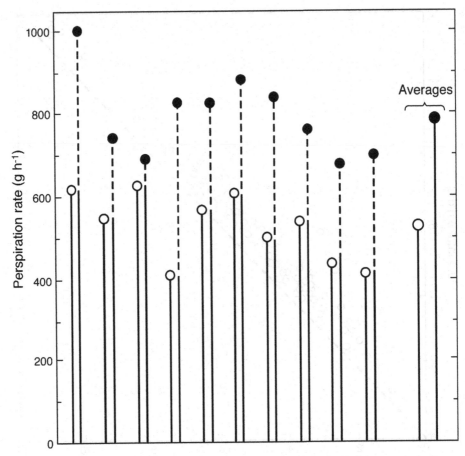

Fig. 19.12 Comparison of evaporative water loss for clothed (open circles) and nearly nude individuals (filled circles), sitting in the sun in the northern Sonoran Desert. Each pair of vertical lines corresponds to measurements on one person. Values are normalized for a 70 kg person. (Adapted from Gosselin 1969.)

be able to travel about three times as far by walking at night rather than during the day, provided that good shade can be found during the day (Adolph 1969). Walking at night will require about one gallon (3.785 l) of water per 20 miles (32 km) traversed. The benefit of walking at night is illustrated in Fig. 19.3. During the day, if the temperature is 43 °C (110 °F), walking in the sun consumes about 1100 g h^{-1} of water, compared with 450 g h^{-1} for sitting in the shade. However, at night if the temperature is 32 °C (90 °F), the difference in the cooling rate between sitting and walking is only about 200 g h^{-1}.

Figure 19.13 illustrates the number of days without water during which night walking would be possible in north Africa. Over most of the Sahara and Sahel, the limit is less than two days. The survival time, if walking is done at night

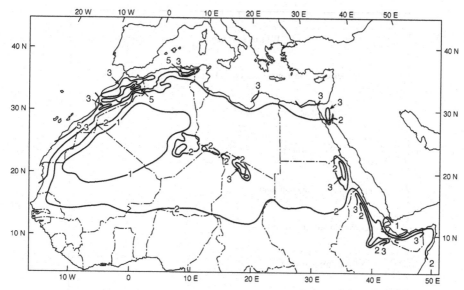

Fig. 19.13 The number of days without water for which night walking is possible. See the text for details. (From Brown 1969b.)

until walking is no longer possible, can be estimated by multiplying the isopleth numbers by 2.3. If no walking is attempted, the survival time can be estimated by multiplying the numbers by 2.6. To obtain the approximate number of kilometers that are walkable at night, before walking is no longer possible, multiply the number by 29. Similar maps are shown in Adolph (1969) for the arid areas of North America, the Middle East, and Australia.

It is worth commenting on the desperate practice of drinking very salty water or urine, when severely dehydrated with no alternative source of water. In the case of the latter, the ingested solutes in the urine will be excreted again, accompanied by the same amount of water, so nothing is gained. This also applies to the drinking of sea water. In order to be physiologically useful, brackish water may be only about one-half as saline as sea water. Very salty water and urine can be put to good use by allowing them to evaporate from the clothing or skin, producing cooling without the use of water from perspiration.

Figure 19.14 represents a good summary of the cooling required, in terms of the perspiration rate, for different types of exposure and exertion in the desert. The cooling required varies by a factor of four for the different conditions.

Physical effects of heat stress and dehydration

The human body consists of about two-thirds water. However, under normal circumstances a departure in the water volume of as little as 1% causes either

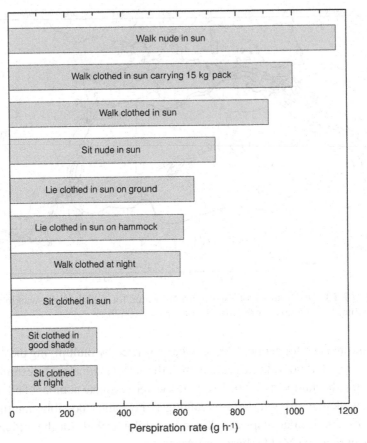

Fig. 19.14 Summary of human perspiration rates for ten conditions of exposure to desert air at 37.8 °C (100 °F). Values are normalized for a 70 kg person. (From Gosselin 1969.)

a removal of the excess through the kidneys or a thirst. At a dehydration of 4% of body mass, the mouth is extremely dry, and at 8% the tongue is swollen and speech is difficult. At 10% loss of body mass, thirst is extreme and mental derangement takes place. But even at this level of dehydration, recovery is complete within an hour of drinking (Adolph 1969). At 12% loss, recovery is generally possible only with medical assistance, and at 14% there is an explosive rise in temperature that is fatal without prompt action. Fig. 19.15 describes most of the major elements of the dehydration–exhaustion syndrome, as a function of water deficit. The symptoms are plotted at the deficit for which they first occur. Also, the displacement in the ordinate direction is roughly proportional to the commonness of occurrence of the symptom. Many of the symptoms are a consequence of the strain on the circulation system. For water deficits of less

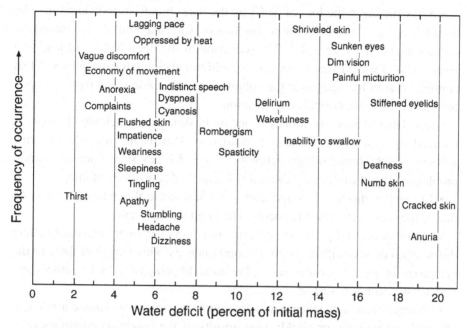

Fig. 19.15 Symptoms associated with different water deficits. Each symptom is plotted for the deficit at which it first occurs, and the position of the symptom name on the ordinate axis is related to its commonness, or frequency, of occurrence. Dyspnea is labored breathing, Rombergism is difficulty balancing, anorexia is loss of appetite, and anuria is absence of urine. (Adapted from Adolph 1969.)

than 10%, the data are from field or laboratory studies with test subjects. For greater deficits, the data are more subjective and result from reports by individuals who have suffered from unavoidable water deprivation, through being lost in the desert, for example.

Adaptation to desert heat can affect the body through two mechanisms. First, the normal heat-regulation process has side effects on a number of body functions. Secondly, if the adaptation to heat fails, there are naturally additional consequences. The following side effects result from the normal heat-regulation process itself (Lee 1968).

• Cardiovascular deficiencies: dilation of the skin blood vessels occurs in order to enable greater flow of blood for cooling at the surface. This increase in the capacity (volume) of the circulatory system reduces its efficiency, and vital organs suffer because of a reduced supply of blood to them.

• Water deficiency: this has been discussed at length above. It should be added that incomplete water intake, to compensate for the loss, reduces the blood volume and contributes to the above threat to the cardiovascular system.

• Salt deficiency: the loss of NaCl by the unacclimatized individual may be as much as 5 g l^{-1} of perspiration. The loss of NaCl may thus easily exceed the normal intake of 10–15 g d^{-1}. The reduction of the body NaCl will lead to a greater loss of water in the urine, causing additional circulation problems. "Heat cramps," cramp-like spasms in the voluntary muscles, also result from a drop in blood NaCl below a critical concentration.

• Renal disturbance: the kidney is one of the organs to which blood supply is reduced as a result of vascular system dilation. This results in a reduced urine volume, which should not go below 0.7 l d^{-1}. Kidney stones are a common problem of desert dwellers, presumably because of low urine volume.

• Alimentary disturbances: reduced blood flow to the alimentary tract may be one of the causes of reduced appetite when under heat stress.

• Disturbances of the nervous system: circulatory deficiency can easily affect the cerebral cortex, with a variety of consequences. This may contribute to the symptoms of loss of consciousness, lassitude, hyperactivity, and sleeplessness that are sometimes observed.

• Changes in the skin: when skin is continuously wet, sweat glands can become plugged and miliaria, or prickly heat, results. If the condition continues or is frequently repeated, the sweat glands can be permanently sealed by scar tissue, which would have disastrous consequences in terms of future maintenance of the heat balance.

Again, the above symptoms are a consequence of the normal heat-regulation processes, and can occur when the regulation process is operating normally and successfully. In contrast, heat stroke (hyperpyrexia) results when the temperature of the body is sufficiently high that the continued functioning of a vital organ is endangered. It can result from failure to perspire, but it also may be a result of a high rate of metabolic heat generation, very high environmental temperatures, etc. The critical body temperature for humans is between 42 and 45 °C (107.6 and 113 °F), beyond which damage is irreversible. This condition indicates the marked failure of the heat regulatory process. Heat stroke is the leading cause of the large-scale loss of life that occasionally results when desert heat escapes its boundaries, or when very hot periods occur in temperate environments. For example, even though residents of the Indian subcontinent should be somewhat acclimatized to heat by late spring, westerly winds transported very hot air from the Rajasthan Desert across the subcontinent in May 2002, and the death toll exceeded 1000. Climate-related deaths are correlated less with generally hot climates than with periods of anomalously high temperature to which inhabitants of an area are not acclimated. For example, Fig. 19.16 shows the daily death rate in Shanghai, China, as a function of the daily maximum temperature in summer.

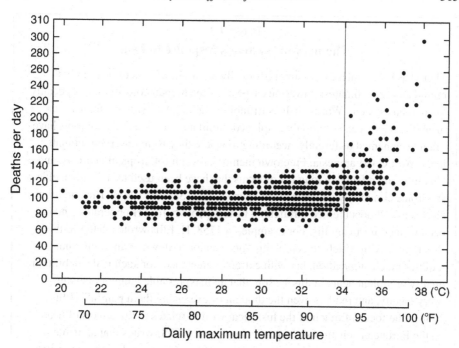

Fig. 19.16 The number of deaths per day during the summer in Shanghai, China, as a function of the daily maximum temperature. The temperature of 34 °C (93 °F) is highlighted because it represents a threshold above which the death rate rises dramatically. Each symbol represents a summer day. (Adapted from Tan 1991.)

Even though the average number of deaths is slightly larger for maximum temperatures near 32–33 °C (90–91 °F) than for lower maxima, it is only for the infrequent days on which the maximum exceeds 34 °C (93 °F) that the death rate rises dramatically. Similar curves, which reflect a critical temperature threshold, could be shown for many other locations in the world, especially those for which technology or poverty do not allow the most sensitive of the population to avoid the full impact of the heat.

Physical effects of desert mineral dust

The inhalation of dust in desert areas can lead to a number of medical conditions in humans. Middleton (1997) provides a brief summary. Bronchitis and emphysema can be aggravated by inhalation of fine dust particles. Bar-Ziv and Goldberg (1974) report high incidences of silicosis and pneumoconiosis in Bedouins in the Negev Desert. In Western Sahara, when the easterly irifi wind blows dust

---

Box 19.2
The nervous system's response to heat

There has been much discussion historically about the effects of heat, in both humid and arid climates, on various aspects of the human constitution, especially the nervous system. Words such as irritability, lethargy, lassitude, and nervous disability have been commonly employed. Similar collections of symptoms have often been named differently, according to where they were observed. Punjab head, West Coast amnesia, Freetown memory, Assam rot, tropical neurosis, and Guamitis, are a few in a long list of terms that have been used to refer to an elusive syndrome commonly called tropical neurasthenia (Sargent 1963). The condition has been well documented throughout history, from before Hippocrates, and has been blamed for many ills. For example, in 1758, H. Ellis wrote a letter to the Royal Society in which he complains "one cannot sit down to anything, that requires much application, but with extreme reluctance; for such is the debilitating quality of violent heats in this season, that an inexpressible languor enervates every faculty, and renders even the thought of exercising them painful" (Ellis 1758). The focus of much of the historical writing relates to the effects of heat on the European constitution in particular. Rasch (1898) writes that an irritable temperament, a stuporous state, and apathy appeared in many Europeans during their first year of residence in Egypt, and that many suicides were reported. Acclimatization by Europeans resulted in "an irreversible weakening of the spirit to live." Sargent (1963) provides a very comprehensive and entertaining account of such historical views of living with heat. Needless to say, there is much present-day evidence for the effects of heat on the nervous system. For example, it is well documented that violent crime rates go up in most cities during periods of unusually high temperatures.

---

from the Sahara, conjunctivitis incidence is high among the nomads (Morales 1946). Airborne dust may also contain fungi that can cause allergic reactions and other problems. "Valley fever" in Arizona, in the United States, is caused by the fungus *Coccidioides immitis* that is carried in dust storms, and is responsible for more than 27 deaths per year in that state (Leathers 1981).

Psychological effects of deserts

There are a range of psychological effects that are associated with life in the desert. Some are related to the physiological response to the heat and dehydration: e.g. apathy and poor performance of skilled tasks. Others seem to be associated with isolation and continued exposure to the harsh environmental conditions. See Box 19.2 for reactions of the nervous system to heat.

Measures of heat stress

A number of different metrics have been developed that are related to thermal comfort and physical wellbeing in hot environments. Some are designed for indoor, industrial situations where intense direct solar radiation is not a factor. Different indices are discussed in Ingram and Mount (1975). Some indices use only temperature and relative humidity (or dew-point temperature). Others, such as the Wet Bulb Globe Temperature (WBGT) take into account air movement and radiant heat (from the sun, and reflected from the ground), as well as air temperature and humidity. Unfortunately, it requires measurements that are sometimes not routinely available, and thus approximations to this indicator are sometimes used.

Electrostatic effects

The production of electrostatic voltage by the impacts of sand grains on each other has long been documented. Such voltages have an effect both on the human body and on equipment. The human impact is likely both physiological and psychological. Some who have traveled on foot in Saharan dust storms have reported intense headaches, which were relieved by carrying a metal staff that served as a better ground. Telephone company employees who were working in a dust storm in the Imperial Valley of the northern Sonoran Desert reported the buildup of an electrostatic charge within a long wire that was lying on the ground; the charge was sufficiently large to knock a person to the ground (Clements *et al.* 1963).

Suggested general references for further reading

Adolph, E. F. (Ed.), 1969: *Physiology of Man in the Desert* – treats virtually all aspects of the physical effects of desert heat and dehydration on humans, based on field and laboratory studies.

Clements, T., *et al.*, 1963: *A Study of Windborne Sand and Dust in Desert Areas* – reviews the causes of dust storms and their properties, with some discussion of their effects on humans, primarily for the northern Sonoran and Mojave Deserts.

Ingram, D. L., and L. E. Mount, 1975: *Man and Animals in Hot Environments* – general discussion of heat-exchange mechanisms, the thermal regulatory system, and adaptations to hot environments.

Kerslake, D. McK., 1972: *The Stress of Hot Environments* – describes all the mechanisms of heat exchange with the environment, the heat balance and how it is maintained, clothing effects, and various indices of heat stress.

Lee, D. H. K., 1968: *Human adaptation to arid environments* – discusses heat regulatory processes, physiological consequences of heat regulation, and adaptive processes.

Monteith, J. L., and L. E. Mount (Eds.), 1974: *Heat Loss From Animals and Man: Assessment and Control* – a series of papers on the physical principles of heat transfer in humans, and the physiology of thermoregulation,

Oke, T. R., 1987: *Boundary Layer Climates* – contains a chapter on the climates of animals.

Tromp, S. W., 1980: *Biometeorology: The impact of the weather and climate on humans and their environment (animals and plants)* – contains a brief discussion of thermoregulation in humans.

Questions for review

(1) Summarize the typical changes that take place to the human body as a result of acclimatization to heat and work in the desert.

(2) Why does a light wind in the "hot" summer in a temperate climate feel cool, whereas in the desert there is no such feeling of great relief?

(3) Explain why region C–D in Fig. 19.5 is referred to as the zone of least thermoregulatory effort.

(4) Summarize the different mechanisms by which heat can be gained or lost by the human body.

(5) In the hot desert, how is heat transported from the core of the body to the skin where the heat is lost through evaporation?

(6) Review the side effects that result from the normal heat-regulation process. What is the common cause of most of them?

(7) What can be the consequences of inadequate heat regulation?

Problems and exercises

(1) Speculate on the physical mechanisms by which clothing influences the rate of evaporation of perspiration.

(2) Assume that your body mass is 70 kg. Make up a fictitious but realistic schedule of recreational hiking and camping during a 24 h period in the desert, and also specify reasonable temperatures for each hour of the day in the desert where you are hiking. How much water do you need to consume each hour in order to replenish the water that is lost through perspiration? (Use Fig. 19.3.)

(3) If very heavy work is undertaken by a 70 kg person in a hot desert, 15 l of perspiration would be produced in a 24 h period in order to maintain a heat balance (Fig. 19.7). If, hypothetically, the heat generated was not lost through evaporation, by how much would the temperature of the body increase during that period? Assume that the entire body has the specific heat capacity of water, of which two-thirds of it is composed.

(4) Explain why the lower saturation vapor pressure over salty perspiration reduces the evaporation rate.

20

Optical properties of desert atmospheres

About a desert sunset:

. . . I sat down and watched a sunset, which grew from grey to pink, and to red; and then to a crimson so intolerably deep that we held our breath in trepidation for some stroke of flame or thunder to break its dizzy stillness.

T. E. Lawrence, British writer and adventurer
Seven Pillars of Wisdom (1926)

Then, after making our purchases, we say good-bye with wishes for safe passage. They go one way and we go the opposite. And our respective images begin to undulate. Soon their camels seem double because of the air. And they themselves, now elongated, now shortened, seem to have two heads apiece, like the kings and queens on decks of cards. Ten o'clock. Ten thirty. It was about this same time when the little fairyland lakes had begun to appear yesterday. Already a few materialize – so cool and so blue! – harbingers no doubt of a larger illusion to come. They still threaten to flood out and swamp you. But on the contrary when you get close – blink! Nothing. The lakes are swallowed by the arid sand or folded up like blue cloth. Then they dissolve rapidly and in silence, like the imagined things they are.

J. Viaud (P. Loti), French writer and adventurer
Le Désert (1895)

Enveloped in the sensual softness of sand-dunes like intertwined bodies that the changing light of the day and of the seasons paints grey and white, ochre and beige, one is tempted to recline and rest for a moment before moving on to other colors, other shapes . . . Leaving behind village, oasis or encampment for a landscape of overwhelming immensity, one is transported by the play of light and shadow to another world, one in which dreams are as accessible as mirages.

Mona Zaalouk, Egyptian painter
A painter's paradise (1994)

Even though atmospheric optics may seem like an esoteric subject, it really is not. In fact the appearance of everything we see is controlled to some degree by what the atmosphere does to light that is transmitted through it. This subject has special relevance for deserts because the extreme atmospheric conditions there produce especially significant effects. We are all aware of the commonness of mirages in deserts; of the scintillation, or shimmering appearance, of the atmosphere as we look horizontally through it near the ground; of the spectacular sunsets; and of the greatly reduced visibility through a dust-laden atmosphere. All of these phenomena are more prevalent in the desert, and will be the subject of this chapter.

Mirages

Light is bent, or **refracted**, when it passes through a medium of varying density, at an angle to the surfaces of constant density that is less than the perpendicular. In particular, it will be bent from a straight path in the direction of higher density. The net effect of atmospheric temperature decreasing with height at normal lapse rates, and of pressure decreasing with height, is for the density to decrease with height also. Thus, light is virtually always refracted as it passes through the atmosphere. The only exception is when the lapse rate is large and equal to 34 K km^{-1}, a value known as the **autoconvective lapse rate**, wherein the density is constant with height. In this special case light travels with a straight path.

At the beginning of the diurnal heating cycle, the lapse rates will likely have negative values (temperature increasing with height) associated with a nocturnal inversion. As the heating of the surface begins, lapse rates near the ground will become positive, and increase in value as the heating continues. For lapse rates smaller than the autoconvective value, the density will decrease with height and light will be bent toward the higher density air near the surface. Fig. 20.1 shows an observer at point O, seeing the surface at point C. However, because of the curvature of the light path from C, the observer is actually looking "over" the horizon (which is actually at B), and furthermore C is seen in the direction C$'$ because that is the direction from which the C image arrives at the observer. Thus, for sub-autoconvective lapse rates, images are elevated above their true position: **superior mirages**. As the heating of the surface continues, the lapse rate near the surface will become approximately equal to the autoconvective value, and an observer at location O sees the horizon at B through light that travels the straight path BO (left side of figure). But in the desert, or over other hot surfaces, lapse rates within a shallow layer near the ground can become much larger than autoconvective, i.e. super-autoconvective. For example, it was noted in Chapter 5 that lapse rates very near the desert surface can be over 10 000 times the dry adiabatic value (or about 3000 times greater than autoconvective).

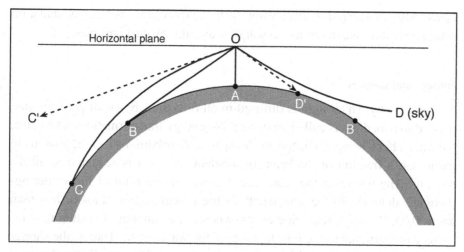

Fig. 20.1 Schematic of light paths under different conditions of atmospheric temperature stratification. The observer's eye is located at point 0. For a constant density atmosphere, with an autoconvective lapse rate (34 K km^{-1}), light travels in a straight path and the observer sees the horizon at its correct geometric location at B. For lapse rates that are more stable than the autoconvective value, the observer can see over the geometric horizon to point C, for example. But, the image of C is seen at C'. This is a superior mirage. When lapse rates are super-autoconvective, the light path is convex toward the ground. The observer looking toward D' actually sees light from the sky at D, which has travelled the curved path DO. This is called an inferior mirage.

For lapse rates that are super-autoconvective, the density increases with height and light paths will curve upward toward the higher density. This situation is illustrated with light path DO. Here, the observer at O is looking toward D', which is below the true horizon; however, the light arriving from that direction is originating at D. Thus, what is seen at D' is the sky, and because the image is seen below its true position it is called an **inferior mirage**. This is the typical mirage that we think of in deserts, where the gray or blue image of the sky is seen below the horizon and is mistaken for water. Mountains or other features in the background of the scene may appear normal because the light path from them is sufficiently far above the surface so that the super-autoconvective layer is avoided.

Effects of atmospheric dust

There are two twilight phenomena discussed here that result from atmospheric scattering, which is affected by dust in the atmosphere. Under the right conditions, both of these processes can be more dramatic in the atmosphere over arid lands. Also considered in this section is the near-surface visibility degradation that can result in the desert as a result of scattering by dust particles.

Sunsets and sunrises

The process by which light is diffused in all directions by small particles suspended in a medium is called scattering. No energy transformations take place, the only effect being a change in the spatial distribution of the radiation. In particular, a fraction of the beam of incident radiation is scattered in all directions. The nature of the scattering depends on the ratio of the scattering-particle's diameter to the wavelength of the radiation. For ratios of less than about 0.03, Rayleigh scattering exists, wherein the amount of scattering is inversely proportional to the fourth power of the wavelength. That is, the shorter the wavelength of the light, the greater the scattering. Very small particles such as air molecules scatter approximately according to this law. Thus, the shorter wavelengths in the visible part of the sun's spectrum, such as those of blue light, are scattered the most, and this is what is responsible for the blue color of the sky when the air molecules are the dominant scatterers. A more complex type of scattering, called Mie scattering, occurs for larger particle sizes. In the latter situation, all wavelengths of visible light are redirected approximately equally.

When the sun is near the horizon, the light has a much longer path through the atmosphere than when it is in a more overhead position. This means that much more blue light is scattered, with the spectrum that reaches the viewer dominated by the remaining colors of the sun's visible spectrum. When the atmosphere is relatively dust-free, the sun near the horizon appears yellow, as do the clouds on which it shines. When there are fine dust particles, more of the short wavelengths are scattered, which shifts the color of the remaining spectrum more toward the orange and red. Without small suspended dust particles, a perceptible red color is not possible. Particles that are larger than fine dust scatter according to the Mie theory, and produce a milky or hazy appearance to the sky and the sun.

The desert atmosphere is often far removed from sources of industrial pollution having large particle sizes. Thus, when substrate conditions and low wind speeds preclude the suspension of dust from the surface, the atmosphere may be quite pristine. This results in bright yellow sunsets and sunrises that are not "washed out" by the haze caused by large dust particles. An exception might exist for deserts near coastlines, where there could be large suspended salt particles. Under conditions of some suspended fine dust from the desert surface, the low sun would have more orange and red. For extremely dusty conditions, and especially when the winds are sufficiently strong to elevate large dust or sand particles, the sun

---

Box 20.1
The colors of the desert surface

Desert beauty is often extolled in terms of the myriad pastel colors of the surface. The color of a surface results, of course, from the light that is reflected by it, or, conversely, by the light that is not absorbed. During non-twilight hours, the sun light is white (containing all colors), and thus we see the "true" color of the surface. Even though there are some brightly colored substrates in the desert, as well as the subtle pastels, much of the surface is actually fairly light-colored, reflecting evenly across the visible spectrum. This has been mentioned before in the context of the large desert albedoes. Thus, when yellow- to red-colored light from the sun near the horizon strikes the surface, it will be reflected rather than absorbed, and the surface will take on the color of the light. In more-temperate latitudes, where the surface is largely covered by vegetation, red light from a sunset will fall on the green foliage (where only green is reflected) and be absorbed. Thus, even though the desert substrate has beautiful inherent colors of its own, it is often the twilight colors of the surface that are praised in prose and poetry.

---

would become yellow or perhaps even obscured completely. Box 20.1 explains how the sun's light colors the surface.

Crepuscular rays

Crepuscular rays are shadows seen in the atmosphere, as a result of the presence of dust. Where the sun is shining on a volume of atmosphere, dust scatters the light in all directions and this air serves as a source of light. If the sun is near the horizon or below it, objects that block the sun, such as deep cumulus clouds or high and narrow terrain prominences, cause narrow bands of atmosphere to be in their shadow. There is no scattering of the direct rays of the sun in these bands, and thus their contrast with the surrounding sunlit sky is obvious. These parallel bands of shadow (they are parallel because the sun's rays are parallel) seem to converge toward the direction of the sun for the same reason that parallel railroad tracks converge in the distance: perspective (Fig. 20.2). Less frequently, for the same reason they can be seen to converge toward the anti-solar point in the sky, the direction opposite to that of the sun. If there is a brightly colored sunset, the rays are dark or bluish against the colored background. If there is no bright sunset coloration to the sky, the rays will appear to be dark blue or gray against a lighter blue background. The above description portrays the narrow shadow bands existing against an illuminated background, so the bands are darker than the background. The opposite situation is even more spectacular, and is one in

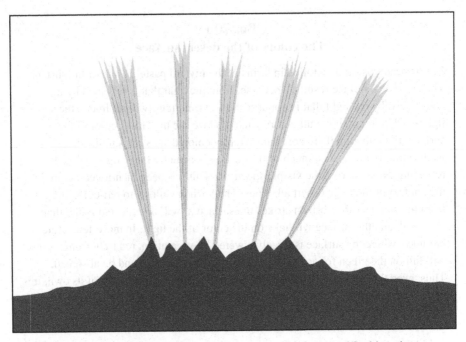

Fig. 20.2 Crepuscular rays caused by terrain prominences blocking the sun and causing shadows in the atmosphere.

which the sun near the horizon is largely blocked by clouds. Here, where the sunlight finds passage between clouds, brightly lit bands radiate across the sky against a dark background.

Even though dust is sometimes more concentrated in the desert atmosphere than elsewhere, there is always some dust everywhere. Thus, crepuscular rays are not unique to deserts. However, they are often more dramatic because of the greater amount of dust. Also, when the dust exists in a deep layer above the surface, the rays will be visible for a longer period after the sun passes below the horizon. When clouds that shade the sunlight are scarce in deserts, the shadows must be created by peaks in surrounding mountains.

Visibility degradation

In previous discussions, references have been made to the "visibility" within dust storms, without any attempt to be rigorous about what affects visibility and how it is defined. In temperate latitudes, visibility degradation more commonly results from rain or snow, and has great impact on many activities such as surface or air transportation. Visibility reduction by dust has similar impacts in arid areas, as described in Chapter 16 relative to dust storms. In general, light passing through a medium can be affected by scattering as well as absorption, for which there are separate coefficients. The sum of the scattering coefficient and the absorption coefficient is known as the attenuation (or extinction) coefficient, and quantifies the total effect of the medium on light passage. In terms of dust and sand, the attenuation coefficient depends on the number of particles per unit volume and the size of the particles. The definition of visibility, or visual range, during the day is the greatest distance at which an object can be distinguished from the background. In sandstorms, with strong winds, even though it is claimed that visibilities can approach zero, the act of opening one's eyes to test the visibility would be challenging. Nevertheless, it is very likely that the visibility in such conditions must approach near zero.

Scintillation

A scene that is viewed in the distance, when the intervening surface of Earth is hot, often appears to "shimmer" and is generally not clearly visible. This is because the air near the ground contains intense convective turbulence, and the density contrasts between the moving eddies of air refract the light. This phenomenon is **scintillation**, sometimes called optical haze. It is characterized by one or more of the following fluctuations (Neuberger 1966): (1) apparent vibratory motion that causes distant objects to appear to move irregularly; (2) shadow bands that move across extended objects, causing contrasts in the light intensity from them and distortions in internal structural features; (3) color changes of sources of white light. The frequencies of the oscillations range from one to several hundred cycles per second, but the common range is five to ten. Turbulence near the observer is more responsible for the scintillation than is the turbulence near the object being viewed. In fact, the first 10% of the air column nearest the observer is 19 times as effective in producing the scintillation as is the 10% of the column nearest the target (Neuberger 1966). The effect is most intense over hot surfaces, such as prevail in the desert. Fig. 20.3 illustrates how density irregularities in the atmosphere can cause variations in the intensity of the light coming from a distant source. In this simple, conceptual example, light passes from a medium of one temperature into a medium of another temperature through an irregular boundary. In some areas, convergent light rays would cause an increase in light intensity, while in areas of divergent rays the light intensity would be less. Clearly an image defined by the original light rays would be distorted in the process of passing through the interface. In reality, these density irregularities would be changing with time, thus causing an oscillation in the light intensity and the position of features defined by the source of light.

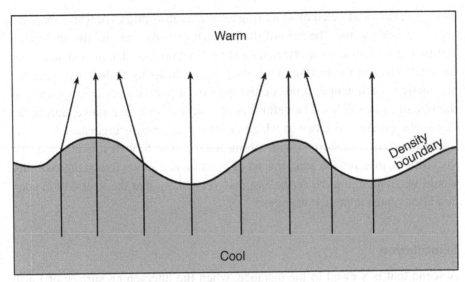

Fig. 20.3 Horizontal density irregularities in the atmosphere, and the resulting variation in the paths of light rays that produce scintillation.

Suggested general references for further reading

Meinel, A., and M. Meinel, 1983: *Sunsets, Twilights, and Evening Skies* – a non-technical discussion of atmospheric optical processes and phenomena that prevail at twilight.

Neuberger, H., 1966: *Introduction to Physical Meteorology* – describes condensation processes in the atmosphere, visibility, solar and terrestrial radiation, meteorological acoustics, meteorological optics, and atmospheric electricity.

Tricker, R. A. R., 1970: *Introduction to Meteorological Optics* – a technical description of refraction, rainbows, the corona, scattering, and visibility.

Questions for review

(1) Use a diagram to draw the light rays for superior and inferior mirages.
(2) Explain the term autoconvective lapse rate, and what it means physically.
(3) What is the physical cause of scintillation?

Problems and exercises

(1) Why does the appearance of water on the ground associated with the inferior mirage disappear as we approach the location? That is, why do we see the ground, rather than sky, close to our location? Suggest how the strength of the super-autoconvective lapse rate might relate to how close the mirage extends toward the observer's location.

(2) Obtain a good text on atmospheric optics, and summarize the various types of mirage that one might encounter in deserts. For example, in this chapter we did not discuss mirages in which the image is inverted (top to bottom).

Appendix A

Glossary of meteorological and land-surface terms

absolute vorticity The vorticity of a fluid particle defined with respect to an absolute reference frame. The vertical component (perpendicular to the horizontal plane) of the absolute vorticity, the one to which reference is normally made, is defined as the sum of the vertical component of the vorticity of the fluid with respect to Earth's surface (the relative vorticity) and the vertical component of the vorticity of Earth (equal to the Coriolis parameter).

absorptivity The ratio of the amount of radiant energy absorbed by a given substance to the total amount incident upon it, at a given wavelength. Absorption may occur at the surface if the substance is opaque, or during the transit of the radiation through a translucent substance.

adiabatic Describes a process in which the thermodynamic state of a system changes without the transfer of heat or mass across the boundaries of the system.

advection A process by which an atmospheric property is transported from one location to another by the motion of the air. A distinction is often made between horizontal and vertical advection.

advection fog A fog that is formed by the horizontal advection of moist air over a surface that is cooler than the air, and the consequent cooling of the air near the surface to its dew-point temperature.

aeolian Windborne.

aerosol scattering An aerosol, as the term is applied to the atmosphere, is a system in which solid or liquid particles are dispersed in the air. The particles must be sufficiently small that they remain stably dispersed in the air. Smoke, haze, or dust particles, and the air in which they are dispersed, would be referred to as an aerosol. Scattering is the process by which the aerosol diffuses part of the incident radiation in all directions.

albedo The ratio of the amount of electromagnetic radiation that is reflected by a surface to the amount that is incident upon it. The value may be given as a fraction or a percentage. The term reflectivity is commonly used to refer to the fraction reflected in only a particular wavelength or wavelength band, whereas the term albedo is generally used to refer to the fraction of the energy reflected in a broad band such as the visible band or the entire solar spectrum.

alluvial fan A fan-shaped accumulation of rock waste washed downslope by the flow of water from elevated terrain.

almost intransitive Describes a special type of transitive system in which the climate remains within a regime for a finite time, with the system then migrating into another equally acceptable regime without any change in the external forcing.

altitude oasis   An area in an arid zone with terrain elevations that are higher than those of the surroundings, with a correspondingly wetter microclimate.

annual plant   A plant that lives for one year or one season, renewing itself through seeds.

autoconvective lapse rate   The lapse rate of temperature at which density is constant with height; about 34 K km^{-1}.

Autumnal Equinox   The time during each year at which the sun passes directly over the equator, moving from the Northern to the Southern Hemisphere.

bajada   A field of rock waste formed when alluvial fans become laterally confluent, forming a continuous apron of waste against a mountain (see **alluvial fan**).

barchan   A crescent-shaped sand dune, the points of which are oriented in the direction of the dune movement and the mean wind direction.

bedload transport   Lateral movement of soil particles through direct action of the wind, or through the impact of other particles.

black body   A hypothetical substance or body that absorbs all of the electromagnetic energy that is incident upon it. That is, none is transmitted or reflected. It also emits energy at the maximum possible rate.

bolsone   An enclosed drainage basin, usually surrounded by mountains, that is often associated with a salt lake or playa.

boundary layer   The layer of air that is immediately adjacent to a solid bounding surface. The context in this book is in terms of the planetary boundary layer, or the layer of air near the ground with either **turbulent** or **laminar** flow.

capillary water movement   The movement of water through soil pores, resulting from surface-tension effects. The water rise in a capillary tube or straw illustrates this effect.

climate   Average weather conditions, including deviations from the average.

contact transport   Movement of soil particles through the impact of other moving particles.

continental-interior deserts   Deserts that are in the interior of continental land masses, and distant from maritime moisture sources. During the transit of the maritime air from the coast to the interior, water vapor is condensed through a variety of mechanisms such as large-scale storms and mountains, and the removal of the cloud water through precipitation causes the air that reaches the interior to be relatively dry.

convection   Primarily vertical atmospheric motions that result in the vertical transport of atmospheric properties. Free convection is the vertical motion caused by density differences in the atmosphere (buoyancy), and forced convection is the vertical transport by the turbulence that is caused by the friction at Earth's surface.

cool coastal deserts   Deserts that exist near coastlines that have an adjacent cool ocean current. Advection fogs are usually present over the near-coastal ocean and land. Air temperatures are lower in these deserts because of the advection over the land of the cool air from the ocean and the interception of some of the direct solar energy by the fog droplets.

creep   Lateral movement of soil particles through the direct action of the wind.

crepuscular rays   Shadows or streaks of light seen in the atmosphere as a result of the presence of dust. If the sun is near the horizon or below it, objects that block the sun, such as deep cumulus clouds or high and narrow terrain prominences, cause narrow dark bands. Or most of the atmosphere may be in shadow, and streaks of light are isolated against the dark background.

cumulus   A cloud type that generally takes the form of singular distinct elements with sharp edges. They often form in clear air as a result of convection within air of sufficiently high water-vapor content.

density current   Movement of air caused by horizontal contrasts in air density, with the term used in this text in the context of cold, dense air spreading out across the surface ahead of a thunderstorm.

desert pavement A sheet of pebbles or gravel that is cemented to the substrate below.

desert varnish A shiny reddish or blue-black coating on rock, created by bacterial action and water-borne deposition of manganese and iron oxides.

desertification The development of any characteristic of the climate or land surface that is more typical of a desert, whether the change is natural or anthropogenic.

desiccation The process of long-term reduction in soil moisture, especially on time scales of decades.

dew-point temperature The temperature to which a parcel of air must be cooled in order for saturation to occur.

diffuse solar radiation Solar radiation that has been scattered from the direct solar beam by molecules or suspended solid or liquid particles in the atmosphere. This includes the blue light of the clear sky, scattered by air molecules, or the white or colored light scattered by cloud or dust in the atmosphere.

direct solar radiation The radiation that is received directly from the sun.

downburst Rapidly descending cold air from the base of a thunderstorm, with associated strong winds as the cold air spreads horizontally near the ground.

dry adiabatic lapse rate The rate of decrease (increase) of temperature of a parcel of air that is lifted (caused to subside) adiabatically through the atmosphere. By definition, no condensation or evaporation take place.

dry fog Fine dust distributed throughout the boundary layer by the turbulence scales of air motion, especially in the Sahara.

electromagnetic spectrum A continuum of wavelengths of radiation, or radiant energy, that propagates through a vacuum or a medium such as the atmosphere, and takes the form of an advancing disturbance in the magnetic and electric fields. The electromagnetic spectrum is illustrated in Fig. 4.2.

emissivity For a given wavelength, the ratio of the radiant energy emitted, per unit time per unit area, from a surface to the radiant energy emitted by a black body at the same temperature.

equinoxes The two times per year that the sun passes above the equator, from one hemisphere to the other.

erg A sandy desert area, especially within the Sahara and Middle East.

evapotranspiration The combination of evaporation and transpiration, with transpiration being the process by which water is transferred from liquid form in the soil, through the plant, and into the atmosphere as water vapor.

extratropical cyclones Mid- or high-latitude storms, with cyclonically rotating winds, that form along the polar front and derive energy from the horizontal temperature contrast.

flux The rate of flow of any quantity, either mass or energy, through a unit surface area. This is sometimes called a flux density.

fossil water Water that has resided in aquifers for a long period of time, with generally a very low rate of natural recharge by rainfall.

free atmosphere The atmosphere above the planetary boundary layer.

groundwater mining Consumption of groundwater at a rate that is faster than it is naturally replenished.

gust front The leading edge of a thunderstorm outflow boundary, with gusty winds.

gypsophyte A plant that can tolerate gypsum soils.

haboob A dust storm caused by strong winds from a thunderstorm gust front.

halophyte Vegetation type with a tolerance for saline soils and water.

hammada (hamada) A rocky desert area, especially within the Sahara and Middle East.

hardpan A layer of soil, at the surface or below, that is relatively impervious to water and vegetation roots.

heat capacity　The ratio of the heat absorbed (released) by a substance to the corresponding temperature rise (fall); also called **thermal capacity**.

heat low　An area of low surface pressure caused by the heating of Earth's surface. Heat lows remain stationary over the continental subtropics in summer, and have a diurnal variability in strength.

hydrograph　A plot of river discharge as a function of time, commonly with dimensions of volume per unit time.

hydrostatic　Describes an equilibrium in a fluid that is a condition of balance between the vertical pressure gradient and gravitational forces. The atmosphere is never in exact hydrostatic balance, but the assumption under some conditions allows useful simplifications of the governing equations.

hygroscopic　Having an affinity or attraction for water. Examples of hygroscopic substances are $NaCl$ (table salt) or AgI (silver iodide, used in cloud seeding).

indirect solar radiation　The same as **diffuse solar radiation**. The illumination received from the sky through the scattering of sunlight by air molecules, clouds, and dust particles. In some usages this may include solar radiation reflected from the substrate and vegetation.

inferior mirage　An image of an object that appears below its actual position.

infiltration　The entry of water from the surface into a substrate.

Inter-Tropical Convergence Zone (ITCZ)　The area separating the northeasterly trade winds of the Northern Hemisphere and southeasterly trade winds of the Southern Hemisphere. This is a zone of confluence between the two large trade-wind regimes, and is sometimes referred to as the meteorological equator.

intransitive　Describes a climate system that has more than one possible stable climate, for a particular set of external forcing parameters.

inviscid　Without viscosity or frictional effects within a fluid. This condition is never realized, but the assumption is sometimes useful for simplifying the governing equations.

laminar　Non-turbulent. The laminar flow of a fluid is smooth, without turbulence.

laminar sublayer　A very shallow, laminar, boundary layer of a few molecules to a few millimeters in depth above the surface, through which transfers of heat, moisture, and surface-frictional effects occur through molecular processes.

latent heat　The heat released or absorbed per unit mass during a change of phase. In atmospheric science, the latents heats of vaporization (or condensation), fusion (or melting), and sublimation of water are important.

latent heat flux　A way of expressing the flux of water vapor, generally at an evaporation surface, in terms of the latent heat that would be released upon the condensation of the water vapor.

latent heat of condensation/evaporation　The heat released (consumed) during the condensation (evaporation) of a unit mass of water.

latent heat of fusion/melting　The heat released (consumed) during the freezing (melting) of a unit mass of water.

loam　A mixture of clay, silt, and sand.

loess　A loamy material that has been chiefly deposited by the wind.

long-wave radiation　Infrared electromagnetic radiation, with wavelengths of Earth's emission spectrum, in the range 3.0–100 µm.

mesoscale　The range of horizontal space scales in the atmosphere between the synoptic scale and the microscale, with wavelengths from 2 km to 2000 km.

microclimate　Small-scale (regional) climate that is determined by local factors such as topography, water-table depth, and substrate type.

mid-latitudes　Latitudes between the subtropical and polar latitudes, from about 35° to 60°.

mixed layer A layer of atmosphere, generally in contact with Earth's surface, that is well mixed by turbulence.

mixing fog A fog produced when two unsaturated volumes of air, with different temperatures and vapor pressures, mix. If the resulting mixture is saturated, a fog forms.

moist adiabatic lapse rate The rate of decrease (increase) of temperature of a parcel of air that is lifted (caused to subside) through the atmosphere while condensation (evaporation) is occurring. The rate varies from about 2 to 10 K km^{-1}, depending on temperature and pressure. The value is about 6 K km^{-1} at 800 mb and 0 °C.

molecular diffusion The transport of a property of the air, such as temperature or water-vapor content, through molecular motion.

molecular scattering The process by which air molecules diffuse some of the incident radiation in all directions. This is called Rayleigh scattering, and is responsible for the blue sky.

monsoon Seasonal winds. The much larger annual variation in temperature over large land areas compared with adjacent seas causes large-scale seasonal swings in the pressure contrast. When the resulting landward winds during summer and seaward winds during winter are not obscured by the planetary circulation, monsoon winds result. Sometimes the associated rainfall is also referred to as the monsoon.

nephelophyte A plant that can absorb water from dew and fog on leaf surfaces.

oasis effect An effect seen when dry desert air is advected over a moist surface. The extremely high evaporation rate dominates the heat budget, with the result that both the net radiation and a downward sensible heat flux are required to supply the necessary energy.

optically active Describes the property of some atmospheric gases such that they selectively absorb electromagnetic energy at certain wavelengths.

orography The nature of the elevation of terrain above sea level.

osmotic (osmosis) The diffusion of a fluid through a porous medium under the influence of a gradient in the salt concentration.

outflow boundary The leading edge of a density current that is flowing outward from a thunderstorm.

percolation The vertical movement of water within a substrate.

perennial vegetation Vegetation that lives for more than one year.

phreatophyte A plant that obtains its water from the water table, generally by sending down deep roots.

phreatophyte mound An area of elevated sandy soil caused by a phreatophyte blocking the flow of windborne soil.

playa Spanish for beach; generally a flat basin-floor area with a high water table and a high salt content in the near-surface substrate.

pluvial Refers to a period in the past when the rainfall was greater than it is presently.

polar front The boundary between polar and tropical air masses, generally located in mid-latitudes.

porosity The fraction of the volume of a substrate that can be occupied by water or air.

potential evaporation The evaporation rate (mm d^{-1}) from a surface of water, under existing atmospheric conditions, that has the same temperature as the atmosphere with which it is in contact.

potential evapotranspiration (PET) The sum of the evaporation and the transpiration for conditions in which soil-water supply is unlimited.

potential temperature The temperature that a parcel of air would have if it were moved adiabatically to 1000 mbar.

potential vorticity The product of the absolute vorticity and the static stability of the atmosphere at a point, defined in Eqn. 2.2, that is conserved in adiabatic flow.

psammophyte   A plant that is typical of sandy soils.

radiant energy   Electromagnetic energy of any wavelength.

radiation fog   A fog that forms when the land cools at night, and the air near the ground is, in turn, cooled by the surface to its dew-point temperature.

radiative temperature   The temperature computed based on measurements of the wavelengths of emitted long-wave radiation.

rain shadow   When precipitation occurs as an airstream is lifted over a mountain barrier, the air on the lee side of the mountain is drier and warmer than that on the upwind side. This modified air is less likely to support subsequent precipitation. This is called the rain-shadow effect.

rain-shadow desert   A desert that is caused by the rain-shadow effect.

reflectivity   The ratio of the amount of electromagnetic radiation that is reflected by a surface to the amount that is incident upon it, for a particular wavelength.

refraction   The bending of the path of a propagating wave as it passes through a medium of varying density.

reg   A stony desert area, especially within the Sahara and Middle East.

relative vorticity   The vorticity defined in terms of a system of coordinates that is fixed to Earth's surface.

reptation   The process by which grains of sand or soil, elevated by the wind, impact other grains at the surface that take only a single hop.

residual layer   The layer above the stable, nocturnal boundary layer where there exists residual turbulence from the daytime boundary layer, with the intensity decaying with time as a result of internal friction within the fluid.

saltation   The process by which sand or soil grains are elevated from the grain bed, gain horizontal momentum through exposure to the higher-velocity airstream above the ground, descend back to the grain bed and impact other grains, and bounce upward again into the airstream.

saltation speed   The speed of the wind at the surface such that particulate matter of a certain size begins to move.

sand   Fine particles of rock with diameters greater than 0.05 mm and less than about 2.0 mm.

sand sea   A large expanse of sand, including flat areas as well as dunes.

sand sheet   A large area of sand, especially a flat area.

scattering   The process by which aerosol particles or molecules diffuse part of the incident radiation in all directions.

scintillation   Air near the ground over hot surfaces during the day contains intense convective turbulence, and the density contrasts between the moving eddies of air refract light. This causes the images of distant objects to appear to "shimmer", and this phenomenon is called scintillation.

sea-/lake-breeze circulation   The land surface has larger diurnal swings in surface temperature than does an adjacent water surface, and the resulting pressure contrasts cause a wind to develop in the coastal region that is landward during the day and seaward at night.

sensible heat   Heat that can be sensed, in contrast to **latent heat.**

serir   Same as **reg**.

shear   A spatial variation in the wind speed or direction.

short-wave radiation   Electromagnetic radiation, with wavelengths of the **solar spectrum** from 0.15 to 3.0 μm.

siefs   Linear sand dunes, formed parallel to the mean wind and named after the Arabic word for sword.

soil-moisture content   The percent of the volume of a soil that is occupied by water.

solar spectrum The intensity of electromagnetic radiation in the different wavelengths emitted by the sun.

solstices Either of the two times of the year when the sun is displaced the farthest north or south of the equator.

stratus A cloud type, usually with large horizontal extent and fairly uniform bases.

sublimation A phase change of a substance directly between the solid and the vapor phases. The change from solid to vapor is the customary one here.

subsidence Downward movement of air.

substrate The material of which Earth's surface is composed. The term is sometimes used interchangeably with soil, but the latter implies an ability to sustain plant life.

subtropical desert A desert that exists primarily because of the large-scale subsidence of the global circulation in subtropical latitudes.

subtropics The belts of latitudes in both hemispheres between the mid-latitudes and the tropics. The boundaries of the belt are indistinct and depend on continental influences, but 30° may be considered as an average central latitude.

Summer Solstice The time of the year when the sun is displaced farthest (poleward) from the equator within that hemisphere.

superior mirage An image of an object that appears above its actual position.

surface layer The lower 50–100 m of the mixed layer, where the turbulent transport of heat, moisture, and momentum vary relatively little compared with the rest of the mixed layer above.

suspended dust Fine dust that remains suspended in the atmosphere after it has been elevated from the surface.

synoptic scale The range of horizontal space scales in the atmosphere between the planetary scale and the mesoscale.

tank A bedrock depression or cavity in the desert, filled with rain water; also called **tinaja**.

temperature inversion A departure from the normal decrease of temperature with height in the troposphere. That is, the temperature increases with height within an inversion layer. This is often simply referred to as an inversion.

thermal capacity Another name for **heat capacity**.

thermal conductivity A property that reflects the ability of a substance to transport heat through molecular motion. The conductivity is the constant that relates the conductive heat flux to the temperature gradient in the direction opposite to the flux (Eqn. 5.1).

thermally direct circulation A closed thermal circulation, in a vertical plane, in which the rising motion occurs at a higher potential temperature than does the subsidence. A sea-breeze circulation is an example, with rising motion over the warmer land during the day.

therophyte Another name for **annual plant**.

tinaja Another name for **tank** (from the Spanish for jar).

trade winds A wind system that prevails over much of the tropics in which the winds are oriented from the subtropical highs to the equatorial low-pressure trough. In the Northern Hemisphere they are northeasterly and in the Southern Hemisphere they are southeasterly.

transitive Describes a climate system in which there is only one permitted set of long-term climate statistics: that is, given a particular set of external forcing parameters for the atmosphere, such as the orography, the solar input, Earth's rotation rate, etc., there is only one stable long-term climate.

transmissivity A measure of the amount of electromagnetic radiation that is transmitted through a medium. It is the ratio of the amount of radiation transmitted to the amount that is incident, at a given wavelength.

transpiration The process by which water is transferred from liquid form in the soil, through the plant, and into the atmosphere as water vapor.

tropopause The boundary between the troposphere and stratosphere, the upper boundary of the troposphere.

troposphere The layer of atmosphere above the surface wherein the temperature lapse rate is generally positive.

turbulent A type of fluid flow in which the motion are apparently random, and describable only in terms of statistical properties.

Vernal Equinox The time at which the sun passes directly over the equator, moving from the Southern to the Northern Hemisphere.

visible spectrum Light with wavelengths from 0.4 to 0.7 μm.

vorticity A vector measure of the local rotation in a fluid.

wadi A desert water course that is often dry.

water-vapor density The mass per unit volume of water vapor in the atmosphere.

Winter Solstice The time of the year when the sun is displaced farthest (poleward) from the equator in the opposite hemisphere.

xerophytes Drought-resistant plants that exhibit various methods for conserving water.

zenith angle The angle between the zenith, or vertical, and the direction of the incoming direct radiation from the sun.

Appendix B

Abbreviations

Abbreviation	Meaning
AGL	above ground level
ASL	above sea level
BP	before the present
DOY	day of year
EML	elevated mixed layer
ENSO	El Niño – Southern Oscillation
ITCZ	inter-tropical convergence zone
LT	local time
NDVI	normalized difference vegetation index
SAL	Saharan air layer
SST	sea-surface temperature
UTC	Greenwich time
WBGT	wet bulb globe temperature

Appendix C

Units, numerical constants, and conversion factors

Basic units

Quantity	Name of unit	symbol
Length	micrometer	μm
	millimeter	mm
	centimeter	cm
	meter	m
	kilometer	km
Mass	kilogram	kg
Time	second	s
Temperature	degree Celsius	°C
	degree Kelvin	K
	degree Fahrenheit	°F

Derived units

Energy	joule	J
Power	watt	W
Volume	liter	l
	cubic meter	m^3
Pressure	pascal	Pa
	millibar	mbar
Area	hectare	ha

Numerical constants

Water

Specific heat of liquid water: $4218 \; J \; kg^{-1} \; K^{-1}$, at 0 °C
Latent heat of vaporization of water: $2.50 \times 10^6 \; J \; kg^{-1}$, at 0 °C: $2.25 \times 10^6 \; J \; kg^{-1}$, at 100 °C
Latent heat of fusion: $3.34 \times 10^5 \; J \; kg^{-1}$, at 0 °C
Density of liquid water: $1.000 \times 10^3 \; kg \; m^{-3}$

536

Air

Thermal conductivity of dry air: 2.40×10^{-2} J m^{-1} s^{-1} K^{-1}

Earth

Average radius: 6.37×10^6 m
Acceleration due to gravity at Earth's surface: 9.81 m s^{-2}
Angular velocity of rotation: 7.292×10^{-5} radian s^{-1}
Average distance from sun to surface of Earth: 1.50×10^{11} m
Solar irradiance on a plane normal to the light at the top of the atmosphere: 1.38×10^3 W m^{-2}

Universal

Stefan–Boltzmann constant: 5.6696×10^{-8} W m^{-2} K^{-4}

Conversion factors and formulae

acres to hectares:	hectares = acres $\times$ 0.4047
acres to square kilometers:	square kilometers = acres $\times$ 0.004047
degrees Celsius to degrees Fahrenheit:	°F = (9/5) °C + 32
degrees Fahrenheit to degrees Celsius:	°C = (5/9)(°F − 32)
degrees Celsius to degrees Kelvin	K = °C + 273
degrees Kelvin to degrees Celsius	°C = K − 273
feet to meters:	meters = feet $\times$ 0.3048
hectares to acres:	acres = hectares $\times$ 2.471
hectares to square kilometers:	square kilometers = hectares $\times$ 0.01
kilocalories to joules:	joules = kilocalories $\times$ 4185
kilometers to miles:	miles = kilometers $\times$ 0.6214
meters to feet:	feet = meters $\times$ 3.281
miles to kilometers:	kilometers = miles $\times$ 1.609
square kilometers to hectares:	hectares = square kilometers $\times$ 100

Appendix D

Symbols

Symbol	Quantity	Common units
Capital letters		
C	thermal capacity or heat capacity	$J\ m^{-3}\ K^{-1}$
D	soil-water diffusivity	$m^2\ s^{-1}$
E	evaporation rate	$kg\ m^{-2}\ s^{-1}$
G	sensible heat flux between the surface and subsurface	$W\ m^{-2},\ J\ s^{-1}\ m^{-2}$
H	sensible heat flux between the surface and the atmosphere	$W\ m^{-2},\ J\ s^{-1}\ m^{-2}$
I	long-wave radiation intensity	$W\ m^{-2},\ J\ s^{-1}\ m^{-2}$
$I{\downarrow}$	downward-directed long-wave radiation intensity	$W\ m^{-2},\ J\ s^{-1}\ m^{-2}$
$I{\uparrow}$	upward-directed long-wave radiation intensity	$W\ m^{-2},\ J\ s^{-1}\ m^{-2}$
k	molecular diffusivity	$m^2\ s^{-1}$
K	eddy diffusivity	$m^2\ s^{-1}$
K_Θ	hydraulic conductivity	$m\ s^{-1}$
L	latent heat of evaporation	$J\ kg^{-1}$
P	wave period	s
Q	direct solar radiation intensity	$W\ m^{-2},\ J\ s^{-1}\ m^{-2}$
	heat flux into soil	$W\ m^{-2},\ J\ s^{-1}\ m^{-2}$
Q_n	the nth quartile in a frequency distribution	same as variable
R	net radiation intensity	$W\ m^{-2},\ J\ s^{-1}\ m^{-2}$
RH	relative humidity	$\%$
T	temperature	K
U	mean wind speed	$m\ s^{-1}$
Z	potential vorticity	s^{-1}
	zenith angle of sun	degrees, radians
Lower case letters		
c	specific heat	$J\ kg^{-1}\ K^{-1}$
e	water vapor pressure	millibars
f	Coriolis parameter	s^{-1}

k	thermal conductivity	W m^{-1} K^{-1}, J s^{-1} m^{-1} K^{-1}
	molecular diffusivity	m^2 s^{-1}
	von Karman constant	dimensionless
p	pressure	millibars
q	indirect solar (diffuse) radiation intensity	W m^{-2}, J s^{-1} m^{-2}
	soil moisture flux	m s^{-1}
	specific humidity	kg m^{-3}
t	time	s
u	east–west component of wind speed	m s^{-1}
	total horizontal wind speed	m s^{-1}
u_*	friction velocity, or shear velocity	m s^{-1}
v	north–south component of wind speed	m s^{-1}
z	vertical space coordinate (distance above or below surface of substrate)	m
z_0	roughness length	m

Greek capital letters

Δ	change or difference in some quantity	operator
Θ	potential temperature	K
	volumetric soil-moisture content	dimensionless
Ψ	soil-moisture potential	m
Ω	rotational frequency of Earth	radians^{-1}

Greek lower case letters

α	albedo	% or fraction of one
	reflectivity	fraction of one
ε	emissivity	dimensionless
ζ	relative vorticity	s^{-1}
	absorptivity	fraction of one
λ	wavelength of electromagnetic radiation	μm
π	pi	dimensionless
ρ	density	kg m^{-3}
σ	Stefan–Boltzmann constant	W m^{-2} K^{-4}
ϕ	latitude	degrees
ψ	transmissivity	fraction of one
μ	thermal admittance	J m^{-2} s$^{-1/2}$ K^{-1}

Common subscripts

a	atmosphere
g	ground
Ha	refers to heat in the atmosphere
max	maximum value
p	applies at constant pressure
s	saturation
	surface
	substrate or soil

sat	saturation
v	water vapor
Wa	refers to water vapor in the atmosphere
z	applies at height z in atmosphere or depth z in substrate
θ	derivative is at constant potential temperature
λ	applies at a particular wavelength
0	applies at $z = 0$
n	applies at $z = $ level n

Appendix E
Maps of the world

Abu Dhabi - 27   Biskra - 9   In Salah - 32   Nouakchott - 15   Tel Aviv - 26
Aozou - 10   Cairo - 18   Khartoum -12   Quseir - 20   Tripoli - 30
Atar - 17   Dakhla Oasis - 19 Kidira - 24   Sabhah - 33   Wadi Halfa - 13
Baghdad - 7   Dhahran - 29   Luanda - 4   Saint Helena Bay - 3   Walvis Bay - 1
Beirut - 25   Djibouti - 22   Lüderitz - 2   Sharjah - 11   Windhoek - 6
Berbera - 21   Dongola - 14   Medina - 28   Swakopmund - 5
Bilma - 16   Faya - 23   Nouadhibou - 31   Tamanrasset - 8

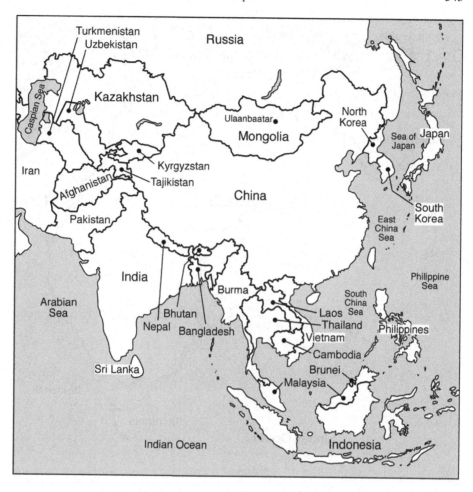

Hints to solving some problems and exercises

Chapter 2

Problem (2) Consider how a surface heat-flux maximum influences the pressure distribution near the surface and above the surface.

Problem (5) This is the reverse of the situation where a mountain extends upward into the atmosphere, and the heating contrast on a horizontal surface causes mountain–valley circulations.

Chapter 5

Problem (2) Table 5.1 shows the admittance of rock (concrete also) to be four times greater than that of dry sand. Recall that the admittance is a measure of the ability of a surface to accept or release heat. In this case, the daytime surfaces are hotter than your feet, so heat is being released from the substrate to you, with greater efficiency from the rock. Not only does the higher thermal conductivity of the rock allow the surface–foot temperature contrast to be maintained, but the higher heat capacity of the rock means that more heat has to be conducted to your foot before the surface temperature is reduced by a degree. Thus, for rock and sand surfaces *at the same temperature*, the concrete should feel hotter to the touch, quite the opposite from our experience. Thus, the surface temperature of the sand must actually be much higher than that of the concrete. This indeed makes sense because the ability of the rock to conduct the solar heat downward away from the surface is about a factor of ten greater than for the sand (Table 5.1). But, notice that the admittance of sand and clay are similar, so sand is not unique in this respect among the porous substrates. We are just more likely to be walking on a bare dry surface of it (at the beach) with our bare feet, and thus it has developed its painful reputation.

Chapter 19

Problem (3) To solve this you need the latent heat of vaporization for water (0.58 kcal g^{-1}) in order to obtain the heat consumed in the evaporation of 15 l of water. You then need the specific heat of water in order to determine how much the body temperature would increase if the heat actually lost through evaporation was used to raise the temperature. Conveniently, the kilogram-calorie, or kilocalorie, is the amount of heat required to raise the temperature of one kilogram of water from 14.5 °C to 15.5 °C. This specific-heat value applies only at 15 °C, but as an approximation assume that it is constant. You will find that the body temperature would have reached the boiling point of water long before the end of the 24 h period without any heat loss.

References

Abbs, D. J., 1986: Sea-breeze interactions along a concave coastline in southern Australia: observations and numerical modeling study. *Mon. Wea. Rev.*, **114**, 831–48.

Abd El Rahman, A. A., 1986: The deserts of the Arabian Peninsula. In *Hot Deserts and Arid Shrublands, Ecosystems of the World Volume 12B.* M. Evenari, I. Noy-Meir, and D. W. Goodall (Eds.), Elsevier Scientific Publishing Company, New York, 29–54.

Abdulaziz, A., 1994: A study of three types of wind-blown dust in Kuwait: Duststorms, rising dust and suspended dust. *J. Meteor.*, **19**, 19–23.

Abrahams, A. D., and A. J. Parsons, 1991: Relation between infiltration and stone cover on a semiarid hillslope, Southern Arizona. *J. Hydrology*, **122,** 49–59.

Ackerman, S. A., and H. Chung, 1992: Radiative effects of airborne dust on regional energy budgets at the top of the atmosphere. *J. Appl. Meteor.*, **31**, 223–33.

Ackerman, S. A., and S. K. Cox, 1982: The Saudi Arabian heat low: Aerosol distribution and thermodynamic structure. *J. Geophys. Res.*, **87**, 8991–9002.

Ackerman, S. A., and T. Inoue, 1994: Radiation energy budget studies using collocated AVHRR and ERBE observations. *J. Appl. Meteor.*, **33**, 370–8.

Acocks, J. P. H., 1952: *Veld Types of South Africa.* Department of Agriculture and Water Supply, Pretoria, South Africa.

Adams, D. K., and A. C. Comrie, 1997: The North American monsoon. *Bull. Amer. Meteor. Soc.*, **78**, 2197–213.

Adolph, E. F. (Ed.), 1969: *Physiology of Man in the Desert.* Hafner Publishing Company, New York.

Agassi, M., Y. Benyamini, J. Morin, S. Marish, and E. Henkin, 1996: Runoff and erosion control in Israel. In *Runoff, Infiltration and Subsurface Flow of Water in Arid and Semi-arid Regions.* A. S. Issar and S. D. Resnick (Eds.), Kluwer Academic Publishers, Dordrecht, pp. 63–120.

Agnew, C. T., 1988: Soil hydrology in the Wahabi sands. *J. Oman Studies*, Special Report no. 3, pp. 191–200.

1990: Spatial aspects of drought in the Sahel. *J. Arid Environments*, **18**, 279–93.

Agnew, C. T., and E. Anderson, 1992: *Water Resources in the Arid Realm.* Routledge, London.

Ahrens, C. D., 2000: *Meteorology Today: An Introduction to Weather, Climate, and the Environment.* Brooks/Cole, Pacific Grove.

Aizenshtat, B. A., 1960: *The heat balance and microclimate of certain landscapes in a sandy desert.* U. S. Department of Commerce, Weather Bureau (translated from Russian by George S. Mitchell), Washington, D.C.

Al-Shalash, A. H., 1966: *The Climate of Iraq*. The Cooperative Printing Presses Workers Society, Amman, Jordan.

Albertson, J. D., M. B. Parlange, G. G. Katul, C.-R. Chu, H. Stricker, and S. Tyler, 1995: Sensible heat flux from arid regions: A simple flux-variance method. *Water Resour. Res.*, **31**, 969–73.

Allison, G. B., and C. J. Barnes, 1985: Estimation of evaporation from the normally "dry" Lake Frome in South Australia. *J. Hydrology*, **78**, 229–42.

Allison, R. F., 1997: Middle East and Arabia. In *Arid Zone Geomorphology: Process Form and Change in Drylands*. D. S. G. Thomas (Ed.), John Wiley and Sons, Chichester, pp. 507–21.

Alonso, S., A. Portela, and C. Ramis, 1994: First considerations on the structure and development of the Iberian thermal low-pressure system. *Ann. Geophys.*, **12**, 457–68.

Alpert, P., Y. J. Kaufman, Y. Shay-El, D. Tanre, A. da Silva, S. Schubert, and J. H. Joseph, 1998: Quantification of dust-forced heating of the lower troposphere. *Nature*, **395**, 367–70.

Ambroggi, R. P., 1966: Water and the Sahara. *Scientific American*, **214**, 21–31.

Amiran, D. H. K., and A. W. Wilson (Eds.), 1973: *Coastal Deserts: Their Natural and Human Environments*. University of Arizona Press, Tucson.

Anderson, R. S., 1987: Eolian sediment transport as a stochastic process: The effects of a fluctuating wind on particle trajectories. *J. Geology*, **95**, 497–512.

Andrews, R. C., 1921: *Across Mongolian Plains*. Blue Ribbon Books, New York.

1932: *The New Conquest of Central Asia*. The American Museum of Natural History, New York.

Anthes, R. A., 1984: Enhancement of convective precipitation by mesoscale variations in vegetative covering in semi-arid regions. *J. Climate Appl. Meteor.*, **23**, 541–54.

Arakawa, H. (Ed.), 1969: *Climates of Northern and Eastern Asia*. Elsevier Scientific Publishing Company, Amsterdam.

Armstrong, R. W. (Ed.), 1983: *Atlas of Hawaii*. University of Hawaii Press, Honolulu.

Arritt, S., 1993: *Deserts*. The Readers Digest Association, Pleasantville.

Ash, J. E., and R. J. Wasson, 1983: Vegetation and sand mobility in the Australian desert dunefield. *Zeitschrift für Geomorphologie*, suppl. **45**, 7–25.

Ashburn, E. V., and R. G. Weldon, 1956: Spectral diffuse reflectance of desert surfaces. *J. Optical Soc. Amer.*, **46**, 583–6.

Atkinson, B. W., 1981: *Meso-scale Atmospheric Circulations*. Academic Press, London.

Aubreville, A., 1949: *Climats, forêts et désertification de l'Afrique tropicale*. Société de éditions géographiques, maritimes et coloniales, Paris.

AUSLIG (Australian Surveying and Land Information Group), 1992: *The Ausmap Atlas of Australia*. Cambridge University Press, Cambridge.

Austin, M., 1903: *The Land of Little Rain*. University of New Mexico Press, Albuquerque (second printing, 1976).

1924: *The Land of Journeys' Ending*. University of Arizona Press, Tucson (second printing, 1983).

Avissar, R., and H. Pan, 2000: Simulations of the summer hydrometeorological processes of Lake Kinneret. *J. Hydrometeor.*, **1**, 95–109.

Axelrod, D. I., 1983: Paleobotanical history of the western deserts. In *Origin and Evolution of Deserts*. S. G. Wells and D. R. Haragan (Eds.), University of New Mexico Press, Albuquerque, pp. 113–29.

Ayyad, M. A., and S. I. Ghabbour, 1986: Hot deserts of Egypt and the Sudan. In *Hot Deserts and Arid Shrublands, Ecosystems of the World Volume 12B*. M. Evenari, I.

Noy-Meir, and D. W. Goodall (Eds.), Elsevier Scientific Publishing Company, New York, pp. 149–202.

Bader, D. C., and T. B. McKee, 1983: Dynamical model simulation of the morning boundary layer development in deep mountain valleys. *J. Climate Appl. Meteor.*, **22**, 341–51.

Bagnold, R. A., 1935: *Libyan Sands: Travel in a Dead World*. Michael Haag Limited, London.

1954: *The Physics of Blown Sand and Desert Dunes*. Methuen, London.

1990: *Sand, Wind and War*. University of Arizona Press, Tucson.

Bahre, C. J., and D. E. Bradbury, 1978: Vegetation change along the Arizona-Sonora boundary. *Ann. Assoc. Amer. Geographe.*, **68**, 145–65.

Bailey, H. P., 1958: A simple moisture index based upon a primary law of evaporation. *Geograf. Ann.*, **1**, 1–16.

1981: Climate features of deserts. In *Water in Desert Ecosystems*. D. D. Evans and J. L. Thames (Eds.), Dowden, Hutchinson, and Ross, Stroudsburg, pp. 13–41.

Bailey, R. W., 1941: Climate and settlement of the arid region. In *Climate and Man: Yearbook of Agriculture – 1941*, United States Department of Agriculture, Washington, D. C., pp. 188–96.

Balland, D., 1994: Hidden waters. *The UNESCO Courier: A Window Open on the World*, **47–1**, 30–3.

Balling, R. C., Jr., 1988: The climatic impact of a Sonoran vegetation discontinuity. *Climatic Change*, **13**, 99–109.

1989: The impact of summer rainfall on the temperature gradient along the United States-Mexican border. *J. Appl. Meteor.*, **28**, 304–8.

Balling, R. C., Jr., and S. W. Brazel, 1989: High-resolution nighttime temperature patterns in Phoenix. *J. Arizona-Nevada Academy of Science*, **23**, 49–53.

Balling, R. C., Jr., and R. S. Cerveny, 1987: Long-term associations between wind speeds and the urban heat island of Phoenix, Arizona. *J. Climate Appl. Meteor.*, **26**, 712–16.

Bar-Ziv, J., and G. M. Goldberg, 1974: Simple siliceous pneumoconiosis in Negev Bedouins. *Archives of Environmental Health*, **29**, 121–6.

Barlow, M., S. Nigam, and E. H. Berbery, 1998: Evolution of the North American monsoon system. *J. Climate*, **11**, 2238–57.

Barnes, C. J., G. B. Allison, and M. W. Hughes, 1989: Temperature gradient effects on stable isotopes and chloride profiles in dry soils. *J. Hydrology*, **112**, 69–87.

Barnston, A. G., and P. T. Schickedanz, 1984: The effect of irrigation on warm-season precipitation in the Southern Great Plains. *J. Climate Appl. Meteor.*, **23**, 865–88.

Bass, D. E., 1963: Thermoregulatory and circulatory adjustments during acclimatization to heat in man. In *Temperature, its Measurement and Control in Science and Industry*, Vol. 3, Pt. 3., C. M. Hertzfeld and J. D. Hardy (Eds.), Reinhold, New York, pp. 299–305.

Beals, E. W., 1965: The remnant cedar forests of Lebanon. *J. Ecology*, **53**, 679–94.

Beard, J. S., 1949: *The Natural Vegetation of the Windward and Leeward Islands*. Oxford University Press, London.

Beebe, R. C., 1974: Large scale irrigation and severe storm enhancement. *Proceedings, Symposium on Atmospheric Diffusion and Air Pollution*, 9–13 September, Santa Barbara, California, American Meteorological Society (co-sponsored by World Meteorological Organization), pp. 392–5.

Beljaars, A. C. M., P. Viterbo, and M. J. Miller, 1996: The anomalous rainfall over the United States during July 1993: Sensitivity to land surface parameterization and soil moisture anomalies. *Mon. Wea. Rev.*, **124**, 362–83.

Bender, G. L. (Ed.), 1982: *Reference Handbook on the Deserts of North America.* Greenwood Press, Westport.

Berger, A. (Ed.), 1981: *Climate Variations and Variability: Facts and Theories.* D. Reidell Publishing Company, Dordrecht.

Berger, A., 1984: Accuracy and frequency stability of the Earth's orbital elements during the Quaternary. In *Milankovitch and Climate: Understanding the Response to Astronomical Forcing*, Vol. 1. A. Berger, J. Imbrie, J. Hays, G. Kukla, and B. Saltzman (Eds.), D. Reidel Publishing Company, Dordrecht, pp. 3–53.

Berger, A., J. Imbrie, J. Hays, G. Kukla, and B. Saltzman (Eds.), 1984: *Milankovitch and Climate: Understanding the Response to Astronomical Forcing*, Vols. 1 and 2. D. Reidel Publishing Company, Dordrecht.

Berger, I. A., 1997: South America. In *Arid Zone Geomorphology: Process Form and Change in Drylands.* D. S. G. Thomas (Ed.), John Wiley and Sons, Chichester, pp. 543–62.

Berkofsky, L., 1976: The effect of variable surface albedo on atmospheric circulation in desert regions. *J. Appl. Meteor.*, **15**, 1139–44.

Berrah, M., 1994: Screenplays in the sand. *The UNESCO Courier: A Window Open on the World*, **47–1**, 21–4.

Betancourt, J. L., C. Latorre, J. A. Rech, J. Quade, and K. A. Rylander, 2000: A 22,000-year record of monsoonal precipitation from northern Chile's Atacama Desert. *Science*, **289**, 1542–6.

Betzer, P. R., K. L. Carder, R. A. Duce, and 9 others, 1988: Long-range transport of giant mineral aerosol particles. *Nature*, **336**, 568–71.

Bingham, S., 1996: *The Last Ranch: A Colorado Community and the Coming Desert.* Harcourt Brace, San Diego.

Bitan, A., 1974: The wind regime in the north-west section of the Dead Sea. *Arch. Met. Geophys. Bioklim.*, B**22**, 313–35.

1977: The influence of the special shape of the Dead Sea and its environment on the local wind system. *Arch. Met. Geophys. Bioklim.*, B**24**, 283–301.

1981: Lake Kinneret (Sea of Galilee) and its exceptional wind system. *Bound. Layer Meteor.*, **21**, 477–87.

Bitan, A., and H. Sa'aroni, 1992: The horizontal and vertical extension of the Persian Gulf pressure trough. *Intl. J. Climatol.*, **12**, 733–47.

Black, J. F., 1963: Weather control: Use of asphalt coatings to tap solar energy. *Science*, **139**, 226–7.

Black, J. F., and B. I. Tarmy, 1963: The use of asphalt coatings to increase rainfall. *J. Appl. Meteor.*, **2**, 557–64.

Blackmore, C., 1995: *The Worst Desert on Earth: Crossing the Taklamakan.* John Murray, London.

Blake D. W., T. N. Krishnamurti, S. Low-Nam, and J. S. Fein, 1983: Heat low over the Saudi Arabian desert during May 1979 (Summer MONEX). *Mon. Wea. Rev.*, **111**, 1759–75.

Boffi, J. A., 1949: Effects of the Andes Mountains on the general circulation over the southern part of South America. *Bull. Amer. Meteor. Soc.*, **30**, 242–7.

Bond, W. J., W. D. Stock, and M. T. Hoffman, 1994: Has the Karoo spread? A test for desertification using stable carbon isotopes. *South African J. Science*, **90**, 391–7.

Borchert, J., 1950: The climate of the central North American grassland. *Ann. Assoc. Amer. Geographs.*, **40**, 1–39.

Bosilovich, M. G., and S. D. Schubert, 2002: Water vapor tracers as diagnostics of the regional hydrologic cycle. *J. Hydrometeor.*, **3**, 149–65.

Boudet, G., 1972: Désertification de l'Afrique tropicale sèche. *Adansonia*, Ser. 2, **12**, 505–24.

Bounoua, L., and T. N. Krishnamurti, 1991: Thermodynamic budget of the five-day wave over the Saharan Desert during summer. *Meteor. Atmos. Phys.*, **47**, 1–25.

Bouwer, H., 1975: Predicting reduction in water losses from open channels by phreatophyte control. *Water Resour. Res.*, **11**, 96–101.

Bowers, J. E., 1986: *Dune Country*. University of Arizona Press, Tucson.

Bowers, S. A., and R. J. Hanks, 1965: Reflection of radiant energy from soils. *Soil Science*, **100**, 130–8.

Bowker, D. E., and R. E. Davis, 1992: Influence of atmospheric aerosols and desert reflectance properties on satellite radiance measurements. *Int. J. Remote Sensing*, **13**, 3105–26.

Bowler, J. M., and R. J. Wasson, 1984: Glacial age environments of inland Australia. In *Late Cenozoic Palaeoclimates of the Southern Hemisphere*. J. C. Vogel (Ed.), Balkema, Rotterdam, pp. 183–208.

Bowles, P. F., 1949: *The Sheltering Sky*. The Ecco Press, Hopewell.

Bowman, I., 1924: *Desert Trails of Atacama*. American Geographical Society, Special Publication no. 5, New York.

Bowman, I., 1935: Our expanding and contracting "desert". *Geographical Rev.*, **25**, 43–61.

Brandt, C. J., and J. B. Thornes (Eds.), 1996: *Mediterranean Desertification and Land Use*. John Wiley and Sons, Chichester.

Braud, I., J. Noilhan, P. Bessemoulin, P. Mascart, R. Haverkamp, and M. Vauclin, 1993: Bare-ground surface heat and water exchanges under dry conditions: Observations and parameterization. *Bound. Layer Meteor.*, **66**, 173–200.

Braudel, F., 1972: *The Mediterranean and the Mediterranean World in the Age of Philip II*, Vol. 1. Harper and Row, New York.

Brazel, A. J., and S. Hsu, 1981: The climatology of hazardous Arizona dust storms. In *Desert Dust*, T. L. Péwé (Ed.), Geological Society of America, Special Paper 186, pp. 293–303.

Brazel, A. J., W. G. Nickling, and J. Lee, 1986: Effect of antecedent moisture conditions on dust storm generation in Arizona. In *Aeolian Geomorphology*. W. G. Nickling (Ed.), Allen and Unwin, Boston, pp. 261–71.

Breckle, S. W., 1983: The temperate deserts and semi-deserts of Afghanistan and Iran. In *Temperate Deserts and Semi-Deserts, Ecosystems of the World Volume 5*. N. E. West (Ed.), Elsevier Scientific Publishing Company, New York, pp. 271–319.

Breed, C. S., S. C. Fryberger, S. Andrews, C. McCauley, F. Lennartz, D. Gebel, and K. Horstman, 1979: Regional studies of sand seas using Landsat (ERTS) imagery. In *A Study of Global Sand Seas*. E. D. McKee (Ed.), U. S. Geological Survey Professional Paper 1052, pp. 305–98.

Bridenbaugh, C., and R. Bridenbaugh, 1972: *No Peace Beyond the Line – The English in the Caribbean 1624–1690*. Oxford University Press, New York.

Broccoli, A. J., and S. Manabe, 1992: The effects of orography on midlatitude Northern Hemisphere dry climates, *J. Climate*, **5**, 1181–201.

Brody, L. R., 1977: *Meteorological Phenomena of the Arabian Sea*. US Naval Environmental Prediction Research Facility, Monterey, NEPRF Application Report AR-77-01.

Brown, A. H., 1969a: Water requirements of man in the desert. In *Physiology of Man in the Desert*. E. F. Adolph (Ed.), Hafner Publishing Company, New York, pp. 115–35.

1969b: Survival without drinking water in the desert. In *Physiology of Man in the Desert*. E. F. Adolph (Ed.), Hafner Publishing Company, New York, pp. 271–9.

Brown, B. G., R. W. Katz, and A. H. Murphy, 1985: Exploratory analysis of precipitation events with implications for stochastic modeling. *J. Climate Appl. Meteor.*, **24**, 57–67.

Brubaker, K. L., D. Entekhabi, and P. S. Eagleson, 1993: Estimation of continental precipitation recycling. *J. Climate*, **6**, 1077–89.

Bruintjes, R. T., 1999: A review of cloud seeding experiments to enhance precipitation and some new prospects. *Bull. Amer. Meteor. Soc.*, **80**, 805–20.

Brutsaert, W. H., 1982: *Evaporation Into the Atmosphere: Theory, History and Applications*. Reidel, Dordrecht.

Bryant, N. A., L. F. Johnson, A. J. Brazel, R. C. Balling, C. F. Hutchinson, and L. R. Beck, 1990: Measuring the effect of overgrazing in the Sonoran Desert. *Climatic Change*, **17**, 243–64.

Bryson, R. A., 1978: The albedo pattern of the La Joya dunes. In *Exploring the World's Driest Climate*. H. H. Lettau and K. Lettau (Eds.), IES Report 101, Univ of Wisconsin, Madison, pp. 54–6.

Bryson, R. A., and D. A. Baerreis, 1967: Possibilities of major climatic modification and their implications: Northwest India, a case study. *Bull. Amer. Meteor. Soc.*, **48**, 136–42.

Bryson, R. A., and F. K. Hare (Eds.), 1974: *Climates of North America*. Elsevier Scientific Publishing Company, Amsterdam.

Bryson, R. A., C. A. Wilson, III, and P. M. Kuhn, 1964: Some preliminary results of radiation sonde ascents over India. *Proceedings WMO-IUGG Symposium Tropical Meteorology*, Rotorua, New Zealand, November 1963. J. W. Hutchings (Ed.), Wellington, New Zealand Meteorological Service, pp. 507–16.

Büdel, J., 1957: The Ice Age in the tropics. *Universitas*, **1**, 183–91.

Budyko, M. I., 1958: *The Heat Balance of the Earth's Surface*. US Department of Commerce, Weather Bureau (translated from Russian by Nina A. Stepanova), Washington, D.C.

1986: *The Evolution of the Biosphere*. Reidel, Dordrecht.

Bullard, J. E., 1997: Vegetation and dryland geomorphology. In *Arid Zone Geomorphology: Process Form and Change in Drylands*. D. S. G. Thomas (Ed.), John Wiley and Sons, Chichester, pp. 109–31.

Bullock, P., and L. H. Le Houérou, 1996: Land degradation and desertification. In *Climate Change 1995: Impacts, Adaptations and Mitigation of Climate Change*. R. T. Watson, M. C. Zinyowera, R. H. Ross, and D. J. Dokken (Eds.), Cambridge University Press, Cambridge, pp. 171–89.

Burk, S. D., and W. T. Thompson, 1996: The summertime low-level jet and marine boundary layer structure along the California coast. *Mon. Wea. Rev.*, **124**, 668–86.

Butterfield, G. R., 1991: Grain transport rates in steady and unsteady turbulent airflows. *Acta Mechanica*, suppl., **1**, 97–122.

1993: Sand transport response to fluctuating wind velocity. In *Turbulence, Perspectives on Flow and Sediment Transport*. N. J. Clifford, J. R. French, and J. Hardisty (Eds.), John Wiley and Sons, Chichester, pp. 305–35.

Butzer, K. W., R. Stuckenrath, A. J. Bruzewics, and D. M. Helgren, 1978: Late Cenozoic palaeoclimates of the Gaap Escarpment, Kalahari margin, South Africa. *Quaternary Research*, **10**, 310–39.

Cable, M., and F. Francesca, 1942: *The Gobi Desert*. Hodder and Stoughton, London.

Cahill, A. T., and M. B. Parlange, 1998: On water vapor transport in field soils. *Water Resour. Res.*, **34**, 731–9.

Campbell, G. S., and G. A. Harris, 1981: Modeling soil-water-plant-atmosphere systems of deserts. In *Water in Desert Ecosystems*. D. D. Evans and J. L. Thames (Eds.), Dowden, Hutchinson and Ross, Stroudsburg, pp. 75–91.

Campbell, J. M., 1997: *Few and Far Between: Moments in the North American Desert.* Museum of New Mexico Press, Santa Fe.

Capehart, W. J., and T. N. Carlson, 1997: Decoupling of surface and near-surface soil water content. *Water Resour. Res.*, **33**, 1383–95.

Carlson, T. N., and S. G. Benjamin, 1980: Radiative heating rates for Saharan dust. *J. Atmos. Sci.*, **37**, 193–213.

Carlson, T. N., and J. M. Prospero, 1972: The large-scale movement of Saharan air outbreaks over the northern equatorial Atlantic. *J. Appl. Meteor.*, **11**, 283–97.

Carlson, T. N., and P. Wendling, 1977: Reflected radiance measured by NOAA 3 VHRR as a function of optical depth for Saharan dust. *J. Appl. Meteor.*, **16**, 1368–71.

Cess, R. D., and I. L. Vulis, 1989: Intercomparison and interpretation of satellite-derived directional albedos over deserts. *J. Climate*, **2**, 393–407.

Chadwick, O. A., L. A. Derry, P. M. Vitousek, B. J. Huebert, and L. O. Hedin, 1999: Changing sources of nutrients during four million years of ecosystem development. *Nature*, **397**, 491–7.

Chang, J.-H., 1972: *Atmospheric Circulation Systems and Climates.* Oriental Publishing Company, Honolulu.

Chang, J.-T., and P. J. Wetzel, 1991: Effects of spatial variations of soil moisture and vegetation on the evolution of a prestorm environment: A numerical case study. *Mon. Wea. Rev.*, **119**, 1368–90.

Changnon, S. A., Jr., and R. G. Semonin, 1979: Impact of man upon local and regional weather. *Rev. Geophys. Space Physics*, **17**, 1891–900.

Chao, S. C., 1984a: The sand deserts and the Gobis of China. In *Deserts and Arid Lands.* F. El Baz (Ed.), Martinus Nijhoff, The Hague, pp. 95–113.

1984b: Analysis of the desert terrain in China using Landsat imagery. In *Deserts and Arid Lands.* F. El Baz (Ed.), Martinus Nijhoff, The Hague, pp. 115–32.

Chapman, V. J., 1960: *Salt Marshes and Salt Deserts of the World.* Leonard Hill Books, London.

Charlson, R. J., and J. Heintzenberg, 1995: *Aerosol Forcing of Climate.* John Wiley and Sons, Chichester.

Charney, J., 1975: Dynamics of deserts and drought in the Sahel. *Quart. J. Roy. Meteor. Soc.*, **101**, 193–202.

Charney, J., W. J. Quirk, S. H. Chow, and J. Kornfield, 1977: A comparative study of the effects of albedo change on drought in semiarid regions. *J. Atmos. Sci.*, **34**, 1366–85.

Charney, J., P. H. Stone and W. J. Quirk, 1975: Drought in the Sahara: A biogeophysical feedback mechanism. *Science*, **187**, 434–5.

Chase, T. N., R. A. Pielke, Sr., T. G. F. Kittel, J. S. Baron, and T. S. Stohlgren, 1999: Potential impacts on Colorado Rocky Mountain weather due to land use changes on the adjacent Great Plains. *J. Geophys. Res.*, **104**, 16673–90.

Chelhod, J., 1990: Islands of welcome in a sea of sand. *The UNESCO Courier: A Window Open on the World*, **43–2**, 11–15.

Chen, F., and R. Avissar, 1994: The impact of shallow convective moist processes on mesoscale heat fluxes. *J. Appl. Meteor.*, **33**, 1382–401.

Chen, F., and J. Dudhia, 2001: Coupling an advanced land surface – hydrology model with the Penn State – NCAR MM5 modeling system. Part I: Model implementation and sensitivity. *Mon. Wea. Rev.*, **129**, 569–85.

Chen, F., T. T. Warner, and K. Manning, 2001: Sensitivity of orographic moist convection to landscape variability: A study of the Buffalo Creek, Colorado, flash flood case of 1996. *J. Atmos. Sci.*, **58**, 3204–23.

Chen, S., Y. Kuo, W. Ming, and H. Ying, 1995: The effect of dust radiative heating on low-level frontogenesis. *J. Atmos. Sci.*, **52**, 1414–20.

Chen, Z., C. D. Elvidge, and D. P. Groeneveld, 1998: Monitoring seasonal dynamics of arid land vegetation using AVIRIS data. *Remote Sensing Environ.*, **65**, 255–66.

Chepil, W. S., 1945: Dynamics of wind erosion: 1. Nature of movement of soil by wind. *Soil Science*, **60**, 305–20.

Chepil, W. S., and N. P. Woodruff, 1957: Sedimentary characteristics of dust storms, II. Visibility and dust concentration. *Amer. J. Science*, **255**, 104–14.

Childs, C., 2000: *The Secret Knowledge of Water*. Sasquatch Books, Seattle.

2002: *The Desert Cries: A Season of Flash Floods in a Dry Land*. Arizona Highways Books, Phoenix.

Chou, M.-D., J. Guoliang, K.-N. Liou, and S.-C. S. Ou, 1992: Calculation of surface radiation in arid regions – A case study. *J. Appl. Meteor.*, **31**, 1084–95.

Chowdhury, A., A. K. Banerjee, and S. S. Gokhale, 1980: Meteorological aspects of arid and semi-arid regions of India with special reference to Thar desert. *Mausam*, **31**, 111–18.

Chudnovskii, A. F., 1966: Plants and light. I. Radiant energy. In *Fundamentals of Agrophysics*. A. F. Ioffe and I. B. Revut (Eds.), Israel Program for Scientific Translations, Jerusalem, pp. 1–51.

Clements, T., J. F. Mann, Jr., R. O. Stone, and J. L. Eymann, 1963: *A Study of Windborne Sand and Dust in Desert Areas*. United States Army Natick Laboratories, Earth Sciences Division, Natick, Massachusetts, Technical Report ES-8.

Clements, T., R. H. Merriam, R. O. Stone, J. L. Eymann, and H. L. Reade, 1957: *A Study of Desert Surface Conditions*. Headquarters, United States Army Quartermaster Research and Development Command, Environmental Protection Research Division, Natick, Massachusetts, Technical Report EP 53.

Cloudsley-Thompson, J. L., 1977: *Man and the Biology of Arid Zones*. Edward Arnold, London.

1984: *Sahara Desert*. Pergamon Press, Oxford.

1993: The future of the Sahara. *Environmental Conservation*, **20**, 335–8.

Coakley, J. A., Jr., and R. D. Cess, 1985: Response of the NCAR community climate model to the radiative forcing by the naturally occurring tropospheric aerosol. *J. Atmos. Sci.*, **42**, 1677–92.

Cogger, H. G., and E. E. Cameron (Eds.), 1984: *Arid Australia*. Australian Museum, Sydney.

Cole, A. E., A. Court, and A. J. Kantor, 1965: Model atmospheres. In *Handbook of Geophysics and Space Environments*. S. L. Valley (Ed.), McGraw-Hill, New York, pp. 2.1–2.22.

Conrad, G., 1969: *L'évolution Continental Posthercynienne du Sahara Algerien*. CNRS, Paris.

Cooke, R. U., and A. Warren, 1973: *Geomorphology in Deserts*. University of California Press, Berkeley.

Cooke, R., A. Warren, and A. Goudie, 1993: *Desert Geomorphology*. University College London Press, London.

Cooley, R. L., G. W. Fiero, L. H. Lattman, and A. L. Mindling, 1973: *Influence of Surface and Near Surface Caliche Distribution on Infiltration Characteristics and Flooding, Las Vegas Area, Nevada*. Desert Research Institute, University of Nevada System, Project Report No. 21.

Cornett, J. W., 1985: *Atacama, Desert of Chile and Peru*. Palm Springs Desert Museum, Palm Springs.

Costa, J. E., 1987: Hydraulics and basin morphometry of the largest flash floods in the conterminous United States. *J. Hydrology*, **93**, 313–38.

Courel, M. F., R. S. Kandel, and S. I. Rasool, 1984: Surface albedo and the Sahel drought. *Nature*, **307**, 528–31.

Croke, J., 1997: Australia. In *Arid Zone Geomorphology: Process Form and Change in Drylands*. D. S. G. Thomas (Ed.), John Wiley and Sons, Chichester, pp. 563–73.

Crooks, G. A., and G. R. C. Cowan, 1993: Duststorm, South Australia, November 7th, 1988. *Bull. Aust. Meteor. Oceanog. Soc.*, **6**, 68–72.

Cunnington, W. M., and P. R. Rowntree, 1986: Simulation of the Saharan atmosphere – dependence on moisture and albedo. *Quart. J. Roy. Meteor. Soc.*, **112**, 971–99.

Darlington, D., 1996: *The Mojave: A Portrait of the Definitive American Desert*. Henry Holt and Company, New York.

Darwin, C., 1845: An account of the fine dust which often falls on vessels in the Atlantic Ocean. *Proc. Geol. Soc. Lond.* 4 June 1845, pp. 26–30.

Darwin, C., 1860: *Journal of Researches into the Natural History and Geology of the Countries Visited During the Voyage of the H.M.S. Beagle Round the World*, London. 1952, Hafner Publishing Company, New York.

Davis, C., T. Warner, E. Astling, and J. Bowers, 1999: Development and application of an operational, relocatable, mesogamma-scale weather analysis and forecasting system. *Tellus*, **51A**, 710–27.

de Martonne, E., 1927: Regions of interior basin drainage. *Geographical Rev.*, **17**, 397–414.

de Martonne, E., and L. Aufrère, 1927: Map of interior basin drainage. *Geographical Rev.*, **17**, 414.

de Saint-Exupéry, A., 1939: Wind, Sand and Stars. In *Airman's Odyssey*. Harcourt Brace, San Diego.

de Villiers, M., 2000: *Water: The Fate of Our Most Precious Resource*. Houghton Mifflin Company, Boston.

de Vreede, M., 1977: *Deserts and Men: A Scrapbook*. Government Publishing Office, The Hague.

de Vries, D. A., 1959: The influence of irrigation on the energy balance and the climate near the ground. *J. Appl. Meteor.*, **16**, 256–70.

Deacon, E. L., 1953: *Vertical Profiles of Mean Wind in the Surface Layers of the Atmosphere*. Geophysical Mem. No. 91, Meteorological Office, Air Ministry, London.

Dean, W. R. S., and I. A. W. McDonald, 1994: Historical changes in stocking rates of domestic livestock as a measure of semi-arid and arid rangeland degradation in the Cape Province. *J. Arid Environments*, **28**, 281–98.

Deering, D. W., T. F. Eck, and J. Otterman, 1990: Bidirectional reflectances of selected desert surfaces and their three-parameter soil characterization. *Agricultural and Forest Meteor.*, **52**, 71–93.

Dennis, A. S., 1990: Water augmentation in arid lands through weather modification. In *Human Intervention in the Climatology of Arid Lands*. D. R. Haragan (Ed.), University of New Mexico Press, Albuquerque, pp. 61–99.

Denton, G. H., and W. Karlén, 1973: Holocene climate variations – Their pattern and possible cause. *Quaternary Research*, **3**, 155–205.

Depierre, D., and H. Gillet, 1971: Désertification de la zone sahélienne du Tchad. *Bois et Forêts des Tropiques*, **139**, 2–25.

Derbyshire, E., and A. S. Goudie, 1997: Asia. In *Arid Zone Geomorphology: Process Form and Change in Drylands*. D. S. G. Thomas (Ed.), John Wiley and Sons, Chichester, pp. 487–506.

Dey, B., 1982: Nature and possible causes of droughts in the Canadian prairies – case studies. *J. Climatol.*, **2**, 233–49.

Diaz de Villegas, J., A. Marin, A. Ochoa, F. Hernandez, G. Bullon, J. de Lizaury, J. Bonelli, and M. Lombardero, 1949: *España en Africa*. Consejo Superior de Investigaciones Científicas, Instituto de Estudios Africanos, Madrid.

Dib, M., 1992: *Le Désért Sans Detour*. Sindbad, Paris.

Dickinson, R. E., 1983: Land surface processes and climate – Surface albedos and energy balance. *Adv. Geophys.*, **25**, 305–53.

Dincer, T., A. Al-Mugrin, and U. Zimmermann, 1974: Study of the infiltration and recharge through the sand dunes in arid zones with special reference to the stable isotopes and thermonuclear tritium. *J. Hydrology*, **23**, 79–109.

Dodd, A. V., and H. S. McPhilimy, 1959: *Yuma Summer Microclimate*. Headquarters, Quartermaster Research and Engineering Command, Environmental Protection Research Division, US Army, Natick, Massachusetts, Technical Report EP-120.

Dolman, A. J., A. D. Culf, and P. Bessemoulin, 1997: Observations of boundary layer development during the HAPEX-Sahel observation period. *J. Hydrology*, **188–9**, 998–1016.

Douglas, M. W., and S. Li, 1996: Diurnal variation of the lower-tropospheric flow over the Arizona low desert from SWAMP-1993 observations. *Mon. Wea. Rev.*, **124**, 1211–24.

Douglas, M. W., R. A. Maddox, K. Howard, and S. Reyes, 1993: The Mexican monsoon. *J. Climate*, **6**, 1665–77.

Douglas, M. W., M. Peña, N. Ordinola, L. Flores, and J. Boustead, 2001: Synoptic and spatial variability of the rainfall along the northern Peruvian coast during the 1997–8 El Niño event. In *Symposium on Precipitation Extremes: Prediction, Impacts and Responses*, Albuquerque, New Mexico, American Meteorological Society, pp. 170–173.

Doyle, J. D., 1997: The influence of mesoscale orography on a coastal jet and rainband. *Mon. Wea. Rev.*, **125**, 1465–88.

Dregne, H. E., 1977: Desertification of arid lands. *Economic Geography*, **3**, 322–31.

1983: *Desertification of Arid Lands*. Harwood Academic Publishers, New York.

1985: Aridity and land degradation. *Environment*, **27**, 16–33.

1986: Desertification of arid lands. In *Physics of Desertification*. F. El-Baz and M. H. A. Hassan (Eds.), Martinus Nijhoff Publishers, Dordrecht, pp. 4–34.

1987: Reflections on the PACD. *Desertification Control Bulletin*, United Nations Environment Program, Special Tenth Anniversary of UNCOD Issue (November 15, 1987), pp. 8–11.

Dregne, H. E., and C. J. Tucker, 1988: Desert encroachment. *Desertification Control Bulletin*, United Nations Environment Program, No. 16, pp. 16–19.

Driese, K. L., and W. A. Reiners, 1997: Aerodynamic roughness parameters for semi-arid natural shrub communities of Wyoming, USA. *Agricultural and Forest Meteor.*, **88**, 1–14.

Droegemeier, K. K., and R. B. Wilhelmson, 1987: Numerical simulation of thunderstorm outflow dynamics. Part I: Outflow sensitivity experiments and turbulence dynamics. *J. Atmos. Sci.*, **44**, 1180–210.

Dubief, J., 1959: *Le Climat du Sahara*, Vol. 1. University of Algeria, Algiers.

1963: *Le Climat du Sahara*, Vol. 2. University of Algeria, Algiers.

Duce, R. A., 1995: Sources, distributions, and fluxes of mineral aerosols and their relationship to climate. In *Aerosol Forcing of Climate*. R. J. Charlson and J. Heintzenberg (Eds.), John Wiley and Sons, Chichester, pp. 43–72.

Duce, R. A., P. S. Liss, J. T. Merrill, and 19 others, 1991: The atmospheric input of trace species to the world ocean. *Global Biogeochemical Cycles*, **5**, 193–259.

Dunkerley, D. L., and K. F. Brown, 1997: Desert soils. In *Arid Zone Geomorphology: Process Form and Change in Drylands*. D. S. G. Thomas (Ed.), John Wiley and Sons, Chichester, pp. 55–68.

Dutton, J. A., 1976: *The Ceaseless Wind: An Introduction to the Theory of Atmospheric Motion*. McGraw Hill, New York.

Dzerdzeevskii, B. L., 1958: On some climatological problems and microclimatological studies of arid and semi-arid lands in the U.S.S.R. In *Climatology and Microclimatology*, Vol. 11, Proceedings Canberra Symposium, UNESCO, Paris, pp. 315–25.

Eckholm, E. P., 1976: *Losing Ground: Environmental Stress and World Food Prospects*. W. W. Norton and Company, New York.

Eddy, J. A., 1976: The Maunder Minimum. *Science*, **192**, 1189–202.

  1977: Climate and the changing sun. *Climatic Change*, **1**, 173–90.

Edholm, O. G., 1972: The effect in man of acclimatization to heat on water intake, sweat rate and water balance. In *Advances in Climatic Physiology*. S. Ito, K. Ogata, and H. Yoshimura (Eds.), Shoin, Tokyo, pp. 144–55.

Ek, M., and R. H. Cuenca, 1994: Variation in soil parameters: Implications for modeling surface fluxes and atmospheric boundary-layer development. *Bound. Layer Meteor.*, **70**, 369–83.

El-Baz, F., 1983: A geological perspective of the desert. In *Origin and Evolution of Deserts*. S. G. Wells and D. R. Haragan (Eds.), University of New Mexico Press, Albuquerque, pp. 163–83.

  1986: On the reddening of quartz grains in dune sand. In *Physics of Desertification*. F. El-Baz and M. H. A. Hassan (Eds.), Martinus Nijhoff Publishers, Dordrecht, pp. 191–209.

El-Fandy, M. G., 1949: Dust – an active meteorological factor in the atmosphere of northern Africa. *J. Appl. Phys.*, **20**, 660–6.

Ellis, H., 1758: An account of the heat of the weather in Georgia. *Phil. Trans. R. Soc. Lond.*, **50**, 754–6.

Ellsaesser, H. W., M. C. MacCracken, G. L. Potter, and F. M. Luther, 1976: An additional model test of positive feedback from high desert albedo. *Quart. J. Roy. Meteor. Soc.*, **102**, 655–66.

Eltahir, E. A. B., 1996: Role of vegetation in sustaining large-scale atmospheric circulations in the tropics. *J. Geophys. Res.*, **101**, 4255–68.

Ely, L. L., Y. Enzel, V. R. Baker, and D. R. Cayan, 1993: A 5000-year record of extreme floods and climate change in the southwestern United States. *Science*, **262**, 410–12.

Enfield, D. B., 1981: Thermally driven wind variability in the planetary boundary layer above Lima, Peru. *J. Geophys. Res.*, **86**, 2005–16.

Evans, D. D., T. W. Sammis, and D. R. Cable, 1981: Actual evapotranspiration under desert conditions. In *Water in Desert Ecosystems*. D. D. Evans and J. L. Thames (Eds.), Dowden, Hutchinson and Ross, Stroudsburg, pp. 195–218.

Evans, D. D., and J. L. Thames (Eds.), 1981: *Water in Desert Ecosystems*. Dowden, Hutchinson and Ross, Stroudsburg.

Evenari, M. I., 1985a: The desert environment. In *Hot Deserts and Arid Shrublands, Ecosystems of the World Volume 12A*. M. Evenari, I. Noy-Meir, and D. W. Goodall (Eds.), Elsevier Scientific Publishing Company, New York, pp. 1–22.

  1985b: Adaptations of plants and animals to the desert environment. In *Hot Deserts and Arid Shrublands, Ecosystems of the World Volume 12A*. M. Evenari, I. Noy-Meir, and D. W. Goodall (Eds.), Elsevier Scientific Publishing Company, New York, pp. 79–92.

Evenari, M., I. Noy-Meir, and D. W. Goodall (Eds.), 1985: *Ecosystems of the World, Volume 12A: Hot Deserts and Arid Shrublands*. Elsevier Scientific Publishing Company, New York.

1986: *Ecosystems of the World*, Volume 12B: *Hot Deserts and Arid Shrublands*. Elsevier Scientific Publishing Company, New York.

Evenari, M., L. Shanan, and N. Tadmor, 1971: *The Negev: The Challenge of a Desert*. Harvard University Press, Cambridge, Massachusetts.

Faus-Pujol, M. C., and A. Higueras-Arnal, 1998: Two examples of environmental transformation in dry Spain. In *Population and Environment in Arid Regions*, Man and the Biosphere Series, Vol. 19. J. Clarke and D. Noin (Eds.), The Parthenon Publishing Company, Pearl River, pp. 105–31.

Ferrari, M., 1996: *Deserts*. Smithmark Publishers, New York.

Findlater, J., 1966: Cross-equatorial jet streams at low levels over Kenya. *Meteor. Mag.*, **95**, 353–64.

1967: Some further evidence of cross-equatorial jet streams at low levels over Kenya. *Meteor. Mag.*, **96**, 216–19.

1969: A major low-level air current near the Indian Ocean during the northern summer., *Quart. J. Roy. Meteor. Soc.*, **95**, 362–80.

Fleagle, R., and J. Businger, 1963: *An Introduction to Atmospheric Physics*. Academic Press, New York.

Flohn, H., 1964: On the causes of the aridity of northeastern Africa. *Würzburger Geograph. Arb.*, **12**, 1–17.

Folland, C. K., T. R. Karl, and K. Y. A. Vinnikov, 1990: Observed climate variations and change. In *Climate Change: The IPCC Scientific Assessment*. J. T. Houghton, G. J. Jenkins, and J. J. Ephraums (Eds.), Cambridge University Press, Cambridge, pp. 195–238.

Folland, C. K., T. N. Palmer, and D. E. Parker, 1986: Sahel rainfall and world-wide sea temperatures 1901–85; Observational, modelling, and simulation studies. *Nature*, **320**, 602–7.

Fontaine, B., and S. Bigot, 1993: West African rainfall deficits and sea surface temperatures. *Intl. J. Climatol.*, **13**, 271–85.

Fouquart, Y., B. Bonnell, G. Brogniez, J. C. Buriez, L. Smith, J. J. Morcrette, and A. Cerf, 1987: Observations of Saharan aerosols: Results of ECLATS field experiment. II Broadband radiative characteristics of the aerosols and vertical radiative flux divergence. *J. Climate Appl. Meteor.*, **26**, 38–52.

Fowler, W. B., and J. D. Helvey, 1974: Effects of large-scale irrigation on climate in the Columbia Basin. *Science*, **184**, 121–7.

Freeman, M. H., 1952: *Duststorms of the Anglo-Egyptian Sudan*. Meteorological Report No. 11, Meteorological Office, London.

Furlong, C. W., 1914: *The Gateway to the Sahara*. Charles Scribner and Sons, New York.

Gaertner, M. A., C. Fernández, and M. Castro, 1993: A two-dimensional simulation of the Iberian summer thermal low. *Mon Wea. Rev.*, **121**, 2740–56.

Gamo, M., 1996: Thickness of the dry convection and large-scale subsidence above deserts. *Bound. Layer Meteor.*, **79**, 265–78.

Garratt, J. R., 1992a: *The Atmospheric Boundary Layer*. Cambridge University Press, Cambridge.

1992b: Extreme maximum land surface temperatures. *J. Appl. Meteor.*, **31**, 1096–105.

Gash, J. H. C., P. Kabat, B. A. Monteny, and 17 others, 1997: The variability of evaporation during the HAPEX-Sahel observation period. *J. Hydrology*, **188–9**, 385–99.

Gates, D. M., 1972: *Man and His Environment: Climate*. Harper and Row, New York.

Gautier, E.-F., 1935: *Sahara: The Great Desert*. Columbia University Press, New York.

Gay, L. W., and C. Bernhofer, 1991: Enhancement of evaporation by advection in arid regions. In *Hydrological Interactions Between Atmosphere, Soil and Vegetation*, Proceedings of the Vienna Symposium. G. Kientz *et al.* (Eds.), IAHS Publ. No. 204, pp. 147–56.

Gehlbach, F. R., 1993: *Mountain Islands and Desert Seas*. Texas A&M University Press, College Station.

Geiger, R., 1966: *The Climate Near the Ground*. Harvard University Press, Cambridge, Massachusetts.

Gentilli, J., 1971: *World Survey of Climatology*, Vol. 13, *Climates of Australia and New Zealand*. Elsevier Scientific Publishing Company, Amsterdam.

George, U., 1977: *In the Deserts of This Earth*. Harcourt Brace Jovanovich, New York.

Gerber, H. E., and A Deepak (Eds.), 1984: *Aerosols and Their Climatic Effects*. A. Deepak Publishing, Hampton.

Gibbs, W. J., 1969: Meteorology and climatology. In *Arid Lands of Australia*. R. O. Slatyer and R. A. Perry (Eds.), Australian National University Press, Canberra, pp. 33–54.

Gillette, D. A., D. A. Bolivar, and D. W. Fryrear, 1974: The influence of wind velocity on the size distributions of aerosols generated by the wind erosion of soils. *J. Geophys. Res.*, **79C**, 4068–75.

Glaccum, R. A., and J. M. Prospero, 1980: Saharan aerosols over the tropical North Atlantic – Mineralogy. *Marine Geology*, **37**, 295–321.

Glantz, M. H. (Ed.), 1994: *Drought Follows the Plow: Cultivating Marginal Areas*. Cambridge University Press, Cambridge.

Glantz, M. H., and N. Orlovsky, 1983: Desertification: a review of the concept. *Desertification Control Bulletin*, United Nations Environment Program, No. 9, pp. 15–22.

Glantz, M. H., A. Z. Rubinstein, and I. Zonn, 1993: Tragedy in the Aral Sea basin. *Global Environmental Change*, **3**, 174–98.

Godske, C. L., T. Bergeron, J. Bjerknes, and R. C. Bundegaard, 1957: *Dynamic Meteorology and Weather Forecasting*. American Meteorological Society and Carnegie Institution of Washington, Boston and Washington.

Gosselin, R. E., 1969: Rates of sweating in the desert. In *Physiology of Man in the Desert*. E. F. Adolph (Ed.), Hafner Publishing Company, New York, pp. 44–76.

Goudie, A. S., 1978: Dust storms and their geomorphological implications. *J. Arid Environments*, **1**, 291–310.

1983a: *Environmental Change*. Clarendon Press, Oxford.

1983b: Dust storms in space and time. *Prog. Physical Geog.*, **7**, 502–30.

Goudie, A. S., and N. J. Middleton, 1992: The changing frequency of dust storms through time. *Climatic Change*, **20**, 197–225.

Goudie, A. S., and J. Wilkinson, 1977: *The Warm Desert Environment*. Cambridge University Press, London.

Goutorbe, J. P., T. Lebel, A. Tinga, and 13 others, 1994: HAPEX-Sahel: A large-scale study of land-atmosphere interactions in the semi-arid tropics. *Ann. Geophys.*, **12**, 53–64.

Goutorbe, J. P., T. Lebel, A. J. Dolman, and 8 others, 1997: An overview of HAPEX-Sahel: A study in climate and desertification. *J. Hydrology*, **188–9**, 4–17.

Grainger, A., 1990: *The Threatening Desert*. Earthscan, London.

Great Plains Committee, 1936: *The Future of the Great Plains*. Report of the Great Plains Committeee, Government Printing Office, Washington, D.C., December 1936.

Greco, S., R. H. Moss, D. Viner, and R. Jenne, 1994: *Climate Scenarios and Socioeconomic Projections for IPCC WGII Assessment*. IPCC-WMO and UNEP, Washington, D.C.

Greeley, R., and J. D. Iverson, 1985: *Wind as a Geological Process on Earth, Mars, Venus and Titan*. Cambridge University Press, Cambridge.

Gribbin, J. (Ed.), 1978: *Climate Change*. Cambridge University Press, Cambridge.

Griffin, D. W., V. H. Garrison, J. R. Herman, and E. A. Shinn, 2001: African desert dust in the Caribbean atmosphere: Microbiology and public health. *Aerobiologia*, **17**, 203–13.

Griffiths, J. F., 1966: *Applied Climatology: An Introduction*. Oxford University Press, London.

1972a: General introduction. In *World Survey of Climatology*, Vol. 10, *Climates of Africa*. J. F. Griffiths (Ed.), Elsevier Scientific Publishing Company, Amsterdam, pp. 1–35.

1972b: The Horn of Africa. In *World Survey of Climatology*, Vol. 10, *Climates of Africa*. J. F. Griffiths (Ed.), Elsevier Scientific Publishing Company, Amsterdam, pp. 133–65.

1972c: *Climates of Africa*. Elsevier Scientific Publishing Company, Amsterdam.

Griffiths, J. F. and K. H. Soliman, 1972: The northern desert (Sahara). In *World Survey of Climatology*, Vol. 10, *Climates of Africa*. J. F. Griffiths (Ed.), Elsevier Scientific Publishing Company, Amsterdam, pp. 75–131.

Guerzoni, S., and R. Chester (Eds.), 1996: *The Impact of Desert Dust Across the Mediterranean*. Kluwer Academic Publishers, Dordrecht.

Gupta, R. K., 1986: The Thar Desert. In *Hot Deserts and Arid Shrublands, Ecosystems of the World Volume 12B*. M. Evenari, I. Noy-Meir, and D. W. Goodall (Eds.), Elsevier Scientific Publishing Company, New York, pp. 55–99.

Gutentag, E. D., and J. B. Weeks, 1980: *Water Table in the High Plains Aquifer in 1978 in Parts of Colorado, Kansas, Nebraska, New Mexico, Oklahoma, South Dakota, Texas and Wyoming*. Hydrologic Investigations Atlas HA-642, U. S. Geological Survey, Reston, Virginia.

Hacker, J. M., 1988: The spatial distribution of the vertical energy fluxes over a desert lake area. *Aust. Meteor. Mag.*, **36**, 235–43.

Hahn, D. G., and S. Manabe, 1975: The role of mountains in the south-Asian monsoon circulation. *J. Atmos. Sci.*, **32**, 1515–41.

Hahn, D. G., and J. Shukla, 1976: An apparent relationship between Eurasian snow cover and Indian monsoon rainfall. *J. Atmos. Sci.*, **33**, 2461–2.

Hansen, J. E., A. A. Lacis, P. Lee, and W. -C. Wang, 1980: Climatic effects of atmospheric aerosols. *Ann. New York Acad. Sci.*, **338**, 575–87.

Harazono, Y., J. Shen, S. Liu, and S. Li, 1992: Micrometeorological characteristics of a sand dune in the eastern part of Inner Mongolia, China in Autumn. *J. Agric. Meteor.*, **47**, 217–24.

Hartmann, W. K., 1989: *Desert Heart: Chronicles of the Sonoran Desert*. Fisher Books, Tucson.

Hassan, M. A., 1990: Observations of desert flood bores. *Earth Surface Processes and Landforms*, **15**, 481–5.

Hastenrath, S., 1994: *Climate Dynamics of the Tropics*. Kluwer Academic Publishers, Dordrecht.

Hastings, J. R., and R. M. Turner, 1965: *The Changing Mile*. University of Arizona Press, Tucson.

Hayden, J. D., 1998: *The Sierra Pinacate*. University of Arizona Press, Tucson.

Heathcote, R. L., 1983: *The Arid Lands: Their Use and Abuse*. Longman, London.

Hellden, U., 1988: Desertification monitoring. *Desertification Control Bulletin*, United Nations Environment Program, No. 17, pp. 8–12.

1991: Assessing desertification. *Ambio*, **20**, 372–83.

Hellwig, D. H. R., 1973a: Evaporation of water from sand: 1. Experimental setup and climate influences. *J. Hydrology*, **18**, 93–108.

1973b: Evaporation of water from sand: 2. Diurnal variations. *J. Hydrology*, **18**, 109–18.

1978: Evaporation of water from sand: 6. The influence of the depth of the water table on diurnal variations. *J. Hydrology*, **39**, 129–38.

Henderson, S. T., 1977: *Daylight and its Spectrum*. Adam Hilger, Bristol.

Henderson-Sellers, A., 1981: The effects of land clearing and agricultural practices upon climate. In *Blowing in the Wind: Deforestation and Long Range Implications*. Studies in Third World Societies, Pub. No. 14, Dept. of Anthropology, College of William and Mary, pp. 443–85.

Henning, D., and F. Flohn, 1977: *Climate Aridity Index Map*. United Nations Conference on Desertification, Nairobi, Kenya.

Herman, J. R., P. K. Bhartia, O. Torres, C. Hsu, C. Seftor, and E. Celarier, 1997: Global distribution of UV-absorbing aerosols from Nimbus 7/TOMS data. *J. Geophys. Res.*, **102**, 16911–22.

Hershfield, D. M., 1968: Rainfall input for hydrologic models. *Intl. Assoc. Sci. Hydrol.*, **78**, 177–88.

Hillel, D., 1998: *Environmental Soil Physics*. Academic Press, San Diego.

Hills, E. S. (Ed.), 1966: *Arid Lands: A Geographical Appraisal*. Methuen and Company, London.

Hinds, B. D., and G. B. Hoidale, 1975: *Boundary Layer Dust Occurrence*, Vol. 2, *Atmospheric Dust Over the Middle East, Near East and North Africa*. Technical Report, US Army Electronics Command, White Sands Missile Range, New Mexico.

Hipps, L. E., 1989: The infrared emissivities of soil and *Artemisia tridentata* and subsequent temperature corrections in a shrub-steppe ecosystem. *Remote Sensing Environ.*, **27**, 337–42.

Hjalmarson, H. W., 1984: Flash flood in Tanque Verde Creek, Tucson, Arizona. *J. Hydraulic Eng.*, *ASCE*, **110**, 1841–52.

Hobbs, P. V., and M. P. McCormick (Eds.), 1988: *Aerosols and Climate*. A. Deepak Publishing, Hampton.

Hoffman, M. T., and R. M. Cowling, 1990: Desertification in the Lower Sundays River Valley, South Africa. *J. Arid Environments*, **19**, 105–17.

Holcombe, T. L., T. Ley, and D. A. Gillette, 1997: Effects of prior precipitation and source area characteristics on threshold wind velocities for blowing dust episodes, Sonoran Desert 1948–1978. *J. Appl. Meteor.*, **36**, 1160–75.

Holm, D. A., 1960: Desert geomorphology in the Arabian Peninsula. *Science*, **132**, 1369–79.

Holmes, R. M., 1969: Note on low-level airborne observations of temperature near prairie oases. *Mon. Wea. Rev*, **97**, 333–9.

Holt, T. R., 1996: Mesoscale forcing of a boundary layer jet along the California coast. *J. Geophys. Res.*, **101**, 4235–54.

Holton, J. R., 1992: *An Introduction to Dynamic Meteorology*. Academic Press, San Diego.

Hong, S.-Y., and E. Kalnay, 2000: Role of sea surface temperature and soil-moisture feedback in the 1998 Oklahoma-Texas drought. *Nature*, **408**, 842–4.

Hong, X., M. J. Leach, and S. Raman, 1995: Role of vegetation in generation of mesoscale circulation. *Atmos. Environ.*, **29**, 2163–76.

Hoogmoed, W. B., 1986: *Analysis of rainfall characteristics relating to soil management from some selected locations in Niger and India*. Wageningen Agricultural University, ICRISAT Sahelian Center Report 86–3, Wageningen, The Netherlands.

Hora, S. L., 1952: The Rajputana Desert: Its value in India's national economy. Proceedings of the Symposium on the Rajputana Desert. *Bull. National Institute Sci. India*, **1**, 1–11.

Hoskins, B. J., H. H. Hsu, I. N. James, M. Masutani, P. D. Sardeshmukh, and G. H. White, 1989: *Diagnostics of the Global Atmospheric Circulation Based on ECMWF Analyses, 1979–1989*. WCRP, WMO/TD No. 326, World Meteorological Organization, Geneva.

Houseman, J., 1961: Dust haze at Bahrain. *Meteor. Mag.*, **19**, 50–2.

Hovis, W. A., Jr., 1966: Infrared spectral reflectance of some common minerals. *Appl. Optics*, **5**, 245–8.

Howe, G. H., L. J. Reed, J. T. Ball, G. E. Fisher, and G. B. Lassow, 1968: *Classification of World Desert Areas*. United States Army Natick Laboratories, Materiel Command, Technical Report 69-38-ES, Natick, Massachusetts.

Hulme, M., and M. Kelly, 1993: Exploring the links between desertification and climate change. *Environment*, **35**, 4–19.

Humes, K. S., W. P. Kustas, and D. C. Goodrich, 1997: Spatially distributed sensible heat flux over a semiarid watershed. Part I: Use of radiometric surface temperatures and a spatially uniform resistance. *J. Appl. Meteor.*, **36**, 281–92.

Humes, K. S., W. P. Kustas, M. S. Moran, W. D. Nichols, and M. A. Weltz, 1994: Variability of emissivity and surface temperature over a sparsely vegetated surface. *Water Resour. Res.*, **30**, 1299–310.

Hunt, C. B., 1983: Physiographic overview of our arid lands in the western U. S. In *Origin and Evolution of Deserts*. S. G. Wells and D. R. Haragan (Eds.), University of New Mexico Press, Albuquerque, pp. 7–63.

Husar, R. B., J. M. Prospero, and L. L. Stowe, 1997: Characterization of tropospheric aerosols over the oceans with the NOAA advanced very high resolution radiometer optical thickness operational product. *J. Geophys. Res.*, **102**, 16 889–909.

Idso, S. B., 1976: Dust storms. *Scientific American*, **235**, 108–14.

1977: A note on some recently proposed mechanisms of genesis of deserts, *Quart. J. Roy. Meteor. Soc.*, **103**, 369–70.

Idso, S. B., and J. W. Deardorff, 1978: Comments on "The effect of variable surface albedo on atmospheric circulation in desert regions" (Berkofsky, L., 1976). *J. Appl. Meteor.*, **17**, 560.

Ingram, D. L., and L. E. Mount, 1975: *Man and Animals in Hot Environments*. Springer-Verlag, New York.

Issar, A. S., and S. D. Resnick (Eds.), 1996: *Runoff, Infiltration and Subsurface Flow of Water in Arid and Semi-Arid Regions*. Kluwer Academic Publishers, Dordrecht.

Ives, R. L., 1962: Kiss tanks. *Weather*, **17**, 194–6.

Jackson, R. D., and S. B. Idso, 1975: Surface albedo and desertification. *Science*, **189**, 1012–13.

Jaeger, E. C., 1957: *The North American Deserts*. Stanford University Press, Stanford.

Jahns, R. H., 1949: Desert floods. *Engineering and Science* (California Institute of Technology, Alumni Association), **12**, 10–14.

Joseph, J. H., 1984: The sensitivity of a numerical model of the global atmosphere to the presence of desert aerosol. In *Aerosols and Their Climatic Effects*. H. E. Gerber and A. Deepak (Eds.), A. Deepak Publishing, Hampton, pp. 215–26.

Joshi, P. C., and P. S. Desai, 1985: The satellite-determined thermal structure of heat lows during Indian south-west monsoon season. *Adv. Space Res.*, **5**, 57–60.

Junning, L., Q. Zhengan, and S. Fumin, 1986: An investigation of the summer lows over the Qinghai-Xizang plateau. In *Proceedings of the International Symposium on the Qinghai-Xizang Plateau and Mountain Meteorology*. 20–24 March 1984, Beijing, American Meteorological Society, Boston, pp. 369–86.

Jury, W. A., W. R. Gardner, and W. H. Gardner, 1991: *Soil Physics*. John Wiley and Sons, New York.

Jury, W. A., J. Letey Jr., and L. H. Stolzy, 1981: Flow of water and energy under desert conditions. In *Water in Desert Ecosystems*. D. D. Evans and J. L. Thames (Eds.), Dowden, Hutchinson and Ross, Stroudsburg, pp. 92–113.

Justice, C. O., and P. H. Y. Hiernaux, 1986: Monitoring the grasslands of the Sahel using NOAA AVHRR data: Niger 1983. *Int. J. Remote Sensing*, **7**, 1475–97.

Kahana, R., B. Ziv, Y. Enzel, and U. Dayan, 2002: Synoptic climatology of major floods in the Negev Desert, Israel. *Intl. J. Climatol.*, **22**, 867–82.

Kamra, A. K., 1969: Electrification in an Indian dust storm. *Weather*, **24**, 145–6.

   1972: Measurements of the electrical properties of dust storms. *J. Geophys. Res.*, **77**, 5856–69.

Karyampudi, V. M., and T. N. Carlson, 1988: Analysis and numerical simulations of the Saharan air layer and its effects on easterly wave disturbances. *J. Atmos. Sci.*, **45**, 3102–36.

Karyampudi, V. M., S. P. Palm, J. A. Reagan, H. Fang, W. B. Grant, R. M. Hoff, C. Moulin, H. F. Pierce, O. Torres, E. V. Browell, and S. H. Melfi, 1999: Validation of the Saharan dust plume conceptual model using lidar, Meteosat, and ECMWF data. *Bull. Amer. Meteor. Soc.*, **80**, 1045–75.

Karyampudi, V. M., J. Simpson, S. Palm, and H. Pierce, 1997: Lidar observations of Saharan dust layer and its influence on tropical cyclogenesis. In *Preprints, 22nd Conf. on Hurricanes and Tropical Meteorology*, Fort Collins, American Meteorological Society, pp. 59–60.

Katsoulis, B. D., and J. M. Tsangaris, 1994: The state of the Greek environment in recent years. *Ambio*, **23**, 274–9.

Katz, R. W., and M. H. Glantz, 1977: Rainfall statistics, droughts, and desertification in the Sahel. In *Desertification: Environmental Degradation in and Around Arid Lands*, M. H. Glantz (Ed.), Westview Press, Boulder, pp. 81–102.

Kellogg, W. W., 1978: Global influences of mankind on the climate. In *Climate Change*, J. Gribbin (Ed.), Cambridge University Press, Cambridge, pp. 205–27.

   1979: Trends in anthropogenic aerosols and their effects on climate. Papers presented at the *WMO Technical Conference on Regional and Global Observations of Atmospheric Pollution Relative to Climate, Boulder, 20–24 August*. WMO – No. 549, WMO, Geneva, pp. 125–32.

Kellogg, W. W., and S. H. Schneider, 1977: Climate, desertification, and human activities. In *Desertification: Environmental Degradation in and Around Arid Lands*. M. H. Glantz (Ed.), Westview Press, Boulder, pp. 141–63.

Kershaw, A. P., 1978: Record of last interglacial-glacial cycle from northeastern Queensland. *Nature*, **272**, 159–61.

Kerslake, D. McK., 1972: *The Stress of Hot Environments*. Cambridge University Press, Cambridge.

Khalaf, F. I., and M. K. Al-Hashash, 1983: Aeolian sedimentation in the northwest part of the Arabian Gulf. *J. Arid Environ.*, **6**, 319–32.

Kidron, G. J., 2000: Analysis of dew precipitation in three habitats within a small drainage basin, Negev Highlands, Israel. *Atmos. Res.*, **55**, 257–70.

Kidron, G. J., A. Yair, and A. Danin, 2000: Dew variability within a small arid drainage basin in the Negev Highlands, Israel. *Quart. J. Roy. Meteor. Soc.*, **126**, 63–80.

Kiefert, L., G. H. McTainsh, and W. G. Nickling, 1996: Sedimentological characteristics of Saharan and Australian dusts. In *The Impact of Desert Dust Across the Mediterranean*. S. Guerzoni and R. Chester (Eds.), Kluwer Academic Publishers, Dordrecht, pp. 183–90.

Kiehl, J. T., and K. E. Trenberth, 1997: Earth's annual global mean energy budget. *Bull. Amer. Meteor. Soc.*, **78**, 197–208.

Kimura, F., and T. Kuwagata, 1995: Horizontal heat fluxes over complex terrain computed using a simple mixed-layer model and a numerical model. *J. Appl. Meteor.*, **34**, 549–58.

Kironchi, G., S. M. Kinyali, and J. P. Mbuvi, 1992: Effect of soils, vegetation and landuse on infiltration in a tropical semi-arid catchment. *E. Afr. Agric. For. J.*, **57**, 177–85.

Knighton, D., and G. Nanson, 1997: Distinctiveness, diversity and uniqueness in arid zone river systems. In *Arid Zone Geomorphology: Process Form and Change in Drylands*. D. S. G. Thomas (Ed.), John Wiley and Sons, Chichester, pp. 185–203.

Kondratyev, K. Ya., V. I. Korzov, V. V. Mukhenberg, and L. N. Dyachenko, 1981: The shortwave albedo and the surface emissivity. In *Study Conference on Land Surface Processes in Atmospheric General Circulation Models*. Greenbelt, 5–10 January 1981, pp. 463–514.

Köppen, W., 1931: *Grundriss der Klimakunde*. Walter de Gruyter Company, Berlin.

Krusel, N., 1987: Surface temperature – A case study at The Salt Lake, New South Wales. Thesis, Monash University.

Krutch, J. W., 1951: *The Desert Year*. University of Arizona Press, Tucson.
    1956: *The Voice of the Desert*. William Sloane, New York.

Kuo, L. T. C., 1976: *Agriculture in the People's Republic of China*. Praeger Publishers, New York.

Kustas, W. P., and K. S. Humes, 1997: Spatially distributed sensible heat flux over a semiarid watershed. Part II: Use of a variable resistance approach with radiometric surface temperatures. *J. Appl. Meteor.*, **36**, 293–301.

Kuwagata, T., and F. Kimura, 1995: Daytime boundary layer evolution in a deep valley. Part I: Observations in the Ina Valley. *J. Appl. Meteor.*, **34**, 1082–91.

Labed, J., and M. P. Stoll, 1991: Spatial variability of land surface emissivity in the thermal infrared band: Spectral signature and effective surface temperature. *Remote Sensing Environ.*, **38**, 1–17.

Lahey, J. F., 1973: On the origin of the dry climate in northern South America and the southern Caribbean. In *Coastal Deserts: Their Natural and Human Environments*. D. H. K. Amiran and A. W. Wilson (Eds.), University of Arizona Press, Tucson, pp. 75–90.

Lakhtakia, M. N., and T. T. Warner, 1987: A real-data numerical study of the development of the precipitation along the edge of an elevated mixed layer. *Mon. Wea. Rev.*, **115**, 156–68.

Lamb, H. H., 1972: *Climate: Present, Past, and Future*, Vol. 1, *Fundamentals and Climate Now*. Methuen and Company, London.
    1977: *Climate: Present, Past, and Future*, Vol. 2, *Climatic History and the Future*. Methuen and Company, London.

Lamb, P. J., 1978a: Case studies of tropical Atlantic surface circulation patterns during recent sub-Saharan weather anomalies: 1967 and 1968. *Mon. Wea. Rev.*, **106**, 482–91.
    1978b: Large-scale tropical Atlantic surface circulation patterns associated with Sub-Saharan weather anomalies. *Tellus*, **30**, 240–51.

Lamprey, H., 1988: Report on the desert encroachment reconnaissance in northern Sudan: 21 October to 10 November 1975. *Desertification Control Bulletin*, United Nations Environment Program, No. 17, pp. 1–7.

Lancaster, N., 1995: *The Geomorphology of Desert Dunes*. Routledge, London.

Landsberg, H. E., 1951: Statistical investigation into the climatology of rainfall in Oahu. In *On the Rainfall of Hawaii: A Group of Contributions*. Meteorological Monograph, American Meteorological Society, Boston, pp. 7–23.
    1970: Man-made climate changes. *Science*, **197**, 1265–74.

Lane, B. C., 1998: *The Solace of Fierce Landscapes: Exploring Desert and Mountain Spirituality*. Oxford University Press, New York.

Lang, A. R. G., G. N. Evans, and P. Y. Ho, 1974: The influence of local advection on evapotranspiration from irrigated rice in a semi-arid region. *Agricultural Meteor.*, **13**, 5–13.

Langbein, W. B., and S. A. Schumm, 1958: Yield of sediment in relation to mean annual precipitation. *Trans. Amer. Geophys. Union*, **39**, 1076–84.

Lare, A. R., and S. E. Nicholson, 1994: Contrasting conditions of surface water balance in wet years and dry years as a possible land surface – atmosphere feedback mechanism in the West African Sahel. *J. Climate*, **7**, 653–68.

Lashof, D. A., B. J. DeAngelo, S. R. Saleska, and J. Harte, 1997: Terrestrial ecosystem feedbacks to global climate change. *Ann. Rev. Energy Environ.*, **22**, 75–118.

Lattman, L. H., 1983: Effects of caliche on desert processes. In *Origin and Evolution of Deserts*. S. G. Wells and D. R. Haragan (Eds.), University of New Mexico Press, Albuquerque, pp. 101–9.

Laval, K., and L. Picon, 1986: Effect of a change of the surface albedo of the Sahel on climate. *J. Atmos. Sci.*, **43**, 2418–29.

Lawrence, T. E., 1926: *Seven Pillars of Wisdom*. Doubleday, Garden City.

Laymon, C., and D. Quattrochi, 2003: Estimating spatially-distributed surface fluxes in a semi-arid Great-Basin Desert using Landsat TM data. In *Thermal Remote Sensing in Land Surface Processes*. D. Quattrochi and J. Luvall (Eds.), Taylor and Francis, London. (In press.)

Le Clézio, J. M. G., 1980: *Désert*. Gallimard Publishers, Paris.

Le Houérou, H. N., 1968: La désertisation du Sahara septentrional et des steppes limitrophes. *Ann. Algér. de Géogr.*, **6**, 2–27.

1976: The nature and causes of desertification. *Arid Zone Newsletter*, Office of Arid Land Studies, Tucson, Arizona, **3**, 1–7.

1979: La désertisation des régions arides. *La Recherche*, **99**, 336–44.

1986: The desert and arid zone of northern Africa. In *Hot Deserts and Arid Shrublands, Ecosystems of the World Volume 12B*. M. Evenari, I. Noy-Meir, and D. W. Goodall (Eds.), Elsevier Scientific Publishing Company, New York, pp. 101–47.

1992: An overview of vegetation and land degradation in the world arid lands. In *Degradation and Restoration of Arid Lands*. H. E. Dregne (Ed.), International Center for Arid and Semi-Arid Lands Studies, Texas Tech University Press, Lubbock, Texas, pp. 127–63.

Le Houérou, H. N., G. F. Popov, and L. See, 1993: *Agrobioclimatic Classification of Africa*. Agrometeorology Series, Working Paper no. 6, FAO, Rome, Italy.

Le Roy Ladurie, E., 1967: *Histoire du Climat depuis l'an mil*. Flammarion, Paris. (Translated by B. Bray, Doubleday and Company, 1971.)

Leathers, C. R., 1981: Plant components of desert dust in Arizona and their significance for man. In *Desert Dust*. T. L. Péwé (Ed.), Geological Society of America, Special Paper 186, pp. 191–206.

Lee, B. E., and B. F. Soliman, 1977: An investigation of the forces on three dimensional bluff bodies in rough wall turbulent boundary layers. *Trans. ASME, J. Fluids Eng.*, **99**, 503–10.

Lee, D. H. K., 1964: Terrestrial animals in dry heat: Man in the desert. In *Handbook of Physiology – Environment*. D. B. Hill, E. F. Adolph, and C. G. Wilber (Eds.), American Physiological Society, Washington, D.C., pp. 551–82.

1968: Human adaptation to arid environments. In *Desert Biology*. G. W. Brown Jr. (Ed.), Academic Press, New York, pp. 517–56.

Lee, J. A., 1991a: The role of desert shrub size and spacing on wind profile parameters. *Phys. Geogr.*, **12**, 72–89.

1991b: Near-surface wind flow around desert shrubs. *Phys. Geogr.*, **12**, 140–6.

Lee, R., 1978: *Forest Microclimatology*. Columbia University Press, New York.

Leighton, R., and R. Deslandes, 1991: *Monthly anticyclonicity and cyclonicity in the Australian region: 23-year averages*. Technical Report No. 64, Bureau of Meteorology, Australia.

Leithead, C. S., and A. R. Lind, 1964: *Heat Stress and Heat Disorders*. Cassell, London.

Leslie, L. M., 1980: Numerical modeling of the summer heat low over Australia. *J. Appl. Meteor.*, **19**, 381–7.

Lettau, H. H., 1978a: Introduction. In *Exploring the World's Driest Climate*. H. H. Lettau and K. Lettau (Eds.), IES Rep. 101, Univ of Wisconsin, Madison, pp. 12–29.

1978b: Explaining the world's driest climate. In *Exploring the World's Driest Climate*. H. H. Lettau and K. Lettau (Eds.), IES Rep. 101, Univ of Wisconsin, Madison, pp. 182–248.

Lewis, H. E., A. R. Foster, B. J. Mullan, R. N. Cox, and R. P. Clark, 1969: Aerodynamics of the human microenvironment. *The Lancet*, **1** (7609), 1273–7.

Lieman, R., and P. Alpert, 1992: Investigation of the temporal and spatial variations of PBL height over Israel. In *Air Pollution Modeling and its Application*, IX. G. Kallos (Ed.), Plenum, pp. 231–9.

Lindesay, J. A., and P. D. Tyson, 1990: Thermo-topographically induced boundary layer oscillations over the central Namib, Southern Africa. *Intl. J. Climatol.*, **10**, 63–77.

Lines, G. C., 1979: *Hydrology and Surface Morphology of the Bonneville Salt Flats and the Pilot Valley Playa, Utah*. Geological Survey Water-Supply Paper 2057, US Government Printing Office.

List, R. J., 1966: *Smithsonian Meteorological Tables*. Smithsonian Institution Press, Washington, D.C.

Littmann, T., 1991: Dust storm frequency in Asia: climatic control and variability. *Intl. J. Climatol.*, **11**, 393–412.

Logan, R. F., 1960: *The Central Namib Desert*. National Research Council Publication 785, National Academy of Sciences, Washington, D.C.

1968: Causes, climates and distribution of deserts. In *Desert Biology*, Vol. 1. G. W. Brown Jr. (Ed.), Academic Press, New York, pp. 21–50.

Longley, R. W., 1975: Precipitation in valleys. *Weather*, **30**, 294–300.

Lopez-Bermudez, F., A. Romero-Diaz, G. Fisher, C. Francis, and J. B. Thornes, 1984: Erosion y ecologia en la Espana semi-arida (Cuenca de Mula, Murcia). *Cuadernos de Investig. Geofra*, **10**, 113–26.

Lorca, F. G., 2002: *Selected Verse*. Farrar, Straus and Giroux, New York.

Lorenz, E. N., 1968: Climate determinism. In *Causes of Climatic Change*. J. M. Mitchell (Ed.), Meteorological Monograph 8, American Meteorological Society, Boston, pp. 1–3.

Lowdermilk, W. C., 1953: *Conquest of the Land Through 7000 Years*. US Department of Agriculture, Natural Resources Conservation Service, Agriculture Information Bulletin No. 99.

Lusigi, W., 1981: *Combatting Desertification and Rehabilitating Degraded Production Systems in Northern Kenya*. IPAL Technical Report A-4, MAB Project No. 3, Impact of Human Activities and Land-Use Practices on Grazing Lands, UNESCO, Nairobi, Kenya.

Lydolph, P. E., 1964: The Russian sukovey. *Ann. Assoc. Amer. Geographs*, **54**, 291–309.

1973: On the causes of aridity along a selected group of coasts. In *Coastal Deserts: Their Natural and Human Environments.* D. H. K. Amiran and A. W. Wilson (Eds.), University of Arizona Press, Tucson, pp. 67–72.

1977: *Climates of the Soviet Union.* Elsevier Scientific Publishing Company, Amsterdam.

Lyford, F. P., and H. K. Qashu, 1969: Infiltration rates as affected by desert vegetation. *Water Resour. Res.,* **5**, 1373–6.

Mabbutt, J. A., 1977: *Desert Landforms.* ANU Press, Canberra.

1978a: The impact of desertification as revealed by mapping. *Environmental Conservation,* **5**, 45.

1978b: *Desertification in Australia.* Water Research Foundation of Australia, Report No. 4, Canberra, Australia.

1984: A new global assessment of the status and trends of desertification. *Environmental Conservation,* **11**, 100–13.

MacMahon, J. A., 1985: *Deserts.* Alfred A. Knopf, New York.

MacMahon, J. A., and F. H. Wagner, 1985: The Mojave, Sonoran and Chihuahuan Deserts of North America. In *Hot Deserts and Arid Shrublands, Ecosystems of the World Volume 12A.* M. Evenari, I. Noy-Meir, and D. W. Goodall (Eds.), Elsevier Scientific Publishing Company, New York, pp. 105–202.

Mainguet, M., 1991: *Desertification: Natural Background and Human Mismanagement.* Springer-Verlag, Berlin.

Malek, E., and G. E. Bingham, 1997: Partitioning of radiation and energy balance components in an inhomogeneous desert valley. *J. Arid Environments,* **37**, 193–207.

Malek, E., G. E. Bingham, and G. D. McCurdy, 1990: Evapotranspiration from the margin and moist playa of a closed desert valley. *J. Hydrology,* **120**, 15–34.

Malek, E., G. E. Bingham, D. Orr, and G. D. McCurdy, 1997: Annual mesoscale study of water balance in a Great Basin heterogeneous desert valley. *J. Hydrology,* **191**, 223–44.

Maley, J., 1982: Dust, clouds, rain types, and climatic variations in tropical North Africa. *Quaternary Research,* **18**, 1–16.

Man, J., 1999: *Gobi: Tracking the Desert.* Yale University Press, New Haven.

Manabe, S., and A. J. Broccoli, 1990: Mountains and arid climates of middle latitudes. *Science,* **247**, 192–5.

Mares, M. A. (Ed.), 1999: *Encyclopedia of Deserts.* University of Oklahoma Press, Norman.

Mares, M. A., J. Morello, and G. Goldstein, 1985: The Monte Desert and other subtropical semi-arid biomes of Argentina, with comments on their relation to North American arid areas. In *Hot Deserts and Arid Shrublands, Ecosystems of the World Volume 12A.* M. Evenari, I. Noy-Meir, and D. W. Goodall (Eds.), Elsevier Scientific Publishing Company, New York, pp. 203–37.

Marshall, J. K., 1970: Assessing the protective role of shrub-dominated rangeland vegetation against soil erosion by wind. *Proceedings, 11th International Grasslands Conference,* Surfer's Paradise, Queensland, Australia Grassland Society of Victoria, Parkville, pp. 19–23.

Marston, R. A., 1986: Maneuver-caused wind erosion impacts, South Central New Mexico. In *Aeolian Geomorphology.* W. G. Nickling (Ed.), Allen and Unwin, Boston, pp. 273–90.

Marticorena, B., G. Bergametti, D. Gillette, and J. Belnap, 1997: Factors controlling threshold friction velocity in semiarid and arid areas of the United States. *J. Geophys. Res.,* **102**, 23 277–87.

Mather, G. K., D. E. Terblanche, F. E. Steffens, and L. Fletcher, 1997: Results of the South African cloud seeding experiments using hygroscopic flares. *J. Appl. Meteor.*, **36**, 1433–47.

McCurdy, G. G., 1989: Radiation balance of a desert salt playa. M.S. Thesis, Utah State University, Logan.

McGinnies, W. G., B. J. Goldman, and P. Paylore (Eds.), 1968: Introduction to *Deserts of the World: An Appraisal of Research Into Their Physical and Biological Environments*. University of Arizona Press, Tucson.

McIntyre, D. S., 1958: Permeability measurements of soil crusts formed by raindrop impact. *Soil Sci.*, **85**, 185–9.

McKee, E. D. (Ed.), 1979: *A Study of Global Sand Seas*. United States Geological Survey Paper 1052, Washington, D.C.

McKee, E. D., C. S. Breed, and S. G. Fryberger, 1977: Desert sand seas. In *Skylab Explores the Earth*. NASA SP-380, National Aeronautics and Space Administration, Washington, D.C., pp. 5–47.

McKenna-Neumann, C., and W. G. Nickling, 1994: Momentum extraction with saltation: implications for experimental evaluation of wind profile parameters. *Bound. Layer Meteor.*, **68**, 35–50.

McNamee, G. (Ed.), 1995: *The Sierra Club Desert Reader: A Literary Companion*. Sierra Club Books, San Francisco.

Medinger, M., 1961: La crue de décembre 1960 de l'Oued Mya. *Inst. de Rech. Sahariennes, Travaux*, **20**, 203–6.

Meehl, G. A., G. W. Branstator, and W. M. Washington, 1993: Tropical Pacific interannual variability and CO_2 climate change. *J. Climate*, **6**, 42–63.

Meigs, P., 1952: Arid and semiarid climate types of the world. In *Proceedings, Eighth General Assembly and Seventeenth International Congress, International Geographical Union*, Washington, D.C., pp. 135–8.

    1953: World distribution of arid and semi-arid homoclimates. In *Reviews of Research on Arid Zone Hydrology*. UNESCO, Paris, Arid Zone Programme 1, pp. 203–10.

    1957: Arid and semi-arid climate types of the world. In *Proceedings, International Geographical Union, 17th Congress, 8th General Assembly*, Washington, D.C., pp. 135–8.

    1966: *Geography of Coastal Deserts*. UNESCO, Paris, Arid Zone Research XXVIII.

Meinel, A., and M. Meinel, 1983: *Sunsets, Twilights, and Evening Skies*. Cambridge University Press, Cambridge.

Membery, D. A., 1985: A gravity wave haboob. *Weather*, **40**, 214–21.

    1997: Unusually wet weather across Arabia. *Weather*, **52**, 166–74.

Menenti, M., 1984: Physical aspects and determination of evaporation in deserts applying remote sensing techniques. Dissertation, Netherlands Agricultural University, Wageningen, The Netherlands.

Menenti, M., and J. C. Ritchie, 1994: Estimation of effective aerodynamic roughness of Walnut Gulch watershed with laser altimeter measurements. *Water Resour. Res.*, **30**, 1329–37.

Middleton, N., 1986: The geography of dust storms. Dissertation, University of Oxford.

    1997: Desert dust. In *Arid Zone Geomorphology: Process Form and Change in Drylands*. D. S. G. Thomas (Ed.), John Wiley and Sons, Chichester, pp. 413–36.

Middleton, N. J., A. S. Goudie, and G. L. Wells, 1986: The frequency and source areas of dust storms. In *Aeolian Geomorphology*. W. G. Nickling (Ed.), Allen and Unwin, Boston, pp. 237–59.

Mikesell, M. W., 1960: Deforestation in northern Morocco. *Science*, **132**, 441–54.

1969: The deforestation of Mount Lebanon. *Geographical Rev.*, **59**, 1–28.

Milford, J. R., 1987: Problems of deducing the soil water balance in dryland regions from Meteosat data. *Soil Use and Management*, **3**, 51–7.

Miller, A. A., 1961: *Climatology*. Methuen, London.

Miller, D. H., 1981: *Energy at the Surface of the Earth: An Introduction to the Energetics of Ecosystems*. Academic Press, New York.

Miller, H., 1945: *The Air-conditioned Nightmare*. New Directions Books, New York.

Miller, R. L., and I. Tegen, 1998: Climate response to soil dust aerosols. *J. Climate*, **11**, 3247–67.

Mitchell, J. F. B., 1989: The "greenhouse" effect and climate change. *Rev. Geophys.*, **27**, 115–39.

Mohalfi, S., H. S. Bedi, T. N. Krishnamurti, and S. D. Cocke, 1998: Impact of shortwave radiative effects of dust aerosols on the summer season heat low over Saudi Arabia. *Mon. Wea. Rev.*, **126**, 3153–68.

Monod, T., 1973: *Les Déserts*. Horizons de France, Paris.

1986: The Sahel zone north of the Equator. In *Hot Deserts and Arid Shrublands, Ecosystems of the World Volume 12B*. M. Evenari, I. Noy-Meir, and D. W. Goodall (Eds.), Elsevier Scientific Publishing Company, New York, pp. 203–43.

1994: Deserts. *The UNESCO Courier: A Window Open on the World*, **47-1**, 4–9.

Monteith, J. L., 1963: Dew: facts and fallacies. In *The Water Relations of Plants*. A. J. Rutter and F. H. Whitehead (Eds.), Blackwell Scientific Publications, Oxford, pp. 37–56.

Monteith, J. L., and L. E. Mount (Eds.), 1974: *Heat Loss From Animals and Man: Assessment and Control*. Butterworths, London.

Monteith, J. L. and M. H. Unsworth, 1990: *Principles of Environmental Physics*. Edward Arnold, London.

Mooney, H. A., S. I. Gulmon, P. W. Rundel, and J. Ehleringer, 1980: Further observations on the water relations of *Prosopis tamarugo* of the Northern Atacama desert. *Oecologia*, **44**, 177–80.

Moore, J. K., S. C. Doney, D. M. Glower, and I. Y. Fung, 2001: Iron cycling and nutrient limitation patterns in surface waters of the world ocean. *Deep-Sea Research II*, **49**, 463–507.

Moore, R. M., and R. A. Perry, 1970: Vegetation of Australia. In *Australian Grasslands*. R. M. Moore (Ed.), Australian National University Press, Canberra, pp. 59–73.

Morales, A. F., 1946: *El Sahara Español*. Alta Comisaria de España en Marruecos, Madrid.

Moriarty, W. W., 1955: *Large-scale effects of heating over Australia. I. The synoptic behavior of the summer low*. Technical Report No. 7, Div. Meteorol. Phys., CSIRO, Australia.

Morrison, R., 1964: *Lake Lahontan: Geology of Southern Carson Desert, Nevada*. United States Geological Survey Professional Paper 401.

Mossman, R. C., 1910: The climate of Chile. *J. Scottish Meteor. Soc.*, **15**, 313–46.

Motts, W. S., 1965: Hydrologic types of playas and closed valleys and some relations of hydrology to playa geology. In *Geology, Mineralogy and Hydrology of U. S. Playas*. J. T. Neal (Ed.), Air Force Cambridge Research Laboratories Report 65–226, Environmental Research Paper 96, Bedford, Massachusetts, pp. 31–72.

Motts, W. S. (Ed.), 1970: *Geology and Hydrology of Selected Playas in Western United States*. University of Massachusetts, Amherst.

Moulin, C., C. E. Lambert, F. Dulac, and U. Dayan, 1997: Control of atmospheric export of dust from North Africa by the North Atlantic oscillation. *Nature*, **387**, 691–4.

Mount, L. E., 1974: The concept of thermal neutrality. In *Heat Loss From Animals and Man*. J. L. Monteith and L. E. Mount (Eds.), Butterworth, London, pp. 425–39.

Muturi, H. R., 1994: *Temperature and Rainfall Trends in Kenya During the Last 50 Years*. Department of Research and Development; Ministry of Research, Technology and Training, Nairobi, Kenya.

Nagel, J. F., 1962: Fog precipitation measurements of Africa's southwest coast. *South Africa Weather Bureau, Notos* **11**, 51–60.

Namias, J., 1989: Cold waters and hot summers. *Nature*, **338**, 15–16.

1991: Spring and summer 1988 drought over the contiguous United States – Causes and prediction. *J. Climate*, **4**, 54–65.

Neal, J. T. (Ed.), 1965: *Geology, Mineralogy and Hydrology of U.S. Playas*. Air Force Cambridge Research Laboratories Report 65-226, Environmental Research Paper 96, Bedford, Massachusetts.

1975: *Playas and Dried Lakes*. Dowden, Hutchinson and Ross; Stroudsburg.

Neihardt, J. G., 1925: *The Twilight of the Sioux*. University of Nebraska Press, Lincoln.

Neruda, P., 1974: *Confieso que he Vivido: Memorias*. Editorial Seix Barral, Barcelona.

Neuberger, H., 1966: *Introduction to Physical Meteorology*. The Pennsylvania State University Press, University Park.

Neuberger, H., and J. Cahir, 1969: *Principles of Climatology*. Holt, Rinehart and Winston, New York.

New, M. G., M. Hulme, and P. D. Jones, 2000: Representing 20th century space-time climate variability. II: Development of 1901–1996 monthly terrestrial climate fields. *J. Climate*, **13** 2217–38.

Nicholson, S. E., 1989a: Long term changes in African rainfall. *Weather*, **44**, 46–56.

1989b: African drought: characteristics, causal theories, and global teleconnections. In *Understanding Climate Change*. A. Berger, R. E. Dickinson, and J. W. Kidson (Eds.), American Geophysical Union, Washington, D.C., pp. 79–100.

Nicholson, S. E., and T. J. Farrar, 1994: The influence of soil type on the relationships between NDVI, rainfall, and soil moisture in semiarid Botswana. I. NDVI response to rainfall. *Remote Sensing Environ.*, **50**, 107–20.

Nicholson, S. E., and H. Flohn, 1980: African environmental and climatic changes and the general atmospheric circulation in Late Pleistocene and Holocene. *Climatic Change*, **2**, 331–48.

Nickling, W. G. (Ed.), 1986: *Aeolian Geomorphology*. Allen and Unwin, Boston.

Nickling, W. G., and S. A. Wolfe, 1994: The morphology and origin of Nabkhas, region of Mopti, Mali, West Africa. *J. Arid Environments*, **28**, 13–30.

Nir, D., 1974: *The Semi Arid World: Man on the Fringe of the Desert*. Longman, London.

Niu, G. -Y., S. -F. Sun, and Z. -X Hong, 1997: Water and heat transport in the desert soil and atmospheric boundary layer in western China. *Bound. Layer Meteor.*, **85**, 179–95.

NOAA, 1983: *Climatic Atlas of the United States*. US Department of Commerce, National Oceanic and Atmospheric Administration, Washington, D.C.

Novak, M. D., 1986: Theoretical values of daily atmospheric and soil thermal admittances. *Bound. Layer Meteor.*, **34**, 17–34.

Novak, M. D., and T. A. Black, 1985: Theoretical determination of the surface energy balance and thermal regime of bare soils. *Bound. Layer Meteor.*, **33**, 313–33.

Noy-Meir, I., 1985: Desert ecosystem structure and function. In *Hot Deserts and Arid Shrublands, Ecosystems of the World Volume 12A*. M. Evenari, I. Noy-Meir, and D. W. Goodall (Eds.), Elsevier Scientific Publishing Company, New York, pp. 93–103.

NSF, 1977: The land of the empty bucket. In *The Study of Aridity: A Mosaic Special*. MOSAIC, **8** (1), National Science Foundation, Washington, D.C., pp. 2–5.

O'Carroll, C. M., and M. Carlyon, 2001: Dust from Africa leads to large toxic algae blooms in Gulf of Mexico, study finds. *The Earth Observer*, **13** (No. 4), 18.

Odingo, R. S., 1990: Review of UNEP's definition of desertification and its programmatic implications. In *Desertification Revisited*. UNEP, Nairobi, pp. 7–44.

Oglesby, R. J., and D. J. Erickson III, 1989: Soil moisture and the persistence of North American drought. *J. Climate*, **2**, 1362–80.

Oke, T. R., 1987: *Boundary Layer Climates*. Methuen, London.

Oldeman, L. R., R. T. A. Hakkeling, and W. G. Sombroek, 1990: *World map of Status of Human-Induced Soil Degradation: An Explanatory Note*. UNEP, Nairobi, and ISRIC, Wageningen.

Oliver, F. W., 1945: Dust storms in Egypt and their relation to the war period, as noted in Maryut, 1939–1945. *Geographical J.*, **106**, 26–49.

Ookouchi, Y., M. Segal, R. C. Kessler and R. A. Pielke, 1984: Evaluation of soil moisture effects on the generation and modification of mesoscale circulations. *Mon. Wea. Rev.*, **112**, 2281–92.

Orshan, G., 1986: The deserts of the Middle East. In *Hot Deserts and Arid Shrublands, Ecosystems of the World Volume 12B*. M. Evenari, I. Noy-Meir, and D. W. Goodall (Eds.), Elsevier Scientific Publishing Company, New York, pp. 1–28.

Otterman, J., 1974: Baring high-albedo soils by overgrazing: A hypothesized desertification mechanism. *Science*, **186**, 531–3.

1977a: Anthropogenic impact on surface characteristics and some possible climatic consequences. *Israel Meteorological Research Papers*, Israel Meteor. Serv., Bet Dagen, pp. 31–41.

1977b: Monitoring surface albedo change with LANDSAT. *Geophys. Res. Lett.*, **4**, 441–4.

1981: Satellite and field studies of man's impact on the surface of arid regions. *Tellus*, **33**, 68–77.

1989: Enhancement of surface-atmosphere fluxes by desert-fringe vegetation through reduction of surface albedo and of soil heat flux. *Theor. Appl. Climatol.*, **40**, 67–79.

Otterman, J., and R. S. Fraser, 1976: Earth-atmosphere system and surface reflectivities in arid regions from LANDSAT MSS data. *Remote Sensing Environ.*, **5**, 247–66.

Otterman, J., M. D. Novak, and D. O'C. Starr, 1993: Turbulent heat transfer from sparsely vegetated surface: Two-component representation. *Bound. Layer Meteor.*, **64**, 409–20.

Otterman, J., and D. Sharon, 1979: Day/night partitioning of rain in an arid region: Computational approaches, results for the Negev and meteorological/climatological implications. *J. Recherches Atmosphériques*, **13**, 11–20.

Otterman, J., and D. O'C. Starr, 1995: Alternative regimes of surface and climate conditions in sandy arid regions: possible relevance to Mesopotamian drought 2200–1900 B.C. *J. Arid Environments*, **31**, 127–35.

Overpeck, J., D. Rind, A. Lacis, and R. Healy, 1996: Possible role of dust-induced regional warming in abrupt climate change during the last glacial period. *Nature*, **384**, 447–9.

Owens, M., and D. Owens, 1984: *Cry of the Kalahari*. Houghton Mifflin, Boston.

Pabot, H., 1962: *Comment Briser le Cercle Vicieux de la Désertification dans les Régions Sèches de l'Orient*. FAO, Rome.

Pachur, H. J., H. P. Roper, S. Kropelin, and M. Groschin, 1987: Late Quaternary hydrography of the Eastern Sahara. *Berliner Geowissenschaften Abhandlungen* (A), **75**, 331–84.

Paegle, J., J. Nogues-Paegle, and K. C. Mo, 1996: Dependence of simulated precipitation on surface evaporation during the 1993 United States summer floods. *Mon. Wea. Rev.*, **124**, 345–61.

Palmen, E., and C. W. Newton, 1969: *Atmospheric Circulation Systems*. Academic Press, New York.

Palmer, T. N., and C. Brancovic, 1989: The 1988 US drought linked to anomalous sea surface temperature. *Nature*, **338**, 54–7.

Paltridge, G. W., and C. M. R. Platt, 1976: *Radiative Processes in Meteorology and Climatology*. Elsevier Scientific Publishing Company, New York.

Parungo, F., Z. Li, X. Li, D. Yang, and J. Harris, 1994: Gobi dust storms and the Great Green Wall. *Geophys. Res. Lett.*, **21**, 999–1002.

Pedgley, D. E., 1972: Desert depressions over north-east Africa. *Meteor. Mag.*, **101**, 228–44.

1974: Winter and spring weather at Riyadh, Saudi Arabia. *Meteor. Mag.*, **103**, 225–36.

Peel, R. F., 1974: Insolation and weathering: Some measures of diurnal temperature changes in exposed rocks in the Tibesti region, central Sahara. *Zeitschrift für Geomorphologie*, suppl., **21**, 19–28.

1975: Water action in desert landscapes. In *Processes in Physical and Human Geography*. R. F. Peel, M. Chisolm, and P. Haggett (Eds.), Heinemann Educational Books, London, pp. 110–29.

Perrolf, K., and K. Sandstrom, 1995: Correlating landscape characteristics and infiltration – A study of surface sealing and subsoil conditions in semi-arid Botswana and Tanzania. *Geografiska Annaler*, **77A**, 119–33.

Perrone, T. J., 1979: *Winter Shamal in the Persian Gulf*. US Naval Environmental Prediction Research Facility, Technical Report TR 79–06, Monterey, California.

Perry, A. H., 1984: Recent climate change – is there a signal amongst the noise? *Progress in Physical Geography*, **8**, 111–17.

Perry, K. D., T. A. Cahill, R. A. Eldred, D. D. Dutcher, and T. E. Gill, 1997: Long-range transport of North African dust to the eastern United States. *J. Geophys. Res.*, **102**, 11 225–38.

Perry, R. A., 1977: The evaluation and exploitation of semi-arid lands: Australian experience. *Phil. Trans. R. Soc. Lond.*, **B278**, 493–505.

Petit, J. R., M. Briat, and A. Royer, 1981: Ice age aerosol content from East Antarctic ice core samples and past wind strength. *Nature*, **293**, 391–4.

Petrov, M. P., 1976: *Deserts of the World*. Keter Publishing House, Jerusalem.

Péwé, T. L., 1981: *Desert Dust: Origin, Characteristics, and Effect on Man*. Geological Society of America, Special Paper 186, Boulder, Colorado.

Philip, J. R., 1960: Advection and the arid zone: Theoretical. In *Proceedings Australian Arid Zone Technical Conference*, CSIRO, Melbourne, pp. 41.1–41.9.

Physick, W. L., and N. J. Tapper, 1990: A numerical study of circulations induced by a dry salt lake. *Mon. Wea. Rev.*, **118**, 1029–42.

Pichi-Sermolli, R. E. G., 1955a: Tropical east Africa (Ethiopia, Somaliland, Kenya, Tanganyika). In *Plant Ecology Reviews of Research*. UNESCO, Paris, *Arid Zone Research*, **6**, 302–60.

1955b: The arid vegetation types of tropical countries and their classification. In *Plant Ecology, Proceedings of the Montpellier Symposium*. UNESCO, Paris, *Arid Zone Research*, **5**, 29–32.

Pielke, R. A., 1984: *Mesoscale Meteorological Modeling*. Academic Press, Orlando, Florida.

Pike, J. G., 1970: Evaporation of groundwater from coastal playas (sabkhah) in the Arabian Gulf. *J. Hydrology*, **11**, 79–88.

Pittock, A. B., 1983: Solar variability, weather and climate: an update. *Quart. J. Roy. Meteor. Soc.*, **109**, 23–55.

Portela, A., and M. Castro, 1996: Summer thermal lows in the Iberian Peninsula: A three-dimensional simulation. *Quart. J. Roy. Meteor. Soc.*, **122**, 1–22.

Prell, W. L., 1984: Monsoonal climate of the Arabian Sea during the Late Quaternary: A response to changing solar radiation. In *Milankovitch and Climate: Understanding the Response to Astronomical Forcing*, Part 1. A. Berger, J. Imbrie, J. Hays, G. Kukla, and B. Saltzman (Eds.), D. Reidel Publishing Company, Dordrecht, pp. 349–66.

Prell, W. L., and J. E. Kutzbach, 1992: Monsoon variability over the past 150,000 years. *J. Geophys. Res.*, **92**, 8411–25.

Price, M. R. S., A. bin H. al-Harthy, and R. P. Whitcombe, 1985: Fog moisture and its ecological effects in Oman. In *Proceedings of the Conference on Arid Lands: Today and Tomorrow, Tucson*. E. E. Whitehead (Ed.), Belhaven Press, London, pp. 69–88.

Priestley, C. H. B., 1959: *Turbulent Transfer in the Lower Atmosphere*. University of Chicago Press, Chicago.

Prospero, J. M., 1996a: The atmospheric transport of particles to the ocean. In *Particle Flux in the Ocean*. V. Ittekkott, S. Honjo, and P. J. Depetris (Eds.), John Wiley and Sons, New York, pp. 19–52.

1996b: Saharan dust transport over the North Atlantic Ocean and Mediterranean: An overview. In *The Impact of Desert Dust Across the Mediterranean*. S. Guerzoni and R. Chester (Eds.), Kluwer Academic Publishers, Norwell, pp. 133–51.

1999: Long-term measurements of the transport of African mineral dust to the southeastern United States: Implications for regional air quality. *J. Geophys. Res.*, **104**, 15 917–27.

Prospero, J. M., and T. N. Carlson, 1972: Vertical and areal distribution of Saharan dust over the western equatorial North Atlantic Ocean. *J. Geophys. Res.*, **77**, 255–65.

Prospero, J. M., R. A. Glaccum, and R. T. Nees, 1981: Atmospheric transport of soil dust from Africa to South America. *Nature*, **289**, 570–2.

Prospero, J. M., and R. T. Nees, 1986: Impact of the North African drought and El Niño on mineral dust in the Barbados trade wind. *Nature*, **320**, 735–8.

Prospero, J. M., M. Uematsu, and D. L. Savoie, 1989: Mineral aerosol transport to the Pacific Ocean. In *Chemical Oceanography*. J. P. Riley, R. Chester, and R. A. Duce (Eds.), Vol. 10, Academic Press, London, pp. 188–218.

Prueger, J. H., L. E. Hipps, and D. I. Cooper, 1996: Evaporation and the development of the local boundary layer over an irrigated surface in an arid region. *Agricultural and Forest Meteor.*, **78**, 223–37.

Puigdefabregas, J., and C. Aguilera, 1996: The Rambla Honda field site: Interactions of soil and vegetation along a catena in semi-arid southwest Spain. In *Mediterranean Desertification and Land Use*. C. J. Brandt and J. B. Thornes (Eds.), John Wiley and Sons, Chichester, pp. 137–68.

Pye, K., 1987: *Aeolian Dust and Dust Deposits*. Academic Press, London.

Pye, K., and H. Tsoar, 1990: *Aeolian Sand and Sand Dunes*. Unwin Hyman, London.

Quézel, P., 1965: *La végétation du Sahara du Tchad à la Mauritanie*. Gustav Fisher Verlag, Stuttgart.

Quijano, A. L., I. N. Sokolik, and O. B. Toon, 2000: Radiative heating rates and direct radiative forcing by mineral dust in cloudy atmospheric conditions. *J. Geophys. Res.*, **105**, 12 207–19.

Quine, T. A., A. Navas, D. E. Walling, and J. Machin, 1994: Soil erosion and redistribution on cultivated and uncultivated land near Las Bardenas in the central Ebro River Basin, Spain. *Land Degradation and Rehabilitation*, **5**, 41–55.

Rácz, Z., and R. K. Smith, 1999: The dynamics of heat lows. *Quart. J. Roy. Meteor. Soc.*, **125**, 225–52.

Ramage, C. S., 1971: *Monsoon Meteorology*. Academic Press, New York.

Ramaswamy, C., 1965: On a remarkable case of dynamical and physical interaction between middle and low latitude weather systems over Iran. *Indian J. Meteor. Geophys.*, **16**, 177–200.

Rasch, C., 1898: Uber die Einfluss des Tropenklimas auf das Nervensystem. *Allg. Z. Psychiat.*, **54**, 745–75.

Rasmussen, K. R., M. Sorensen, and B. B. Willetts, 1985: Measurement of saltation and wind strength on beaches. In *Proceedings of International Workshop on Physics of Blown Sand*. Memoirs **8**, Department of Theoretical Statistics, Aarhus University, Denmark, pp. 301–25.

Ratcliffe, R. A. S., 1981: Meteorological aspects of the 1975–76 drought in westerrn Europe. In *Climate Variations and Variability: Facts and Theories*. A. Berger (Ed.), D. Reidel Publishing Company, Dordrecht, pp. 355–67.

Rauh, W., 1985: The Peruvian-Chilean Deserts. In *Hot Deserts and Arid Shrublands, Ecosystems of the World Volume 12A*. M. Evenari, I. Noy-Meir, and D. W. Goodall (Eds.), Elsevier Scientific Publishing Company, New York, pp. 239–67.

1986: The arid region of Madagascar. In *Hot Deserts and Arid Shrublands, Ecosystems of the World Volume 12B*. M. Evenari, I. Noy-Meir, and D. W. Goodall (Eds.), Elsevier Scientific Publishing Company, New York, pp. 361–77.

Raupach, M. R., A. S. Thom, and I. Edwards, 1980: A wind tunnel study of turbulent flow close to regularly arrayed rough surfaces. *Bound. Layer Meteor.*, **18**, 373–97.

Reeves, C. C., Jr., 1976: *Caliche*. Estacado Books, Lubbock.

Reeves, R. G., A. Anson, and D. Landen (Eds.), 1975: *Manual of Remote Sensing*, Vols. I and II. American Society of Photogrammetry, Falls Church.

Reid, I., and L. E. Frostick, 1997: Channel form, flows and sediments in deserts. In *Arid Zone Geomorphology: Process Form and Change in Drylands*. D. S. G. Thomas (Ed.), John Wiley and Sons, Chichester, pp. 205–29.

Reining, P., 1978: *Handbook on Desertification Indicators*. American Association for the Advancement of Science, Washington, D.C.

Reisner, M., 1986: *Cadillac Desert: The American West and its Disappearing Water*. Penguin, New York.

Riehl, H., M. A. Alaka, C. L. Jordan, and R. J. Renard, 1954: *The Jet Stream*. Meteorological Monographs, American Meteorological Society, Vol. 2, no. 7.

Rife, D. L., T. T. Warner, F. Chen, and E. A. Astling, 2002: Mechanisms for diurnal boundary-layer circulations in the Great Basin Desert. *Mon. Wea. Rev.*, **130**, 921–38.

Rijks, D., 1971: Analysis of rainfall reliability in the Senegal River Basin. In *Agroclimatology in the Semi-arid Areas South of the Sahara*, Proceedings of the Regional Technical Conference, Dakar, 8–20 February 1971, World Meteorological Organization, WMO – No. 340, pp. 32–41.

Ripley, E. A., 1976a: Comments on the paper "Dynamics of deserts and drought in the Sahel" (Charney 1975). *Quart. J. Roy. Meteor. Soc.*, **102**, 466–7.

1976b: Drought in the Sahara: Insufficient biogeophysical feedback. *Science*, **191**, 100.

Roberts, N., M. Taneb, P. Barker, B. Damnati, M. Toole, and D. Williamson, 1993: Timing of the younger Dryas event in East Africa from lake level change. *Nature*, **366**, 146–8.

Robinove, C. J., P. S. Chavez Jr., D. Gehring, and R. Holmgren, 1981: Arid land monitoring using Landsat albedo difference images. *Remote Sensing Environ.*, **11**, 133–56.

Robinson, S., 1963: Circulatory adjustments of men in hot environments. In *Temperature, its Measurement and Control in Science and Industry*, Vol. 3, Pt. 3. C. M. Hertzfeld and J. D. Hardy (Eds.), Reinhold, New York, pp. 287–97.

Robinson, T. W., 1958. *Phreatophytes*. United States Geological Survey Water Supply Paper 1423.

Rodwell, M. J., and B. J. Hoskins, 1996: Monsoons and the dynamics of deserts. *Quart. J. Roy. Meteor. Soc.*, **122**, 1385–404.

Rognon, P., 1987: Late Quaternary climate reconstruction for the Maghreb (north Africa). *Palaeogeography, Palaeoclimatology, Palaeoecology*, **58**, 11–34.

Ropelewski, C. F., and M. S. Halpert, 1987: Global and regional scale precipitation patterns associated with the El Niño/Southern Oscillation. *Mon. Wea. Rev.*, **115**, 1606–26.

Rose, C. W., 1968: Water transport in a soil with a daily temperature wave, II. Analysis. *Aust. J. Soil Res.*, **6**, 45–57.

Rosen, M. R., 1994: The importance of groundwater in playas: A review of playa classifications and the sedimentology and hydrology of playas. In *Paleoclimate and Basin Evolution of Playa Systems*. M. R. Rosen (Ed.), The Geological Society of America, Special Paper 289, pp. 1–18.

Rosenan, N., 1963: Changes of Climate. In *Proceedings of the Rome Symposium*. UNESCO, Paris, and World Meteorological Organization, Geneva, pp. 67–73.

Rossby, C. G., 1941. The scientific basis of modern meteorology. In *Climate and Man: Yearbook of Agriculture – 1941*. United States Department of Agriculture, Washington, D.C., pp. 599–655.

Rossignol-Strict, M., 1983: African monsoons, an immediate climate response to orbital insolation. *Nature*, **304**, 46–9.

Rowell, D. P., and C. Blondin, 1990: The influence of soil wetness distribution on short-range rainfall forecasting in the West African Sahel. *Quart. J. Roy. Meteor. Soc.*, **116**, 1471–85.

Rowson, D. R., and S. J. Colucci, 1992: Synoptic climatology of thermal low-pressure systems over south-western North America. *Intl. J. Climatol.*, **12**, 529–45.

Rozanov, B., 1990: Global assessment of desertification: status and methodologies. In *Desertification Revisited* Proceedings of an ad hoc consultative Meeting on the Assessment of Desertification. UNEP DC/PAC, Nairobi, pp. 45–122.

Rubio, J. L., 1987: Desertificacion en la communidad valenciana antecedantes historicos y situacion actual de erosion. *Rev Valenc. Estud. Autonomicos*, **7**, 231–58.

Rumney, G. R., 1968: *Climatology and the World's Climates*. Macmillan, New York.

Russell, R. J., 1936: The desert-rainfall factor in denudation. In *International Geological Congress, Report of the XVI Session, 1933*, Vol. 2, pp. 753–63.

Rutllant, J., and P. Ulriksen, 1979: Boundary-layer dynamics of the extremely arid northern part of Chile: The Antofagasta Field Experiment. *Bound. Layer Meteor.*, **15**, 41–55.

Sagan, C., O. B. Toon, and J. B. Pollack, 1979: Anthropogenic albedo changes and the earth's climate. *Science*, **206**, 1363–8.

Said, F. J. L., B. Benech, A. Druilhet, P. Durand, M. H. Marciniak, and B. Monteny, 1997: Spatial variability in airborne surface flux measurements during HAPEX-Sahel. *J. Hydrology*, **188–9**, 878–911.

Sammis, T. W., and L. W. Gay, 1979: Evapotranspiration from an arid zone plant community. *J. Arid Environments*, **2**, 313–21.

Sanger, R. H., 1954: *The Arabian Peninsula*. Cornell University Press, Cornell.

Sargent, F., II, 1963: Tropical neurasthenia: Giant or windmill. In *Environmental Physiology and Psychology in Arid Conditions*. Arid Zone Research XXII, UNESCO, Paris, pp. 273–314.

Sarnthein, M., 1978: Sand deserts during glacial maximum and climatic optimum. *Nature*, **272**, 43–6.

Sarre, R. D., 1987: Aeolian sand transport. *Prog. Physical Geog*, **11**, 157–82.

Satchell, J. E., 1978: Ecology and environment in the United Arab Emirates. *J. Arid Environments*, **1**, 201–26.

Schlesinger, W. H., J. F. Reynolds, G. L. Cunningham, L. F. Huenneke, W. M. Jarrell, R. A. Virginia, and W. G. Whitford, 1990: Biological feedbacks in global desertification. *Science*, **247**, 1043–8.

Schreiber, H. A., and D. R. Kincaid, 1967: Regression models for predicting on-site runoff from short-duration convective storms. *Water Resour. Res.*, **3**, 389–95.

Schulze, B. R., 1972: South Africa. In *World Survey of Climatology*, Vol. 10, *Climates of Africa*. J. F. Griffiths (Ed.), Elsevier Scientific Publishing Company, Amsterdam, pp. 501–86.

Schütz, L., 1980: Long range transport of desert dust with special emphasis on the Sahara. *Ann. N. Y. Acad. Sci.*, **338**, 515–32.

Schwerdtfeger, W. (Ed.), 1976: *Climates of Central and South America*. Elsevier Scientific Publishing Company, Amsterdam.

Schwikowski, M., P. Seibert, U. Baltensperger, and H. W. Gäggeler, 1995: A study of an outstanding Saharan dust event at the high-Alpine site Jungfraujoch, Switzerland. *Atmos. Environ.*, **29**, 1829–42.

Scoging, H., 1991: Desertification and its management. In *Global Change and Challenge. Geography for the 1990s*. R. Bennet and R. Estall (Eds.), Routledge, London, pp. 57–79.

Sears, P., 1935: *Deserts on the March*. University of Oklahoma Press, Norman.

Seely, M. K., 1978: Grassland productivity: The desert end of the curve. *S. African J. Sci.*, **74**, 295–7.

Segal, M., J. H. Cramer, R. A. Pielke, J. R. Garratt, and P. Hildebrand, 1991: Observational evaluation of the snow breeze. *Mon. Wea. Rev.*, **119**, 412–24.

Segal, M., W. E. Schreiber, G. Kallos, J. R. Garratt, A. Rodi, J. Weaver, and R. A. Pielke, 1989: The impact of crop areas in northeast Colorado on midsummer mesoscale thermal circulations. *Mon. Wea. Rev.*, **117**, 809–25.

Sellers, W. D., 1965: *Physical Climatology*. University of Chicago Press, Chicago.

Sellers, W. D., and R. H. Hill, 1974: *Arizona Climate, 1931–1972*. University of Arizona Press, Tucson.

Sepaskhah, A. R., and L. Boersma, 1979: Thermal conductivity of soils as a function of temperature and water content. *J. Soil Science Soc. Amer.*, **43**, 439–44.

Servant, M., and S. Servant-Vidary, 1980: L'environment quaternaire du basin du Tchad. In *The Sahara and the Nile*. M. A. J. Williams and H. Faure (Eds.), Balkema, Rotterdam, pp. 133–62.

Shantz, H. L., 1956: History and problems of arid lands development. In *The Future of Arid Lands*. G. F. White (Ed.), American Association for the Advancement of Science, Washington, D.C, pp. 3–25.

Sharon, D., 1972: The spottiness of rainfall in a desert area. *J. Hydrology*, **17**, 161–75.

1974: The spatial pattern of convective rainfall in Sukumaland, Tanzania – a statistical analysis. *Arch. Met. Geophys. Biokl.*, Ser. B, **22**, 201–18.

1979: Correlation analysis of the Jordan Valley rainfall field. *Mon. Wea. Rev.*, **107**, 1042–7.

1981: The distribution in space of local rainfall in the Namib Desert. *J. Climatol.*, **1**, 69–75.

Shaw, P. A., 1997: Africa and Europe. In *Arid Zone Geomorphology: Process Form and Change in Drylands*. D. S. G. Thomas (Ed.), John Wiley and Sons, Chichester, pp. 467–85.

Shaw, P. A., and H. J. Cooke, 1986: Geomorphic evidence for the Late Quaternary palaeoclimate of the Middle Kalahari of Northern Botswana. *Catena*, **13**, 349–59.

Shaw, P. A., and D. S. G. Thomas, 1997: Pans, playas and salt lakes. In *Arid Zone Geomorphology: Process Form and Change in Drylands*. D. S. G. Thomas (Ed.), John Wiley and Sons, Chichester, pp. 293–317.

Shaw, R. H., and A. R. Pereira, 1982: Aerodynamic roughness of a plant canopy: A numerical experiment. *Agricultural Meteor.*, **26**, 51–65.

Shephard, M., 1995: *The Great Victoria Desert*. Reed Books, Chatswood.

Sheridan, D., 1981: *Desertification of the United States*. Council on Environmental Quality, US Government Printing Office, Washinton, D.C.

Shinn, E. A., G. W. Smith, J. M. Prospero, P. Betzer, M. L. Hayes, V. Garrison, and R. T. Barber, 2000: African dust and the demise of Caribbean coral reefs. *Geophys. Res. Lett.*, **27**, 3029–32.

Shmida, A., 1985: Biogeography of the desert flora. In *Hot Deserts and Arid Shrublands, Ecosystems of the World Volume 12A*. M. Evenari, I. Noy-Meir, and D. W. Goodall (Eds.), Elsevier Scientific Publishing Company, New York, pp. 23–77.

Shuhua, L., H. Zichen, and L. Lichao, 1997: Numerical simulation of the influence of vegetation cover factor on boundary layer climate in semi-arid region. *Acta Meteor. Sinica*, **11**, 66–78.

Sikka, D. R., 1997: Desert climate and its dynamics. *Current Science*, **72**, 35–46.

Sinclair, P. C., 1966: A quantitative analysis of the dust devil. Dissertation, University of Arizona, Tucson.

Singh, G., R. D. Joshi, and A. B. Singh, 1972: Stratigraphic and radiocarbon evidence for the age and development of three salt lake deposits in Rajasthan, India. *Quaternary Res.*, **2**, 496–505.

Slatyer, R. O., and J. A. Mabbutt, 1965: Hydrology of arid and semi-arid regions. In *Handbook of Applied Hydrology*. V. T. Chow (Ed.), McGraw Hill, New York, pp. 241–46.

Slatyer, R. O., and I. C. McIlroy, 1961: *Practical Micrometeorology*. CSIRO-UNESCO, Australia.

Smagorinsky, J., 1974: Global atmospheric modeling and the numerical simulation of climate. In *Weather and Climate Modification*. W. N. Hess (Ed.), John Wiley and Sons, New York, pp. 633–86.

Smiley, T. L., and J. H. Zumberge (Eds.), 1974: *Polar Deserts and Modern Man*. University of Arizona Press, Tucson.

Smith, E. A., 1986a: The structure of the Arabian heat low. Part I: Surface energy budget. *Mon. Wea. Rev.*, **114**, 1067–83.

1986b: The structure of the Arabian heat low. Part II: Bulk tropospheric heat budget and implications. *Mon. Wea. Rev.*, **114**, 1084–102.

Smith, H. T. U., 1968: Geologic and geomorphic aspects of deserts. In *Desert Biology: Special Topics on the Physical and Biological Aspects of Arid Regions*. G. W. Brown (Ed.), Academic Press, New York, pp. 51–100.

Smith, R. B., 1979a: The influence of mountains on the atmosphere. In *Advances in Geophysics*, Vol. 21. B. Saltzman (Ed.), Academic, New York, pp. 87–230.

1979b: Some aspects of the quasi-geostrophic flow over mountains. *J. Atmos. Sci.*, **36**, 2385–93.

Sokolik, I. N., and G. Golitsyn, 1993: Investigation of optical and radiative properties of atmospheric dust aerosols. *Atmos. Environ.*, Ser. A, **27**, 2509–17.

Sokolik, I. N., and O. B. Toon, 1996: Direct radiative forcing by anthropogenic airborne mineral aerosols. *Nature*, **381**, 681–3.

Sokolik, I., D. Winker, G. Bergametti, D. Gillette, G. Carmichael, Y. Kaufman, L. Gomes, L. Schuetz, and J. Penner, 2001: Introduction to special section: Outstanding problems in quantifying the radiative impacts of mineral dust. *J. Geophys. Res.*, **106**, 18 015–27.

Sokolik, N. I., 1961: *The effect of irrigation on the heat regime of surrounding areas.* US Department of Commerce, Weather Bureau (translated from Russian by George S. Mitchell), Washington, D.C.

Soriano, A., 1983: Deserts and semi-deserts of Patagonia. In *Temperate Deserts and Semi-Deserts, Ecosystems of the World Volume 5.* N. E. West (Ed.), Elsevier Scientific Publishing Company, New York, pp. 423–60.

Stearns, C. R., 1967: Micrometeorological studies in the coastal desert of southern Peru. Thesis, University of Wisconsin, Madison.

Stebbing, E. P., 1935: The encroaching Sahara: The threat to the West African colonies. *Geographical J.*, **85**, 506–24.

  1938: The man-made desert in Africa: erosion and drought. *J. Roy. African Soc., Suppl.* **37**, 1–40.

Steedman, R. A., and Y. Ashour, 1976: Sea breezes over north-west Arabia. *Tellus*, **28**, 299–306.

Steiert, J., 1995: *Playas: Jewels of the Plains.* Texas Tech University, Lubbock.

Stevenson, C. E., 1969: The dust fall and severe storm of 1 July 1968. *Weather*, **24**, 126–32.

Stowe, C. D., 1969: Dust and sand storm electrification. *Weather*, **24**, 134–40.

Strahler, A. N., and A. H. Strahler, 1973: *Environmental Geoscience: Interactions Between Natural Systems and Man.* Hamilton, Santa Barbara.

Stranz, D., 1974: A strong Shamal in the Arabian Gulf (December 1970). *Rivista Italiana Geofisica*, **6**, 296–8.

Street, F. A., and A. T. Grove, 1979: Global maps of lake-level fluctuations since 30,000 yr B.P. *Quaternary Res.*, **12**, 83–118.

Strydom, N. B., and C. H. Wyndham, 1963: Natural state of heat acclimatization of different ethnic groups. *Fed. Proc.*, **22**, 801–9.

Stull, R. B., 1988: *An Introduction to Boundary Layer Meteorology.* Kluwer Academic, Dordrecht.

Sud, Y. C., and M. Fennessey, 1982: A study of the influence of surface albedo on July circulation in semi-arid regions using the GLAS-GCM. *J. Climatol.*, **2**, 105–25.

Sud, Y. C., and W. E. Smith, 1985: Influence of local land-surface processes on the Indian monsoon: A numerical study. *J. Climate Appl. Meteor.*, **24**, 1015–36.

Sullivan, R. J., 1987: Comparison of aeolian roughness measured in a field experiment and in a wind tunnel simulation. Thesis, Arizona State University, Tempe.

Swap, R., M. Garstang, S. Greco, R. Talbot, and P. Kallberg, 1992: Saharan dust in the Amazon Basin. *Tellus*, **44B**, 133–49.

Swap, R., S. Ulanski, M. Cobbett, and M. Garstang, 1996: Temporal and spatial characteristics of Saharan dust outbreaks. *J. Geophys. Res.*, **101**, 4205–20.

Tajadod, M., 1994: A place of trials. *The UNESCO Courier: A Window Open on the World*, **47–1**, 16–17.

Takahashi, K., and H. Arakawa (Eds.), 1981: *Climates of Southern and Western Asia.* Elsevier Scientific Publishing Company, Amsterdam.

Talbot, M. R., 1980: Environmental response to climate change in the West African Sahel over the past 20,000 years. In *The Sahara and the Nile.* M. A. J. Williams and H. Faure (Eds.), Balkema, Rotterdam, pp. 37–62.

Talbot, R. W., R. C. Harriss, E. V. Browe, G. L. Gregory, D. I. Sebacher, and S. M. Beck, 1986: Distribution and geochemistry of aerosols in the tropical North Atlantic troposphere: Relationship to Saharan dust. *J. Geophys. Res.*, **91**, 5173–82.

Tan, G., 1991: Weather-related human mortality in Shanghai and Guangzhou, China. In *Preprints, Tenth Conference on Biometeorology and Aerobiology, Salt Lake City, Utah, 10–13 September 1991*. American Meteorological Society, Boston, pp. 15–17.

Tang, M., and E. R. Reiter, 1984: Plateau monsoons of the Northern Hemisphere: A comparison between North America and Tibet. *Mon. Wea. Rev.*, **112**, 617–37.

Tao, G., L. Jingtao, Y. Xiao, K. Ling, F. Yida, and H. Yinghua, 2002: Objective pattern discrimination model for dust storm forecasting. *Meteor. Appl.*, **9**, 55–62.

Tapper, N. J., 1988: Some evidence for a mesoscale thermal circulation at The Salt Lake, New South Wales. *Aust. Meteor. Mag.*, **36**, 101–2.

   1991: Evidence for a mesoscale thermal circulation over dry salt lakes. *Palaeogeography, Palaeoclimatology, Palaeoecology*, **84**, 259–69.

Taylor, C. A., 1934: Transpiration- and evaporation-losses from areas of native vegetation. *Trans. Amer. Geophys. Union*, **15**, 554–9.

Taylor, C. M., and T. Lebel, 1998: Observational evidence of persistent convective-scale rainfall patterns. *Mon. Wea. Rev.*, **126**, 1597–607.

Taylor, C. M., F. Said, and T. Lebel, 1997: Interactions between the land surface and mesoscale rainfall variability during HAPEX-Sahel. *Mon. Wea. Rev.*, **125**, 2211–27.

Tchakerian, V. P., 1997: North America. In *Arid Zone Geomorphology: Process Form and Change in Drylands*. D. S. G. Thomas (Ed.), John Wiley and Sons, Chichester, pp. 523–41.

Tegen, I., and I. Fung, 1994: Modeling of mineral dust in the atmosphere: Sources, transport and optical thickness. *J. Geophys. Res.*, **99**, 22 897–914.

Tegen, I., A. A. Lacis, and I. Fung, 1996: The influence on climate forcing of mineral aerosols from disturbed soils. *Nature*, **380**, 419–22.

Terjung, W. H., S. O. Ojo, and S. W. Swarts (1970): A nighttime energy and moisture budget in Death Valley, California in mid August. *Geog. Ann.*, **52(A)**, 160–73.

Thames, J. L., and D. D. Evans, 1981: Desert systems: An overview. In *Water in Desert Ecosystems*. D. D. Evans and J. L. Thames (Eds.), Dowden, Hutchinson and Ross, Stroudsburg, Pennsylvania, pp. 1–12.

Thesiger, W., 1964: *Arabian Sands*. Penguin Books, London.

Thomas, D. S. G. (Ed.), 1997a: *Arid Zone Geomorphology: Process Form and Change in Drylands*. John Wiley and Sons, Chichester.

   1997b: Sand seas and aeolian landforms. In *Arid Zone Geomorphology: Process Form and Change in Drylands*. D. S. G. Thomas (Ed.), John Wiley and Sons, Chichester, pp. 373–412.

   1997c: Science and the desertification debate. *J. Arid Environments*, **37**, 599–608.

   1997d: Reconstructing ancient arid environments. In *Arid Zone Geomorphology: Process Form and Change in Drylands*. D. S. G. Thomas (Ed.), John Wiley and Sons, Chichester, pp. 577–605.

Thomas, D. S. G., and N. J. Middleton, 1994: *Desertification: Exploding the Myth*. John Wiley and Sons, Chichester.

Thomas, D. S. G., and P. A. Shaw, 1991: *The Kalahari Environment*. Cambridge University Press, Cambridge.

Thompson, R. D., 1975: *The Climatology of the Arid World*. Reading Geographical Paper 35, Department of Geography, University of Reading.

Thornthwaite, C. W., 1945: Report of the committee on transpiration and evaporation, 1943–1944. *Trans. Amer. Geophys. Union 1944*, pp. 686–93.

   1948: An approach toward a rational classification of climate. *Geographical Rev.*, **38**, 55–94.

Tivey, J., 1993: *Biogeography, a Study of Plants in the Ecosphere*. Longman, Essex.

Tomaselli, R., 1977: The degradation of the Mediterranean Maquis. *Ambio*, **6**, 356–62.

Trenberth, K. E. (Ed.), 1992: *Climate System Modeling*. Cambridge University Press, Cambridge.

Trenberth, K. E., 1999: Atmospheric moisture recycling: role of advection and local evaporation. *J. Climate*, **12**, 1368–81.

Trewartha, G. T., 1961: *The Earth's Problem Climates*. University of Wisconsin Press, Madison.

1968: *An Introduction to Climate*. McGraw-Hill, New York.

Tricker, R. A. R., 1970: *Introduction to Meteorological Optics*. American Elsevier Publishing Company, New York.

Tromp, S. W., 1980: *Biometeorology: The Impact of the Weather and Climate on Humans and their Environment (Animals and Plants)*. Heydon and Son, London.

Tsoar, H., and K. Pye, 1987: Dust transport and the question of desert loess formation. *Sedimentology*, **34**, 139–54.

Tsukamoto, O., 1993: Comments on "Peculiar downward water vapor flux over Gobi Desert in the daytime". *J. Meteor. Soc. Japan*, **71**, 161–2.

Tsukamoto, O., J. Wang, and Y. Mitsuta, 1992: A significant evening peak of vapor pressure at an oasis in the semi-arid region. *J. Meteor. Soc. Japan*, **70**, 1155–60.

Tucker, C. J., H. E. Dregne, and W. W. Newcomb, 1991: Expansion and contraction of the Sahara Desert from 1980 to 1990. *Science*, **253**, 299–301.

Tweit, S. J., 1992: *The Great Southwest Nature Factbook*. Alaska Northwest Books, Anchorage.

1995: *Barren, Wild and Worthless: Living in the Chihuahuan Desert*. University of New Mexico Press, Albuquerque.

Tyson, P. D., 1981: Atmospheric circulation variations and occurrence of extended wet and dry spells over Southern Africa. *J. Climatology*, **1**, 115–30.

1984: The atmospheric modulation of extended wet and dry spells over South Africa, 1958–1978. *J. Climatology*, **4**, 621–35.

1986: *Climate Change and Variability in Southern Africa*. Oxford University Press, Oxford.

UCAR, 1994: *El Niño and Climate Prediction*. University Corporation for Atmospheric Research, Boulder, Colorado.

Uman, M. A., 1987: *The Lightning Discharge*. Academic Press, Orlando, Florida.

UN, 1977: *World Map of Desertification*. United Nations Conference on Desertification, Nairobi, Kenya, 29 August–9 September 1977, UN Document A/CONF.74/2.

1978: United Nations Conference on Desertification: *Roundup, Plan of Action and Resolutions*, United Nations, New York.

UNEP, 1992: *World Atlas of Desertification*. UNEP, Nairobi, and Edward Arnold Press, London.

Unland, H. E., P. R. Houser, W. J. Shuttleworth, and Z.-L Yang, 1996: Surface flux measurement and modeling at a semi-arid Sonoran Desert site. *Agricultural and Forest Meteor.*, **82**, 119–53.

Uriarte, A., 1980: Rainfall on the northern coast of the Iberian peninsula. *J. Meteor.*, **5**, 138–44.

van de Griend, A. A., M. Owe, M. Groen, and M. P. Stoll, 1991: Measurement and spatial variation of thermal infrared surface emissivity in a savanna environment. *Water Resour. Res.*, **27**, 371–9.

van Hylckama, T. E. A., 1970: Water use by salt cedar. *Water Resour. Res.*, **6**, 728–35.

van Loon, H., R. L. Jenne, and K. Labitzke, 1973: Zonal harmonic standing waves. *J. Geophys. Res.*, **78**, 4463–71.

van Wijk, W. R. (Ed.), 1963: *Physics of Plant Environment*. North Holland Publishing Company, Amsterdam.

Vehrencamp, J. E., 1951: *An Experimental Investigation of Heat and Momentum Transfer at a Smooth Air-Earth Interface*. Department of Engineering, University of California, Los Angeles.

 1953: Experimental investigation of heat transfer at an air-earth interface. *Trans. Amer. Geophys. Union*, **34**, 22–30.

Viaud, J. (P. Loti), 1895: *Le Désert*. (Republished 1993; Jay Paul Minn, translator; University of Utah Press; Salt Lake City.)

Vonder Haar, T. H., 1980: *Earth radiation budget measurements from satellites and their interpretation for climate modeling and studies*. Contract NAS1–15553, Research Institute of Colorado, Fort Collins.

Wakimoto, R. M., 1982: The life cycle of thunderstorm gust fronts as viewed with Doppler radar and radiosonde data. *Mon. Wea. Rev.*, **110**, 1060–82.

Waldmeier, M., 1961: *The Sunspot Activity in the Years 1610–1960*. Schulthess and Company, Zürich.

Walker, A. S., 1982: Deserts of China. *Amer. Scientist*, **70**, 366–76.

Walker, J., and P. R. Rowntree, 1977: The effects of soil moisture on circulation and rainfall in a tropical model. *Quart. J. Roy. Meteor. Soc.*, **103**, 29–46.

Wallace, J. S., and C. J. Holwill, 1997: Soil evaporation from tiger bush in south west Niger. *J. Hydrology*, **188–9**, 426–42.

Wallin, J. R., 1967: Agrometeorological aspects of dew. *Agricultural Meteor.*, **4**, 85–102.

Walsh, R. P. D., M. Hulme, and M. D. Campbell, 1988: Recent rainfall changes and their impact on hydrology and water supply in the semi-arid zone of the Sudan. *Geographical J.*, **154**, 181–98.

Walter, H., 1986: The Namib Desert. In *Hot Deserts and Arid Shrublands, Ecosystems of the World Volume 12B*. M. Evenari, I. Noy-Meir, and D. W. Goodall (Eds.), Elsevier Scientific Publishing Company, New York, pp. 245–82.

Walter, H., and E. O. Box, 1983a: Caspian lowland biome. In *Temperate Deserts and Semi-Deserts, Ecosystems of the World Volume 5*. N. E. West (Ed.), Elsevier Scientific Publishing Company, New York, pp. 9–41.

 1983b: Semi-deserts and deserts of central Kazakhstan. In *Temperate Deserts and Semi-Deserts, Ecosystems of the World Volume 5*. N. E. West (Ed.), Elsevier Scientific Publishing Company, New York, pp. 43–78.

 1983c: Middle Asian deserts. In *Temperate Deserts and Semi-Deserts, Ecosystems of the World Volume 5*. N. E. West (Ed.), Elsevier Scientific Publishing Company, New York, pp. 79–104.

 1983d: The Karakum Desert, an example of a well-studied eu-biome. In *Temperate Deserts and Semi-Deserts, Ecosystems of the World Volume 5*. N. E. West (Ed.), Elsevier Scientific Publishing Company, New York, pp. 105–59.

 1983e: The orobiomes of middle Asia. In *Temperate Deserts and Semi-Deserts, Ecosystems of the World Volume 5*. N. E. West (Ed.), Elsevier Scientific Publishing Company, New York, pp. 161–91.

 1983f: The deserts of central Asia. In *Temperate Deserts and Semi-Deserts, Ecosystems of the World Volume 5*. N. E. West (Ed.), Elsevier Scientific Publishing Company, New York, pp. 193–236.

 1983g: The Pamir – An ecologically well-studied high-mountain desert biome. In *Temperate Deserts and Semi-Deserts, Ecosystems of the World Volume 5*. N. E. West (Ed.), Elsevier Scientific Publishing Company, New York, pp. 237–69.

Walters, K. R., and W. F. Sjoberg, 1988: *The Persian Gulf Region: A Climatological Study*. US Air Force Environmental Technical Applications Center, Technical Note TN-88/002.

Walton, K., 1969: *The Arid Zones*. Aldine, Chicago.

Wang, J., and Y. Mitsuta, 1990: Peculiar downward water vapor flux over Gobi Desert in the daytime. *J. Meteor. Soc. Japan*, **68**, 399–401.

1992: Evaporation from the desert: Some preliminary results of HEIFE. *Bound. Layer Meteor.*, **59**, 413–18.

Warner, T. T., 1989: Mesoscale atmospheric modeling. *Earth-Science Rev.*, **26**, 221–51.

Warner, T. T., and R.-S. Sheu, 2000: Multiscale local forcing of the Arabian Desert daytime boundary layer, and implications for the dispersion of surface-released contaminants. *J. Appl. Meteor.*, **39**, 686–707.

Warren, A., and J. K. Maizels, 1977: Ecological change and desertification. In *Desertification: Its Causes and Consequences*. Pergamon Press, New York, pp. 169–260.

Warren, A., Y. C. Sud, and B. Rozanov, 1996: The future of deserts. *J. Arid Environments*, **32**, 75–89.

Wasson, R. J., S. N. Rajagura, V. N. Misra, D. P. Agrawal, R. P. Ohir, A. K. Singhvi, and K. K. Rao, 1983: Geomorphology, Late Quaternary stratigraphy and palaeoclimatology of the Thar Desert. *Zeitschrift für Geomorphologie*, suppl. **45**, 117–52.

Webb, R. H., and H. G. Wilshire (Eds.), 1983: *Environmental Effects of Off-Road Vehicles*. Springer-Verlag, New York.

Wells, S. G., 1983: Preface to *Origin and Evolution of Deserts*. S. G. Wells and D. R. Haragan (Eds.), University of New Mexico Press, Albuquerque.

Wendler, G., and F. Eaton, 1983: On the desertification of the Sahel zone. *Climate Change*, **10**, 380.

Went, F. W., 1975: Water absorption in Prosopis. In *Physiological Adaptation to the Environment*. F. J. Vernberg (Ed.), Intext Educational Publishers, New York, pp. 67–75.

Werger, M. J. A., 1986: The Karoo and southern Kalahari. In *Hot Deserts and Arid Shrublands, Ecosystems of the World Volume 12B*. M. Evenari, I. Noy-Meir, and D. W. Goodall (Eds.), Elsevier Scientific Publishing Company, New York, pp. 283–359.

West, N. E., 1983a: Overview of North American temperate deserts and semi-deserts. In *Temperate Deserts and Semi-Deserts, Ecosystems of the World Volume 5*. N. E. West (Ed.), Elsevier Scientific Publishing Company, New York, pp. 321–30.

1983b: Great Basin-Colorado Plateau sagebrush semi-desert. In *Temperate Deserts and Semi-Deserts, Ecosystems of the World Volume 5*. N. E. West (Ed.), Elsevier Scientific Publishing Company, New York, pp. 331–49.

1983c: Western intermountain sagebrush steppe. In *Temperate Deserts and Semi-Deserts, Ecosystems of the World Volume 5*. N. E. West (Ed.), Elsevier Scientific Publishing Company, New York, pp. 351–74.

1983d: Intermountain salt-desert shrubland. In *Temperate Deserts and Semi-Deserts, Ecosystems of the World Volume 5*. N. E. West (Ed.), Elsevier Scientific Publishing Company, New York, pp. 375–97.

1983e: Colorado Plateau-Mohavian blackbrush semi-desert. In *Temperate Deserts and Semi-Deserts, Ecosystems of the World Volume 5*. N. E. West (Ed.), Elsevier Scientific Publishing Company, New York, pp. 399–411.

Westcot, D. W., and P. J. Wierenga, 1974: Transfer of heat by conduction and vapor movement in a closed soil system. *Proc. Soil Sci. Soc. Amer.*, **38**, 9–14.

Westing, A. H., 1994: Population, desertification and migration. *Environmental Conservation*, **21**, 109–14.

Westphal, D. L., O. B. Toon, and T. N. Carlson, 1987: A two-dimensional numerical investigation of the dynamics and microphysics of Saharan dust storms. *J. Geophys. Res.*, **92**, 3027–49.

1988: A case study of mobilization and transport of Saharan dust. *J. Atmos. Sci.*, **45**, 2145–75.

Whalley, W. B., J. P. McGreevy, and R. I. Ferguson, 1984: Rock temperature observations and chemical weathering in the Hunza region, Karakoram: preliminary data. In *The International Karakoram Project*, Vol 2. K. J. Miller (Ed.), Cambridge University Press, Cambridge, pp. 616–33.

Wheater, H. S., C. Onof, A. P. Butler, and G. S. Hamilton, 1991: A multivariate spatial-temporal model of rainfall in southwest Saudi Arabia. II. Regional analysis and long-term performance. *J. Hydrology*, **125**, 201–20.

Whiteman, C. D., 1982: Breakup of temperature inversions in deep mountain valleys: Part I. Observations. *J. Appl. Meteor.*, **21**, 270–89.

2000: *Mountain Meteorology: Fundamentals and Applications.* Oxford University Press, New York.

Whitlock, C. H., and S. R. LeCroy, 1987: *Surface Bidirectional Reflectance Properties of Two Southwestern Arizona Deserts for Wavelengths Between 0.4 and 2.2 Micrometers.* NASA Technical Paper 2643, National Aeronautics and Space Administration, Washington, D.C.

Wiggs, G. F. S., 1992: Airflow over barchan dunes: field measurements, mathematical modelling and wind tunnel testing. Dissertation, University of London.

1997: Sediment mobilisation by the wind. In *Arid Zone Geomorphology: Process Form and Change in Drylands.* D. S. G. Thomas (Ed.), John Wiley and Sons, Chichester, pp. 351–72.

Wilhite, D. A., 2000a: *Drought – A Global Assessment*, Volume 1. Routledge, London.

2000b: *Drought – A Global Assessment*, Volume 2. Routledge, London.

Wilks, D. S., 1995: *Statistical Methods in Atmospheric Sciences.* Academic Press, San Diego.

Williams, M., D. Dunkerley, P. de Deckker, P. Kershaw, and J. Chappell, 1998: *Quaternary Environments.* Arnold, London.

Williams, M. A. J., 1975: Late Pleistocene tropical aridity synchronous in both hemispheres? *Nature*, **253**, 617–18.

Williams, M. A. J., and D. A. Adamson, 1980: Later Quaternary depositional history of the Blue and White Nile rivers in central Sudan. In *The Sahara and the Nile.* M. A. J. Williams and H. Faure (Eds.), Balkema, Rotterdam, pp. 281–304.

Williams, O. B., and J. H. Calaby, 1985: The hot deserts of Australia. In *Hot Deserts and Arid Shrublands, Ecosystems of the World Volume 12A.* M. Evenari, I. Noy-Meir, and D. W. Goodall (Eds.), Elsevier Scientific Publishing Company, New York, pp. 269–312.

Wilson, J. W., and R. M. Wakimoto, 2001: The discovery of the downburst: T. T. Fujita's contribution. *Bull. Amer. Meteor. Soc.*, **82**, 49–62.

Wolfe, S. A., and W. G. Nickling, 1993: The protective role of sparse vegetation in wind erosion. *Progress in Physical Geography*, **17**, 50–68.

Xie, P., and P. A. Arkin, 1996: Analyses of global monthly precipitation using gauge observations, satellite estimates, and numerical model predictions. *J. Climate*, **9**, 840–58.

1997: Global precipitation: A 17-year monthly analysis based on gauge observations, satellite estimates and numerical model outputs. *Bull. Amer. Meteor. Soc.*, **78**, 2539–58.

Xue, Y., and J. Shukla, 1993: The influence of land surface properties on Sahel climate: Part I. Desertification. *J. Climate*, **6**, 2232–45.

Yair, A., 1990: Runoff generation in a sandy area – the Nizzani sands, western Negev, Israel. *Earth Surf. Proc. and Landforms*, **15**, 597–609.

Yan, H., and R. A. Anthes, 1988: The effects of variations in surface moisture on mesoscale circulations. *Mon. Wea. Rev.*, **116**, 192–208.

Yan, Z., and N. Petit-Maire, 1994: The last 140 ka in the Afro-Asian arid/semi-arid transitional zone. *Palaeogeography, Palaeoclimatology, Palaeoecology*, **110**, 217–33.

Yang, S., P. J. Webster, and M. Dong, 1992: Longitudinal heating gradient: Another possible factor influencing the intensity of the Asian summer monsoon circulation. *Adv. Atmos. Sci.*, **9**, 397–410.

Yarham, E. R., 1958: Singing sands; a strange concert heard in the desert. *The UNESCO Courier: A Window Open on the World*, **11–6**, 26–7.

Yeh, T.-C., R. T. Wetherald, and S. Manabe, 1984: The effect of soil moisture on the short-term climate and hydrology change – A numerical experiment. *Mon. Wea. Rev.*, **112**, 474–90.

Yinqiao, H., and G. Youxi, 1996: Some new understandings of land-surface processes in arid area from HEIFE. *Acta. Meteor. Sin.*, **10**, 233–46.

Yoshino, M. M., 1975: *Climate in a Small Area: An Introduction to Local Meteorology*. University of Tokyo Press, Tokyo.

Yu, B., P. P. Hesse, and D. T. Neil, 1993: The relationship between antecedent rainfall conditions and the occurrence of dust events in Mildura, Australia. *J. Arid Environments*, **24**, 109–24.

Zaalouk, M., 1994: A painter's paradise. *The UNESCO Courier: A Window Open on the World*, **47–1**, 18–19.

Zheng, X., and E. A. B. Eltahir, 1997: The response of deforestation and desertification in a model of the West African monsoon. *Geophys. Res. Lett.*, **24**, 155–8.

Zittel, K. A. von, 1875: *Briefe aus der Libyschen Wüste*. R. Oldenbourg, Munich.

Zones, C. P., 1961: *Ground water reconnaissance of Winnemucca Lake Valley, Pershing and Washoe Counties, Nevada*. Geological Survey Water-Supply Paper 1539-C, US Government Printing Office, Washington, D.C.

Zonn, I., M. H. Glantz, and A Rubinstein, 1994: The virgin lands scheme in the former Soviet Union. In *Drought Follows the Plow: Cultivating Marginal Areas*. M. H. Glantz (Ed.), Cambridge University Press, Cambridge, pp. 135–50.

Index